Twentieth Century Mouse Genetics

A Historical and Scientific Review

Twentieth Century Mouse Genetics

A Historical and Scientific Review

Robert P. Erickson
Department of Pediatrics University of Arizona

ELSEVIER

ACADEMIC PRESS

An imprint of Elsevier
elsevier.com/books-and-journals

Academic Press is an imprint of Elsevier
125 London Wall, London EC2Y 5AS, United Kingdom
525 B Street, Suite 1650, San Diego, CA 92101, United States
50 Hampshire Street, 5th Floor, Cambridge, MA 02139, United States
The Boulevard, Langford Lane, Kidlington, Oxford OX5 1GB, United Kingdom

British Library Cataloguing-in-Publication Data
A catalogue record for this book is available from the British Library

Library of Congress Cataloging-in-Publication Data
A catalog record for this book is available from the Library of Congress

ISBN: 978-0-12-824016-8

For Information on all Academic Press publications visit our website at
https://www.elsevier.com/books-and-journals

Publisher: Andre Gerhard Wolff
Acquisitions Editor: Peter B. Linsley
Editorial Project Manager: Susan Ikeda
Production Project Manager: Omer Mukthar
Cover Designer: Matthew Limbert

Typeset by Aptara, New Delhi, India

Dedication

In honor of my mentors: Tahir M. Rizki, undergraduate inspiration for genetics; Leonard A Herzenberg, medical school basic science guide; Salome Gluecksohn-Waelsch, guide to mouse genetics and scientific mother; Bryan Hall, teacher of dysmorphology; and Charles J. Epstein, clinical mentor

For my wife, Sandra nee De'Ath, and children: Andrew Ian, Colin De'Ath, Tanya Nadene, and Christoph Philipe.

Tahir, Andrew

Contents

Preface

Evolution and development can be said to be the defining features of life, and genetics is the science that underlies our understanding of them. In many ways, the study of genetics can be considered the central core of biological research in the 20th Century.

For millennia mice were a pest. Robert Burn's "wee little beastie" was a Romantic notion not shared by the Scottish crofters. Cats, which pursued this vermin, were highly regarded. For physiological experiments rats were favored since surgery to prepare them for physiological experiments, such as cannulating the thoracic duct (the small lymphatic vessel returning lymphatic fluid to the blood circulation) were difficult to perform in mice but the genetic study or rats came much later. The small size of mice was a benefit for the space needed, their adaptability to life in a circumscribed environment, and their rapid breeding time made them ideal. Their use as the major mammalian genetics model is the story of this book. Unlike earlier treatments, notably Rader's "Making Mice", the contribution of European and Japanese scientists is more thoroughly discussed.

In writing this book, I originally set out to be more of an historian than scientist but, as I wrote more about developments in mouse genetics which occurred in the last quarter of the 20th Century, many of which I had played a small part in, I realized that in many cases I was writing a historical, scientific review. I still did archival research when it added to the story but it was the scientific literature to which I mostly turned. In taking this turn, I realized that most historians wouldn't be interested but, hopefully many scientists will be. I believe that they will enjoy the "pure" history, which is also included. For my part, little "gems" like finding out who Nadine Dobrovalskaia-Zavadskaia's husband really was or what Ernst Caspari thought about the one gene/one enzyme theory, which some credited his early work to have found, have fascinated me.

Given the explosion of mouse genetics in the second half of the 20th Century, it would be impossible to fully cover the field in even a few volumes, let alone one. Thus, I have chosen, as a *leit motif,* the *T/t* complex. The *T/t* complex—a series of inversions preventing gene segregation by crossing-over, accumulating lethal mutations, and being maintained by segregation distortion—is a phenomenon rarely found in other mammals, especially humans. It is chosen as a central paradigm since it is a problem that has attracted the attention of many of the leading mouse geneticists over much of the 20th Century. As one of the participants put it:

"With all the exuberant information gathered worldwide, from the analysis of the various T/t locus phenotypes (in particular the lethality of the t/t embryos), we had the feeling that some sort of master gene had been discovered that was able to do many, many important things, in particular, regulating the first critical events in embryo's development. I had the

feeling that studying it would lead to the discovery of new principles of development and was puzzled by the extraordinary importance of these 0.5 cM of chromosome 17 when compared with the ~1,499.5 other cM representing the rest of the genome[1]"

For others, this complexity was something to be avoided. "The appropriately named T complex spawned an entire field of study that for years was plagued by complex genetics, bizarre interactions, contradictions, and misinterpretations.... [some] might have been frightened out of the field early on, as I was for many years" (Papaioannou, 1999). But most of the mouse geneticists did not avoid the topic. As John Scimenti pointed out in a commentary on one of the major advances in studying the complex at the end of the Century, "many of the world's most prominent mouse geneticists and developmental biologists investigated *t* haplotypes at one time or another" (Scimenti, 2000). The general interest among biologists is reflected in the very large number of articles concerning it that were published in high impact journals such as *Nature, Science,* and *Cell.* Adam Wilkins, in his 1986 book, **Genetic Analysis of Development,** about one third of which is devoted to each to *C. elegans, Drosophila,* and *Mus musculus,* makes the *T/t* complex a major focus of the latter. The unraveling of its unique molecular structure required the major advances that occurred in molecular biology, especially DNA cloning, and provides a basis for showing how these advances changed mouse genetics towards the end of the 20th Century. Segregation distortion was (and is) the subject of many evolutionary studies. The study of the developmental lethals which the *T/t* complex contains was a major spur to mammalian embryology and developmental genetics. In addition, because of its centrality for that period, I also found it a convenient and beautiful example for many of the developments in mouse genetics. These include early mouse genetics (chapter one), mammalian evolutionary genetics (chapter two), gene mapping (chapter three), developmental genetics (chapter four), reproductive genetics including the genetics of gametes (chapter eight), the era of gene cloning (chapter nine), and population genetics (chapter five). It is also highly relevant to mutational studies (chapter six) and imprinting (chapter 10). It is an aside for immunogenetics (chapter seven) and is only slightly relevant to sex determination (chapter 11). Thus, it readily suits the purpose of providing a central theme.

If one adopts the Whiggish view of history, which would only describe the most central events that led to the advancement of science, the *T/t* complex would not be chosen as a *leit motif.* In the non-Whiggish view, the fact that so many leading geneticists worked on the problem, and so many papers were published concerning it, makes the history of the *T/t* complex worthy of description as part of the history of mouse genetics in the 20th Century. In addition, I believe that there are major concepts that came from studies on this complex: the usefulness of genetic lethals as "experiments" in development, the occurrence of postmeiotic gene expression in mammals (which was previously thought not to occur, since selection should act on the zygote, not gametes), the importance of segregation distortion in gene drive, and the contribution of gene inversions in "fixing" gene associations.

I have provided "minibiographies" of some contributors to the field. These are usually not the most famous investigators, since the latter have had multiple biographies. Rather, they are of individuals whose story many not have been adequately told. For instance, the one previous short biography in English of Nadine Dobrovalskaia-Zavadskaia misidentifies her husband, who was an important writer, while Lucien Cuenot has not had much biographical information provided in English. Anne McLaren and Mary Lyon were individuals that I had contact with and thus, choose to share some memories along with their "formal" brief, biographies. I am sure book length biographies will one day be written for each of them although it will be difficult in the case of Mary since she was such a private person.

The history of mouse genetics is hardly a neglected field. It is the subject of many articles [e.g. a series on historical episodes in genetics, with many in mouse genetics, titled "Anecdotal, Historical and Critical Commentaries on Genetics" edited by James F. Crow and William F. Dove in the journal *Genetics*, which included two articles on "One Hundred years of Mouse Genetics: An Intellectual History" by Kenneth Paigen (2003a, 2003b)]. There are also several books on the origins of mouse genetics [**Morris, 1978; Rader, 2003**—the former a multi-author compendium (books are cited in bold and are listed in the Bibliography, articles are in the references at the end of each section)]. However, these books are very oriented to mouse genetics in the U.S.A. In addition, there should be mentioned several overviews by scientists heavily involved in the research: Elizabeth S. ("Tibby") Russell (1985) and Mary F. Lyon (2002). However, these too are heavily Anglo-American centric and, in fact, Willison and Lyon's (2000) history of the T/t-complex is titled "A UK-centric history of studies on the mouse t-complex".

In writing about this subject, I have mainly used secondary sources and many hundreds of published scientific papers. Where I needed to delve deeper, I found, or was helped by primary sources. . For earlier periods there are extensive archival materials, for example 126 boxes of L.C. Dunn's papers at the American Philosophical Society. For later periods, there is much less archival material but saved personal interviews are abundant. The latter frequently suffer from imperfect memories. I agree strongly with Benjamin's (1999) statement "To write history is to cite it" and, thus, there are many citations. This book has many sources and inspirations: the books in the bibliography and the papers cited (which have been a pleasure to re-read), my own experience working with the T/t complex in the decades of the 1970s and19 80s, relationships with workers in the field ranging from close friends to acquaintances, and the foment of ideas presented at meetings and seminars. Most chapters cite some of my laboratory's work, having been a fox, rather than a hedgehog. This was part of the motivation for writing the book since I had worked on, and knew many of the players in, multiple areas of mouse (and human) genetics. Thus, the book probably gives too much exposure to my lab's contributions, for which I apologize—one knows one's own work more thoroughly than that of others. My many acknowledgments are listed separately; the biographical sidebars and footnotes sometime reveal my personal involvement with individuals.

These historical reviews take one major subfield of mouse genetics at a time and develop each across the Century. Therefore, the ordering of chapters presents some challenges. The earlier chapters deal with the beginning of the Century and are, appropriately, more historic. I have not provided updates of further results in the 20th Century for them since they were not needed. The origins of mouse genetics in the first quarter of the Century (chapter one) obviously came first as did the discovery of the T/t complex which occurred in this time period. The evolutionary synthesis took place predominantly during the second quarter of the Century as did the discovery of t alleles in the wild and so comes second. Studies of chromosomes and linkage spanned the Century and provided a major tool of mouse genetics and so came third.

Developmental genetics started early in the Century and was a major concern for studies of the T/t complex and, thus, was placed fourth. Population genetics also developed early and continued through the Century and is more thoroughly reviewed in chapter five. With the 1940s, the advent of atomic bombs led to studies of radiation effects and mutation and became a major stimulus to mouse genetics with the development of the Harwell and Oakridge centers (see chapter one). The mutation studies also provided another tool and source of material for mouse studies, even the T/t complex, and are described in chapter six. Immunogenetics also started its major development at the time of WWII and continued; it provided an interesting "false alley" for the T/t complex (chapter seven). An intense study of reproductive genetics and the mechanism of transmission ratio distortion also had a mid-century origin and continued—the subject is treated as chapter eight. Many of the remaining areas of mouse genetics only came to prominence in the last quarter of the century and their ordering is more arbitrary. The cloning era provided the major tools used then and now and is reviewed in chapter nine. This period is closer to us, there was a massive increase in the number of scientists working with mice and in the number of published papers. Thus, the reviews are less historical and more scientific. They include "X-inactivation, epigenetics, and imprinting", chapter 10; "Sex Determination, Sex Differentiation, and the Y chromosome", chapter 11; and "Pharmacogentics", chapter 12. These chapters which review far from finished work are provided with, albeit sketchy, updates to the time of writing (2020). I realize that there are several areas of mouse genetics which I have neglected. Neurogenetics which Richard Sidman greatly developed and behavioral genetics being two obvious topics of neglect.

Note

1) Jean-Louis Guenet, personal communication 2012.

References

Benjamin, W, 1999. **The Arcades Project,** trans. By Eiland, H. and McLaughlin, K. Harvard University Press, London.

Lyon, M.F., 2002. A personal history of the mouse genome. Annu. Rev. Genomics Hum. Genet. 3, 1–16.

Paigen, K., 2003a. One hundred years of mouse genetics: an intellectual history I. The classical period (1902-1980). Genetics 163, 1–7.

Paigen, K., 2003b. One hundred years of mouse genetics: an intellectual history II. The molecular revolution. Genetics 163, 1227–1235.

Papaioannou, V.E., 1999. The ascendency of developmental genetics, or how the t complex educated a generation of developmental biologists. Genetics 151, 421–425.

Russell, E.S., 1985. A history of mouse genetics. Ann. Rev. Genet. 19, 1–28.

Scimenti, J., 2000. Segregation distortion of mouse t haplotypes: the molecular basis emerges. Trends Genetics 16, 240–243.

Willison, K.R., Lyon, M.F., 2000. A UK-centric history of studies on the mouse t-complex. Int. J. Dev. Biol. 44, 57–63.

Acknowledgements

As is always the case, a work like this requires a lot of assistance. Two *ancient eleves* of the Institut Pasteur, Paris deserve special thanks: 1) Dr. Gabriel Gachelin, scientist-turned-historian helped me find archival material when I had only found primary published articles and secondary sources for portions of this history related to France and searched some of these archives for me [he became very interested in the story of Nadine Zavadskaia-Dobrovalskaia and Russian émigré scientists and was to go on to write an article about her (Pigeard-Micalut and Gachelin, 2017, see Nadine Zavadskaia-Dobrovalskaia mini-biography)]; 2) Dr. Jean-Louis Guénet, by allowing me to read his drafts of his coauthored textbook of mouse genetics (**Guénet et al., 2015),** helped me to know how many aspects of mouse genetics I needed to cover. He also edited my translations from French and provided useful comments on the early history of mouse genetics in France. Archivists who helped me find material include Natalie Kolotilova, Moscow State University and Daniel Demellier, Institut Pasteur, Paris. Professor Emeritus Stan R. Blecher, University of Guelph, kindly reviewed and red-lined edited the first 4 chapters with their associated minibiographies before illness forced him to stop. Later chapters inevitably are diminished by his lack of editing. My very good friends, Dr. Franco Mangia, Professor Emeritus of Universita Sapienza di Roma provided a developmental biologist's views of the all the book while Dr. Murray Brilliant, retired Head of Human Genetics at the Marshfield Clinic Institute of Research, similarly provided a geneticist's perceptions. Dr. Robert Karn and Dr. Christina Laukaitis of the Universiiy of Arizona, experts on the salivary proteins and their population genetics in *Mus musculus* thoughtfully reviewed Chapter 2. Useful comments have included those of Dr. Asangla Ao, McGill University; Dr. Nabeel Affara, University of Cambridge; Dr. Walter Bodmer, Stanford University; Dr. Paul Edwards, University of Cambridge;, Dr. Peter J. I. Ellis, University of Kent; and Dr. Marilyn Monk, University College of London. I am grateful to Asangla Ao, Murray Brilliant, Marilyn Monk, and Peter Waelsch for sharing photographs. Special thanks are due to Dr. Jasna Markovac, my former doctoral student turned medical/scientific editor who led me to Elsevier Press, in essence serving as my "agent". Ms. Susan Ikeda, Elsevier publication editor, has patiently and kindly guided me through the publication process. Ms. Roopa Lingayath and Ms. Indhumanthi Mani of Elsevier Press, have been very helpful in obtaining figures. My wife, Sandra Doris nee De'Ath has encouraged me through the many years of writing and been my strongest supporter. As always, the errors which I, and others, have missed remain my responsibility.

Introduction

This book on the history of mouse genetics over the 1900s uses the problem of the T/t complex as a focus. It is amazing that the first mouse genetic studies were published within 2 years of the rediscovery of Mendel's laws (Cuenot, 1902). It is probably apocryphal that Mendel had started his genetic research looking at coat color variation in mice but was forbidden to continue because they "had sex" (**Henig, pp.15–16**).[1] Perhaps, it was just as well that Mendel did not use mice, as he would never have had the number of progeny (near thousands for his seeds) to convince him, with his background in physics, of the reality of the genetic ratios he found.[2] On the other hand, Thomas Hunt Morgan did start his genetic research in mice.

The book should primarily be of interest to mammalian geneticists, including human geneticists and, thus, I have assumed some genetic knowledge on the part of readers. However, the intricacies of the T/t complex deserve some further introduction. The dominant mutation, T, causes a short tail but, because of variable penetrance, greatly dependent on genetic background, full tailed and tailless mice may rarely occur. This dominant mutation is also known as the T-locus. Its discovery is described in Chapter One. T interacts with a series of recessive loci, t^n, to cause a tailless phenotype. These recessive loci are abundant in wild mice, the discovery of which is briefly described in Chapter One and amplified in Chapter Two. Initially these were thought to be alleles of the T-locus, but, with the discovery that they are part of a large chromosomal segment with 4 inversions (described in Chapter Three), t-haplotypes became the correct designation. While T is a homozygous lethal at about day 10, the various t^ns are homozygous lethals at various stages, many much earlier. This allows the maintenance of "balanced lethal lines" where crosses of T/t^n with T/t^n results in only T/t^n progeny. The description of the developmental abnormalities of the various homozygotes was the origin of mammalian developmental genetics and is described in Chapter Four. The population dynamics which maintains these t-haplotypes, which are homozygous lethal in nature is discussed in Chapter Five.

Since they are at complementing loci, heterozygotes for two different t^ns, such as, t^0 and t^1, are usually viable but sterile. Studies on reproductive genetics are presented in Chapter Eight. These include the much studied segregation distortion which can usually occurs: a heterozygote of a wild-type with a t^n may result in more than 90% of the progeny carrying the t^n. This property produces some very interesting population genetic outcomes (see Chapter Five).

The development of our current understanding of immunogenetics has depended heavily on studies on mice, and the development of this field is covered in Chapter Seven. For a while, it was thought that genes in the T/t complex controlled cell surface antigens and this is also discussed in that chapter. Many advances in mouse genetics involved creating new mutations and the T/t complex was one of the targets, as discussed in Chapter Six. Imprinting was an important discovery in

---

[1] This is almost certainly apocryphal. According to Mendel scholar Daniel Fairbanks (personal communication, Mar., 2016), the problems with the story are: a) there is absolutely no record of Mendel ever having mice (and we have a fair number of records from, and concerning, him); b) the story only appears after his death; c) Mendel was a very independent, even rebellious, individual (especially as shown in his relationship to government officials when he became abbot) and it **is** unlikely that he would have accepted such a stricture; and d) his abbot was very supportive of him and would also have opposed the bishop.

[2] The fiction that Mendel "fudged" his data has also been laid to rest (**Franklin, et al, 2008**).

mouse and humans. The T/t complex is imprinted and this is discussed in Chapter 10. When cloning and sequencing of DNA became available, many of the intricacies of the T/t complex were resolved, and, as each chapter follows its theme to the end of the century (and sometimes beyond), the relevant molecular genetics will be reported in many chapters but amplified, with the story of *TBOX* genes in man, in Chapter Nine.

No history of mouse genetics in the 20th Century would be complete without a discussion of the story of sex determination and, although the relationship of the T/t complex to sex determination is slight and not direct, the subject is covered in Chapter 11. Similarly, pharmacogenetics, although primarily an endeavor of human genetics, received much attention in mice and this is the subject of Chapter 12.

I have used the convention of citing books of the bibliography in **bold,** references are gathered at the ends of chapter, including for associated mini-biographies.

Reference

Cuénot, L, 1902. La loi de Mendel et l'hérédité de la pigmentation chez les Souris. Compte. Rend. Acad. Sci. Paris, 134, 779–781.

1

The origins of mouse genetics and the discovery of the *T*-locus

Introduction

The history of mouse genetics is the story of diverse individuals with shared curiosity about the mechanisms of living organisms. That they choose mice as their object of study was the result of many accidents of history; that they choose genetics was the imperative of a powerful new scientific paradigm. The story of mouse genetics generated myths, such as the notion that Mendel started his genetic research with mice, and controversies, such as whether Mary Lyon was the first to enunciate random X-inactivation, which became known as the "Lyon hypothesis." The field has blossomed to such an extent that each of the twelve chapters could become a book of thier own by providing greater detail and continuing the story more fully to the present.

The house mouse, as the name suggests, is a commensal species that has developed the ability to live alongside the human species. This coevolution is usually thought to have started with grain domestication in the early Neolithic. However, molecular systematics demonstrates that domestic mice of the *Mus musculus* and *domesticus* species separated from other mouse species more than 500,000 years ago (Salcedo, et al, 2007). This raises the interesting question as to whether the relationship between human and mouse started with the hominid ancestors who first utilized fire and softened food. It is usually assumed that migrant hunter-gatherers would not be in one place long enough for mice to settle in with them. Thus, this deviation of *Mus musculus/domesticus* from other species is unlikely to be based on an association with humans but the species rapidly became commensal with humans when stable dwelling developed. The finding of many *Mus musculus* skeletons in neolithic towns such as Çatal Hüyuk (7,400 BC-6,000 BC) bespeaks the association (**Hodder,p. 59**) and it has been suggested that the species even had religious significance at those times(Berry, 1987). The long association with humans is reflected in the etymology of the name: *mus* in Latin, *mus* or *mys* in Greek and *mush* in Sanskrit, which are derived from a root meaning "to steal"(**Keeler, p. 7**). It is thought that the "human-adapted" *Mus domesticus* of Asia followed Neolithic farmers into Europe, creating the zone of hybridization between *musculus* and *domesticus* (see Chapter 5. These two subspecies of commensal mice followed humans to the New World on their ships, with *M. m. musculus* travelling with Northern explorers and settlers to most of North America, while *M. m. domesticus* travelled with the Spanish and Portuguese to Latin America. The devastating effects of the introduction of rodents (rats more than mice) to new "worlds" has been reviewed by **Crosby (1986)**. A preceding colonization within extended Europe occurred with the Vikings (**Manco, p. 250**). A mouse mitochondrial variant has been identified that is associated with the locations of settlement of the Vikings, centered on Orkney Island and named after it (Searle, et al, 2009).

Twentieth Century Mouse Genetics. DOI: https://doi.org/10.1016/B978-0-12-824016-8.00015-5

The variant is found in Ireland, the Scottish Islands, Norway, and even Iceland (Jones, et al, 2012). Thus, even the Viking long boats were inhabited by mice!

Responses to mice differed among ancient Western cultures. In Crete, they were considered beneficial and the mouse god, Apollo Smithens, had a temple (**Lauder, p. 24**). In Egypt, where plagues of mice were more the rule, cats were domesticated and a cat goddess spoke to the cat's beneficence in getting rid of mice. A special interest in these small "house mates" is also evident in the Orient, with mention of them as pets in Chinese histories 3000 years ago (Morse, 1981;, which provides a fuller pre19th century history of the relationship of mice to humans). There was a particular interest in albino mice—"in China...all albino mice caught in the wild had to be turned over to the magistrates, and 26 such albinos are listed in Chinese history as found between 307 and 1641 A.D." (Keeler, 1978). This interest was continued in Japan with the collection and sharing of mice of extravagant colors and behaviors. A Japanese handbook, *Chinganso-date-gusi,* from 1787 is much cited (Tokuda, 1935) but it probably was about rats, not mice. Dr. Kuramoto has convincingly shown that another handbook from that period concerned rats rather than mice and that "if one were to specify the species that the term *nezumi* [my italics] denotes in *Yoso-tama-no-kakehashi*, it would be helpful for identifying the species that are referred to in *Chinganso-date-gusi*" (Kuramoto, 2011). His overall best interpretation is that the latter also concerns rats[1].

A more modern Western interest seems to have arisen with *Japonisme* in the mid-19th century, since part of this influence was the introduction of mice, prized by the Japanese, to the West. Although Japanese ceramics were being imported in the 17th century, it was during the Kali era (1848–1854) that foreign ships again traded abundantly with Japan. The confrontation by U.S. Commodore Matthew Perry, commanding four men-of-war in July, 1853, accelerated the trend and was instrumental in starting the ensuing internal struggles and civil war which led to the Meiji revolution in 1868. This resulted in an orientation of Japan to, and an effort to match, the Western nations (**Jansen, pp. 257-333**). The resulting increased trade led, in France, to Japanese objects became highly influential. The shop, *La Porte Chinoise* opened on rue de Rivoli in 1861, selling Japanese articles, especially woodprints (Ives, 1974). The pursuit of, and even competition for these *objets d'arte* are well described in de Waal's account of the acquisition as a family heirloom, of a large collection of *netsuke* (the small carvings, usually of animals or people, used as weights for the cords of the Japanese clothing) (**de Waal, p. 47**). He evokes the image of rich, well-dressed couples (frequently in *liaisons*) perusing the *objets d'art* in private showings in the galleries. The Sichel brother's gallery was another one with a major influence in Paris. It had "an overflowing morass of everything Japanese. The quantities were overwhelming. Philippe Sichel sent forty-five crates with 5,000 objects back from Yokohama after one buying trip in 1874 alone" (**de Waal, p. 47**). The influence, especially of the woodprints, was so great that *Japanisme* was not just a term of cultural influence but, in France, an art form (Ives, 1974).

The varied colors and behaviors ("waltzing" in particular) of these "table top" pets made them popular. Many histories of mouse genetics emphasize the origins of experimental mice from stocks maintained for mouse fanciers (Morse, 1981, Keeler, 1978). In the U.S.A., these historians particularly emphasize the purchase of mice from Miss Abbie E. C. Lathrop of Granby, Massachusetts. She initially received orders "from mouse fanciers interested in obtaining creamy buffs, red creams, ruby-eyed yellows (very rare) or white English sable" but gradually received orders from research labs (**Morse, 1978**). This Japanese origin of some of the progenitors of much used stocks is confirmed in modern times with the use of molecular markers, both protein and DNA (Blank, et al, 1986; Bonhomme & Guénet, 1996), even to the origin of particular mutants (Brilliant, et al, 1994). In fact, a mouse purchased at a Danish pet market in the 1980s was nearly pure "Japanese"—only 1% of its genome was from other subspecies but this 1% providing strong evidence that it wasn't a recent import (Takada, et al, 2013).

Interest in the inheritance of mouse variation long predated this new wave of mouse fancying. A Swiss pharmacist, M. Coladon (Jean-Pierre Colladon according to **Stubbs, p.151**) is cited by Edwards, in 1829, and quoted by Rostand in 1956 (Rostand, 1956), as having shown the non-blending of gray and white coat colors. It is interesting to speculate as to what this cross could have been. If the "white" mice were classic albino and the gray mice were dilute (which is considered to be "gray" in modern terminology), then there would have been dark mice in the crosses of these two recessives. It is likely that the "gray" was really black agouti. This is the normal color of feral house mice (which can look gray), and which, on the wild-type black background, would give a 3:1 ratio of "gray" to white in the segregation of the color-allowing albino locus. Others who studied heredity in mice before the re-discovery of Mendel's laws included William Haacke, who crossed albino mice with Japanese waltzing mice during the period 1893-1895, mistakenly interpreting waltzing to be inherited through the cytoplasm while he thought gray coat color was controlled by the nucleus (**Stubbs, p. 261**). G. Von Guaita made similar crosses: " he started in 1896 with fifty-five *Japanese waltzing* mice and numerous *white* mice belonging to a race bred by Weissmann since 1888" (Davenport, 1900). He found blending inheritance in the first generation with variable segregation in following generations.

France

It really seems quite amazing that within 2 years of the rediscovery of Mendel's laws, the first paper on mouse genetics confirming those laws was published (Cuénot, 1902; see mini-biography). With delays in publication (perhaps slower in those days, although peer review was not usually involved), Cuénot must have started his work almost immediately after the rediscovery. In this first paper, confirming, for the animal realm, the laws that Mendel found in the plant realm, he states: "*depuis deux ans, j'expérimente sur un matériel très favorable*" (for two years I have experimented with a very favorable material).

It is fitting that this work occurred in France, since France played a major part in *Japonisme* and the introduction of mice as pets/household zoo creatures, as discussed above. In this paper, he confirms the laws of Mendel for the segregation of "gray" with recessive albino, finding an all gray first generation and ¾ grays and ¼ albinos in the second generation. As discussed above, his "gray" must have really been agouti-black and he was observing the segregation at the *albino* locus. He found that the albinos bred pure, and that about 1/3 of the grays bred pure while 2/3s of the offspring again segregated for gray and albino. He ends the paper by stating that he has obtained yellow, black, grays mixed with white, and mixed blacks from his students and that the experimental results from these mice are very different from those following Mendel's laws (my emphasis).

In his continuing study of these more diverse mice, Cuenot came to realize that the determinant he called "gray" in contrast to albinism in his first crosses was essential for any color, and he came to call it *chromogene pigmentaire*, designated C (Cuenot, 1903, 1904). He then worked out that "gray" was dominant to black and brown, that black was dominant to brown, and that yellow was dominant to the other 3 colors. He realized that there were 16 possible combinations of the determinants for these colors but did not show segregation ratios of the many potential crosses (Cuénot, 1905). This is understandable, as 3 unlinked loci are involved (the *albino* locus, the *black/brown* locus and the *agouti* locus[2]) and the ratios would have been difficult to ascertain. He did observe a nearly 3:1 ratio (tending to a 2:1 ratio) of yellow to other colors in crosses of yellow with brown or black, which occurs because *yellow* is a dominant allele at the *agouti* locus. He noted that when he intercrossed albinos purchased from a merchant, he obtained yellows and grays but no browns, although others, whom he cited (Bateson, 1903; Allen, 1904), had. However, he was never able to obtain a yellow homozygote, though 81 were tested (Cuénot, 1905). With this result, and with the ratios which were not quite 3:1 of yellow to gray in crosses with this wild type, he proposed that yellow homozygotes did not occur. His explanation for this deficiency followed findings in plants where self-fertilization is frequently excluded (Cuénot, 1905). It is interesting that he cites W.E Castle's studies on prevention of self-fertilization of gametes from hermaphroditic *Ciona intestinalis* (sea squirt) in his proposal of sperm/egg incompatibility—it would be Castle who correctly showed that homozygous yellow embryos died in utero (Castle & Little, 1910). Importantly he went on to propose that *C* controlled a *ferment,* or *diastase,* two of the names for enzymes at the time. Thus, he was the first to enunciate the one gene one-enzyme hypothesis. This important concept has usually been credited to Garrod (1908); Beadle & Tatum's (1941) work received the Nobel prize for the concept. It is interesting to consider why Cuénot's contribution was neglected. A full interpretation of this discovery and the reasons for its lack of fame are well discussed by Hickman and Cairns (2003), who largely blame Cuénot's use of phenotype symbols, rather than gene symbols, for his variants. Sewall Wright (1917) provided a surprisingly modern, although incorrect, step-by-step pathway for the synthesis of yellow (pheomelanin) compared to brown/black (eumelanin) pigments. His thinking can certainly be characterized as "one gene, one enzyme."

This work was widely recognized in France, and Cuenot was lauded. "Two symbolic events testify to the interest of the French scientific community in Mendelian genetics before 1914. One is the attribution of the Académie des Sciences' prestigious Cuvier Prize to Cuénot in 1911. The official citation accompanying this Prize mentions clearly Cuénot's contributions to 'genetics' (mentioned as such). The other symbolic event is the international Conference of Genetics that was held the same year in Paris." (Gayon and Burian, 2000).

Despite this early example of Mendelism in France, the French were extremely slow to admit Mendelian genetics to their biological view. Many reasons have been given and are thoroughly analyzed by Burian and coauthors (Burian, et al, 1988; Burian and Goyon, 1999). The strength of Lamarckism and the orientation of French biologists towards physiology (à la Claude Bernard) and bacteriology (à la Louis Pasteur), subjects in which processes were visible ("analog") compared to genetics with its "invisible" ("digital") factors controlling phenotypes seem major reasons (Burian, et al, 1988; Burian & Gayon, 1999). Gayon and Burian (2000) also point out that many young scientists, who would have been more likely to accept new ideas, were killed during the first World War. In addition, apparent evidence for the inheritance of acquired characters (Lamarckism) continued to be presented by respected scientists. Perhaps the most famous results were those involved in the rise and fall of Paul Kammerer. His studies on the seeming inheritance of acquired pigment patterns and secondary sexual characters in amphibia were damned when it was found that one case (nuptial pad coloration in *Alytes* , the midwife toad) was due to the injection of india ink. While it is almost certain that this was not done by Kammerer (see **Koestler, pp 106–116**), his suicide shortly after the discovery of the fraud confirmed his guilt in many minds.

Cytoplasmic inheritance was also clearly evident, especially in the plastids of plants, and French scientists saw this as supporting Lamarckism. It was mostly ignored by the Mendelists. For example, Sturtevant reluctantly allows "we must be prepared to admit that the nucleus is not the sole organ involved in heredity" (Sturtevant, 1915). If, following Ludwik Fleck and Thomas Kuhn, one views Mendelism as a new paradigm which needed to overthrow the old, the intolerance by the holders of the old paradigm, especially in France, is easily understood. Jan Sapp has well developed this argument in *Beyond the Gene: Cytoplasmic Inheritance and the Struggle for Authority in the Field of Heredity* (Sapp, 1987).

This opposition to the acceptance of Mendel's laws, despite many presentations of the new genetic results in French, and in journals of high standing, led Cuénot to not allow his graduate students to do theses on genetic subjects (Burian, et al, 1988). After the destruction of his mouse stocks in Nancy during its occupation in the First World War, he did not rebuild them, although it would have been relatively easy to do so. He had been a leader in this new field, but he had lost time with the war and was no longer at the forefront, which was important to him. However, he remained a proponent of Mendelism, wrote several monographs (one on genetics with Rostand), and attacked Lamarckism (Cuénot,1925; see minibiography).

It is interesting to contrast Cuénot's research with the future Nobel laureate geneticist Thomas Hunt Morgan's studies at a similar time. He, too, attempted genetic studies in mice but "when [he] crossed white-bellied yellow-flanked house mice (not the albino mice Cuenot had used) with the wild type, his results were erratic and indicated that the germ cells carried other colors. Because he could not confirm Mendel's finding himself, Morgan had come by 1909 to feel quite strongly that Mendelian theory was being given more credence than it merited" (**Shine & Wrobel, p. 51**). In fact, he had a "ten-year refusal to espouse the Mendelian genetics"(Gilbert, 1978). "Until 1911, he pitched his writing against those who saw the nucleus as the center of hereditary activity, and he actively championed the cytoplasm to have that role. When he inadvertently proved the gene theory, one of the chief embryologists, and the person who had been the persecutor of the geneticists became their champion." (Gilbert, 1998). Perhaps his later devotion to, and success in, studying the genetics of *Drosophila* owes much to the zeal which frequently occurs in late converts to causes, especially religion.

Thus, although the first Mendelian results with mice (and any mammal) and the formulation of developmental genetics occurred in France, there was a tremendous resistance to

Mendelian genetics. As we contrast these developments with those in England and the U.S., it is important to note the major differences in the organization of science in France contrasted to the latter two countries: "recognition of a new discipline in France entailed much more than a budgetary commitment for a new chair. It necessitated provisions for a national system of examinations in the subject and a staff to prepare students for the new examinations, in many, if not all, of the universities" (**Sapp, p. 127**). Thus, it was not until the late '50s that the exams for the *License d'Enseignement en Science Naturelle* included genetics (**Sapp, p. 188**).

Lucien Cuénot (1866–1951)

15 Novembre 1921

Louis Cuenot at his prime, age 55.

Lucien Cuénot was born in Paris, Oct. 21, 1866.[1] His father fought the Germans during the Franco-Prussian War and was away 3 years, necessitating a reintroduction between father and son when he returned. After a private education, he entered the Lycée Chaptal in 1882 on a scholarship from his local government (*Mairie de Quartier*, city hall). He was a devoted naturalist and found much to observe in the greener Paris of the time, and in the nearby suburbs, sometimes explored on class excursions. His stay at Lycée Chaptal was short and he entered the Sorbonne in 1883 at the age of 17. He received a "first" for his degree in 1885, followed by a month at the Roscoff Marine Station (Brittany) that helped determine his career in research. After his obligatory military service (spent at Beauvais), he returned to Roscoff to study echinoderms. He received his Doctor of Natural Sciences after defending his thesis on the anatomy of *Asteriidae*. He was named a "préparateur" (an assistant in charge of preparing the hands-on activities) in anatomy and physiology at the Sorbonne and commenced the study of medicine. This was cut short in 1890 when he received an appointment as *chargé de cours complémentaire* (equivalent to an assistant professorship) in zoology at the University of Nancy where he moved with his parental family.

In Nancy, he was a popular teacher. There was much joking during his lectures but, at the same time, he encouraged a critical approach. In the early 1890s, he spent summers at Roscoff but gradually spent more time in the lab of his "mentor", Henri de Lacaze-Duthiers. His research during the decade was very diverse and included histological studies (on lymph nodes in relationship to the formation of blood and phagocytosis), anatomy (echinoderms), physiology (of crabs, other decapods, and insects) parasitism, and sex determination. His theorizing also started to appear (Cuénot, 1895a). His early book length writings (Cuénot, 1892, 1895b) were already written with Darwinian interpretations in contrast to the Lamarckism which was dominant (at that time) among French naturalists. However, he dropped the Neo-Darwinist interpretations in the mid-90's. "In 1897, [in] his review article on the blood corpuscles and lymphoid organs of invertebrates, Cuénot passed over in silence all Darwinian interpretation of phagocytosis, contenting himself with a simple description of the phenomenon" (Limoges, 1976). This was also a decade of travel, taking advantage of his bachelorhood, especially around France. In 1898 he was promoted to the Chair of Zoology in Nancy. The same year he visited London and Cambridge as part of a French delegation.

The decade ended with his marriage in 1900 to Geneviève de Maupassant, no relative to the author but from an aristocratic family which had made a fortune in railroads. He and Genevieve had 6 children, most of whom had professional or academic careers. The Maupassants had a villa in Arcachon, a seaside resort at the tip of Cap Ferret, 50 miles from Bordeaux (Mr. Maupassant had built the railroad from Bordeaux to Arcachon). Cuénot was to spend long summers there,

---

[1]The major source on the life of L. Cuénot is the biography by A CHOMARSD-LEXA (2004). She had access to his papers and communication with his last son.

building a villa at the side of his in-laws which he named after his daughter Nelly. The marine station of the University of Bordeaux was located here and he continued both laboratory research and describing the fauna of the region during these "vacations." These interludes continued until 1919.

In 1900, with the rediscovery of Mendel's laws, Cuénot took up mouse genetics. Perhaps his enthusiasm was prompted by the then current discussions in his department in Nancy about telegony, or "infection of the germ line." This was the notion that the first male to impregnate a female contributes characters to subsequent pregnancies. Even Weisman, who, of course, defined germ plasm, allowed, after reviewing published putative examples, that it might sometimes occur: "We may, however, at any rate suppose that this so-called 'infection', if not altogether deceptive, only occurs in rare instances, and by no means regularly, or at most only in some cases" **(Weisman, p. 385)**.

Cuénot's genetic studies started by finding a room which he called *la chambre des souris* in the basement. It consisted of large glass "tubs" covered with screens on trestles. Boards laid in the tubs were the "bedding", cigar boxes provided hide-a-ways, and bits of cotton were for the nests. *"La faiblesse des moyens était inversement proportionnelle à l'importance de la découverte"* (The feebleness of the means were inversely proportional to the importance of the discovery [**Chomard-Lexa, p. 88**]).

His important discoveries, including the first demonstration of Mendel's laws and the first enunciation of the one gene, one enzyme hypothesis are described in the text. Some of the reasons that he gave up mouse genetics are also described there but can be amplified. *"Primés au congrès de Boston en 1907, ses travaux furent mal compris, minimisés, ignorés et même critiqués dans le milieu scientifique français"* (First, at the congress in Boston in 1907 [the 7th International Zoological Congress, Boston, Mass.], his work was badly understood, minimized, ignored, and even criticized by the French scientific community [**Chomard-Lexa, p. 31**]). He was invited to London in 1908 to present his work on heredity but his speech was a mixture of English and German (which he involuntarily mixed) and it didn't go over well (**Chomard-Lexa, p. 32**). He had, meanwhile, turned to studies on the genetics of cancer in mice. Sadly, his research facilities were small and his mouse colony infected him with some parasite from which he suffered from for years. *"Il avait l'habitude de venir observer ses souris et déposait négligemment sa cigarette sur le couvercle des cages* (He had the habit of coming to observe his mice and negligently laying his cigarette on the cage covers [**Chomard-Lexa, p. 32**])—an obvious route for fecal-oral transmission. Thus, with poor acceptance of his results at home and abroad and difficulties in performing mouse genetic research, it was easy to give it up when World War I started.

Although near the age limit for military service, and with 5 children at the time (a sixth was to come), he was called up. He was ordered (with two other professors of the University of Nancy) to conduct conscripted horses from Nancy to Toul. He was still here with the horses awaiting further orders when he was discovered in a "military stable" by a visiting general [**Chomard-Lexa, p. 32**]. He was then sent to Paris for an "important post" but this seems not to have materialized. Meanwhile, his family had moved to Arcachon. He ended up spending the war teaching at the Lycée Poincaré and pursued his research in Arcachon. By happenstance, he was wounded by the explosion of a bridge in 1917. When he returned to Nancy at the end of the war, all that remained of his mice was *"un ensemencement de tout le quartier en souris brunes et en souris blanches"* a sample in the street of brown and white mice (correspondence of A. Cuénot to Dr. Michon, Sept. 23, 1954; **Chomard-Lexa, p. 32**)—remnants of his colony since the wild mice would be "gray" (agouti).

His post-war years were consumed with zoological research and "philosophical" writings. "In later years he covered practically the whole field of invertebrates, wrote monographs on sea urchins, comatula, phascolosoma; contributed monographs on sipunculids, priapulids, and tardigrades to the fauna of France; wrote chapters of a handbook of comparative physiology and others on animal geography; wrote the chapters on echinoderms, onychophores, tardigrades, and linguatulids for another handbook. All this and many other papers would have sufficed to establish him as a great invertebrate zoologist" (Goldschmidt, 1951). Having amassed this amazing zoological knowledge, he used it as material for philosophical consideration. *"Le biologiste éclaire un autre abime, la signification de la vie et il met au jour de grandes idées qui font maintenant partie du patrimoine intellectuel de l'humanité"* (The biologist lights up another depth, the significance of life and updates grand ideas that now are the basis of part of the intellectual patrimony of humanity" (Cuénot, *Discours*, 1898, cited by **Chomard-Lexa, p. 31.**)

His writings were strongly anti-Lamarckian and he carefully analyzed misinterpretations and alternative explanations of popular examples of the inheritance of acquired characteristics in his book: *L'adaptation* (Cuénot, 1925). Here he fully developed his non-Darwinian theories of evolution. His hypothesis of preadaptation proposed that organisms with pre-existing mutations would occupy "by chance or attraction" new ecological niches. *"Pour Lucien Cuénot, ce n'est pas la caverne qui fait le cavernicole mais l'animal aveugle se rend dans la caverne et son infériorité le gêne moins."* For Lucien Cuénot, it isn't the cave that makes the cave-dweller but the blind animal which enters the cave and its inferiority is less of a hindrance (Guénet, preface, Diligent, 2002). "His huge knowledge of zoological facts helped him to assemble and discuss the most interesting examples and to prove his point abundantly" (Goldschmidt, 1951). Thus, he accepted mutationist theory but used it in a non-Darwinian fashion because he couldn't accept the notion that small changes potentially useful later could be selected. "There is no reason why individuals should be selected because they possess a rudiment of an electric organ, or a little harder skin, or a color approaching that of their usual background" (Limoges, 1976). However, he also could not accept the new synthesis, the mathematical population genetics of Fisher, Haldane and Wright which united genetics and Darwin-Wallace's evolutionary hypothesis. *"Il ne peut adhérer à la théorie synthétique tout simplement parce qu'il ne peut abandoner l'étude de l'organisme individuel alors que la théorie synthétique privilégiait la population."* He wasn't able to accept the synthetic theory simply because he could not abandon the study of the individual organism while the synthetic theory preferenced the population (**Chomard-Lexa, p. 308**). He realized that the new synthesis explained micro-evolution but not macroevolution and this is still a bone of contention between "gradualists" and "saltationists." As Chomard-Lexa well summarized his views:

> *"Cuénot resta toute sa vie face à cette alternative:*
> *Soit supposer l'existence d'une volonté immanente, responsable de l'inventivité au sein de la matière vivante, et dont l'intervention serait plus sensisble à certains moment capitaux de l'évolution, comme la mise en place des grands plans d'organisation (le tout-venant se suffisant de la théorie darwinienne), Soit des facteurs mécaniques non encore découverts, actullement inconnus, et qui permettraient de lever le voile sur tout ce qui est obscur."*

> Cuénot rested all his life faced with this alternative:
> Granted to suppose the existence of an Immanent Will, responsible for the invention in the bosom of living material, and whose intervention will be more perceptible at certain capital moments of evolution like putting in place the grand plans of organization (the all coming to be sufficient for the Darwinian theory).
> Granted mechanical factors not yet discovered, actually unknowable, and that would allow one to lift the cloak over all that is obscure (**Chomard-Lexa, p. 308**).
> Thus, in the end, he was not a materialist.[2][3]

---

[2]Andrée Tétry (1907-1992), a student and collaborator of Cuénot, had written a memoir of L. Cuénot which was not published, after broken commitments and despite numerous attempts, at the time of her death (Diligent, 2002). She had no heirs and, thus, her papers were donated to the *L'Académie de Metz*. Her apartment in Paris had to be cleared out for sale and papers were hastily gathered to take to Metz. Among these was the manuscript of her memoir on Cuénot (over 800 pages ....!). A short version of this was eventually published in the Annales de l'Académie de Nancy (Diligent, 2002) with a foreword by Marie-Bernard Diligent and a preface by Jean-Louis Guénet. These writing provided a complementary view of the development of his scientific thought. They particularly emphasize his humor, his civility, his enjoyment of theatre and films and life.

[3]The quote is as obscure in French (testified to by Jean-Louis Guenet) as it is in English. I use it to show how obscure some of his deistic writings were.

United Kingdom

While the development of mouse genetics in France, after Cuenot and Dobrovalskia-Zavadskia, was delayed by a prejudice against genetics, the development of mouse genetics in England had to overcome the resistance of the quantitative trait-oriented Galtonists. In some ways this is surprising, since Galton had recognized the importance of single gene changes (sports) for evolution: "A mere variety can never establish a sticking-point in the forward course of evolution, but each new sport, affords one. A substantial change of type is effected, as I conceive, by a succession of small changes of typical centre, each more or less stable, and each being in its turn favored and established by natural selection, to the exclusion of its competitors. The distinction between a mere variety and a sport is real and fundamental" [Galton, 1892, cited in **Forrest, p. 219**]. Thus, it might have been anticipated that he would have recognized the importance of Mendel's work on the inheritance of such large changes, but none of his publications after 1900 took note of the rediscovery of Mendel's work. However, his disciples, especially Pearson and Weldon, took a very negative note of the rediscovery. The story of the conflict between Mendelism and the biometricians has been often told (Cock, 1973; Farrall, 1975; Kevles, 1980), as has the role of William Bateson, the bulldog for Mendelism as Huxley was for Darwinism.

Bateson was ready to rapidly accept Mendelism—his *Material for the Study of Variation*(1894) provided many examples that were more easily understood in terms of dominants and recessives. Although he provides several examples of such variation from mice, including color, silky hair, a black variety, and naked, most of the examples were botanical and his personal Mendelian experiments were botanical. Thus, it was a biometrician who published the first mouse genetics paper from England which used these terms, but concluded that dominance wasn't expressed in the crosses (which involved *piebald*, which has variable expressivity). Darbishire's short note (published in the biometrician's journal, "Biometrika") did not have the results of segregation ratios in the crosses which he had set up [Darbishire, 1902]. Darbishire had been a student of the biometrician Weldon at Oxford. When he saw the results of his mouse crosses, he "gradually came to an acceptance of Mendelism" [Kevles, 1980].

Although the genetics of mouse coat color made progress in England in the first half decade of the 20th Century, initially it was Cuenot's results which were confirmed and extended. Thus, in a note in *Nature* entitled "Mendel's Principles of Heredity in Mice", only vague generalities were given [Bateson, 1903]. Although Bateson was a fellow of St. John's College and a junior member of the Department of Animal Morphology, he had almost no research funds (small amounts of support came from the Evolution Committee of the Royal Society) but, as supervisor of the St. John's College farm and gardens, it was easy to do experiments with plants. He gradually attracted students from the women's colleges (mostly Newnham) and a group best characterized as a family developed [Richmond, 2006]. One of these students, his sister-in-law from Girton College, Florence Margaret Durham, was put to work on mice (and canaries). By 1908, Durham [1908] had extended Cuenot's findings on *dilute* and found a new example of *piebald*, but Bateson's 1902 book, *Mendel's Principles of Heredity*[Bateson, 1909], which summarized Mendelian experiments in many species, relied heavily on Cuenot for mouse work.

Miss Durham bred her mice "in a kind of attic over the Museums" (a collection of Natural History museums off Tennis Court Road [Richmond, 2001]). As her work progressed, she extended Cuenot's interpretation of gene interaction in coat color using the notion of presence or absence of gene activity. Thus, she enunciated the importance of epistasis (Durham, 1911). "A factor, then, is epistatic to another, when by its presence it conceals the existence of the other factor, although not allelomorphic to it" [cited in Richmond, 2001]. She also corrected Cuenot by showing that chocolate is a dilution form of black. Her work in mouse genetics ended when Bateson gave up his professorship in genetics, which did not include research support, and which was intended for 5 years (**Bateson W, Mendel G., 1909**)—he left after two years and became director of the privately funded John Innes Horticultural Institute. Miss Durham went on to work for the Medical Research Council and was involved in some human genetics; "she [also] conducted a nine-year investigation of the genetics of guinea pigs that suggested parental alcoholism was not an inheritable trait" (**Clayton, p. 223**).

Mouse genetics did not become well developed in the United Kingdom during the first half of the 20th century. There were no "focusing" institutions such as the ones in the United States—the Bussey Institute at Harvard, then the Cold Spring Harbor Laboratory, and, finally, the Jackson Laboratory. Each of these, mostly in succession, created a gathering and training place for mouse geneticists (see section on the development of mouse genetics in the U.S.A.). Douglas Falconer, in his note on the Institute of Animal Genetics in Edinburgh in 1947, reflects that "there were then (I think) only three university departments of genetics in the United Kingdom: London's University College (where J.B.S. HALDANE was Professor), Cambridge (with R.A. FISHER), and Edinburgh (with F. A. E. CREW). Edinburgh's department was the first, established in 1919 as the Animal Breeding Research Department, with CREW as its Director but no other staff and no building" [Falconer, 1993].

In an attempt to further locate mouse geneticists working in the United Kingdom in the first half of the 20th century, one can turn to the second edition of Hans Gruneberg's compendium of mouse genetics [**Gruneberg, 1952**]. Gruneberg worked at University College, London and his first edition of the volume, in 1943, was somewhat of a "bible" for mouse geneticists [**Gruneberg, 1943**]. It competed with the Jackson Laboratory's multiauthored *Biology of the Laboratory Mouse* which had its first edition in 1941. Gruneberg's massive compilation in the second edition [**Gruneberg, 1952**] thoroughly surveyed what was known of mouse genetics and physiology from the literature, mostly in 3 languages (French, German, and English). It contains 1751 references, many to abstracts, which cover all aspects of mouse biology. Making the assumption that the language indicated the source (English for the US and UK and, if the author's affiliation was known, other countries, French for France, and German for Germany), the distribution of researchers was ascertained[4]. As a measure of the location of mouse genetical research in the U.K. compared to the U.S., the source of many of the references was determined in a semirandom fashion[5] with the assumption that the locations of mouse biological research would also be the locations of mouse genetical research (since many of the references did not concern genetics). Fifty-eight of the subset of these 105, semirandomly chosen articles came from the United States, sixteen from the United Kingdom, eleven from France, ten from Germany or were written in German, four from Canada and six from 4 other countries with 1 or 2 each. Indeed, of the sixteen from England, the genetical ones were from Edinburgh, Cambridge or London while Leeds provided several articles, which used inbred strains of mice.

Falconer gives the reason for the founding of the Institute of Animal Genetics in Edinburgh as the need to improve agricultural output [Falconer, 1993]. It was initially funded by the Agricultural Research Institute which was set up as an animal breeding and research organization in 1945. Of the original 14 associated researchers, five (Falconer, R. A. Beatty, N. Bateman, A. L. McLaren and R. C. Roberts) worked with mice. It was a very active center for 33 years, producing many doctoral theses and having many foreign and local visiting scientists. Five of its members became Fellows of the Royal Society (including Ann McLaren, shortly after she left to found the Medical Research Council's Mammalian Development Unit in London—see her minibiography with Chapter 11).

The first institution in Great Britain which one can compare to the Jackson Laboratory was established when a Radiobiological Research unit was inaugurated at the Atomic Energy Research Establishment at Harwell in 1947 [de Chadarevian, 2006]. The motivation was to perform large scale studies of radiation-induced mutation rates in mice, since the whole world was concerned about the long term effects of exposure to radiation. "The scientists involved in the research were two recruits from Ronald A. Fisher's Department of Genetics at Cambridge, Tobias Carter and Mary Lyons, and Rita Phillips, a young geneticist from Liverpool." [ibid.]. The experiments conducted at Harwell were meant to supplement the much larger scale experiments by the Russells at Oak Ridge National Laboratories in Knoxville, Tenn. "For every British cage there existed 16 American cages." The "collaboration between the two groups remained difficult for reasons that are hard to pin down. Researchers at Harwell put it down to the reluctance of the Russells to share information with the British researchers as well as to keep to agreements." [ibid.]. Some other mouse geneticists also felt that the Russells kept things to themselves and did not fairly acknowledge the contributions of others. The unit was down sized in 1968, after the 1963 test ban treaty. Charles E. Ford's cytogenetic unit then moved to Oxford. However, the Harwell unit remained a key center of genetic research in the United Kingdom, and Mary Lyon's work with the *T/t*-complex, which will loom large in this book, was done at Harwell (Lyon, 2002). The center developed a particular interest in mouse models of human genetic diseases.

Another center of mouse genetics in the United Kingdom was the Medical Research Council's (MRC) National Institute of Medical Research (NIMR) at Mill Hill, a part of Northern London.

The institute also played a leading role in helping the MRC to set up the Laboratory Animals Bureau (LAB) in 1947, renamed as the Laboratory Animals Centre (LAC) in 1958. The LAB provided a standardized source of information on production, supply, maintenance and use of laboratory animals—much of it based on the MRC Animal Care Programme developed at NIMR (**Clayton, p. 213***)*.

The institute was an early center for maintaining Specific Pathogen Free (SPF) stocks which were important for exciting advances in mouse immunogenetics (see Chapter 7) and, later, making Mill Hill an important source of transgenics (see Chapter 9). The developmental biology unit at the MRC NIMR is also where *T* was cloned (see Chapter 9).

It is surprising that so little mouse genetics was being performed in England compared to the U.S. in these first decades of the 1900's. Kevles [1980] has suggested several reasons for the difference in overall amounts of genetic research between the U.S. and the U.K. While the opposition of the biometricians slowed the development of Mendelian genetics in the U.K., there also was conflict between biometricians and Mendelists in the United States, although the conflict was much less vitriolic. However, in the United Kingdom, there were fewer geneticists concentrated in a much smaller number of centers than in the United States. In addition, there was a closer relationship of academics with agriculturalists in the U.S. and the agricultural experimental stations found at land grant universities and colleges welcomed geneticists [Kevles, 1980].

Germany

As mentioned above, in the analysis presented of the references in Gruneberg's second edition [**Gruneberg, 1952**], only about 10% of his references were to German articles and, since he had emigrated from Germany, it is likely that he knew this literature as well as he knew the English literature. This relative paucity of German work in mouse genetics is confirmed by the nearly complete lack of mention of mice in Harwood's survey of genetics in Germany, which has the subtitle *The German Genetics Community 1900-1933* (**Harwood, 1993**). While Harwood's is primarily an intellectual history with a focus on thought processes as influenced by other scientists, society, politics and finances, he reviews the genetic material studied for almost everybody mentioned. His only mention of mice (they don't make it to the index) is that to Paula Hertwig, one of Bauer's assistants at the Agricultural College of Berlin, (all of whom were female and had Ph.D.s). she worked on mutagenesis and radiation-induced genetic damage in mice (**Harwood, p. 201**). The flour moth, *Ephestia,* seemed to be the most popular organism from the animal kingdom. Other insects (Goldschmidt worked on the gypsy moth and bees were studied), many plants, protozoa, and fungi are much discussed, but the only other mammals mentioned were pigs and rabbits.

Of 48 German geneticists whose social class and type of education are given in a table by **Harwood (pp. 289-290) (**in order to determine whether they could be classified as "prag-matists" or " comprehensives" on this basis), two were later to become mouse geneticists. Hans Gruneberg worked on *Drosophila* for his doctorate and then could only get a position in anatomy. He was invited by J.B. S. Haldane to London where his career in mouse genetics developed. Ernst Caspari (see his minibiography attached to Chapter 7) did exciting work in physiological genetics in the flour moth but, as a Jew, could not *habilitate* in Germany and went into exile in Turkey. This was the result of the Turks taking advantage of the antisemitism in Germany. "A progressive Turkish government, advised by a prominent Swiss educator, Malche, decided to take advantage of the unusual situation to build up a modern university in Istanbul" (**Goldschmidt, p. 292**). Caspari was later invited to the United States by L.C. Dunn and took up mouse genetics there.

A useful source of information on early mouse genetics in Germany is a perusal of the bib-liographies of **Searle (1968)** and **Silvers (1979).** These books on the genetics of coat color deal with the most popular early subject in mouse genetics, since coat color variation is the most obvious phenotypic variation in mice. Searle's compendium on mammalian coat color genet-ics has about 750 citations, with many in German and some in French. The earliest citation to mouse genetics in German is 1936 which contains a citation to an early German reference for cat coat color genetics in 1907. Silvers' compendium is limited to mice and includes citations in French, German, Italian, and Spanish among the roughly 800 in his bibliography. There are 2 papers in German on mouse pigmentation in 1916 and 1917 but nothing further until 1939.

Although genetics was much more studied in Germany than in France, there are some paral-lels in the relative underdevelopment compared to the United States. "Until 1945, there was but a single chair—at the Berlin Agricultural College—devoted exclusively to genetics in the twenty-two German universities and four agricultural colleges" (**Harwood, p. 143**). There was also a strong interest in cytoplasmic genetics, as there was in France: "German geneticists were interested in the role which heritable cytoplasmic structures might play in regulating nuclear genes' activities" (**Harwood, p. 6**). Nonetheless, as we have seen, there was active genetic research in many species but a paucity of research in mouse genetics. One can wonder if the anti-French attitudes of the previous century had prevented the spread of *Japanisme* to Germany. If so, this could have resulted in pet stores and homes not frequently having held the many variants of coat color and behavior (e.g., waltzers) that were available in France and the United States.

The United States

The history of mouse genetics in the United States has been extensively reviewed (**Morse, 1978**; Rader, 2004) so it will not be covered in great detail.

A major part of the success of mouse genetics in the United States was the preexistence of an institution, the Bussey Institute of Harvard University, which was ready to embrace mammalian genetics, including that of mice (Rader, 1998). This institute, located 6 miles from the main campus, had been established in 1871 as a setting "to foster teaching and research on agriculture and applied sciences" (Rader, 1998). It was reorganized as the Graduate School of Applied Biology in 1908. Of the 3 researchers hired, William Castle had already been doing mouse genetics (Snell and Read, 1993—see his contributions to coat color genetics mentioned under "France"). He had been on the faculty at Harvard since 1897 and had started work in genetics in 1900, at the time of the rediscovery of Mendel's work. It is an interesting asymmetry to note that he was, following the suggestion of a student (Davenport, 1941), the first to use *Drosophila* as a genetic material (Snell & Read, 1993), whereas T. H. Morgan, who pursued *Drosophila* so thoroughly, had started, and failed, with mice (see above).

> *From 1901 to 1905, CASTLE, with the cooperation of four students in successive years, followed the effects of inbreeding in Drosophila melanogaster through 59 generations of brother-sister mating. After moving to the Bussey, no one used Drosophila for thesis research, but it became the favorite material for Harvard undergraduate teaching in Cambridge so the graduate students were familiar with its usefulness*
>
> *(Weir, 1994).*

At the beginning of his career, Castle was a very strong defender of Mendelism,. He had started his work on mice, rats, rabbits, and guinea pigs in a Harvard basement and moved his animals with him to the Bussey Institute. Luckily, he "brought with him a Carnegie Foundation grant specifically dedicated for work on mammalian heredity" (Rader, 1998).

The Bussey Institute, far enough from Cambridge to encourage local autonomy, developed a horizontal organization. Many of the students lived on the Bussey grounds in a house which was turned into a dormitory in 1915. Students took turns making meals and sharing chores, although eventually a cook was hired (Rader, 1998). Each mouse genetics graduate student supervised the mouse colonies during his thesis mentorship and worked with the mutants then available. These were primarily from the mouse fanciers:

Castle often obtained new strains through his dealings with American mouse fanciers, whose extensive breeding programs and careful observational tradition turned up many unusual organisms. Clyde Keeler remembered that in addition to buying mice from fancier suppliers like Abbie Lathrop's Granby Mouse Farm [mentioned above], 'mouse researchers always cooperated with the Mouse Fancy' by exhibiting their special mice at local Boston mouse shows in return for samples of new or existing mutants (Rader, 1998).

As mentioned, Castle worked extensively on the genetics of coat color in a variety of species of mammals. He was another pioneer in the notion of "one gene, one enzyme": he "said that the general color factor *C* was acted upon by specific substance, perhaps color enzymes… producing specific pigments such as black, brown, or yellow. The color enzymes were somehow produced by the genetic information" (**Provine, p. 57**). However, his studies of hybrids and attempts at selection led him to some distinctly un-Mendelian ideas.: "he continued to maintain that it was possible to create permanent blends of contrasting Mendelian characters, and he spent much of his time between 1907 and 1914 trying to prove that selection played a role in this process" (Rader, 1998). Thus, his major contribution was the training of a strong group of mammalian geneticists who were to develop the field in the United States.

The communal atmosphere of the Bussey, and the minimum 4 years that it took to do a thesis in mammalian genetics (given breeding times), had the potential to create a strong sense of community among these students. At the same time, as mentioned, mouse genetics graduate student's supervised the mouse colonies during their time in Castle's laboratory and under his thesis mentorship. Mouse care was much more laborious then than now—self-drinking water bottles had not been developed and small dishes were filled with milk each day. The boxes were wooden and gnawing mice led to frequent replacements. Of course, the students reeked of mouse urine. Thus, "this culture stressed an ethos of independence and individualism more than cooperation and communality. Nevertheless, Mouse People also achieved a strong sense of community through their shared experience of working and living at the Bussey" (Rader, 1998). Part of this ethos was sharing of strains and mutants, an important tradition among geneticists that, in later generations, was not always emulated by molecular biologists.

One of the important developments at the Bussey was the establishment of inbred strains (earlier than Nadine Dobrovalskia-Zavadskia started in France). Although Mendel's laws showed that "pure bred" plants bred true, "Johannsen's classic paper published in 1903 on 'Heredity in populations and pure lines [in beans]' may be regarded as the first paper which demonstrated the properties of inbred strains" (**Festing, p. 14**). Although initial litters were small and some lines of brother-sister matings were lost, it became clear that mice were amenable to such inbreeding (as were other mammals—Sewall Wright inbred guinea pigs at the same time). As Elizabeth Russell describes it:

At least three former Bussey graduate students have described to me discussions they had about inbreeding, its genetic consequences, and the types of characteristics established in inbred strains. These discussions must have been persuasive, because almost all Castle students have produced and studied their own inbred strains

(Russell, 1985).

The major purpose was to facilitate cancer genetics. "The majority of inbred strains, from the most recent back to DBA [C.C. Littles's first inbred strain], were developed for use in cancer research, to prove or disprove the existence of genetic factors influencing the incidence of cancer and the independence of inheritance of different types of cancers" (Staats, 1966).

Of the mouse geneticist's Castle trained, Clarence C. Little (see minibiography) became the king pin of the continuation of mouse genetics in the United States while L.C. Dunn will figure importantly in later chapters. The students did not just learn mouse genetics—Castle encouraged broad discussions of results and the Bussey Institute was shared with plant geneticists with whom seminars occurred.

Little remember[ed] learning the hard way that he was expected to take these Bussey faculty interactions as a model for his own success as a mouse geneticist: he failed his initial Ph.D. qualifying exams precisely because he could not demonstrate and argue about the broad knowledge of biological and genetic phenomena that Castle so valued

(Rader, 1998).

Sadly, the Bussey Institution was closed by the Harvard Board of Overseers in 1938 when they would not approve of Raymon Pearl as a new director (Goldman, 2002).

Following the Bussey Institute, the Station for Experimental Evolution at Cold Spring Harbor became an important center for mouse genetics. Castle's student, E. C. McDowell was recruited, after he graduated in 1912, by the station's founder, Charles Davenport. Little joined him after his military service in 1919. Little brought his inbred strains and continued to cultivate them as an important resource, an effort which McDowell shared. L. C. Dunn, William Gates, and George Snell were recruited as summer visitors. "Just as T. H. Morgan would send his flies to Woods Hole every summer, Dunn initiated a regular practice of sending his entire mouse colony to Cold Spring Harbor for the warmer months" (**Rader, p. 52**). The work at the Station for Experimental Evolution was also supported by the Carnegie Institution and allowed the colony to grow to about 10,000 mice. In the early 20s the summer groupings grew to the extent that "Mouse Club of America" was started; it's newsletter became an important organ for dissemination of experimental results and availability of various inbred strains. **Rader** (p. 170) states that it continued, and that Snell made it a more formal publication starting in 1941, although its text was typewritten (? mimeographed) for many years. Michael Festing provides a different description of its origins: George Snell "proposed that a *Mouse Genetic News* should be produced. The first edition was published in November 1941, but because of the war it could only be distributed in North America. It contained the rules on gene nomenclature, lists of mutants and inbred strains, and lists of laboratories holding mouse stocks. A second edition was published in the *Journal of Heredity* in 1948. Although this publication was useful, it was recognized by many that a regular news sheet was needed. Eventually, *Mouse News Letter* was started, and the first edition was produced in July 1949, under the editorship of Drs L. C. Dunn and S. Gluecksohn-Schoenheimer, although the editorship was immediately handed over to Dr. T. C. Carter." (**Festing, p. 18**).

At Cold Spring Harbor, Little and Davenport developed a conflict and Little left for the University of Maine (see Little biography). Davenport continued his eugenic research and

Little's move to Maine eventually led to the establishment of the Jackson Laboratory, the institution that guaranteed a major role for mice in genetic research.

As already emphasized, the acceptance of genetic researchers by the agricultural research stations (Dunn spent 9 years at the Storrs Agricultural Experiment Station in Connecticut) and universities helped establish the pre-eminence of mouse genetic research in the United States. However, the role of strong personalities and sympathetic funding organizations (the Carnegie Institution especially) allowed the development of other institutions (the Bussey, the Station for Experimental Evolution) which also played a huge role in the "institutionalization of mice."

Clarence Cook Little, 1888–1971

C. C. Little was born a Boston Brahmin (i.e., a member of the Boston elite class), a valuable distinction for one who eventually needed to raise money for work with his mice[1] . He attended Harvard, where he excelled in track, and while there fell under the influence of W. E. Castle, the grandfather of early mammalian genetics in the United States. He spent time at the Bussey Institute, the Harvard "out-station" for agriculture and horticulture where the students helped with the "chores" (see Chapter 1) and this involved taking care of the mice. His major interest in mouse husbandry must have been influenced by these experiences. While an undergraduate, Little perceived the utility of inbred stocks for genetic uniformity and started such inbreeding despite Castle's negative views on inbreeding. He also insisted on a more rigorous nomenclature than was usual with the mouse fanciers from whom many of the initial stocks came. Thus, his first inbred strain was designated dba (now DBA; the lower case was used to designate the recessive nature of the d [dilute], b [brown], and a [non-agouti] alleles that it carried). Little stayed on at the Bussey Institute as a graduate student, working mostly on coat color genetics and graduating in 1914[2].

The world of mouse genetics was much smaller in the second decade of the 20th Century and personal relationships less diluted—disagreements over results occurred between CC Little and A. Sturtevant about linkage of coat colors and, in particular, between Little and Maude Slye about the genetics of cancer. Little had found employment with E. E. Tyzzer, one of the early medical scientists who found mice interesting for the study of cancer , in a position analogous to a present-day post-doctoral fellowship. Little's work with Tyzzer was particularly fruitful in the area of tumor transplantation (which he thought for decades was relevant to oncogenesis, while it really was relevant to transplantation immunology). For example, their F2 of the dba inbred strain and the less inbred Japanese waltzer ,mice resulted in only 3/182 mice that would accept a waltzer

Clarence Cook Little when President of the University of Michigan.
Source: Used with permission from University of Michigan, 1927 Michiganensian (University of Michigan yearbook), p. 3 (published by the University of Michigan in 1927).

transplanted tumor. Little correctly argued that mouse tumor susceptibility (read "transplantation acceptance") might require as many as 15 segregating factors (Little and Tyzzer, 1916), which was a reasonable estimate of the number

---

[1]A good source is *Making Mice* by Karen Rader which is mostly a biography of C.C. Little and a history of the Roscoe B. Jackson Memorial Laboratory.

[2]Little's contributions to the genetics of coat colors are well documented in Searle's *Comparitive Genetics of Coat Color in Mammals*, Silver's *The Coat Colors of Mice,* and Little's review (1958).

of segregating transplantation antigens. These experiments made him very prejudiced against Slye's claims of simple Mendelian recessive inheritance of cancer in mice. The objections were published in *Science*. The argument between these two was to go on .

After 3 years with Tyzzer in Boston, Little spent nearly 3 years in the U. S. Army Air Corps (Air Force), remaining in the United States during the First World War. Following his military service, he moved to the Station for Experimental Evolution at Cold Spring Harbor Laboratory. This institution was headed by Charles Davenport, a now notorious eugenicist who was part of a group of "closed-mind individuals whose social views greatly influenced their interpretation of scientific evidence" **(Ludmerer, p. 48)**. Davenport valued mouse genetics, and supported the Cold Spring Harbor lab in becoming a major center for it. In the summer, some investigators (e.g., L.C. Dunn) even brought their mouse colonies with them. In the 3 years (1919-1922) that Little was there, the exchanges between mouse geneticists created a cooperative group which launched the Mouse Club Newsletter. The newsletter continued, according to **Rader [p.170]**, until it became a more formal publication in 1941, (see text) which provided a rapid method of information exchange.

Although made Assistant Director of the genetics research facility, Little soon had a falling out with Davenport. His increased national prominence, partly from the debates with Maude Slye on the genetics of cancer in mice, led to an offer to be President of the University of Maine. He had been introduced to academic politics as a paid secretary to the Harvard Corporation, the university's governing body, while a graduate student. He was attracted to the possibilities of implementing progressive approaches to higher education which he valued, as he did scientific research. He accepted the offer, after the university agreed to provide money for his personal research. However, this agreement was soon withdrawn and his mouse research made little progress during his 3 years in this job. Perhaps the most important career development while at the University of Maine was the opportunity to meet and get to know rich industrialists who summered in Maine, in particular Roscoe B. Jackson. These contacts, many from Detroit, also probably helped in his being chosen for the Presidency of the University of Michigan in 1925. Again, his scientific pre-eminence was an important factor, and arrangements were made for him to continue his mouse research. This increasingly focused on cancer research and had two important outcomes: many public speaking opportunities to physicians and potential philanthropists, and funding from the latter for a mammalian genetics/cancer research lab. However, his attempts to alter the curriculum and, possibly, the perceptions of his extra-marital affair with the Dean of Women (Beatrice Johnson, a former student of his at the University of Maine whom he married after his divorce) led to conflicts with the Regents, and he resigned in early 1929.

Although he had no job lined up when he resigned the University of Michigan Presidency, his efforts to build a large research institution at Michigan and his contacts with industrialists led to a gift of land on Mt. Desert Island by George B. Dorr, a friend of his mother, to initiate a major mouse research lab. With the death of one of the major donors, it became the Roscoe B. Jackson Memorial Laboratory.

While well-funded, by the day's standards, at its start, the Oct., '29 "crash" soon caused economic hardships as 3 year funding commitments came to an end. In the meantime, Little's very active schedule of public speaking (at which he excelled) had made him a major figure in cancer circles, and he was chosen as managing director of the American Society for the Control for Cancer. This organization later became the American Cancer Society after the Laskers, also the funders of the "pre-Nobel" Lasker Prizes, transformed it from "a professional organization of doctors and a few scientists, [which] was self-contained and moribund, an ossifying Manhattan social club" to "an intense, well-oiled fund-raising machine" from which Little was forced-out in 1945[**Mukherjee, pp. 111, 113**]) While in this position, Little had an excellent podium from which to advertise the importance of standardized (inbred) stocks of mice. Although Little was not initially in favor of selling mice (mouse geneticists had a well-established tradition of sharing mice), the request from United States Public Health Service (USPHS—later to establish the National Institutes of health [NIH]) cancer researchers for the lab to supply all of their mice stimulated a gradual shift in perception of the need for selling mice. By 1932, major economies in breeding mice (fox chow by the ton instead of seeds by the hundred-weight, self-regulating water bottles instead of a daily exposure to a dish of milk) allowed cheaper production, but sales were still limited to the one group of USPHS researchers. However, by 1934, the desperate lack of funding led to Little's development of a catalog for mouse sales targeted to a variety of medical researchers, and purchases gradually increased. At the same time, the Rockefeller Foundation was reorganizing its priorities with genetics high on the list. An "emergency" grant of $11,000 in 1934 was followed by a 3 times larger commitment starting in 1935.

Over the next several years, Little was heavily involved in "pushing" genetically standardized mice for research, with articles in magazines such as *Good Housekeeping* and even with a cover photo of mice in *Life*. This publicity, and the fact that mice were "low on the screen" of the antivivisectionists, who were then prominent in the public eye, helped him to secure annual support for the Jackson laboratory when the National Cancer Institute act was passed in 1937, and the Rockefeller Foundation contributed $40,000 for a new building. Thus, in the last half of the 4th decade of the 20th Century there was money to hire more staff scientists, including George Snell (a future Nobel Laurate). The Russells, Bill and

Tibby, were also particularly outstanding recruits. Another recruit was J.J. Bittner,who worked on the mouse milk factor which led to the discovery of Murine Mammary Tumor Virus. In his case, concerns about duplicate funding (paying for the mice and then paying for the research on the mice) and duplicate efforts (Bittner's work at the Jackson Lab and Howard Andervont's similar work for the National Cancer Institute, first at Harvard and then in Bethesda) resulted in decreasing research funding and Bittner left for the University of Minnesota in 1943.

The era of the '30s and '40s, thus, was a period in which Jackson Lab research was "struggling", and this was blamed by Tibby Russell on Little's refusal to consider any research that "might overthrow the genetic theory of cancer" (**Rader, p. 177**). The analysis of one reviewer (Walter Loomis of the Rockefeller Institute) of a research grant application from the Jackson Lab was succinct: "It is probably significant that Little sends out vast quantities of typewritten reports rather than sending reprints. It can probably be honestly said that the valuable export of the Jackson Memorial Laboratory is in terms of boxes of mice rather than scientific publications"[3].

The fire that destroyed most of the Jackson Lab in Oct., 1947 allowed Little to shine as a leader in the rebuilding effort, which led to a second birth of the lab. Mouse geneticists showed their desire for the Jackson Lab to remain a center for mouse genetics by sending back breeding pairs of lost strains. "The 1947 fire came at a propitious time for the scientific community. Just as large number of researchers were coming to depend on animals from outside suppliers, disruption by the fire focused attention on the importance of selecting the right animals for a particular project...[the staff] now wanted to apply genetic know-how to guarantee ready availability and continuity of pertinent, genetically uniform , well-characterized mice for the growing biomedical research community (Russell, 1987) There was also great public financial support attendant to national publicity about the fire.

The better-than-resurrected Roscoe B. Jackson Memorial Laboratory celebrated its 25th anniversary in 1954 with C. C. Little presiding over a well-attended event. Little retired from the Directorship in 1956 and wrote a few scientific reviews (little, 1958). That same year he became director of the Tobacco Industry Research Committee[4], which was a "smokescreen" to promulgate the idea that there was no cigarette smoke toxicity. In 1957, Little testified at a Federal Trade Commission hearing on limiting cigarette advertising that "efficacy of filters was immaterial because, after all, there was nothing harmful to be filtered anyway" (**Mukherjee, p. 263**). He died of a heart-attack in 1971.

[3]William Loomis Diary excerpt, 28–29 March, 1951, cited in **Rader, p. 216**.

[4]That legacy lived on, as until the mid-80's, smoking was permitted in the Jackson Laboratory – even at seminars in the C. C. Little auditorium (although confined to one side).

The discovery of the *T*-locus

Despite these antagonisms to genetics discussed above, some genetic research was going on in France. "Most genetic research between the two wars was performed outside the universities in the Institut du Radium [and 6 other locations]. For the most part, those places produced only sporadic and routine work..." (Burian, et al, 1988). I would beg to disagree and would argue that Nadine Dobrovolskaia-Zavadskaia's work at the Institut du Radium, and her discovery of the *T*-locus went far beyond these dismissals. As mentioned in the minibiography of Dobrovolskaia-Zavadskaia, her department head, Professor Regaud, suggested that she study the effects of X-radiation, not just on the morphology of the testes, but on its effects on heredity. At this time (1923), the concept of mutation was still in flux. There was a tension between beliefs about the unit of inheritance, named "gene" by Johannsen in 1909 , to account both for the notion of its permanence in passage through generations, and the need for variability in the unit for evolution to occur. A name was needed for changes in genes which phenotypically were called "sports." De Vries introduced the word "mutation" for the process by which apparently new species of *Oenothera* had appeared (**Mayr, p. 743**). "Morgan... adopted de Vries's term 'mutation' for the origin of a new allele. This transfer of the term was rather unfortunate in view of de Vries's evolutionary mutation theory [see Chapter 2]....Consequently, it resulted in considerable confusion during the ensuing twenty or thirty years" (**Mayr, p.752**).

At the time Dobrovolskaia-Zavadskaia started her studies, others had sought to generate mutations with heat, chemicals, and radiation with both X-rays and with radium: "As early as 1920 some strictly genetic experiments with X-rays and radium were started. The most conclusive results [for the latter were]...obtained by Nadson and Philippov [1911], who succeeded in inducing new stable races of fungi" (Timofeeff-Ressovsky, 1934). However, it was Muller's definitive work in *Drosophila* that marked the particular chromosomes exposed to radiation, and the stages of gametic development during which various rates of X-ray induced mutation occurred that was well accepted. The work also showed increasing mutation rates with increasing exposure, which convincingly supported the hypothesis of cause and effect (and won him the Nobel prize) (Muller, 1928).

The results of Nadine Dobrovolskaia-Zavadskaia's work on heritable genetic changes were reported in 1927. In the descendants of irradiated mice she found tail abnormalities which are now known as *Brachury (T)* (Dobrovalskaia-Zavadskaia, 1927b). She was impressed that sometimes a normal length tail was replaced by a thin filament, as if dry gangrene had occurred, and thought that short tails without the filament may have lost it. She named the character "non viable" (in contrast to lethal) because only one organ (the tail) was incapable of life. She thought that this nonviable character was accompanied by a lethal factor, since homozygotes could not be obtained; there were always some normal tailed mice among the offspring.

The quantitative results of these genetic crosses were presented in an accompanying paper, and were compared to results obtained by two previous authors (Lang, Duboscq) who had bred lines of tailless mice (Doborvolskaia-Zavadskaia & Kobozieff, 1927). She found a ratio of 2:1 for tailless and short-tailed to normal tailed mice and compared this to the 2:1 ratio of yellow to albino Cuénot had found with the yellow mice he had studied, concluding that a lethal factor was involved and that homozygotes did not survive.

She asked herself whether the appearance of this character was due to the X-ray exposure to antecedents, and concluded *"Nous croyons que non, tout d'abord pour cette raison que, sur un nombre considérable de Souris irradiées et mises en reproduction, nous n'avons trouvé ce phénomène que dans deux lignées"* (Dobrovolskaia-Zavadskaia, 1927b). (We believe not, first for the reason that, from a number of mice that were irradiated and mated, we only found two lines that had this phenomenon). In stating this, she seems to believe that a mutagen would produce only one kind of mutation multiple times—perhaps more like the effects of a teratogen which can cause a repetitive defect of development, an interpretation of her work that Lee Silver has presented (Silver, 1990). The gradation in severity, also seen by M. Vallois in the mice studied by O. Duboscq, suggested other factors than X-rays to her. the concept of variable expressivity having not yet been developed). This first paper concluded that the observed abnormality wasn't absence of the organ, but defective development, and that it *"présente un grand intérêt au point de vue de l'origine des mutations"* (it has great interest from the point of view of the origin of mutations, Dobrovolskaia-Zavadskaia, 1927b). This again suggests her view of mutations as a developmental defect rather than a change in the hereditary material.

FIG. 1.1 The building that the *Institut du Radium* relocated to in 1935. Author's photograph.

Dobrovolskaia-Zavadskaia persisted in her denial that the anury/brachury factor, *T,* was an induced mutation during all her studies on the material. This is of interest because, when the gene was cloned, and her initial *T* mutant sequenced, it was found to be due to a deletion, strongly supporting the likelihood that it was induced by radiation (Herman, et al, 1990). In 1934 she restated that her own attempts to induce mutations by X-rays resulted in only 3 "*mutations*", (1 waltzer and 2 short tails, the latter from 2 different irradiated males) among 3,000- offspring from 500 litters across 3 or 4 generations, obtained from 23 irradiated males (Dobrovolskaia-Zavadskaia, et al 1934). Another reason for rejecting mutations as the cause of her *T* mice had been reported in her earlier review of mutational studies (1929): one of the irradiated males had "a tail a little shorter than normal and having a kink on the end....this male was also irradiated twice and mated with three normal females." These matings resulted in three litters, and in two of the litters, *i.e.* in the F$_1$ generations, some frankly *brachury* descendants were found. However, she convinced herself that a pre-existing variation was not the cause of her new mutants "since we have enlarged our stock of mice with the somatically normal tail, taking our breeders from the ancient stock and also from some new source. We have now more than three thousand mice, in several pedigree lines, which after careful examination have failed to show any abnormality" (Dobrovolskaia-Zavadskaia, 1929). This argument was not repeated in 1934 (see below). She had not only convinced herself, but the mouse community at large, that the short tail form was pre-existing. As Gruneberg states in his definitive book: "The short-tail gene was discovered by Dobrovolskaia-Zavadskaia ... in the course of X-ray experiments. Its appearance had nothing to do with the action of the X-rays, as one of the males originally treated proved to be heterozygous for the gene; another case occurred in the second generations after treatment" [**Gruneberg, 1953**].

It is interesting that in Nadine Dobrovolskaia-Zaavadskaia's review of the increased mutation rates found by Mueller and others, these were interpreted by her as the enhancing of a latent possibility. She noted the variability in response to X-rays in Muller's and other 's work and interpreted her work in this light. In particular, in her overview in 1929, after Mueller's work that led to his Nobel prize had been published, she noted the variability in mutation rates that Muller found, quoting him as having obtained "variation of the order of 1000 per cent"). This finding , and the overall low rate of production of mutations by X-rays led her to conclude "*that the rays are unable to produce any new form* (her italics)" (Dobrovolskaia-Zavadskaia, 1929). Thus, she thought that the radiation merely revealed forms already potentially present, but her use of "mutation" and "potential mutants" discloses an ambiguity about the source of phenotypic variability. It is as if she did not locate it in the gene. Perhaps she was influenced by L. J. Stadler's views on mutation in maize: "Stadler accepted only those cases where the irradiation of the new mutant could produce a reversion to the pre-irradiation character. Such cases, at least in maize, were very rare" (**Mayr**, p. 804).

The developmental implications of the tailless/short-tailed mutant was developed further in a third paper (Dobrovolskaia-Zavadskaia, 1928). Here she concluded that the material demonstrated that separate genes for the "*parties molles de la queue*" (soft parts of the tail) and for their "*squllette*" (skeleton) existed. It is these developmental speculations that have led many to consider her the founder of developmental genetics (Korzh & Grunwald, 2001, Silver, 1990).

She continued studies on these anury/brachury (*T*, tailless, short-tailed) mice into the next decade. She soon reported segregation ratios on much larger numbers of mice, firmly establishing the 2:1 ratio of *T*:normal in crosses of *T*/+ by *T*/+ (Dobrovolskaia-Zavadskaia & Kobozieff, 1930). In this study, a small number of pregnant females were dissected and the approximately expected number of fetal deaths (resorbing fetuses) were found. This was confirmed with a much larger group of dissections by Chesley (1932).

Her discovery, with the assistance of another Russian *émigré* scientist turned mouse geneticist, Kobozieff, of recessive alleles which bred true for no tail, was announced in 1932 (Dobrovolskaia-Zavadskaia & Kobozieff, 1932). They developed 1 line from the mating of a short-tailed mouse with a female from another stock "*qui n'a aucune parenté avec notre stock muté*" (which did not have any parent from a mutated stock). The mating produced tailless mice that bred true for 7 generation. Another line which bred true for taillessness came from the cross of a short-tailed female with a normal tailed male from one of Dobrovolskaia-Zavadskaia's long-established lines, which she characterized as "normal." A third line that bred true for taillessness was obtained from the cross of a tailless female from their established stock with a wild male, caught near Saint-Cyr (20 Km West of Paris). It is interesting that the American workers who studied these stocks, Chesley & Dunn (1936), thought of them as all being from crosses to wild mice ("From the original short-tailed (Brachy) stock following outcrosses to wild mice...."). This is an interesting "slip", perhaps due to the ease of finding recessive *t*-alleles which interact with *T* to only produce tailless progeny in the wild (see next chapter). Dobrovolskaia and Kobozieff (1932) actually deliberately bred mice from the tailless lines to wild mice to show that normal-tailed mice could result, i.e. the dominant *T* seen on a "wild type" background. They noted that there was a difference in how *T* was expressed, coming from the same females, with the 2 males of quite different provenance. One male from Banyuls, which was later determined to be *Mus spicilegus hispanicus* Miller, gave short tails, while the other male from the Paris region gave mostly tailless progeny—in both cases segregating 1:1 with normal tails. This variation was not interpreted in terms of modifying genes, perhaps because the term was just entering the vocabulary, but she did discuss modifiers in her 1934 review (see below). Dobovolskaia-Zavadskaia and Kobozieff correctly interpreted these "tailless" stocks as balanced lethal lines.

Her work on *T*, and an extensive review/history of mouse genetics in general, was summarized in a 108 page article, with 4 plates, showing 153 radiographs of skeletons, mostly just tails, and interspersed with 25 photos of mice and their tails (Dobrovolskaia-Zavadskaia, et al, 1934)[3]. By this time, she could review 17 genes in 3 linkage groups. Chromosomes were listed (the correct number of autosomes [20] was known) but assignments of linkage groups to chromosomes was arbitrary [see Chapter 3]. After her already presented arguments as to why X-rays had not caused the "mutants", much of this massive article provides anatomical and radiological characterizations of the mutant tails. Extensive lists of almost every mouse's deviation from normal add little to her correct finding that usually the deformation of the tail depends on underlying deformation of the vertebrae. Quantitative inheritance of differing lengths of tails, with and without varying numbers of vertebrae, were described. In the lengthy discussion, she compares *T* to the *Drosophila* mutant *Notch*. Both share a loss of substance of the organ affected (wings in the case of *Notch*). Secondly, they both show great variability. Since Mohr (1919) had described a chromosomal deficiency in a *Notch* allele (shown by nearby recessive genes being partially expressed, and by male lethality for *Notch* recipients (*Notch* is X-linked), she made the case for *T* being a deficiency (which, indeed, her mutant was, as already discussed, and almost certainly due to X-rays!).

She followed Brachet (1927) in emphasizing that development is not only of form (morphology) but also of size. Thus, she feels *T* is a good choice of gene symbol for anury/brachury (first used by her in this review) because "*Le symbol T présente cet avantage d'être la lettre initiale du mot français 'taille' qui traduit pour nous l'essence de ce facteur; en même temps T est la lettre initiale du mot anglais "tail" queue, et elle est adoptée par la nomenclature génétique américaine pour désigner la mutation 'queue courte'*" (The symbol *T* presents this advantage of being the initial letter of the French word "size", which translates for us the essence of this factor, and at the same time T is the letter initial of the English word tail , and it has been adopted by the American genetic nomenclature to designate the short tail mutation).

She points out that the shortening of the tail always involves a reduction in the number of vertebrae. Since the length of the tail is more similar to that of the parents and "neighbors", she supposes that *t*, controlling the development of the normal tail, is influenced by modifiers, a concept not used in her paper of 1932 but used by Sturtevant and Schultz for variation in bristle number in 1931 (Sturtevant and Schultz, 1931).

Since many of the *T* mice had bent tails, she also discusses the mutation *kink* and follows Hunt (1932) in concluding that it is recessive (others had argued otherwise, with data suggesting incomplete penetrance of a dominant). She concludes that the normal allele, *K*, produces a straight tail, with *kk* causing curved tails due to either deviation in the vertebrae or in the intervertebral cartilage (ankylosis). Although she interpreted *K* as functioning independently of *T*, she realizes that its existence as a separate gene is not yet proven [*Kink* is further discussed in Chapter 3].

Her discussion not only encompasses the developmental implication of *T* (perhaps justifying her status as the founder of developmental genetics), but also the phylogenetic implications of tail loss in some species such as our own. As she discusses gradual loss versus loss in one step, her thinking clearly implies a concept of 1 to 1 relationships of genes to the parts of the body they control: "*En effet, la diminution progressive des vertèbres coccygiennes aurait pu être effectuée en même temps que la suppression de la queue, en raison de la perte de substance du plasma germinatif au cours d'une perturbation chromosomique*" (In effect, the progressive decrease in coccygeal vertebrae might have occurred at the same time as the suppression of the tail, because of the loss of the germinal substance due to a chromosomal perturbation).

She finishes this review, which is her "swan song" to *t* genetics, with an overview of chromosomal mutations (as we would now term them [translocations, deletions, inversions]) in the light of all known radiobiological effects. She gives as the reason that new mutations do not occur "*Cela peut aider à comprendre pourquoi on n'a obtenu dans les gamètes irradiés que des mutations qui apparaissent spontanément; peut être parce que ce n'est que celles-ci qui échappent à la destruction*" (This helps us to understand why we only obtain mutations which occur spontaneously from irradiated gametes, perhaps because only these escape destruction). It is hard to follow her reasoning here but it seems put spontaneous mutations on a different chemical footing than induced mutations and a belief that they would survive the radiation which caused the chromosomal mutations (Dobrovolskaia-Zavadskaia, et al, 1934).

Her seven final conclusions to this magnus opus include another statement that *T* is a deletion: "*La rapprochement de la mutation "queue courte" des phénomènes* analogues étudiés sur la Drosophile, rend probable l'hypothèse qu'il trouve à sa base *une aberration chromosomique du type de déficience*" (The comparison of the "short tail" mutation with analogous phenomena in *Drosophila* makes probable the hypothesis that it is based on a chromosomal aberration of the deficiency type) (Doobrovolskaia-Zavadsiaia, et al, 1934). She sums up the morphological work with a clear developmental genetic hypothesis:

L'étude morphologique des déviations du type normal a permis d'avancer une hypothèse, d'après laquelle il y aurait des gènes, ou des centres régulateurs spéciaux, pour la taille spécifique, et pour les dispositions formative de l'organe. Ces gènes régulateurs principaux fonctionneraient en collaboration avec des gènes accessoires ou modificateurs" (*Morphological study of deviations from normal allows the advancement of an hypothesis by which there will be genes, or regulatory centers, specifically for the tail, and for the developmental forms of the organ. These regulatory genes function principally in collaboration with accessory genes or modifiers*)

(Dobrovolskaia-Zavadskaia, et al, 1934).

With this hypothesis (which one could make more modern with *Hox* genes as the regulators and signaling molecule genes as the modifiers), she deserves the title suggested by others (Korzh & Grunwald, 2001; Silver, 1990) of the founder of mammalian developmental genetics.

Her work was continued by L.C. Dunn, with whom she had correspondence and visits and who received her mice (see minibiography). According to L.C. Dunn's biographer, D. Bennett, she

came to the United States to lecture for some Russian refugee organizations, and appeared at Dunn's laboratory at Columbia looking for a solution to her puzzle. The best solution she could see was 'to give up the confusing tails and return to my proper field which is cancer research.' Dunn was fascinated with the problem and answered, 'If you are going to give them up, just give them to me"

(Bennett, 1977).

Nadine Dobrovolskaia-Zavadskaia, 1878–1954

We know almost nothing about the early years of Nadine Dobrovolskaia-Zavadskaia's life. There are almost no personal papers; some may have been sent to Russia and lost there[1]. A search of the Russian Academy of Sciences archives in Moscow, courtesy of Natalia Kolotilova, Moscow State University, finds only one letter from her (and even that non-authenticated). She was born in Kiev in 1878 and received her higher education in St. Petersburg[2]. This suggests an upper middle class family— either the family was of a commercial/professional class that could move to St. Petersburg, or they could afford to send her there for her education. We can also guess that she was introduced to foreign languages at a young age— she was fluent in French, and the few of her scientific articles in English (all of which were reviews, presumably to introduce her work to an Anglophone audience) are without linguistic mistakes, and no linguistic help is acknowledged while that of technicians is. Her handwritten letters to L.C. Dunn show both fluent English and a lovely script[10].

Her medical education was at the Women's Medical Institution in St. Petersburg, although one source names the University of St. Petersburg[2] which seems unlikely, see below. In the 1870s, Dmitry Tolstoy's rigid control of the universities kept them closed to women: "From 1882-1897 there were no [university] courses available to them in Russia" (Abisetti, 1982) although Tsar Alexander II allowed women's medical courses at such "medical institutes'" starting in 1876 (**Kassow, p. 23**). A considerable number of Russian women studied medicine in Germany and Switzerland during the

Nadine Dobrovolskaia-Zavadskia from BioEssays, *Nadine Dobrovolskaïa-Zavadskaïa and the dawn of developmental genetics, Volume 23, Issue4, April 2001, Pages 365-371, by permission.*

[1]Korzh and Grunwald (2001) provide an excellent short biography of Nadine Dobrovolskaia-Zavadskaia, with an emphasis on her role in starting murine developmental genetics through her discovery of the *T*-locus. Its major error is in its incorrect identification of her husband. It amplifies Korzh's briefer biography (Korzh, 2001). A fuller biography based on extensive records in Paris is that of Natalie Pigeard-Micault and Gabriel Gachelin (2017). They focus on her cancer research and present more information on *L'Institut du Radium* and its subunit, *Fondation Rosenthal*, which was established to study biological effects of radiation, and where her laboratory was located. They have extensively unearthed details of the networks developed by émigré Russian scientists which allowed Dobovalskaia-Zavadskaia to find a position. They also published a complete list of her publications (Gachelin and Pigeard -Micault, 2017).

[2]Biographical entries in "Russian Emigration in France (1919-2000). Biographical dictionary in three volumes", ed.s L. Mnucine, M. Avril, V. Losskaia. M. Nauka, Museum M. Tsvetaevoi (2008-2010) [(Rissijskoie zarubezhie vo Grantzii), ed. by L. Mnuchine, M. Avril, V. Losskaia, M. Nauka, Dommusei M. Tsvvetaevoi (2008-2010)]. Translations courtesy of Natalie Kolotilova, Moscow State University.

latter part of the 19th Century—Switzerland led the way in accepting women to Medical Faculties and gave the first Ph.D. to a woman in 1865 (**Overbye, p. 31**). After the 1905 Russian Revolution, universities were variably opened to women but the University of St. Petersburg did not have a Department of Medicine (**Kassow, p. 24**).

Nadine Dobrovolskaia-Zavadskaia then trained at the surgical clinic of the Military Academy of Medicine and prepared a doctoral thesis, receiving the degree in 1911 (Pigeard-Micalut and Gachelin, 2017). She continued working in St. Petersburg, both at the surgical clinic and a psychiatric hospital (ibid.). Meanwhile, she was working toward the *privat docent* degree for her research in surgical pathology (the degree required to join the faculty in German-influenced universities) from the Women's University of Petrograd in 1918. She obtained a faculty position at the University of Tartu in Estonia, but was moved to the new university of Voronej in central Russia when the Germans occupied Estonia (ibid.). As the Revolution further disrupted educational institutions, she served as an army surgeon with General Wrangel's army in the Crimea during the Civil War (Korzh and Grunwald, 2001). In April, 1920 General Baron Wrangel had been appointed as the last commander-in-chief of the White forces in the South of Russia. Despite some formidable forays by his army, the Red forces defeated his Southern army in November, ending the Civil War. Wrangel "fled to Constantinople, leaving behind him guns, aeroplanes, tanks, armoured trains, munitions, rifles, cartridges. As for his soldiers, let them shift for themselves" (**Aragon, p. 175**). It was left for staff, such as Nadine Dobrovolskaia-Zavadskaia, to get away on their own. Her route from there to Paris involved Egypt where she briefly studied treatment of typhus with iodine (Pigeard-Micault and Gachelin, 2017). France was freely accepting White Russian refugees and she proceeded to France from Constantinople in 1921 (ibid.).

Upon her arrival in Paris, she was soon involved with many associations with émigré Russian Scientists. Several of her old professors from the Universities of St. Petersburg and Petrograd were already there and, since she had already published 20 papers in mostly French and German journals, she was well known (Pigeard-Micault and Gachelin, 2017). She became a very good friend of Marie Curie (Ornsdoff, 1958) and it is probably through her that she became associated with the *Institut du radium, foundation Curie*, where she worked starting in 1923, and where she joined Professor Regaud's group, which was working on the biological effects of the rays of radium. This was a major change in direction from her previous work—her first publication in France was in the *Presse Medicale.* on the subject of forming surgical collaterals for vascularization of tissue (Kharamenko and Dobrovalskaia, 1922). Regaud's focus was largely on histopathological results, and he had studied the effects of radium on the testes of many species (Regaud, 1900; Regaud and Debreil, 1908). Dobrovolskaia-Zavadskaia's first work with him was also primarily morphological and involved studies of the effects of X-rays on striated muscle and testes (Dobrovolskaia-Zavadskaia, 1924; Dobrovolskaia-Zavadskaia, 1927a). From the number of mice involved, it is obvious that a large mouse colony was available. She reports that "in 1923, Professor Regaud directed my attention to the desirability of studying the influence of the modifications produced by X-rays in the testicles, on heredity in mice" (Dobrovolskaia-Zavadskaia, 1929) and this led to the work relevant to the discovery of the *T*-locus. As discussed in the text, Dobrovolskaia-Zavadskaia never accepted the notion that her abnormal, short-tailed mouse was due to an X-ray induced mutation. She remained convinced that the X-rays are "only a revealer of a pre-existing latent condition", that they may be "profitably utilized to detect the potential mutants in different species" (Dobrovolskaia-Zavadskaia, 1929). She met L.C. Dunn during his trip to Europe in 1927. They shared an interest in Science in Russia, Poland and Soviet Union. In 1930, she was invited to the United States to give lectures to Russian émigré organizations, and she visited L.C. Dunn (Korzh, 2001). She continued a correspondence with him and later sent him some of her *T* mice[3]. As told by Dorothea Bennett (1977) she "came to the United States to lecture for some Russian refugee organizations, and appeared at Dunn's laboratory looking for a solution to her puzzle [complete tailessness in *T/t* heterozygotes—she actually knew of the homozygous lethality and maintained them as balanced lethal lines]. As discussed in the text, she left the problem in his hands, sending him mice.

Nadine Dobrovolskaia-Zavadskaia's major work was on the genetics of cancer, again, a pursuit suggested by Professor Regaud (Dobrovolskaia-Zavadskaia, 1933a). She became a "Chief of Laboratory" in January, 1927 at the the Rosenthal Laboratory (Rosenthal was a banker who subsidized the lab, as the Rothschild's did for the neighboring *Institut de Biologie Physico-chimique*) with Kobozieff as her assistant[4]. Her initial approach was to mate cancerous mice with each other or with apparently noncancerous mice. (She wryly comments that "no one except Miss Maud Slye [the controversial American mouse geneticist who claimed that susceptibility to cancer is a single Mendelian recessive] seems to possess a verified pure non-cancerous strain" (Dobrovolskaia-Zavadskaia, 1933b). She called the progeny of one pair a strain, but parent/offspring or sib matings might be used to create the strain. There were never more than 8 generations, and usually many fewer, but the mice were maintained to an old age. She correctly concluded that cancer

---

[3]Archival material of L. D. Dunn, Archives of the American Philosophical Society.

[4]Archives of the foundation Curie (FC-A1G). These contain no lab notes, only publications and administrative papers related to N D-Z. Archives perused by Dr. Gabriel Gachelin.

susceptibility is not a simple dominant Mendelian character and that multiple genes must exist, since liability to mammary tumors occurred in some lines independently of sarcomas in other lines (Dobrovolskaia-Zavadskaia,1933a). Nonetheless, she hypothesizes a single recessive gene for cancer susceptibility, *no* (for neoplasm), with modifiers for the different kinds of cancer (Dobrovolskaia-Zavadskaia, 1933a). Interestingly, one of her "strains" showed "adenocarcinoma of the mammary gland" only in the occipital region (one wonders about the true histological type of the cancer since, of course, there is no breast tissue in the occipital region) and was descended from a short-tail male (*T*). She supposes that *T* is the modifying factor for the unusual location (Dobrovolskaia-Zavadskaia, 1933a). In another article from the same year, an additional modifier (in addition to that for sarcoma) was posited for epitheliomas, (Dobrovolskaia-Zavadskaia,1933b) and she proposed that perhaps the Slye hypothesis of recessive inheritance required multiple genes, not just modifiers (Dobrovolskaia-Zavadskaia, 1933b). In this second publication for an Anglophone audience, she indicated the kind of family studies which would be needed for the study of cancer in humans. These were previously published in French (Dobrovolskaia-Zavadskaia, 1930) and were followed, at least in a large study from Belgium (Wassink, 1935). She again visited the United States in 1932 to attend the International Congress of Genetics[5].

"She was active in genetics until 1935, then moved to experimental trials of a variety of drugs on mice. The reason for that shift was not scientific. First, all animals were killed when the laboratory left Paris in June 1940, due to the progression of the Wehrmacht. Second, her assistant, Kozobieff, left for the Laboratoire d'évolution des êtres organizes de l'Université de Paris; Kobozief had clearly been in charge of the mating of mice and the follow up of the progeny. Thus, with the exception of those mice which she had previously given to various laboratories, her entire collection of mutants was lost"[6]. Thus, her career in mouse genetics ended, as had that of Lucien Cuenot, in part because of the German invasion.

She married Veniamin Valerianovitch Korsak-Zavadskii, a Russian emigre writer, in 1922. He was born in 1884 in Kalach-on-Don and educated in law at the University of Warsaw, which had been evacuated to Rostov-on-Don during the first world war[2]. He was a prisoner of war in Germany and then participated in the Russian civil war. He, like Dobrovolskaia-Zavadskaia, evacuated to Egypt and then to Paris[2]. This was the place for a Russian émigré writer to be: as of 1937 "there were four hundred thousand Russian emigres in the country, and the main newspapers and journals for the whole emigration were firmly based in Paris **(Boyd, p. 432)**. He wrote under the pseudonym of V. Korsak and published about 40 works, including more than 10 books[7]. We do not know when and where Dobrovolskaia-Zavadskaia and Veniamin met. He was a member of the Union of Russian Writers and Journalists in Paris[8]. He died in Paris in 1944 and, in 1946, the union of Russian Writers and Journalists organized a reunion in his memory in which Dobrovolskaia-Zavadskaia participated[2]. After his death, she published a number of his works.

Dobrovolskaia-Zavadskaia, like her husband, remained involved with the Parisian Russian *emigres* and with Russian scientists—in 1925 she attended conferences of the Society of Russian Chemists, the Metchnikov Society of Russian Medicine, and other Russian academic groups and societies in Russia[2]. She gave courses in radiology at the University of Paris, the Union of Doctors associated with the Movement of Russian Christian Students and the Russian Society of the Red Cross[2]. She received *le prix de l-Academie Francaise* for her investigations in the domain of the heredity of cancer[2].

By 1949, Dobrovolskaia-Zavadskaia considered mammary cancer to be genetic (while it was really due to a maternally transmitted virus), while maintaining that sarcomas and epitheliomas were primarily environmental—a conclusion based on her work with carcinogens (Dobrovolskaia-Zavadskaia, 1940). With her concern for these environmental effects, she developed notions of cancer prevention (Chaine, 1954). Her last works were also concerned with pyruvate metabolism and its role in cancer (Dobrovolskaia-Zavadskaia, 1950). She had hoped to visit New York again in 1952 and give lectures on her current work related to cancer, and to International Peace organizations on the state of the USSR, but this trip did not occur[3]. She died in Milan in 1954 while travelling on a lecture tour in Italy. The cause of death was not given in her obituary (Chaine, 1954) but one wonders, given her radiation and carcinogen exposures, if it was due to cancer.

Nadine Dobrovolskaia-Zavadskaia deserves a special amount of credit for studying mouse genetics during this era in France. Although L. Cuenot made the important contribution discussed in this chapter (and his mini-biography), his experiments stopped in 1914 when his mice were destroyed by the invasion of WWI. Nevertheless he taught genetics

---

[5]Letter dated June 2, 1932 by Regaud in the Institut Pasteur archives, where some of Regaud's papers are located. This, and a few other administrative details related to N D-Z, were found by Daniel Demellier, archivist of the Institut Pasteur, who searched the Institut Pasteur archives for material related to her.

[6]Personal communication from Dr. Gabriel Gachelin, who perused the archives of the foundation Curie for any items relevant to N D-Z.

[7]Some of his correspondence to N D-Z is in the collection of the Union of Russian Writers and Journalists Abroad at Amherst University, Amherst, Mass., U.S.A.

[8]Some of his writing are in the Maria Dmitrievna Vrangel Collection, 1915-1944, Hoover Institution, Stanford University, Palo Alto, Cal., U.S.A.

from the end of the war into the 1930s in Nancy (Burian and Gayon, 1999). However, as well documented by Burian and Gayon, there was, at that time, a paucity of genetic teaching and textbooks in France compared to Germany and England (1999). They blame this lack on multiple factors including the over-emphasis of the Mendelian geneticists on nuclear factors compared to cytoplasmic inheritance, the "unphysiological" characteristics of these genes (as perceived in the context of Bernard's studies), and the lack of a close relationship of biological research to agriculture in France (Burian and Gayon, 1999). Perhaps Dobrovolskaia-Zavadskaia's pursuit of genetics was influenced by her Russian background, in which there was a strong interaction of biology with agriculture, and where genetics flourished until the time of Lysensko (Medvedev, 1969).

Notes

1. Takahashi Kuramoto, personal communication, 2015. The chain of e-mails which led me to Dr. Kuramoto discouraged me from trying to describe more of the history of mouse genetics in Japan. Akihisa Setoguchi, a biologist turned historian said that there was "no tradition of archiving scientist's papers in Japan." Dr. Moriwaki's (he is a major modern contributor to mouse genetics) sister, Dr. Shiroishi, also a historian of science said that there are "no English materials on history of mouse genetics in Japan [sic]."

2. These genes are now designated *Tyr*, *Tyrp1*, and *A*.

3. This represents a page allowance lost long ago with our current use of online supplemental material to shorten the printed article.

4. There didn't seem to bee any in Italian, another language with which I am semi-conversant. According to Dr. Franco Mangia (May 19,2020): "I think you are absolutely right not quoting any Italian geneticist in the first half of XX century, because as far as I know genetics was not considered at all by Italian scientists before 1950. Montalenti was actually a great academic but not a real scientist (having however the merit of introducing genetics in the Italian academic environment). Anyway it is surprising that during that period we had excellent scientists like Grassi and Levi, which in Turin had pupils like Dulbecco and Levi-Montalcini. In other words, Italian biology was mostly addressed to zoological/anatomo-histological (remember Sertoli?)/ embryological issues, but not molecular ones."

5. Every 17th, but occasionally the 16th, starting arbitrarily at number 3, reference was chosen.

6. Personal communication, Salome Gluecksohn-Waelsch. In this she may have reflected a prejudice, shared by many mouse geneticists, that Tibby Russell had been wronged when Liane Brauch, then a graduate student with her husband Bill, became his bride. There are many devotees of LB Russell, too.

References

Albisetti, J.C., 1982. The fight for female physicians in Imperial Germany. Cent. Euro. Hist. 15, 99–123.

Allen, G., 1904. The heredity of coat color in Mice. Proc. Amer. Acad. Of Arts and Sci. 40, 61–163.

Bateson, W., 1903. The present state of knowledge of colour-hereditiy in Mice and Rats. Proc. Zool. Soc. London 44, 74.

Bennett, D., 1977. L. C. Dunn and his contribution to T-locus genetics. Ann. Rev. Genet. 11, 1–12.

Berry, R.J., 1987. The house mouse. Biologist 34, 177–186.

Beadle, G.Q., Tatum, E.L., 1941. Genetic Control of Biochemical Reactions in NeurosporaGenetic Control of Biochemical Reactions in Neurospora. Proc. Nat'l Acac. Sci., U. S. A. 7, 499–506.

Blank, R.D., Campbell, G.R., D'Eustachio, P., 1986. Possible derivation of the laboratory mouse genome from multiple wild *MUS* species. Genetics 114, 1257–1269.

Bonhomme, F., Guénet, J.-L., 1996. The laboratory mouse and its wild relatives. In: Lyon, M.F., Rastan, S., Brown, S.D.M. (Eds.), Genetic Variants and Strains of the Laboratory Mouse. Oxford University Press, Oxford, pp. 1577–1596.

Brachet, A., 1927. La vie créatrice des formes. Felix Alcan, Paris, pp. 178–196 ed.

Brilliant, M.H., Ching, A., Nakatsu, Y., Eicher, E., 1994. The original pink-eyed dilution mutation (p) arose in Asiatic mice; implications for the H4 minor histocompatibility antigen, Myod1 regulation and the origin of inbred srains. Genetics 138, 203–211.

Burian, R.M., Gayon, J., 1999. The French school of genetics: from physiological and population genetics to regulatory molecular genetics. Annu. Rev. Genet. 33, 313–349.

Burian, R.M., Gayon, J., Zallen, D., 1988. The singular fate of genetics in the history of French biology, 1900-1940. J Hist. Biol. 21, 357–402.

Castle, W.E., Little, C.C., 1910. On a modified Mendelian ratio among yellow mice. Science 84, 62.

Chaine, J., 1954. Madame Dobobolskaia-Zavadskaia (1878-1954). Bull du Cancer 41, 31–34.

Chesley, P., 1932. Lethal action on the short-tailed mutation in the house mouse. Proc. Soc. Exp. Biol. And Med. 29, 437–438.

Chesley, P., Dunn, L.C., 1936. The inheritance of taillessness (anury) in the house mouse. Genetics. 21, 525–536.

Cock, A.G., 1973. William Bateson, Mendelism, and biometry. J. Hist. Biol. 6, 1–36.

Cuénot, L., 1902. La loi de Mendel et l'hérédité de la pigmentation chez les Souris. Compte Rend. Acad. Sci. Paris, 134, 779–781.

Cuénot, L., 1892. Les moyens de défense dans la serie animale. Encyclopédie scientifique des aide-mémoires. Paris, 1–184.

Cuénot, L., 1895a. La nouvelle théorie transformiste. Rev. Gen. Sci. 5, 74–79.

Cuénot, L., 1895b. L'influence de milieu sur les animaux. Encyclopédie scientifique des aide-memoires. Paris, 1–176.

Cuénot, L., 1903. L'heredite de la pigmentation chez les Souris. Arch. Zool. Exp et Gen. 4 1, 34–41 Notes et Revue.

Cuénot, L., 1904. ibid. Notes et Revue 2, 45–56.

Cuénot, L., 1905. ibid. Notes et Revue 3, 123–132.

Cuénot, L., 1925. Non hérédité des caractères acquis in L'Adaptation, Ed. Gaston Doin (Encyclopédie Scientifique), librairie Octave Doin, Paris.

Darbishire, A.D., 1902. I. Note on the results of crossing japanese waltzing mice with european albino races. Biometrika 2, 101–104.

Davenport, C.B., 1900. Review of Von Giuata's experiments in breeding mice. Biol. Bull. (Woods Hole, Mass.) 2, 121–128.

Davenport, C.B., 1941. The early history of research with Drosophila. Science 93, 305–306.

De Chadaraevian, S., 2006. Mice and the reactor: the "genetics experiment" in 1950s Britain. J. Hist. Biol. 39, 707–735.

Diligent, M.B., 2002. Lucien Cuénot (1866-1951) par Andrée Tetry (1907-1992): Histoire d'un ouvrage inédit. Acad. Natl de Metz? Memoirs 2002, 75–130. doi:http://hdl.handle.net/2042/33874.

Dobrovolskaia-Zavadskaia, N., 1924. Modifications des fibres striées sous l'influence d'irradiations prolongées au moyen de foyers radifères introduits dans les muscles. J. de Radiol. Et d'Electr. 2, 49–61.

Dobrovolskaia-Zavadskaia, N., 1927a. Etudes sur les effets produits par les rayons X dans le testicule de la Souris. Arch Anat. Micr. 23, 396–438.

Dobrovolskaia-Zavadskaia, N., 1927b. Sur la mortification spontanée de la queue chez la Souris nouveau-née et sur l'existence d'un caractère (facteur) héréditaire "non viable. Compte Rendu. Soc Biol. 97, 114–116.

Dobrovolskaia-Zavadskaia, N., 1928. Contribution à l'étude de la structure génétique d'un organe. Ibid 99, 1140–1143.

Dobrovolskaia-Zavadskaia, N., 1929. The problem of species in view of the origin of some new forms in mice. Biol Rev 4, 327–351.

Dobrovolskaia-Zavadskaia, N., 1930. Quelles conditions devrait réunir, pour être valable, une étude de l'hérédité des cancers dans l'espèce humaine. Ass. Franc. Pour l'étude du Cancer 19, 413.

Dobrovolskaia-Zavadskaia, N., 1933a. Heredity of cancer susceptibility in mice. J. Genetics 27, 181–200.

Dobrovolskaia-Zavadskaia, N., 1933b. Heredity of Cancer. Am. J. Canc. 18, 357–379.

Dobrovolskaia-Zavadskaia, N., 1940. Hereditary and Environmental Factors in the origin of Different Cancers. J. Genetics 40, 157–170.

Dobrovolskaia-Zavadskaia, N., 1950. The possible role of some intermediary metabolites in general pathology. Revue Canadienne de Biol 9, 28–53.

Dobrovolskaia-Zavadskaia, N., Kobozieff, N., 1927. Sur la reproduction des Souris anoures. Ibid. 97, 116–119.

Dobrovolskaia-Zavadskaia, N., Kobozieff, M.N., 1930. Sur le facteur létal accompagnant l'anourie et la brachyurie chez la Souris. Compt. Ibid. 191, 352–355.

Dobrovolskaia-Zavadskaia, N., Kobozieff, M.N., 1932. Les souris anoures et a queue filiforme qui se reproduisent entre elles sans disjunction. Ibid 193, 782–784.

Dobrovolskaia-Zavadskaia, N., Kobozieff, N., Veretennikoff, S., 1934. Etude morphologique et génétique de LA BRACHYOURIE chez les descendants de souris à testicules irradiés. Arch. De Zool. Experiment. et Générale, 76, 249–358.

Durham, F.M., 1908. A preliminary account of the Inheritance of Coat-colour in Mice. Rep. Evol. Comm. IV, 411159–411178.

Durham, F.M., 1911. Further experiments in the inheritance of Coat-colour in Mice. J. Genet.

Falconer, D., 1993. Quantitative Genetics in Edinburgh: 1947-1986. Genetics 137, 137–142.

Farrall, L., 1975. Controversy and Conflict in Science: A Case Study—The English Biometric School and Mendel's Laws. Social Studies of Science 5, 269–301.

Gachelin, G., Pigeard-Micault, N., 2017. La Fondation Curie, lieu singulier de la rechereche en genetique en France entre les deux guerres. Introduction aux publications de Nadine Dobrovolskaia-Zavadskia (1878-1954). Arch. Ourverte en Sciences de l'Homme et de la Societe https://halshs.archives.ouverte.fr/halshs-01535882/document.

Garrod, A., 1908. The Croonian Lectures on inborn errors of metabolism. Lancet 2, 1–7 73-79, 142-148, 214-220.

Gayon, J., Burian, R.M., 2000. France in the era of mendelism (1900-1930). Compt. Rend. Acad. Sci., Paris. Life Sciences 323, 1097–1106.

Gilbert, S.F., 1978. The embryological origins of the gene theory. J. Hist Biol 11, 307–351.

Gilbert, S.F., 1998. Bearing Crosses: A Historiography of Genetics and Embryology. Am. J. Med. Genet. 76, 168–182.

Goldman I, L., 2002. Raymond Pearl, smoking and longevity. Genetics 162, 997–1001.

Goldschmidt, R., 1951. L. Cuénot: 1866-1951. Science 113, 309–310.

Herrmann, B.G., Labeit, S., Poustka, A., King, T., Lehrach, H., 1990. Cloning of the T gene required in mesoderm formation in the mouse. Nature 343, 617–622.

Hickman, M., Cairns, J., 2003. The centenary of the One-Gene One-Enzyme hypothesis. Genetics 163, 839–841.

Hunt H.R. (1932) Proc. VI Intern. Congress of Genetics, T. II:91-93.

Ives, C.F., 1974. The Great Wave: The Influence of Japanese Woodcuts on French Prints". The Metropolitan Museum of Art.

Jones, E.P., Skirnisson, K., McGovern, T.H., Gilbert, M.T.P., Willerslev, E., Searle, J.B., 2012. Fellow travellers: a concordance of colonization patterns between mice and men in the North Atlantic region. BMC Evol. Biol. 12, 35. doi:10.1186/1471-2148-12-35.

Keeler, C., 1978. How it began. In: Morris, H.C. III (Ed.), Origins of inbred mice. Academic Press, New York, pp. 179–192.

Kevles, D.J., 1980. Genetics in the United States and Great Britain, 1890-1930: A review with speculations. ISIS 71, 441–456.

Kharamenko, E., Dobrovolskaia, N., 1922. Etude sur les collatérales artificielles. La Presse médicale 3 (11 janvier), 27.

Korzh, V.P., 2001. N. Dobrovolskaya-Zavadskaya and the discovery of the T gene I. Russian J. of Develop. Biol. 32, 192–195.

Korzh, V., Grunwald, D., 2001. Nadine Dobrovolskaia-Zavadakaia and the dawn of developmental genetics. BioEssays 23, 365–371.

Kuramoto, T., 2011. Yoso-Tama-No-Kakehashi, The FIRST Japanese guidebook on raising rats. Exp. Anim. 60, 1–6.

Limoges, D., 1976. Natural selection, phagocytosis and preadaptation: Lucien Cuénot 1886-1901. J. Hist. Med. And Allied Sci. 31, 176–214.

Little, C.C., 1958. Coat color genes in rodents and carnivores. Quart. Rev. Biol. 33, 103–137.

Little, C.C., Tyzzer, E.E., 1916. Further Experimental Studies on the Inheritance of Susceptibility to a Transplantable Carcinoma (J.W.A.) of the Japanese Waltzing Mice. J. of Med. Res. 33, 393–427.

Lyon, M.F., 2002. A Personal History of the mouse genome. Annu. Rev. Genomics Hum. Genet. 3, 1–16.

Mohr, O.L., 1919. Character changes casued by mutation of an entire region of chromosome in Drosophila melanogaster. Genetics 4, 275–282.

Morse, H.C., 1981. The laboratory mouse—A historical perspective. In: Foster, H.L., Small, J.D., Fox, J.G. (Eds.), The Mouse in Biomedical Research. Academic Press, New York. 1, pp. 1–16.

Muller, H., 1928. The production of mutations by X-rays. Proc. Nat. Acad. Sci., U.S.A. 14, 714–726.

Nadson, G.A., Philippov, G.-S., 1911. Influence des Rayons X sur la sexualité et la formation des mutants chez les Champignons inférieurs (Mucorinées). Compt. Rendu Soc. Biol. 93, 473–475.

Orndoff, B., 1958. An interview with Madame Curie : An Historical Note. Radiol 71, 750–752.

Pigeard-Micault, N., Gachelin, G., 2017. Nadine Dobrovadskaia-Zavadskaia, une Russe a la Fondation Curie, une vision signuliere de la genetiqued dans le cancer (1921-1954). Canadian Bull of Med. Hist. 34. doi:10.3138/cbmh/193-012017.

Rader, K.A., 1998. The Mouse People": Murine Genetics Work at the Bussey Institution, 1909-1936. J. Hist Biol. 31, 327–354.

Regaud, C., Dubreil, G., 1908. Perturbation dans le développement des œufs fécondés des spermatozooïdes roentgenisés chez le Lapin. Compt. Rend. Soc. Biol. 64, 1014.

Regaud, C., 1900. Evolution tératologique des cellules séminales, les spermatides à noyaux multiples, chez les mammifères. Bibl. Anat. 1, 24–42.

Richmond, M.L., 2001. Women in the early history of genetics: William Bateson and the Newnham College Mendelians 1900-1910. ISIS 92, 55–90.

Richmond, M.L., 2006. The 'Domestication' of Heredity: the Familial Organization of Geneticists at Cambridge University, 1895-1910. J. Hist. Biol. 39, 565–605.

Rostand, J., 1956. Un précurseur de Mendel: Le Pharmacien Coladon? In: Rostand, J. (Ed.), L-Atomisme en Biologie. Gallimard, Paris, pp. 203–208.

Russell, E.S., 1985. A History of Mouse Genetics. Ann. Rev. Genet. 19, 1–28.

Russell, E.S., 1987. A mouse phoenix rose from the ashes. Genetics 117, 155–156.

Salcedo, T., Geraldes, A., Nachman, M.W., 2007. Nucleotide variation in wild and inbred mice. Genetics 177, 2277–2291.

Searle, J.B., Jones, C.S., Gündüz, I., Scascitelli, M., Jones, E.P., Herman, J.S., Rambau, R.V., Noble, L.R., Berry, R.J., Giménez, M.D., Jóhannesdóttir, F., 2009. Of mice and (Viking?) men: phylogeography of British and Irish house mice. Proc. Roy. Soc. B 276, 201–207.

Silver, L.M., 1990. At the Crossroads of Developmental Genetics: The Cloning of the Classical Mouse T locus. BioEssays 12, 377–380.

Snell, G.D., Read, S., 1993. WILLIAM ERNEST CASTLE, Pioneer Mammalian Geneticist. Genetics 135, 751–753.

Staats, J., 1966. The Laboratory Mouse. In: Green, E.L. (Ed.), Biology of the Laboratory Mouse2nd Ed. McGraw-Hill, New York.

Sturtevant, A.H., 1915. The behavior of chromosomes as studied through linkage. Zeit. Fur induktive Abstam. Und Vererbung. 13, 234–287.

Sturtevant, A.H., Schultz, J., 1931. The inadequacy of the sub-gene hypothesis of the nature of the scute allelomorphs of Drosophila Proc. Nat. Acad. Sci, USA, T 16, 265–270.

Tjimofeeff-Ressovsky, N.W., 1934. The experimental production of mutations. Biol. Rev. 9, 411–457.

Tokuda, M., 1935. An Eighteenth Century Japanese Guide-Book on Mouse Breeding. J. Hered 26, 481–484.

Tokkuda, T., Ebata, T., Noguchi, H., Keane, T.M., Adams, D.J., 2013. The ancestor of extant Japanese fancy mice contributed to the mosaic genomes of classical inbred strains. Genome Res 23, 1329–1338.

Wassink, W.F., 1935. Cancer et Hérédité. Genetica 17, 103–144.

Weir, J.A., 1994. Harvard, Agriculture, and the Bussey Institution. Genetics 136, 1227–1231.

Wright, S., 1917. Color inheritance in mammals. J. Hered. 8, 224–235.

2

The evolutionary modern synthesis, *t*-haplotypes in the wild, and reproductive isolation

Genetics and natural selection—the search for mutants in the wild

An interesting coincidence (or the result of cultural influences which result in the "time is ripe") is that Darwin and Wallace published their theories of evolution and Mendel his discoveries in genetics within a short time of each other (Wallace, 1858; Darwin, 1858; Mendel 1866). Sadly, while Darwin's (and Wallace's) work was highly applauded and disputed, Mendel's remained "hidden" until 34 years later. Darwin and Wallace needed a theory of inheritance to provide the basis for selection. Darwin's theory of pangenes (gemmules arising in peripheral tissues where they were influenced by environmental forces and which then travelled to the gonads to be transmitted to the next generation) was almost Lamarckian. Thus, an important accomplishment of the second and third decades of the twentieth century was the development of theories that could reconcile Mendelian genetics with transmission of hereditary traits as the material for natural selection. This was termed (by Julian Huxley) the Evolutionary Synthesis, his book was titled "Evolution, The Modern Synthesis" (**Huxley, 1942**). At first, the battle was slowed by the arguments between the biometricians and the Mendelian geneticists discussed in Chapter One. Then, as will be further discussed below, the role of mutations in driving evolution was heavily discussed. However, "by the time of the First World War mutationism had fallen out of favor. For the next two decades many geneticists—though not all—seem to have withdrawn from the evolutionary debate" (**Harwood, 1993, p. 99**). A major problem was that the geneticists and the taxonomists were not communicating with each other, and had different notions of what a species was—lack of interactions between the two groups is one of the problems that needed to be resolved in order for the Modern Synthesis to occur.

Ernst Mayr summarized the historical separation between naturalists and geneticists: "Diversity was almost totally excluded from evolutionary discussion of the presynthesis geneticists. Their focus was entirely on genes and characters and on their changes (transformations) in time. They wrote as if they were unaware that there are taxa, and that they...are the real actors on the evolutionary stage" (**Mayr, 1982, p. 541**). This at least was the situation in America. In Germany, there was more interest in evolution amongst geneticists. "While most American geneticists in the 1920s and 1930s seem to have set aside the complexities of evolutionary genetics in favor of transmission genetics, many of their German counterparts did the opposite" (**Harwood, 1993, p. 103**). In the early 20th century, the Mendelian geneticists working

Twentieth Century Mouse Genetics. DOI: https://doi.org/10.1016/B978-0-12-824016-8.00014-3

with animals were heavily focused on chromosomal mechanisms and cytogenetics; the plant geneticists were more interested in mutation than the animal geneticists at this time. Also, naturalists were finding more evidence for geographical separation as a major process in the creation of new species, but they, too, argued about the amount of variation required to define a new species.

The question of the kind of variation needed for evolution to occur was present before Mendel was re-discovered. Galton "felt that the small, incremental steps by which natural selection supposedly proceeded would be thwarted by a phenomenon he had discovered, which he called regression (or reversion) to the mean. Hence, Galton believed that evolution must proceed via discontinuous steps" (Gillham, 2001). The regression on the mean occurs with height, for instance. When two very tall people mate, their children will be closer to the mean than the parents were if the environment is the same. Bateson, in compiling his vast catalog of material for the study of evolution (**Bateson, 1894**), "concentrated on those variations from the norm that were the magnitude of species differences" (**Mayr, 1982, p. 545**). These views were in marked contrast to those of most naturalists, who saw gradation in the variation between species. One important naturalist differed; Wallace, joining the fray after Darwin's death, also saw the potential for large variations. "Those variations important for evolution were not necessarily 'infinitesimal or even as small as they are constantly asserted to be.' Most species possessed great variability, and natural selection favored only the most fit individuals. But the struggle for existence was an intermittent affair" (Gillham, 2001)—in this he anticipated the notions of "punctuated equilibrium," favored by Stephen Jay Gould and still moot.

While the focus of Mendelian geneticists was not on species variation, they did continue to study mutations with great interest. Here the mutation theory of De Vries, one of the re-discoverers of Mendel, was a great stimulus, as mentioned in Chapter 1. "Unfortunately, de Vries's argument was entirely circular: he called any discontinuous variant a species, hence species originate by any single step that causes a discontinuity" (**Mayr, 1982, p. 546**). While coat color mutants found in mice, and many of the mutants found in *Drosophila,* such as those for eye color or wing shape, did not seem to meet the criteria of material for evolution, attention gradually turned to finding such mutations in nature. This was most thoroughly done in *Drosophila.* Chetverikov, Tiofeeff-Ressovsky, and Dobzhansky were three great Russian geneticists who found mutations in natural populations. In particular, the Russians did extensive field studies and mated captive *Drosophila* to detect recessive mutations. "The Tinofeeff-Ressovskys had conducted some of the first genetic analyses of variation in wild populations during the 1920s and endorsed a theory of evolution via mutation and selection" (**Harwood, 1993, p. 111**). Dobzhansky summarized the findings as "individuals resembling mutations in the laboratory have been repeatedly found in natural populations of Drosophila"(**Dobzhansky, 1937, p. 41**). Several findings emerged from these studies: 1) there were frequent lethal recessive mutations being carried in populations, 2) there was a large variation in minor characters such as bristle number, and 3) these could be found in populations of *Drosophila* from many places in the world. Another important advance was the finding of many reverse mutations. "After Morgan had discovered *eosin* in 1913, a reverse mutation from *white-eye,* more and more reverse mutations were discovered and in many cases the frequency from wild type to mutant was not greater than the reverse from mutant to wild type" (**Mayr, 1982, p. 552**). Finally, as linkage and crossing-over were better understood, the availability of *new combinations of genes* in gametes as material for selection became apparent. These advances led Morgan to reverse his early opposition to the theory of evolution (Allen, 1968). Much of Allen's article reviews Morgan's work with balanced lethal stocks, which were an effective tool for detecting new mutations or combinations of previous mutations (mutations to loss of lethality, mutations to gain of modifiers, or crossing over events resulting in one chromosome free of the lethal factor). "It is possible to produce at will other balanced lethal stocks that will 'mutate' in the sense that they will throw off a small predictable number of a mutant type—a type that we can introduce into the stock for the express purpose of recovering it by an apparent mutation process" (Morgan, 1918). By the end of the decade Morgan was fully in support of evolution by descent.

"There is a predisposition on the part of systematists, paleontologists, and a few other students of 'wild' types to deny that mutants are identical with the variation from which evolution obtained its materials...The chief contention that evolution has been by means of very small changes does not require further attention since we now know that some of the genes that are typically Mendelian in behavior produce even smaller differences than those that distinguish wild varieties and paleontological gradations" (Morgan, 1918).

Castle (see Chapter 1) followed the plant and animal breeders in seeing to what degree selection could be performed in rats. He showed that only a small number of generations was needed to generate nearly completely black rats starting with black and white spotted rats (Castle and Phillips, 1914). "However, the crucial point is that the role of natural selection for Castle was not that of selection in the *Origin* [Darwin's]. Nowhere in his seminal writings did Castle direct his selection experiments to problems of adaptation... Castle's interests were...the mechanisms of heredity" (**Ruse, p. 366**), not evolution.

This was also true of the early studies of Morgan who, as with Castle, had started out as an embryologist (see Chapter 1). Bateson remembered "Morphology was studied because it was the material believed to be most favourable for the elucidation of the problems of evolution, and...embryology [was] the quintessence of morphological truth. The one topic of conversation was evolution....Discussion of evolution came to an end primarily because it was obvious that no progress was being made" Bateson, 1922. Ruse also makes the argument in **Monad to Man, The Concept of Progress in Evolutionary Biology,** that the belief in "progress" with evolution from lower to the higher (human) was a philosophical concept that contaminated the pure scientific study of evolution. He points out that Huxley, "Darwin's Bulldog" made no mention of evolution in his scientific papers and very rarely (perhaps once a year) in his anatomy lectures to school biology teachers during summer courses he taught for them (**Ruse, pp 217–220**). His promulgation of Evolutionary Theory was in highly public lectures. "Huxley was prepared—willing, even—to talk on matters evolutionary in such fora as presidential addresses to learned societies or when he was speaking by virtue of his authority as a scientist to a general audience" (**Ruse, p. 221**).

Of the three "greats" for the mathematical theories needed (Fisher, Haldane and Wright), it was probably Fisher in his book "The Genetical Theory of Natural Selection" **(Fisher, 1930)** who had the greatest influence. To quote the non-theoretician, Huxley "Mathematical analysis showed that only particulate inheritance would permit evolutionary change: blending inheritance as postulated by Darwin, was shown by R. A. Fisher [ref. to his book] to demand mutation-rates enormously higher than those actually found to occur" **(Huxley, 1942, p. 26)**.

It is perhaps surprising that there were few searches for variation in wild populations of *Mus musculus*. Part of the deficiency is the much longer time required to search for recessives by breeding in mice than in *Drosophila*. One early study was by L C Dunn. "Dunn (1921) was the first to apply the above genetic interpretation [a high load of mutations, including lethals, in natural populations] to the aberrant individuals in wild species of rodents" **(Dobzhansky, 1937, p. 45)**. The paper referred to by Dobzhansky consists of Dunn's review of collections of coat color variation in rodents stored at several museums of comparative zoology and natural history (Harvard, Boston, New York). The finding of albino, pink-eye, yellow, white-spotting, and black in a number of rodent species is recorded, but the numbers are very small and he only considered the numbers for black as being very substantial (Dunn, 1921). The occasional mutant was found in wild mice (e.g. *Dilute*, found by an exterminator in a corn crib [Detlefsen, 1921]) while more mutants were being found in some of the large breeding colonies (e.g. *Caracul*, found in a breeding colony of 2000 Swiss mice [Carnochan, 1937]). There was a relative lack of interest in studying natural variation in this species.

In contrast, rather extensive studies on *Peromyscus* were performed in the field by Francis Sumner. Henry Osborn had earlier reviewed Osgood's studies on 27,000 *Peromyscus* specimens and emphasized the continuity of variation between the many subspecies, and remained an "agnostic" as to the mechanisms of evolution: "A natural view is that the invisible germ is being continuously enriched with the visible body by processes of which we can form no conception whatever" (Osborn, 1915). Sumner's views were to develop in a different direction. "The extensive studies of Francis B. Sumner (1932) on the variation in geographic races of mice of the genus Peromyscus...pointed the way toward the kind of studies that were needed. The interpretations given...pointed also to the need for theoretical analysis of the consequences of Mendelian heredity since Sumner had appealed to Lamarckian explanations" **(Dunn, 1965, p. 204)**. Sumner studied *Peromyscus* in the field and made crosses between subspecies. In his initial interpretations, he saw no evidence of segregation of individual factors (Sumner, 1923) but his views gradually changed (Sumner and Huestis, 1925). Some of the quantitative traits he studied, like the extent of a particular coat color, were intermediate in the F_1 and showed an appropriate segregation of near parental and intermediate types in the F_2, with a 1:2:1 ratio appropriate for a single major gene (Sumner, 1929). Thus, it was *Peromyscus* and not *Mus* which provided information for the Modern Synthesis. The latter depended on the mathematical population genetics of Fisher, Haldane and Wright. Mayr dates this as occurring in the 3rd and 4th decades of the 20th Century: "a meeting of the minds came quite suddenly and completely in a period of about a dozen years, from 1936-1947" **(Mayr, 1982, p. 567)**.

Mayr (1973), following **Provine**, does credit mouse genetics with one important contribution to the Modern Synthesis: "Castle's successful selection experiments. The latter were particularly important because they probed three facts which had been vigorously denied by the anti-selectionists: the genetic nature of continuous variation, the response of continuous variation to selection, and the ability of selection to push variation beyond the limits of the 'pure lines' that had previously existed in these populations".

Although their observations were not to contribute to the Modern Synthesis, nineteenth century naturalists recorded variations in coloration of local wild populations of house mice, e.g. Lloyd (1912) summarized many of these. Jameson (1898) had reported a light-colored population on an island in Dublin Bay which **Huxley (1942 p, 187)** used as an example of selection by color camouflage to reduce predation: "The average coloration of the populations was considerably lighter than normal, but with greater variability: the paler animals' colour matched the sandy background. Predator pressure was intense owing to the lack of cover... the island could not have been isolated for more than 100 to 125 years". More modern studies include the experimental study in a Missouri farm where controlled addition of cats was the variable (Brown, 1965). There was variation in the number of light colored mice, rising to 45% until cats were introduced when they were reduced to zero while the overall populations was only reduced to ½.

Wild mice have made many other contributions to mouse genetics in the 20th Century, particularly in the study of antigenic differences among mice. Specific examples will be included in the chapters on Immunogenetics (Chapter 7) and, of course, Genetics of Population Size (Chapter 5). Some general overviews include those of Jean-Louis Guenet and Francois Bonhomme (2003), RD Sage (1981), and the volumes edited by H. Foster, J. Smith, and J. Fox **(Foster, et al, 1981)** and by K Moriwaki, T Shiroishi, and H Yonekawa **(Moriwaki, et al, 1994)**. A thorough summary of mouse evolutionary genetics occurs in the masterful book edited by **Macholan, et al, The Evolution of the House Mouse, 2012.**

The Re-discovery of *t* alleles in wild mice

It is perhaps surprising that Dunn, 1921 who ended his paper with evolutionary speculations, did not find *t*-alleles in the wild until 1953 since Nadene Dobrovalskaia-Zavadskaia had found them in the wild much earlier and he knew her results. Dunn had received three stocks of *t*-locus mice in 1931 from her (see Chapter One). A student in his lab, Paul Chesley, worked with the material while Dunn was on sabbatical in Norway in 1932–1933 (see mini-biography of Dunn), working with both mice and *Drosophila*. His first publication about the *T*-locus, after Chesley's untimely death, was in 1936. This work basically confirms Dobrovalskaia-Zavadskaia's findings on balanced lethal lines with *in utero* deaths of both *t* and *T* homozygotes, while adding the description of segregation distortion (Chesley and Dunn, 1936). Although the source of the mice from Dobovalskaia-Zavadskaia is noted, there is no mention that the origin of 2 of the interacting factors (the first named t^0) was from wild mice (see Chapter One). According to Dorothea Bennett, his motivation to look for *t*-alleles in wild mice was that "he remembered that one of his original lethal *t* alleles had come from a wild Spanish mouse" (Bennett, 1977). This is in contrast to the statements made in the paper reporting the finding of *t* alleles in wild mice (Dunn and Morgan, Jr., 1953). The *T/t*-complex had been shown to "mutate" at very high frequencies in the laboratory stocks (actually these were rare cross-overs in the region—see Chapter Three). Dunn wondered "whether the mutability was determined by factors peculiar to certain laboratory stocks in which this locus is maintained in a balanced lethal condition...or whether it was commonly a property of the locus. If the latter, then the locus should be mutable as well in wild populations of this species." (Dunn and Morgan, Jr., 1953). He was surprised by finding many t lethal factors in the wild mice that he studied (a population from the suburbs of New York and Philadelphia): "This discovery of an unexpectedly high frequency of $+/t^n$ heterozygotes among wild mice caused a change in the plan of the experiment" (Dunn and Morgan, Jr., 1953). The frequency of t^n varied from over 50% in a "confined population" of wild mice at the Rockefeller Institute to 0% in several Wisconsin populations. Their properties were studied by observations of F_2 offspring and crosses to laboratory stocks of t^0 and t^1 (Dunn and Morgan, Jr.,1953). The properties these studies found were those of homozygous lethality (in one t^{wn} *(wild, number unspecified)* studied this time; another was viable in the homozygous condition), male segregation distortion, and quasi-sterility in t^x/t^y compound heterozygotes were found. Only one population (the Rockefeller "confined population") had two different t^{wn}s. Modifiers affecting the expressivity of T were also found, but Nadine Dobovolskaia-Zavadskaia's previous discovery of this property (see Chapter One) was not acknowledged.

Seven wild alleles that had been found were further studied in a second paper and named as t^{w1}-t^{w7} (Dunn and Suckling, 1956). Two of the seven were homozygous viable. The t^{wn}/t^n compound heterozygotes were normal tailed (except in the case of t^{w4}, which seemed allelic to t^0) but males were sterile. Many of the crosses of t^{wx} X t^{wy} also gave viable offspring, suggesting that these alleles were different from each other. The markedly high male segregation distortion (84-99% t^w offspring) led to many population studies (see Chapter 5).

These observations were soon extended to other populations , and t^{wn}s were found in the four subspecies of *Mus* from many parts of the world . These included *M. molossinus* and *M. castaneus* from Japan and Asia (Tutikawa, 1955; Kaneda, 1989), *M. domesticus* from North America and Europe (Lewontin and Dunn, 1960; Dunn, et al 1973), and *M. musculus* from northern Europe and Russia (Dunn, et al, 1973, Klein, et al, 1984). The early studies depended on crossing wild mice to stocks with *T*. The use of *H-2* cDNA probes (see Chapter 7) suggested "that all t haplotypes have a common origin and are not products of independent mutational events" (Shin, et al, 1982). As these and other molecular markers of the T/t complex became available, more extensive surveys were possible since genetic crosses were not required for testing. One study of 58 US, 3 European, and 13 Australian populations found only 46% of the populations to have t-haplotypes, with an overall frequency of 6% of individual mice carrying a t^n (Ardlie and Silver, 1998). A 21st Century study in Denmark using similar methods found a 12% frequency in 186 populations (Dod, et al, 2003). The availability of molecular markers also provided a chance to study the evolutionary history of the T/t-complex (Silver, et el,1987; Delarbre, et al, 1988; Mizuno, et al, 1989; Morita, et al, 1992). Sequencing of the region, not surprisingly given the lack of chromosomal exchange and the chance for selection against them, revealed very high levels of indels and base substitutions, 0.7%, about 7-fold higher than average (Figueroa, et al, 1985; Uehara, et al, 1987). Studies involving species *M. spretus* and *M. abboti*, more distantly related from *M. m. musculus* and *M.m. domestius,* indicate an origin predating the radiation of

Mus 2-4 million years ago (Forejt; et al., 1988; Hammer, et al, 1989). Given this ancient origin, it is not surprising that distinctive alleles occur in different subspecies/populations (Ruvinsky, et al, 1991; Horiuchi, eti al, 1992). The different haplotypes are mosaics of wild-type and *T/t* complex chromosomes (Erhart, et al, 1989) which apparently occurred by gene conversion events (Hammer, et al, 1989). Partial *t* haplotypes were also found in the surveyed wild populations (Fujimoto, et al, 1995). These discoveries were made long after the Modern Synthesis had occurred and became the object of great scrutiny (see Chapters 3 and 5).

Leslie Clarence Dunn, 1893–1974

L. C. Dunn was born in Buffalo, New York to parents who, though not college graduates, valued literature and had a library[1]. He enjoyed nature, and several influences led to an interest in Botany. "From early childhood he had spent Saturday and Sunday afternoons working with his favorite person, his grandmother, in her garden, and had learned the satisfaction of steady application to simple tasks that gave insight into living things. A high school teacher, Dr. Marie Walcott, had amplified his love for gardens into an interest in taxonomic botany and through that, had given him an awareness of the powerful beauty of science." (Bennett, 1977). He attended Dartmouth College on a scholarship and was a botany major. Here he discovered genetics through exposure to Punnett's "Mendelism" and Morgan's "Heredity and Sex". He, or his professor, J. H. Gerould (Bennett and Dobzhansky [1978] differ) started a seminar to discuss Mendelian genetics. Dunn even visited Morgan's laboratory seeking graduate studies with him but Morgan had no room for a botanist and he entered Harvard and the Bussey Institute instead.

L. C. Dunn. By permission of the American Philosophical Society.

Dunn became Castle's assistant (see Chapter 1) for mouse care and research when he arrived in 1915. Dunn was trained in this role by Sewall Wright in the summer of 1915, before Wright went to Washington, D.C. He worked on coat color genetics, and his first paper (Dunn, 1916) was concerned with the relationship of red and yellow hair color. His work was interrupted by the U.S. entry into WWI and he volunteered for the Army. He "signed on because of his agitation over Germany's unrestricted submarine warfare" (**Gormley, p. 45**). He joined the Harvard Regiment which went to France in July, 1918, but he wasn't involved in fighting, as their division did supply work. He had married Louise Porter before leaving for France and they were to have 2 sons: the elder, Robert Leslie (born 1921), was a pilot in the Second World War, much to the chagrin of Dunn who was by then a pacifist. His other son, Stephen Porter

---

[1] I am aware of three brief reviews of Dunn's life: Dorothea Bennett's in Annual Reviews of Genetics (Bennett, 1977), Theodosius Dobzhansky's biographical memoir for the U. S. National Academy of Science (Dobzhansky, 1978) and the introduction to the index of his papers at the American Philosophical Society Library (amphilsoc.org). Dunn and Dobzhansky were great friends and collaborators—they had promised each that whoever was the survivor would write the other's obituary. Dobzhansky in fact left Columbia University for the Rockefeller Institute [it wasn't a university then] after Dunn's death because he couldn't stand to be there when Dunn wasn't. A major work on Dunn's life is by Melinda Gormley (**2006**). Her unpublished thesis of over 500 pages focuses on Dunn's political and humanitarian efforts, for which it is comprehensive. The account of his later career as a human geneticist is quite extensive, while his work on mouse genetics is lightly covered. Dunn also left 322 boxes of papers for the American Philosophical Society library, and over 1000 pages of transcripts of oral history at Columbia University. ("The Reminisces of Leslie Clarence Dunn", **Oral History Collection of Columbia University**, interviewed by Saul Benison, 1955-1960).

(born 1928) had severe cerebral palsy but a brilliant intellect and, with immense care from his parents, ultimately became a Ph.D. anthropologist and had a very successful academic career despite always being confined to a wheel chair.

On his return, Dunn continued his Ph.D work on coat color genetics in mice and interpreted an excess of brown-haired mice in the cross of agouti with black as possible linkage of the 2 genes **(Dunn, 1919)**, an interpretation that was eventually found to be false . He had early accepted the major role of the nucleus over other locations for genetic determinants (Dunn, 1917). The 3 papers cited above are among 8 on rodent genetics that he accomplished at Harvard.

Dunn had accepted a position as a poultry geneticist at the Agricultural Experiment Station in Storrs, Conn. even before defending his thesis. Here he was to publish 49 papers on poultry genetics and diet, etc. which established him as the premier poultry geneticist of the time. His work on egg weight and hatching were to be of importance for evolutionary genetics. "Moreover, in the 1940s and 1950s, researchers realized that Dunn's egg weight and hatchability examinations contributed to evolutionary theory. He had demonstrated that natural selection stabilizes and produces intermediate phenotypes" **(Gormley, p. 55)**. It was during his time at the station that his interest in developmental genetics had its birth. He worked with Walter Landauer on *rumplessness* (defective caudal structure development), presaging his work on defective developmental of the tail in *t*-allele mice. The work was very extensive: "Over 3,000 postmortems had shown that...chick embryos died with a variety of deformities" **(Gormley, p. 88)**. He also maintained mice and published one paper on mouse color genetics while there (Dunn and Durham, 1925). An early interest in human genetics occurred as well—he analyzed anthropometric data on Hawaiians that had been collected by Alfred Tozzer (Dunn, 1928).

Another activity while at Storrs was the writing of a text book of genetics with the botanist, E W Sinnott. This textbook (Sinnott & Dunn, "Principles of Genetics") was widely successful, went through many editions and was translated into many languages. "Several elements contributed to the book's success. Their frequent revisions attempted to keep a pace with the changing discipline: they re-wrote sections in order to provide up-to-date information about new developments and deleted outmoded ideas" **(Gormley, p. 60)**. They also included questions at the end of each chapter, which was quite innovative for a biology text (although familiar to engineers, physicists, etc.). The text shifted from being neutral about eugenics to being very negative, as Nazi atrocities became familiar.

Dunn's work against eugenics continued and involved several fronts. "He helped overhaul the Journal of Heredity during the 1930s because the journals pro-eugenic and slapdash content concerned him and other geneticists. He also contributed to the closure of the Eugenics Record Office" **(Gormley, p. 79)**. This effort with the Journal of Heredity was not completely successful—the journal was still publishing pro-eugenic articles in the 1950s and 1960s (Cooke, 1998).[2]

An important event that occurred while he was at the agricultural station was a trip to Europe, which included the Soviet Union, in 1927. His impression of the geneticists and their work made on this short visit (17 days) was to influence his views on the funding of science and the U.S.S.R. He found them and their work stimulating even though he discovered that they lived in poverty. They had funds for research which were not tied to their political views. He was visiting during the period of the New Economic Policy, which allowed "state capitalism". "Freedom to trade...was a backward stage for socialism, but private capital was under the control of the state, which was master of the land, industry, transport and foreign trade. The aim of this retreat towards state capitalism was, under state control, to make even foreign capitalism contribute to the development of socialism" **(Aragon, p. 182)**. This, of course only lasted a short time (it had started in 1921 and ended shortly after his visit) and scientists were soon being purged. Another aspect of this trip was that it represented one of many public services Dunn provided—he was surveying science abroad for the International Education Board.

In 1928, among several competing offers, Dunn accepted the Professorship of Zoology at Columbia University, available because T. H. Morgan had moved to Cal. Tech. New York City, with its active political organizations and as the point of entry for visiting scientists from abroad, became a stimulus to his scientific and social politics. His activities at Columbia included recruiting T. Dobzhansky and re-starting the annual series of lectures in science, the Jessup Lectures, of which he edited ten volumes. He became very active in American scientific societies. "In 1929 Dunn accepted the Chairmanship of the Committee on Organization of the Genetics Sections of the American Society of Zoologists and the Botanical Society of America. This committee established the Genetics Society of America (GSA) and its members voted Dunn as the first president. It is interesting to note that the first 7 presidents of the society were all associated with the Bussey Institute" (Weir, 1994).

Although his first work at Columbia focused on pigmentation genetics in mice, he attempted to work on both *Drosophila* and mice during a sabbatical in 1932-1933 at the University of Oslo, working with Otto Mohr on the former and Christina Bonnevie on the latter. These were to be his last efforts with the fruit fly but his work with Bonnevie

---

[2] This is probably the reason that Salome Gluecksohn-Schoenheimer warned me against publishing in the journal in the 1970s, with the comment that it was "tainted with eugenic notions".

led to his doing embryology in the mouse, working on a mutation that produced a short tail: *Shaker Short*. He had obtained the *T*-mutant and *t*-alleles from Nadine Dobrovalskaia-Zavadskaia in 1931 (see Chapt. One and mini-biography of Dobrovalskaia-Zavadskaia), and a young graduate student was working on them while he was away. Their initial publication appeared after Dunn's return and the student's untimely death (Chesley and Dunn, 1936). Dunn was to be author or co-author of over 50 publications concerned with the *T*-locus during the rest of his career—it remained the topic of his mouse genetical research. The latter will be discussed in Chapters 3, 4, and 5 (as well as chapter 1). His other publications after this, and there were many, were mostly educational, historical, "political", or reviews, although there were some scientific publications in human genetics.

His personal politics were always inclined to socialism and he was semi-active in the liberal organization, the American Association of Scientific Workers. He and another member of his Dept. of Zoology at Columbia (Hecht) gave lectures in the New School for Social Research, a leftist institution promoting education for workers. Starting in 1933, when Nazi laws deprived Jews of academic positions, he became active in helping them re-locate to the United States. He was an executive member of the Fellowship Fund at Columbia, which raised and provided some funds for refugees. He was also a member of the national Emergency Committee in Aid of Displace Foreign Scholars. "In order to bypass immigration restriction laws, the Emergency Committee invited scholars before they arrived to the United States. Jewish immigration quotas did not apply to a refugee with an invitation in hand" (**Gormley, p. 133**). In the mid-1930s, as Lysenko's power grew (see Chapter One) and Vavilov was arrested, his attention further turned to international politics. The 7th International Congress of Genetics had been scheduled for Moscow but was cancelled, as the Soviets rejected Mendelian genetics. Dunn helped the re-scheduling in Edinburgh. This meeting was severely disrupted by the outbreak of the Second World War on its closing day , and many attendees had fled back to Europe during the meeting's last days when the declaration of war was foreseen.

When the United States entered the war, Dunn became concerned with the ways in which biologists could contribute to the war effort, and he got involved in organizational activities which eventually led him to a role in the founding of National Science Foundation. With the Soviets as allies by this time, Dunn presided over the American-Soviet Science Society, which facilitated cooperation between the two countries' scientists. He played an active role in helping Soviet scientists get research articles into American journals, arranging translations and editing them. He and his wife were joint treasurers of the Russian War Relief Fund. Although he had had a heart attack in 1940, he took over the Chairmanship of Zoology at Columbia from 1940-1946. All these efforts, and anxiety about his son in the Army Air Force, were stressful, and his health declined. He took a sabbatical at the end of the war, starting with a "rest cure" in the Arizona desert.

After the war, Dunn at first supported science in the Soviet Union because it provided a model for government support of science. His views were based on his visit of 15 years earlier (Dunn, 1944). However, his outmoded ideas were sharply criticized by Karl Sax, a botanist at Harvard (Sax, 1944). Dunn felt that he needed to personally study Lysenko's claims, and he got Dobzhansky to translate Lysenko's book, Heredity and Its Variabilitiy. When it was published, he arranged for its review in several journals and influenced the review which occurred in the New York Times. His views turned about, and "Krementsov [author of "Stalinist Science", 1997, Princeton University Press, Princeton, NJ] notes that the two most active organizations in coordinating American geneticists' against Lysenko were headed by Dunn and Muller" (**Gormley, p. 340**).

Dunn remained convinced that government funding of science was crucial. He was active in several ways in the effort to get legislation passed to found the National Science Foundation (NSF). He helped write the pro-science legislation of Senator Harley M. Kilgore to mobilize biological scientists for the war effort, and he testified before the Senate in 1945. He also contributed material for Vannevar Bush's Science, the Endless Frontier, which is usually seen as the start of the campaign for the NSF.

With the start of the Cold War, the American-Soviet Science Society was considered a communist front organization, and it was disbanded in 1948 when the government denied it tax exempt status. Dunn's socialist views led him to be attacked in the press, and he was investigated by the House Un-American Activities Committee. In 1948 he was offered an appointment to be Scientific Attache in London, but it was not confirmed for political reasons. He was temporarily denied a passport in 1953 but eventually obtained it. Unhappy with a changed administration in the Dept. of Zoology, he took a visiting professorship at Harvard in fall, 1949 but it didn't become permanent, again, apparently because of his political views.

A new direction of political and scientific activity occurred at this time when he took a strong stance against racism and became more involved in human genetics. He helped UNESCO with their campaign against racism and wrote Heredity, Race and Society (1946, Penguin Press, Harmondsworth, U.K.) as a popular account to explain the lack of a genetic base for meaningful racial distinction (though he still thought race was a useful categorization). He strongly emphasized the role of environment in establishing differences in accomplishments. This new focus led to his recruiting faculty from many departments to work together in an Institute of Study of Human Variation at Columbia University.

Although Dunn and Dobzhansky had acquired an estate North of NYC (the Nevis estate) as a potential site for the institute, it was too expensive to turn it into a modern research institute, and an old building adjacent to the campus was found for the Institute, but only slowly prepared for it. Sadly, the university was not helpful with funding and the institute closed in 1958.

Having previously searched for isolated populations in Italy, Dunn choose the Jewish ghetto in Rome for study. Dunn and his wife and second son, traveled to Italy in 1953 to collect data on an isolated Jewish population in Rome. His wife performed demography in the community while his son used the work for his Ph.D. research in anthropology (Dunn and Dunn, 1957). Blood was collected from 650 individuals for blood group typing, and differences were found from the general Italians, supported the conclusion that the population remained an isolate. After his return,he carried out a similar study on an African-American community living on James Island near South Carolina. These studies contributed greatly to his book, Heredity and Evolution in Human Populations (1959, Harvard University Press,). His work in human genetics led him to be elected president of the American Society of Human Genetics in 1960.[3] In retirement, Dunn kept his mice at the Nevis estate and continued to work on T-locus genetics in collaboration with Dorothea Bennett and others.

It is interesting to speculate as to the reasons that Dunn is not well remembered today (there being no published book length biography, for instance). He did not have many graduate students, perhaps because he was so busy politically, and, thus, did not "found a school" as Castle had done. He was unsuccessful in getting Columbia to start an Institute of Genetics early in the '30s, and the Institute of Human Genetics was short-lived (compared to Little's founding of the Jackson Laboratory). Finally, his expertise was spread over a variety of organisms (mostly chickens, mice and humans). "One of his most damaging faults was his inability to say no. Even when he felt the need to curtail his activities, he could not always follow through with his intentions." (**Gormley, p. 507**). Nonetheless, he was highly revered in his life time and very important for the T-locus story.

[3] I was given a collection of Dunn reprints by M. Tahir Rizki, who had obtained his Ph.D. doing *Drosophila* genetics at Columbia University in the 1950s. Included in the collection is a covered typescript of Dunn's presidential address to the American Society of Human Genetics for May 3, 1961 (Dunn, 1962). Also, though not bound in the cover but enclosed behind this typescript, was a document, also typewritten, titled "The Doughty Street Discussions". (I could not find this in an electronic search of the American Philosophical Society's collection of Dunn papers). This is the beautifully written account of an evening spent discussing science with four foreign visitors: an Italian, a German, a Russian, and an Englishman (the latter may have been L.S. Penrose, who was his host at the time at the Galton laboratory). The evening was obviously all male and took a ribald turn, with discussions of the heterosis that would occur with matings between geneticists from differing countries. The question of choosing the sires came up "and I saw over the head of that burly, deep-voiced Russian, now engaged on his fourth bottle of beer, as clearly as though it had been pricked out in lights like car numbers at the opera—a large bright ONE. He would be the Premier Sire of course." This would almost certainly have Dobzhansky, a close fried of Dunn's. Later in the evening a lament about the paucity of female geneticists for their scheme arose. It is not surprising that this literary piece has not been published with its blatant sexism, but it gives a glimpse of a very engaging man.

Studies on prezygotic species isolation in *Mus musculus* and *domesticus*

Although studies of *Mus musculus* did not contribute much to the field genetics that led to the Modern Synthesis, studies in this species contributed to the solution of the problem of maintaining species (or subspecies) identity in over-lapping zones. Sumner had noted this problem in his studies of *Peromyscus*. "This abrupt transition which was found between one subspecies and the other was surprising in view of the lack of any evident barrier, geographic or ecological, between the ranges of the forms" (Sumner, 1929). In this case the subspecies were fertile in intercrosses while in others a high percentage of off-spring from the intercrosses were sterile (Sumner, 1923).

The problem was well studied in house mice, mostly in the last quarter of the 20th Century. The subspecies of *Mus musculus, M. m. musculus* in Eastern Europe and *M. m. domesticus* in Western Europe are thought to have initially diverged about five hundred thousand to about a million

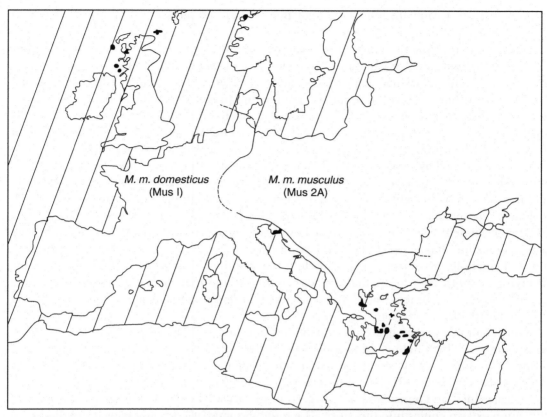

FIG. 2.1 The hybrid zone between Mus musculus musculus and Mus musculus domesticus from Vanlerberghe, et al, 1986., Genetical Research, by permission.

years ago (Suzuki, et al 2004). It is believed that *M. m. domesticus* followed agriculturists from the Middle East into Europe and came into contact with *M. m. musculus* only several thousand years ago (Cucchi, et al 2005). The two sub-species are distinguished by coloring, morphometrics, and behavioral traits (Ursin, 1952, Hunt and Selander, 1973). The zone of contact between the two sub-species is quite narrow and apparently quite stable (Hunt and Selander, 1973), Fig. 2.1. It has been extensively studied in Denmark where it crosses the Jutland peninsula and separates a Northern *musculus* population from the Southern *domesticus* population which continues into Germany (Ursin, 1952; Hunt and Selander, 1973; Dod, et al, 2003) in a transect across S. Germany and Austria (Sage, et al, 1986) and in Czechoslovakia (now the Czech Republic and Slovakia; Munclinger, et al 2002). Early studies of genetic variation across the hybrid zone (HZ) used electrophoretic variation in enzymes as markers. An extensive study across the Denmark zone of separation showed asymmetry of gene transfer with extensive introgression (the "invasion" of genes from one sub-species into the other sub-species at the boundary between them) of *domesticus* alleles into the *musculus* population but little introgression of *musculus* alleles into the *domesticus* population (Hunt and Selander, 1973).

The problem of maintenance of species individuality can broadly be subsumed under the name of "reproductive isolation". There are many aspects of reproductive isolation and they are frequently classified as prezygotic and post-zygotic. One of the earliest studies of a prezygotic mechanism involved odor recognition and pregnancy. Not surprisingly for a nocturnal animal, mice have a very sensitive sense of smell, even recognizing other individuals of the same sex and strain (Crawford, 1934; Bowers and Alexander, 1967). Hilda Bruce, at the National Institute of Medical Research, Mill Hill, discovered that pregnancy could frequently be blocked by removing the sire and replacing it with a different male; the effect was stronger if the new male was of a different strain (Bruce and Parrott, 1960)[1]. Since the male did not need to be in sight or sound of the females, it seemed likely that odor was involved. Hilda Bruce decided to test whether humans could distinguish the odors. "Knowing how skillful perfumers must be in distinguishing between thousands of different odours, they persuaded some Boake [world famous perfumers] representatives to visit NIMR for the purpose of smelling mice" using pieces of cloth that had been exposed to the different strains (**Clayton, p. 208**). They found that the odors from 5 strains of mice could be separated into 2 groups (Parkes and Bruce, 1962)[2]. But what was it in mouse urine that affected the behavior of females (the Bruce Effect)?

One major line of investigation to answer this question, a direction that seems to have been erroneous but which held great prominence for decades, was that odiferous components of urine were influenced by genetic variation at the Major Histocompatability Complex (MHC). This complex of genes, which controls cell surface antigens, is of great importance for immunological studies and transplantation surgery. Its history will be covered in Chapter Seven. The *T/t* complex also comes into this story. The route to the MHC as being relevant to the Bruce effect was a serendipitous one. As stated in their first paper on the subject (Yamazaki, et al, 1976), it was the confluence of Lewis Thomas's speculation that the MHC might impart a characteristic scent to each individual and observations in the mouse room that led to this hypothesis. Thomas had made this speculation while thinking about the possibility of pheromones in humans: "Man's best friend might be used to sniff out histocompatible donors" (**Thomas, p. 19**). The mouse-room worker(s) had noted that, with trio breeders, where 2 females only differed with respect to their MHC[3], and the male matched one of them, there was a mating preference for the female with the dissimilar MHC. This initial work used the preferential mating choice of the female for a male of the same inbred strain, but with a different (congenic) MHC, over a male with her own MHC (Yamazaki, et al 1976). The choice of female by the male was apparently influenced by the odors of the parental mice—male mice fostered with females of a different MHC preferred the odor of females of their own MHC to that of the odor to which they were exposed during rearing (Beauchamp, et al, 1988). These preferences were also shown for MHC haplotypes then found in wild populations (Egid and Brown, 1989) and were shown to maintain diversity of MHC haplotypes in natural settings (Potts, et al, 1991). If the MHC has any effects on odor preference, they may influence MHC diversity (Penn and Potts, 1999) but would not contribute to species isolation.

To pursue the problem, a Y maze was used, in which air directed down both arms flowed over odorant samples towards gates which were opened at intervals, to let the mouse enter one of the arms (Boyse and Thomas, 1979). Mice were rewarded with a drop of water for correct responses after being deprived of water for 23 hours. Initially the odorant was a mouse in the odorant compartment. The trained mice could correctly identify the congenic homozygous seg-regants from the F_2 cross of the 2 lines only differing for their MHC, (congenic, see Chapter 7), thus eliminating other possible causes (e.g. age, previous reproductive experience, etc.) for the discrimination (Boyse and Thomas, 1979).

These apparent MHC discriminations were soon found to be achieved with only urine or bedding as the markers, and were equally distinguishable in germ-free mice (Yamazaki, et al, 1990). This was important as the MHC greatly influences immune responses and could read-ily be responsible for a different bacterial biota, creating different urinary volatiles' in mice of different MHC (a germ-free environment eliminated odorant discrimination in rats; Singh, et al., 1990). Although the MHC proteins could not be detected in serum, they were found after the serum was digested with a protease (pronase) which presumable released peptides which would be more volatile than intact proteins (Yamazaki, et al, 1999) but MHC peptides are pres-ent in urine in very minute quantities.

Besides the MHC, there has been much interest in the idea that hybrids in the zone have a weakened resistance to parasite-born disease due to the breakup of coadapted gene clusters conferring resistance on the allopatric populations on either side of the HZ [Moulia, et al, 1993], although those findings have been disputed by others"[4].

Update

The importance and basis for MHC-based mouse discrimination remains controversial in the Twenty first Century (Overat, et al, 2014). Mouse urinary proteins (MUPs) seem much more important in nature (Cheetham, et al, 2007) and are found at a million-fold higher concentrations in urine than are MHC peptides. They have great genetic diversity and bind many volatile com-pounds (Cheetham, et al, 2007). The problem of replication for the MHC olfaction studies may be due to the paradigm of congenic lines: when mice are maintained in a uniform environment and are genetically uniform except for one small "patch" of chromosome, minute differences may easily become statistically very significant—especially when a long training period for the dis-criminating mice is used[5]. In addition, the role of MHC odorant distinctions in wild mice is very controversial—the most recent conclusion is that they aren't present (Green, et al, 2015).

T/t complex effects on odorant discrimination

It was discovered that the Bruce effect, in which the presence of a foreign male induces pregnancy block, is also influenced by *T/t* complex determined factors (Coopersmith and Lennington, 1998). These authors explored this role of *t*-haplotypes with laboratory-descen-dants of wild mice from 23 distinct populations, but the particular *t*-haplotype involved was not determined. The authors tested for t^{ns} by doing crosses only to *T*. They did not test for allelism by doing crosses to other t^{wn} s. Most of the t^{wn}s would probably be t^{w5}, as that is the most frequently found t-haplotype in wild populations in the USA. They found that +/+ females were more likely to undergo pregnancy block (the loss of pups already in utero) when the male was +/+ than when it was +/t.

These findings would suggest a genotypically negative consequence, since they would select for the recessive lethal t-allele through subsequent mating with the +/t male. However, the results might not have reflected the t-haplotypes effect, but rather that of the tightly linked MHC. (Although the mapping distance between T and MHC is 15 centiMorgans, inversions of the T/t complex prevent recombination –see Chapter 3). Again, the findings may be specious, as is the current belief about MHC prezygotic selection (see above). The general approach of training mice to recognize odorant differences continued, and has been used to detect X and Y chromosomal differences (Yamazaki, et al, 1986). This was done with reciprocal (by sex) F_1s, so that only the sources of the X or Y were genetically distinct, and were tested only in males. The females from either cross had 2 different Xs and, thus, were not distinguishable). The amount of training required suggested that the differences were slighter than those found with the MHC.

Update on other odorant systems and subspecies isolation

Another odorant detection system has been studied which may better serve subspecies isolation. This is the salivary androgen-binding protein (ABP), the alpha subunit of which is coded by *Abpa*. Behavioral studies have shown that females prefer to mate with males sharing their allele of *Abpa* (Laukaitis, et al, 1997; Talley, et al 2001) This gene is now called *Abpa27* (see Laukaitis et al 2008). ABP (a dimer with ABPBG, encoded by *Abpbg*) is only present in saliva and not in urine (Dlouhy, et al, 1986) and only saliva stimulates preference behavior in female subjects in the Y-maze (Talley, et al, 2001). Mice, like most mammals, lick their coats and, thus, the salivary proteins are spread over a large surface area.

Two alleles of *Abpa27* are differently monomorphic in *M. m. musculus* and *M. m. domesticus*, in samples distant from the barrier zone in mid-Northern Europe (Karn & Dlouhy, 1991) which is the reason it was studied as an odorant.[6] Analyses of sequence variation suggest that there has been positive selection for the different forms in these two sub-species (Hwang, et al 1997; Karn and Nachman, 1999). However, other populations of *M. musculus* are more panmictic (freely genetically exchanging) for forms of *Abpa*. A 21st Century study tested wild mice, and first generation from wild, from both sides of the barrier zone in a portion of the Germany/ Czech Republic transect. A slight preference of males for the odor of saliva of their own *Abpa27* allele (tested in an olfactory maze) was found (Bimova, et al, 2005). This was followed by a study that tested preferences for salivas from strains congenic for the alleles in *M. m. domesticus* and *M. m. musculus* (Voslajerova Bimova, et al 2011). That study provided evidence for an ABP-based reinforcement on the European mouse hybrid zone.

Given the overall variable evidence, stronger in the later study (Voslajerova Bimova, et al, 2011) for these prezygotic mechanisms of the Bruce Effect blocking pregnancies and of mate preference via odorants, other mechanisms of sub-species isolation have been studied in mice: "Most of us still feel that the existence of the HZ [hybrid zone] owes primarily to a postzygotic barrier(s) and that reinforcement involving a prezygotic mechanism ... arose secondarily (as predicted by the concept of reinforcement). [Voslajerova Bimova, et al 2011; also reviewed in Laukaitis and Karn, 2012]. The final conclusion of Laukaitis and Karn is "The only chemical cue that has been shown to differentiate the subspecies of *M.m. domesticus* and *M. m. musculusis* is ABP. Although urine has been proposed to have a role in mediating subspecies recognition in the European house mouse hybrid zone, no component with molecular characteristics that potentially could differentiate the two subspecies has yet been described."

Haldane's rule, sex-chromosomal hybrid sterility genes

J. B. S. Haldane, a seminal biochemist and geneticist who is credited as one of the "big three" for the mathematical foundation of the Modern Synthesis (see "Haldane-Mitchison Clan" mini-biography and Chapter 3) formulated an important rule relevant to species isolation in 1922: "When in the F_1 offspring of two different animal races one sex is absent, rare, or sterile, that sex is the heterozygous (heterogametic) sex"(Haldane, 1922). At that time he was generalizing from 45 *Lepidopteran*, 10 bird, 6 mammalian and a few other species. The rule has been extensively confirmed with about 200 examples, many in *Drosophila* and only about 1% exceptions have been found (Laurie, 1997). Obviously such inviability, sterility, or decreased fertility can contribute to species isolation. In fact, Haldane's rule and the large X-effect (substitution of the X of one species by that of another, in hybridization) has an excessively large effect on fitness compared to substituting an autosome) are considered the "two rules of speciation" (Masly & Presgraves, 2007) This 21st Century article soundly confirms these rules with step-by-step replacement of small segments of the genome between two *Drosophila* species for 70% of their genomes.

An early set of results by Muller in *Drosophila* were interpreted to indicate that X:autosome imbalance was responsible for Haldane's rule (Muller, 1942). This explanation was not much questioned until the last quarter of the 20th century, when the molecular/genetical mechanisms involved in Haldane's rule became much debated. A classic experiment by Coyne (1985), also in *Drosophila*, strongly suggested that the responsible interaction is between X and Y chromosomes, as had been previously suggested by White (1945). Wu (1992) pointed out an alternative explanation of Coyne's findings, and Coyne agreed that both X:autosome and X:Y interaction must be involved in many examples of hybrid sterility in *Drosophila* (Coyne, 1992). What is clear is that hybrid sterility evolves rapidly. "These results show that, in the incipient stage of speciation, sterility in hybrid males is the first to evolve" (Wu & Davis, 1993). It is also clear that a mutation on the X with a positive selective value in an XY species will be selected for more rapidly than such a mutation on an autosome, because of its "exposure" in heterogametic males. This should particularly affect sterility genes, but not viability genes, since the former are sex specific (there are different gene programs for male versus female gametogenesis, see Chapt.11). However, X-linked genes of negative effect could result in female sterility when homozygous. These are rare, as Coyne (1985) showed that the presence of 2 Xs from the "foreign" species in homozygous condition did not cause sterility. As Orr (1993) points out, "if the probability, *p*, of picking up a complementary sterile is small (as it must be or speciation would be nearly instantaneous), the probability that a pair of species will pick up *both* an X-linked allele (or set of alleles) causing hybrid male sterility and an X-linked recessive(s) causing hybrid female sterility is far smaller (roughly p^2)". Thus, hybrid male sterility would arise faster than hybrid female sterility. However, there remains the problem of rates of mutation to hybrid sterility. "The mutagenic potential of the *Drosophila* genome for inviability is known to be 10 times that for sterility of either sex…and yet what has been realized in evolution is quite the opposite: male sterility, both from hybridization data and genetic analysis, is at least 10 times more prevalent than inviability" (Wu & Davis, 1993).

Turning to mice, the importance of X and Y chromosomes for reproductive isolation is reflected in their degree of introgression across the *M. m. musculus/M. m. domesticus* hybrid zone. The Y chromosome is introgressed for a much shorter distance from the boundary than are autosomal markers (Vanlerberghe, et al, 1986). This could be due to preferential mating of the females near the boundary with males from within their subspecies compared to the other, across the boundary subspecies, but this would also introgress the X chromosome (Tucker, et al, 1992; Dod, et al, 1993). The *M .m. domesticus* Y is the one found in the standard inbred strains and "is also the type found in a zone of natural hybridization between *M. m. musculus* and *M. m. castaneus,* whether in the form of a loose interaction (central China) or a complex admixture (Japan)" (Boursot,et al, 1993). In crosses of wild-derived mice (sometimes recently inbred), male infertility or sterility maps particularly to the *M. m. musculus* X Chromosome (Britten-Davidian, et al, 2005; Good, et al, 2007). Incompatibilities between subspecies of *Peromyscus* also include X-linked loci (Vrana, et al, 2000). Thus, mapping and cloning experiments were pursued.

One of the first hybrid sterility loci to be mapped on the X chromosome was *Hst-3* (Guenet, et al, 1990). Perhaps this gene is identical to another X-linked locus, *Ihpd* (interspecific hybrid placental dysplasia) mapping to the X (Zechner, et al, 1996, 1997). This latter gene interacts with autosomal loci in causing the malfunctional placental dysplasia. *Iphd* is also influenced by the Y chromosome in these crosses (Hamberger, et al 2001).

Update on sex chromosomes and hybrid sterility

In further research on mutual effects of X and Y chromosomes during the current Century, Oka, et al (2004) studied interactions of the *M. m. molossinus* X chromosome with the *M. m.domesticus* genome. Quantitative trait mapping identified 3 genes involved in sperm head morphology and 1 gene involved in testis weight. However, the degree to which these were due to Y- or autosomal-interactions was not dissected. In crosses involving *M. m. domsesticus* and *M. M. musculus*, the evidence is against a strong contribution of the Y (Campbell, et al, 2012).

In the 21st Century, it has also been found that a likely major basis for hybrid sterility in crosses between subspecies is rapid evolution of X and Y genes involved in X:Y competition, i.e., sex drive. The concept was first broached in 1991 by Frank (1991) and Hurst and Pomiankowski, 1991. The mechanism also involves the degree of inactivation of the X:Y bivalent during spermatogenesis and is further discussed in Chapter 8).

The *T/t* complex and hybrid sterility

Although, as discussed above, the MHC locus linked to *T* on chromosome 17 is not responsible for subspecies isolation, a number of Hybrid sterility genes (*Hsts*) are linked to *T* and are likely to be involved in transmission ratio distortion (see Chapter 8). The first to be identified was *Hst1* (Forejt and Ivanyi, 1975). It maps 3 centiMorgans distal to *T* (Forejt, et al 1991). In a high-resolution back-cross made to fine-localize the gene, 5 testis-specific transcripts were found in the smallest region of non-recombination (Gregorova, et al 1996). Eventual cloning identified *Prdm9*, a histone-3, lysine-4 trimethyltransferase, as the causative gene (Mihola, et al, 2009). Previously described knockouts of this gene (Hayashi, et al, 2005) had shown disrupted formation of the sex-body (the X:Y heterochromatic pair seen during male meiosis). Thus, this hybrid sterility gene also involves X:Y interactions. Three other hybrid sterility loci (*Hst4, Hst5,* and *Hst* 6) map to the distal-most inversion of the *t*-complex *Hst4, Hst5,* and *Hst 6;* Pilder, et al, 1998). The effect of *Hst6* seems to be on sperm morphology.

Other modes of action of Hybrid sterility genes involve sperm flagellar structure. By taking advantage of the fact that *Mus spretus*, as compared to *M. m. domesticus*, carries an inversion in the *t* region which interacts with t^n alleles to cause sterility, fine mapping and cloning of *Hst6* was performed (Fossela, et al, 2000). It was found to encode an axonemal heavy chain required for sperm flagellar formation in *M. m. domesticus* but not in *M. spretus*. Thus, rapid evolution of many genes related to testis formation may contribute to species isolation.

Notes

1. This may appear to be a post-zygotic mechanism, since pregnancy has been blocked, but it represents a form of mate selection.

2. The version of this story that I heard when I was a post-doctoral fellow at NIMR, Mill Hill, in 1969-1970 was different, and might have represented an earlier trial or might be apocryphal. In this version, the 5 strains of mice were in ceramic "battery" jars with wooden lids (thus, the different coat colors were not visible) and the odor tester was a wine expert–he was to lift the lid and smell with his eyes closed. Supposedly he took a deep whiff at the first jar and promptly got sick in the sink!

3. Personal communication, Robert Karn, Mar., 2016. with permission.

4. Congenic lines will be discussed more fully in Chapters 3 and 7 and are the result of back-crossing mice of one inbred stain to another, choosing heterozygotes carrying the donor's allele (in this case of the MHC) for the next back-cross. After a series of usually 10 backcrosses, the mice can be inter-crossed to obtain a congenic inbred strain only differing from the initial strain by the homozygous alleles of the gene chosen and a small number of closely linked genes that have not been eliminated by genetic recombination. Even although 5 generations of backcrossing are calculated to result in mice with 98.5% of the new genetic background, because of genetic linkage, theoretical calculations suggest a near maximum decrease in the amount of linked genes after N16 (16 generations of backcrosses; Naveira and Barbadilla, 1993). Congenics are discussed further in Chapter 3.

5. This is a point I first heard made by the Czechoslovakian immunogeneticist Alena Lengorova but I cannot find a citation. All my experience in personal studies of congenics, and reading the results of others with congenics, support the importance of this observation.

6. "I can confirm, however, that it was the observation of fixed *Abpa* alleles in the three subspecies, as well as the ubiquitous capability of rodent salivas [sic] to bind androgen that caused us to establish our hypothesis (J. Heredity, 1991 paper: 'oral and/or olfactory recognition purposes'[Karn and Dlouhy, 1991]). That work led to the later behavioral work (Evolution '97 [Laukaitis, et al, 1997]) that included observations that ABP could be found on the mouse's pelage". Personal communication, Robert C Karn, with permission.

References

Allen, G.E., 1968. Thomas Hunt Morgan and the problem of natural selection. J. Hist. Biol. 1, 113–139.

Ardlie, K.G., Silver, L.M., 1998. Low frequency of *t* haplotypes in natural populations of house mice (*Mus musculus domesticus*). Evolution 52, 1185–1196.

Bateson, W., 1922. Evolutionary faith and modern doubts. Science 55, 1412.

Beauchamp, G.K., Yamazaki, K., Bard, J., Boyse, E.A., 1988. Preweaning experience in the control of mating preferences by genes in the major histocompatability complex of the mouse. Behavior Genet 18, 537–547.

Bennett D., 1977. L. C. Dunn and his contribution to T-locus genetics. Ann. Rev. Genet. 11 :1–12.

Bimova, B., Karn, R.C., Pialek, J., 2005. The role of the salivary androgen-binding protein in reproductive isolation between two subspecies of house mouse : *Mus musculus musculus* and *Mus musculus domesticus*. Biol. J. Linnean Soc. 84, 349–361.

Bowers, J.M., Alexander, B.K., 1967. Mice : Individual Recogition by Olfactory Cues. Science 158, 1208–1210.

Boursot, P., Auffray, J.-C., Britton-Davidian, J., Bonhomme, F., 1993. The Evolution of House Mice. Annu. Rev. Ecol. Syst. 24, 119–152.

Boyse, E., Thomas, L., 1979. Recognition Among Mice: Evidence from the use of a Y-maze differentially scented by Congenic Mice of Different Major Histocompatibility Types. J. Exp. Med. 150, 755–760.

Britton-Davidian, J., Fel-Clair, F., Lopez, J., Alibert, P., Boursot, 2005. Postzygotic isolation between two European subspecies of the house mouse : estimates from fertility patterns in wild and laboaory-bred hybrids. Biol. J. Linnean Soc. 84, 379–393.

Brown, L.N., 1965. Selection in a population of house mice containing mutant individuals. J. Mammal. 46, 461–465.

Bruce, H.M., Parrott, D.M.V., 1960. Rôle of olfactory sense in pregnancy block by strange males. Science 131, 1526–1528.

Campbell, P., Good, J.M., Dean, M.D., Tucker, P.K., Nachman, M.W., 2012. The contribution of the Y chromosome to hybrid male sterilitiy in house mice. Genetics 191, 1271–1281.

Carnochan, F.G., 1937. A new mutation in the house mouse. J. Hered. 28, 333–334.

Castle, W.E., Phillips, J.D., 1914. Piebald rats and selection : An experimental test of the effectiveness of selection and of the theory of gamete purity in mendelian crosses. Carnegie Institution of Washington, Washington, D.C.

Cheetham, S.A., Thom, M.D., Jury, F., Ollier, W.E.R., Beynon, R.J., Hurst, J.L., 2007. The genetic basis of individual recognition signals in the mouse. Curr. Biol. 17, 1771–1777.

Chesley, P., Dunn, L.C., 1936. The inheritance of taillessness (Anury) in the house mouse. Genetics 21, 525–536.

Cooke, K., 1998. Twisting the ladder of science : pure and practical goals in twentieth century studies of inheritance. Endeavour 22, 12–16.

Coopersmith, C.B., 1998. Pregnancy block in house mice (*Mus domesticus*) as a function of t-complex genotype: Examination of the mate choice and male infanticide hypotheses. J. Comp. Psychol. 112, 82–91.

Coyne, J.A., 1985. The genetic basis of Haldane's rule. Nature 314, 736–773.

Coyne, J.A., 1992. Genetics and speciation. Nature 355, 511–515.

Crawford, S.C., 1934. The habits and characteristics of nocturnal animals. Quart. Rev. Biol. 9, 201–214.

Cucchi, T., Vigne, J.D., Auffray, J.C., 2005. First occurrence of the house mouse (*Mus musculus* domesticus Schwarz & Schwarz, 1943) in the Western Mediterranean : a zooarchaeological revison of subfossil occurrences. Biol J. Linn. Soc. 84, 429–445.

Darwin, C.R., 1858. Extract from an unpublished work on species. Proc. Linnean Soc. 3, 46–50.

Delarbre, C., Kashi, Y., Boursot, P., Beckman, J.S., Kourilsky, P., Bonhomme, F., Gachelin, G., 1988. Phylogenetic distribution in the genus *Mus* of *t*-complex-specific DNA and protein markers :inferences on the origin of *t*-haplotypes. Mol. Biol. Evol. 5, 120–133.

Detlefsen, J.A., 1921. A new mutation in the house mouse. Amer. Natural. 45, 469–473.

Dlouhy, S.R., Nichhls, W.C., Karn, R.C., 1986. Production of an antibody to mouse salivary androgen binding protein (ABP) and its use in identifying a prostate protein produced by a gene distinct from *Abp*. Biochem. Genet. 24, 743–763.

Dobzhansky, T., 1978. Leslie Clarence Dunn : November 2,1983-March 19, 1974. Biographical Memoirs, Nat'l Acad. Sci, USA.

Dod, B., Hermiin, L.S., Boursot, P., Chapman, V.M., Nielsen, J.T., Bonhomme, F., 1993. Counter selection on sex chromosomes in the *Mus musculus* hybrid zone. J. Evol. Bio. 6, 529–546.

Dod, B., Litel, C., Makoundou, P., Orth, A., Boursot, P., 2003. Identification and characterization of *t* haplotypes in wild mice populations using molecular markers. Genet. Res. 81, 103–114.

Dunn, L.C., 1916. The genetic behaviour of mice of the color varieties «Black-and Tan» and « Red». Amer. Naturalist 50, 664–675.

Dunn, L.C., 1917. Nucleus and cytooplasm as vehicles of heredity. Amer. Naturalist 51, 286–300.

Dunn, L.C., 1921. Unit character variation in rodents. J. Mammal. 2, 125–140.

Dunn, L.C., 1928An anthorpomorphic study of Hawaiians. Papers of the Peabody Museum11. Harvard Univ., pp. 20–211.

Dunn, L.C., 1944. Soviet Biology. Science 99, 65–67.

Dunn, L.C., Bennett, D., Cookingham, J., 1973. Polymorphisms for lethal alleles in European population of *Mus musculus*. J. Mammal. 54, 822–830.

Dunn, L.C., Suckling, J., 1956. Studies of genetic variability in populations of wild house mice. I Analysis of eight additional alleles at locus T. Genetics 42, 299–311.

Dunn, L.C., Dunn, S.P., 1957. The Jewish community of Rome. Sci. Amer. 196, 118–128.

Dunn, L.C., Durham, G.B., 1925. The isolation of a pattern variety in piebald house mice. Amer. Naturalist 59, 36–49.

Dunn, L.C., Morgan, Jr., W.C., 1953. Alleles at a mutable locus found in populations of wild house mice (Mus Musculus). Proc. Natl Acad. Sci., USA 39, 391–402.

Egid, K., Brown, J.L., 1989. The Major Histocompatability Complex and Female Mating Preference in Mice. Anim. Behav. 38, 548–549.

Erhart, M.A., Phillips, S.J., Bonhomme, F., Boursot, P., Wakeland, E.K., Nadeau, J.H., 1989. Haplotypes that are mosaic for wild-type and *t* complex-specific alleles in wild mice. Genetics 123, 405–415.

Figueroa, F., Golubic, M., Nizetic, D., Klein, J., 1985. Evolution of mouse major histocompatibility complex genes borne by t chromosomes. Proc. Nati. Acad. Sci. USA 82, 2819–2824.

Forejt, J., Gregorova, S., Jansa, P., 1988. Three new *t*-haplotypes of *Mus musculus* reveal structural similarities to *t*-haplotypes of *Mus domesticus*. Genet. Res., Camb. 51, 111–119.

Forejt, J., Ivanyi, P., 1975. Genetic studies on male sterility of hybrids between laboratory and wild mice (*Mus musculus L.* Genet Res, Camb 24, 189–206.

Forejt, J., Vincek, V., Klein, J., Lehrach, H., Loudova-Mickova, M., 1991. Genetic mapping of the *t*-complex region on mouse chromosome 17 including the *Hybrid sterility-1* gene. Mammal. Genome 1, 84–91.

Fossella, J., Samant, S.A., Silver, L.M., King, S.M., Vaughan, K.T., Olds-Clarke, P., Johnson, K.A., Mikami, A., Vallee, R.B., Pilder, S.H., 2000. An axonemal dynein at the *Hybrid Sterility 6* locus : implications for *t* haplotype-specific male sterility and the evolution of species barriers. Mammal. Genome 11, 8–15.

Frank, S.A., 1991. Divergence of meiotic drive suppression systems as an explanation for sex-biased hybrid sterility and inviability. Evol 45, 262–267.

Fujimoto, A., Wakassugi, N., Tomita, T., 1995. A novel partial *t* haplotype with a Brachyury-independent effect on tail phenotype. Mammal. Genome 6, 396–400.

Gillham, N.W., 2001. Evolution by Jumps : Francis Galton and William Bateson and the Mechanism of Evolutionary Change. Genetics 159, 1383–1392.

Good, J.M., Handel, M.A., Nachman, M.W., 2007. Asymmetry and polymorphism of hybrid male sterility during the early stages of speciation in house mice. Evol 62, 50–65.

Green, J.P., Holmes, A.M., Davidson, A.J., Paterson, S., Stockley, P., Beynon, R.J., Hurst, J.L., 2015. The genetic basis of kin recognition in a cooperatively breeding mammal. Curr. Biol. 25, 2631–2641.

Gregorova, S., Mnukova-Fajdelova, M., Trachtulec, Z., Capkova, J., Loudova, M., Hoglund, M., Hamvas, R., Lehrach, H., Vincek, V., Klein, J., 1996. Sub-milliMorgan map of the proximal part of chromosome 17 including the hybrid sterility 1 gene. Mammal. Genome, 107–113 &.

Guenet, J.-L., Bonhomme, F., 2003. Wild mice : an ever increasing contribution to a popular mammalian model. Trends in Genet 19, 24–31.

Guenet, J.-L., Nagamine, C., Simon-Chazottes, D., Montagutelli, X., Bonhomme, F., 1990. *Hst-3* : an X-linked hybrid sterility gene. Genet. Res., Camb. 56, 161–165.

Haldane, J.B.S., 1922. Sex ratio and unisexual sterilty in hybrid animals. J. Genet. 12, 101–109.

Hamburger, M., Kurz, H., Orth, A., Otto, S., Luttges, A., Elliot, R., Nagy, A., Tan, S.-S., Tam, P., Zechner, U., Fundele, R., 2001. Genetic and developmental analysis of X-inactivation in interspecific hybrid mice suggests a role for the Y chromosome in placental dysplasia. Genetics 157, 341–348.

Hammer, M.F., Schimenti, J., Silver, L.M., 1989. Evolution of mouse chromosome 17 and the origin of inversions associated with *t* haplotypes. Proc. Nat'l Acad. Sci. USA 86, 3261–3265.

Hayashi, K., Yoshida, K., Matsui, Y., 2005. A histone H3 methyltransferase controls epigenetic events required for meiotic prophase. Nature 438, 374–378.

Horiuchi, Y., Agulnik, A., Figueroa, F., Tichy, H., Klein, J., 1992. Polymorphisms distinguishing different mouse species and *t* haplotypes. Genet. Res. Camb. 60, 43–52.

Hunt, W.G., Selander, R.K., 1973. Biochemical genetics of hybridization in european house mice. Heredity 31, 11–33.

Hurst, L.D., Pomiankowski, A., 1991. Causes of sex-ratio bias may account for unisexual sterility in hybrids : a new explanation of haldane's rule and related phenomena. Genetics 128, 841–858.

Hwang, J.M., Hofstetter, J.R., Bonhomme, F., Karn, R.C., 1997. The microevolution of mouse salivary androgen-binding (ABP) paralleled subspeciation of *Mus musculus*. J. Hered. 88, 93–97.

Jameson, H.L., 1898. On a probable case of protective coloration in the house-mouse (*Mus musculus*. Linn.). Zool. J. of the Linnean Soc. 27, 465–473.

Kaneda, H., Maeda, Y.Y., Moriwaki, K., Sakaizumi, M., Taya, C., Watanabe, S., Yanekawa, H., 1989. *t* chromosome found in East Asiatic wild mice, *M m molissinus, M m castaneus*. Mouse News Letter 84, 119.

Karn, R.C., Dlouhy, S.R., 1991. Salivary androgen-binding protein variation in *Mus* and other rodents. J. Hered. 88, 93–97.

Karn, R.C., Nachman, M.W., 1999. Reduced nucleotide variability at an androgen-binding protein locus (*Abpa*) in house mice : Evidence for positive natural selection. Mol. Biol. Evol. 16, 1192–1197.

Klein, J., Sipos, P., Figueroa, F., 1984. Polymorphism of *t*-complex genes in European wild mice. Genet. Res., 3093–3106.

Laukaitis, C.M., Critser, E.S., Karn, R.C., 1997. Salivary androgen-binding protein (ABP) mediates sexual isolation in *Mus musculus*. Evol 51, 2000–2005.

Laukaitis, C.M., Heger, A., Blakley, T.D., Munclinger, P., Ponting, C.P., Karn, R.C., 2008. Rapid bursts of *androgen-binding protein (Abp)* gene duplication occurred independently in diverse mammals. BMC Evolutionary Biol 8 article # 46. doi:doi.org/10.1186/1471-2148-8-46.

Laukaitis, C.M., Karn, R.C., 2012. Recognition of subspecies status mediated by androgen-binding protein (abp) in the evolution of incipient reinforcement on the european house mouse hybrid zone. In: Macholan, M, Baird, SJE, Munclinger, P, Pialek, J (Eds.), Evolution of the House Mouse. Cambridge University Press, London, New York, pp. 150–190.

Laurie, C.C., 1997. The Weaker Sex Is Heterogametic : 75 Years of Haldane's Rule. Genetics 147, 937–951.

Lewontin, R.C., Dunn, L.C., 1960. The evolutionary dynamics of a polymorphism in the house mouse. Genetics 45, 705–722.

Lloyd, R.E., 1912. The Growth of Groups in the Animal Kingdom. Longmans, Green and Co, London.

Masly, J.P., Presgraves, D.C., 2007. High-resolution genome-wide dissection of the two rules of speciation in Drosophils. PLoS Biol 5, e243. doi:10.1371/journal.pbio.0050243.

Mayr, E., 1973. Essay Review : The Recent Historiography of Genetics. J. Hist. Biol. 6, 125–154.

Mendel, J.G., 1866. Versuche uber Pflanzenhybride. Verhand. Naturforsch. Ver Brunn IV, 3–47.

Mihola, O., Trachtulec, Z., Vicek, C., Schimenti, J.C., Forejt, J., 2009. A mouse speciation gene encodes a meiotic histone H3 methyltransferase. Science 323, 373–375.

Morgan, T.H., 1918. Concerning the mutation theory. Sci. Monthly 6, 385–405.

Morita, T., Kubota, H., Murata, K., Nozaki, M., Delarbre, C., Willison, K., Satta, Y., Sakaizumi, M., Takahata, N., Gachelin, G., 1992. Evolution of the mouse *t* haplotype : Recent and worldwide introgression in *Mus musculus*. Proc. Nat'l Acad. Sci., U.S.A. 89, 651–6855.

Mizuno, K., Vincek, V., Figueroa, F., Klein, J., 1989. The *D17Tu5* locus in the *t* complex : implications for the origin of *t* haplotypes and inbred strain. Immunogenet 30, 105–111.

Muller, H.J., 1942. Isolating mechanisms, evolution and temperature. Biol. Symp. 6, 71–125.

Munclinger, P., Bozikova, E., Sugerkova, M., Pialek, J., Macholan, M., 2002. Genetic variation in house mice (*Mus*, Muridae, Rodentia) from the Czech and Slovak republics. Folia Zool 51, 81–92.

Oka, A., Mita, A., Sakurai-Yamatani, N., Yamamoto, H., Takagi, N., Takano-Shimizu, T., Toshimori, K., Moriwaki, K., Shiroishi, T., 2004. Hybrid Breakdown Caused by substitution of the X Chromosome Between Two Mouse Subspecies. Genetics 166, 913–924.

Orr, H.A., 1993. A mathematical model of Haldane's rule. Evol 47, 1606–1611.

Osborn, H.F., 1915. Origin of single characters as observed in fossil and living animals and plants. Amer. Natural. 49, 193–239.

Overath, P., Sturm, T., Rammansee, H.-G., 2014. Of volatiles and peptides : in search for MHC-dependent olfactory signals in social communication. Cll Molec. Life Sci. 71, 2429–2442.

Parkes, A.S., Bruce, H.M., 1962. Pregnancy-block in female mice placed in boxes soiled by males. J. Reprod. Fertil. 4, 303–308.

Penn, D.J., Potts, W.K., 1999. The evolution of mating preferences and major histocompatiability genes. Amer. Naturalist 153, 145–164.

Pilder, S.H., Olds-Clarke, P., Phillips, D.M., Silver, L.M., 1998. *Hybrid sterility-6* : A mouse *t* comples locus controllling sperm flagellar assembly and movement. Dev. Biol. 159, 631–642.

Potts, W.K., Manning, C.J., Wakeland, E.K., 1991. Mating patterns in seminatural populations of mice influenced by MHC genotype. Nature 352, 619–621.

Ruvinsky, A., Polyakov, A., Agulnik, A., Tichy, H., Figueroa, F., Klein, J., 1991. Low density of *t* haplotypes in the eastern fom of the house mouse. Mus musculus L. Genetics 127, 161–168.

Sage, R.D., 1981. « Wild mice ». In: Foster, HL, Small, JD, Fox, JG (Eds.), The Mouse in Biomedical Research, Vol. I, History, Genetics, and Wild Mice. Academic Press, Inc, New York.

Sage, R.D., Heynemann, D., Lim, K.-C., Wilson, A.C., 1986. Wormy mice in a hybrid zone. Nature 324, 60–63.

Sax, K., 1944. Soviet Biology. Science 99, 298–299.

Shin, H.-S., Stavnezer, J., Artzt, J., Bennet, D., 1982. Genetic structure and origin of *t* haplotypes of mice, analyzed with H-2 probes. Cell 29, 969–976.

Silver, L.M., Hammer, M., Fox, H., Garrels, J., Bucan, M., Herrmann, B., Frischauf, A.-M., Lehrach, H., Winnking, H., Figueroa, F., Klein, J., 1987. Molecular Evidence for the Rapid Propagation of Mouse *t* Haplotypes from a Single, Recent, Ancestral Chrommosome. Mol. Biol. Evol. 4, 473–482.

Singh, P.B., Herbert, J., Roser, B., Arnott, L., Tuker, D.K., Brown, R.E., 1990. Rearing rats in a germ-free environment eliminates their odors of individuality. J. Chem. Ecol. 16, 1667–1682.

Sumner, F.B., 1923. Studies of subspecific hybrids in Peromyscus. Proc. Nat'l Acad. Sci., USA 9, 47–52.

Sumner, F.B., 1929. The analysis of a concrete case of intergradation between two subspecies. Proc. Nat'l Acad. Sci., USA 15, 110–120.

Sumner, F.B., Huestis, R.R., 1925. Studies of coat-color and foot pigmentation in subspecific hybrids of Peromyscus eremiticus. Biol. Bull. 48, 37–55.

Suzuki, H., Shimadda, T., Terashima, M., Tsuchiya, K., Aplin, K., 2004. Temporal, spatial, and ecological modes of evolution of Eurasian Mus based on mitochondrial and nuclear gene sequences. Mol. Phylogenet. Evol. 33, 626–646.

Talley, H.M., Laukaitis, C.m., Karn, R.C., 2001. Female preference for male saliva : implications for sexual isolation of *Mus musculus* subspecies. Evol 55, 631–634.

Tucker, P.K., Sage, R.D., Warner, J., Wilson, A.C., Eicher, E.M., 1992. Abrupt cline for sex chromosomes in a hybrid zone between two species of mice. Evol 46, 1146–1163.

Tutikawa, K., 1955. Further studies on *T* locus in the Japanese wild mouse, *Mus musculus molossinus*. Annu. Rep. Nal Inst. Genet. Jpn 5, 13–15.

Uehara, H., Abe, K., Park C-H, T., Shin, H.-S., Bennett, D., Artzt, K., 1987. The molecular organization of the H-2K region of two t-haplotypes: implications for the evolution of genetic diversity. EMBO J 6, 83–90.

Ursin, I., 1952. Occurrence of voles, mice, and rats (*Muridae)* in Denmark, with special note on a zone of intergradation between two subspecies of the house mouse (*Mus musculus* L.) Vid. Medd. Dansk Naturist. Foren 114, 217–244.

Vanderberghe, F., Dod, B., Boursot, P., Bellis, B.o.n.h.o.m.m.e.F., 1986. Absence of Y chromosome introgression across the hybrid zone between *Mus mus domesticus* and *Mus mus musculus*. Genet. Res., Camb. 48, 191–197.

Voslajerova-Bimova, B., Macholan, M., Baird, S., Munclinger, P., Dufkova, P., Laukaitis, C.M., Karn, R.C., Luzynski, K., Tucker, P.K., Pialek, J., 2011. Reinforcement selection acting on the European house mouse hybrid zone. Mol. Ecol. 20, 2403–2424.

Vrana, P.B., Fossella, J.A., Matteson, P., del Rio, T., O'Neill, M.J., Tilgman, S.M., 2000. Genetic and epigenetic incompatibilities under lie hybrid dysgenesis in *Peromyscus*. Nature Genet 25, 120–124.

Wallace, A.R., 1858. On the tendency of varieties to depart indefinitely from the original type. Proc. Linnean Soc. 3, 53–62.

Weir, J.A., 1994. Harvard, Agriculture, and the Bussey Institution. Gentics 136, 1227–1231.

White, M.J.D., 1945Animal Cytology and Evolution1997. Cambridge University Press, London (cited by Laurie.

Wu, C.-I., 1992. A note on Haldane's rule : hybrid inviability vs. Hybrid sterility. Evol. 46, 1584–1587.

Wu, C.-I., Davis, A.W., 1993. Evolution of postmating reproductive isolation : the composite nature of Haldane's rule and its genetic basis. The Amer. Natural. 142, 187–212.

Yamazaki, K., Boyse, E.A., Mike, V., Thaler, H.T., Mathieson, B.J., Abbott, J., Boyse, J., Zayas, Z.A., Thomas, L., 1976. Control of Mating Preference in Mice by Genes of the Major Histocompatibility Complex. J. Exp. Med. 144, 1324–1355.

Yamazaki, K., Beauchamp, G.K., Matsuzaki, O., Bard, J., Thomas, L., Boyse, E.A., 1986. Participation of the murine X and Y chromosomes in genetically determined chemosensory identity. Proc. Nat'l Acad. Sci.,USA 83, 4438–4440.

Yamazaki, K., Beauchamp, G.K., Imai, Y., Bard, J., Phelan, S.P., Thomas, L., Boyse, E.A., 1990. Odortypes determined by the major histocompatibility complex in germ-free mice. Proc. Natl. Acad. Sci. USA 87, 8413–8416.

Yamazaki, K., Beauchamp, G.K., Singer, A., Bard, J., Boyse, E.A., 1999. Odortypes : Their origin and composition. Proc. Nat'l Acad. Sci., USA 96, 1522–1525.

Zechner, U., Reule, M., Orth, A., Bonhomme, F., Strack, B., Guenet, J.-L., Hameister, H., Fundele, R., 1996. An X-chromosome linnked locus contributes to abnormal placental development in mouse interspecific hybrids. Nature Genet 12, 398–400.

Zechner, U., Reule, M., Burgoyne, P.S., Schubert, A., Orth, A., Hameister, H., Fundele, R., 1997. Paternal transmission of X-linked Placental Dysplasia in mouse interspecific hybrids. Genetics 146, 1399–1405.

3

Linkage studies, cytogenetics, and the discovery that *t*-haplotypes consist of a series of inversions—an over-the-Century Project

General linkage studies in mice

When Mendel proposed his laws of independent segregation of, and assortment for, characters he knew nothing about the entities involved. The assignment of these entities (named "genes" by Johannsen in 1909) to chromosomes could be considered to be the true start of genetics. Thus, early in the 20th Century it was established that chromosomes were the bearers of hereditary materials. In 1896, before the rediscovery of Mendel, based on the equivalent number of chromosomes provided by the male and the female (half the somatic number), the great American cytologist E. B. Wilson could state "chromatin is the physical basis of inheritance, and that the smallest visible units of structure by which inheritance is effected are to be sought in the chromatin-granules or chromomeres" **(Wilson, p.135)**. Boveri's experiments, in which enucleated eggs of one species of sea urchin were fertilized with sperm from another species, leading to embryos (plutei) with characteristics of the sperm donor, were not yet considered definitive. There were arguments against the continuity of the chromosomes, as they "disappeared" during interphase. Some scientists considered that the chromosomes were "reconstructed" from scratch simply to make mitosis possible, but they also thought that their contents were not always the same from one cell to the next. Boveri's experiments in the first decade of the 20th Century, with separated cleavage cells of dispermic eggs that showed varying abnormalities with abnormal numbers of chromosomes, were accepted as establishing the individuality of chromosomes (Boveri, 1907). He studied the case in which the extra sperm of dispermic fertilizations made 3-poled, instead of 4-poled asters (the aster being the centriole and the filaments which radiate from it, attaching to chromosomes to pull them apart) In such instances Boveri found that more chromosomes than normal (one haploid set from the egg and one from each sperm) went to each isolated blastula cell, and some developed into normal appearing plutei despite abnormal combinations of some of the chromosomes. Morgan and his coauthors from the "fly room" at Columbia University (a small room where so many of the great discoveries of early genetics were made with *Drosophila*—see **Kohler, 1994**) used the results of these experiments and similar ones by Baltzer and others to reach a firm conclusion on the chromosome theory. They determined that "the evidence in favor of the view that the chromosomes are the

Twentieth Century Mouse Genetics. DOI: https://doi.org/10.1016/B978-0-12-824016-8.00005-2

bearers of hereditary factors comes from several sources and has continually grown stronger, while a number of alleged facts, that seemed opposed to this evidence, have either been disproven, or else their value has been seriously questioned" **(Morgan, et al, p. 108)**.

The realization that the genes could be linearly arranged along the chromosome was not immediately obvious. The 7 "genes" that Mendel studied in the pea were all on separate chromosomes—more likely to have been by choice than be serendipity. Chromosomal crossing-over was not reported in these early cytological studies. This had changed by the time of Wilson's Croonian lecture of 1914 (Wilson, 1914). Now Janssen's (1909) theory of chiasmatypes had been used to explain cross-overs between linked genes.

There is some controversy about who should be credited for the theory of genetic linkage of genes on chromosomes. The textbook of Mouse Genetics by Guenet, et al credits Bateson and Punnett: the "fundamental concept of genetic linkage was introduced in 1906, by Bateson and Punnett, after a series of experiments on the inheritance of comb shape in chickens which shows inheritance patterns that are explained by close linkage of the genes involved (Bateson and Punnett, 1906)" **(Guénet, et al, p. 90)**. Over four years of experimental reports by Bateson to the Royal Society of London, with varying coauthors, under the title "Experimental studies in the physiology of heredity" were reported. Their work explored the inheritance of plumage and other phenotypic features in chickens. These include no text that considers genes to be associated on chromosomes. Instead, in the 1907 summary of the results of 4 years of work on the genetics of the rooster's comb (based on studies of 12,500 birds), Bateson and Punnett conclude "Our experiments upon the inheritance of the walnut comb and its components, the rose, the pea, and the single, are now concluded, and we may briefly recapitulate the explanation we gave in our third report (p. 12). The case is one of simple hybridism in which both the factors concerned affect the same structure, the comb" (Bateson and Punnett, 1907). No mention is made of linkage, coupling, or repulsion as a possible explanation of the unexpected numbers of offspring they saw from the allelic pairs of *Rose (Rr)* and *Pea (Pp)*, which in the combination of the 2 dominant forms *(RP)*, with any of the other alleles, showed the walnut phenotype. The unexplained finding is that rather than the expected ratios of 9/16 walnut, 3/16 rose, 3/16 pea and 1/16 single, which would be expected for separate segregation of the allelic pairs, in a variety of crosses they found ¼ walnut, ¼ rose, ¼ pea and ¼ single *(rp)*, a result explained by close linkage.

Although the terms "coupling" and "repulsion" were sometimes used for anomalous genetic segregations, these terms did not have the modern concept. The meaning of "coupling" in classic genetics is that in the situation of 2 closely linked loci, the respective alleles of the 2 loci on the same homologue stay together more often than expected by chance, while "repulsion" refers to alleles of the 2 loci that are on opposite chromosomes and stay apart more often than expected by chance. In those days, "the terms *coupling* and *repulsion* were coined to account for this unusual finding through some sort of underlying physical force. In a genetics book from 1911, Punnett imagined that alleles of different genes might 'repel one another, refusing, as it were, to enter into the same zygote, or they may attract one another, and becoming linked, pass into the same gamete, as it were by preference'" **(Silver, p. 139)**. Edwards (2012) has argued that

Punnett is the first to discover linkage (1905), since partial linkage was not subject to the kind of biochemical explanation Bateson's statement implies but, as the statement from 1911 shows, a realization that this was due to alleles being located on the same chromosome was not part of his thinking.

Walter Sutton is usually given the credit for the chromosomal theory of the linkage of genes based on his studies of meiosis in the "lubber" grasshopper which has 11 pairs of fairly recognizable chromosomes (and a single accessory chromosome determining sex ; Sutton, 1902). "Sutton was the first to present the idea in the form in which we recognize it today. Moreover, he not only called attention to the fact that both chromosomes and hereditary factors undergo segregation, he also showed that the parallelism between their methods of distribution goes even further than this–the segregation of one pair of chromosomes is probably independent of the segregation of the other pairs" **(Morgan, et al, p. 4)**. However, since Sutton's work had been done in Wilson's laboratory, the latter "sought to propagate the name 'Sutton-Boveri' theory for these ideas. It is perhaps worth mentioning that Sutton was aware of Boveri's work when he wrote his 1902 paper, as was Wilson, who had actually worked with Boveri. In fact Sutton expressly said so, in a passage that does him credit: 'The evidence advanced in the case of the ordinary chromosomes is obviously more in the nature of a suggestion than of proof, but it is offered in this connection as a morphological complement to the beautiful experimental researches of Boveri already referred to'" **(Harris, p. 173)**.

Also, it is controversial as to who was the first to find genetic linkage in the mouse. Many attribute it to the great geneticist, J.B.S. Haldane and coauthors (one of whom was his sister, an undergraduate at a private school in Oxford at the time) in 1915 (see mini-biography, The Haldane-Mitchison Scientific Clan). The note was submitted during WWI with A.D. Sprunt as another coauthor—he already had been killed in France. In this paper, Darbishire's (1904) experiments were cited as indicating "the existence of Reduplication in Mice," and Haldane further states that "this work was undertaken to verify and extend his results" (Haldane, et al, 1915), i.e. Haldane, et al were crediting Darbishire for the finding, although he certainly didn't see it himself. "Reduplication" was Haldane's term for linkage and his mathematical analysis of the result were interpreted as indicating "reduplication of the repulsion type"; ("coupling type" might have sounded more like linkage). He clearly thought that if he had published this result in 1910 when he first reanalyzed Darbishire's data, he would have established chromosomal linkage before it was found in *Drosophila* **(Clark, p. 40)**. Thus, it is perhaps not surprising that LC Dunn credits Castle and Wright, as follows: "There are now four well established cases of linked genes in rats and mice, and these are also the only known cases of linkage (aside from sex-linkage) in mammals. CASTLE and WRIGHT (1915) reported the first of these cases. They found evidence of linkage between the genes for two new color variations of the Norway rat.... The publication of the first report on this linkage was soon followed by a preliminary report by HALDANE, SPRUNT and HALDANE (1915) on the linkage between the genes for pink-eye and albinism in house mice. The authors preferred, however, to explain the results of their experiments in terms of the reduplication hypothesis," again indicating the confusion about the use of this term (Dunn, 1920).

The Haldane-Mitchison Scientific Clan

It is unusual, especially in the modern era when there are so many thousands of highly accomplished scientists, for one family to have four generations of Fellows of the Royal Society (FRS).[1] Yet, the Haldane-Mitchison clan has done so in the mostly modern era—the scientific legacy of J. B. and J. B. S. Haldane continued through J. B. S.'s nephews (one of who, N. Avrion Mitchison, made major contributions to mouse immunogentics) through JBS's sister, Naomi Mitchison, the coauthor of the first mouse genetic linkage paper (see main text) who became a famous authoress.

John Scott Haldane, 1860–1936

John Scott Haldane was born to an influential family that had played major roles in Scottish history since the early 18th Century. The family seat from 1851 was located at Gleneagles, located at the junction of lowlands and highlands in Perthshire (**Maurice, p. 1**). He studied medicine at Edinburgh, with a year abroad at the Friedrich Schiller University of Jena, receiving his medical degree in 1884. "He worked for a while as demonstrator at University College, Dundee, carrying out there a classic investigation of the air in houses, schools and sewers" (**Clark, p. 12**). He then joined his uncle, Sir John Burdon Sanderson at Oxford, becoming a Fellow of New college. He soon had a large "mansion" built at Cherwell with a lab on the ground floor. J. S. "Haldane was the prototype of the Victorian scientist, deeply concerned with the spiritual world into which his work was intruding; dedicated in such a way that normal hours ceased to exist; and of an absent-mindedness that became legendary even at Oxford" (**Clark, p. 23**).

As a scientist he was able to help his more famous, Liberal party politician brother, Viscount Haldane of Cloan. When the latter, who was at first a successful barrister, became interested in explosives from legal cases concerning them, gave "'A Public Lecture on Explosives by Mr. R. B. Haldane, M.P., [it was] illustrated by experiments conducted by Professor J. Haldane'"(**Maurice, p.104**). In 1911, when his brother went on a secret mission to Berlin to try to slow the arms race in

J. B. HALDANE AT OXFORD, 1914.

battleships (*Dreadnoughts*, **Massie, 1991**) under the cover of a visit to look into developments in technical education in Germany, J.S. went along as his brother's personal secretary (**Maurice, p. 293**).

J. S. Haldane shared his brother's interest in philosophy and, like his brother, was invited to give the Gifford Lectures at the University of Glasgow. "These lectures had been founded by Lord Gifford 'to promote a *thinking* consideration of the Nature of God and of his relation to the actual world' " (**Maurice, pp. 119-120**) and were published as *The Sciences and Philosophy* which ended on a deist note. He also popularized his work in a series of essays entitled *The New Physiology and Other Addresses*. In this he was to be followed by his son.

At Oxford, J. S. Haldane became a world-renowned pulmonary physiologist. He discovered the Haldane effect in hemoglobin, the effect that deoxygenation of hemoglobin increases its capacity for carrying carbon dioxide and vice-versa. He developed a decompression chamber to help deep sea divers and developed tables of the lengths of time needed at various depths for decompression. His son, JBS Haldane participated in this work "for in 1908, JBS had taken part in experiments which were to transform the whole diving practice of the British Navy"(**Clark, p. 26**). He studied

---

[1]The Royal Society of London, for Improving of Natural Knowledge was founded in 1662 and its charter "laid down that it was 'to Consist of a President, Council, Fellows, Secretaries, Curators, Operator, Printer, Graver & other officers'" (**Jardine, p. 83**). Of course, the King was an honorary member but a large part of the early officers and members would not be considered as scientists but as "interested noblemen." For instance, even our founder of the Haldane-Mitchison clan's, J.B. Haldane, brother Lord Haldane, a prominent politician (he was Secretary of State for War at the outset of the First World War) was FRS on the basis of his support of universities and of scientific education.

toxic gasses in mines, identifying carbon monoxide and hydrogen sulfide as deadly gases in different situations, and the effects of silicon in the lung (silicosis). During World War I he studied the effects of gas warfare, on himself and his son, invented the gas mask, and developed the oxygen tent for treatment of gas exposure. "John Scott Haldane was one of those very rare men who can train themselves to ignore fear. His son, described how his father disliked experimenting on animals and 'preferred to work on himself or other human beings who were sufficiently interested in the work to ignore pain or fear,' explaining that his father had achieved a state in which he was almost indifferent to pain."(**Clark, p. 15**). J. S. Haldane's efforts on behalf of aiding the military on gas warfare were never recognized. His brother, Lord Haldane, was dismissed as Lord Chancellor in 1915 because of his alleged German sympathies (he had studied in Germany and written admiringly of German philosophy)—no matter that the British and German Royal families were close relatives and got along well with each other. This attitude influenced how J. S. Haldane's work was viewed.

J. S. Haldane received many honorary degrees. Especially significant was the one from the University of Birmingham, received at the same time that his brother did. Of this, his brother Lord Haldane wrote "Yesterday was a fine ceremony. I think the nicest thing they said about anybody was said about Johnny: "There are those living whose lives have been preserved by his scientific work, some of it done at the peril of his own life.'" (**Maurice, p. 257**), J. S. Haldane also received the Royal Honor of Companion of Honor and, of course, was FRS.

John Burdon Sanderson Haldane, 1892–1964

J. B. S. Haldane was born on Guy Fawkes Day, 1892. He was extremely precocious and could read at 3 and spoke German at 5 (taught by a nanny). By the age of 8 he was taking scientific notes for his father. He started his education at the Dragon preparatory school in Oxford where he was first in many subjects at age 12, even over the upper classmen in higher forms. This won him a scholarship to Eton where he was hazed horribly, both because of his vast intellectual superiority and his absolute refusal to "bow under." "Yet in one way Eton made him, and made him more decisively than the rigors of a public school can be expected to make boys who are abnormally intelligent or unusually sensitive. It brought out his determined sense of independence and self-sufficiency; at the same time it both enlarged his self-identification with the persecuted minority and developed this into an unreasoning belief that minorities were always more likely to be right than wrong, and always more likely to be persecuted than not."(**Clark, p. 20**). He followed Eton with his father's college at Oxford where he had 1st class honors in modern mathematics in just one year. He then continued in classics in which he found relaxation and developed a honed writing skill.

J. B. S. HALDANE. PHOTO BY KLAUS PATAU, COURTESY OF JURGEN PATAU.

It was as an undergraduate at Oxford that he and his sister worked on mammalian genetics.[2] In "1901, John Scott Haldane had taken his son to an evening *conversazione* of the Oxford University Junior Scientific Club at which A. D. Darbishire had lectured on the recently rediscovered work of Gregor Mendel" (**Clark, p. 29**). Naomi Haldane Mitchison's comment on she and her brother's

---

[2]Both J.B. S. and Naomi were very active and falls, in J.B. S.'s case from his father's bike at the age of 8 and in his sister's case from a horse, may have influenced their approaches to genetics. J.B. S. suffered a basal skull fracture and mental damage and/or deafness were likely. Instead, "'After the accident' says one of his schoolfellows, 'Jack blossomed out into a brilliant mathematician, and rumor had it that he actually taught the maths master." (**Clark, p. 17**). This, of course presaged his brilliant mathematical models in population genetics (see Chapter 2). Naomi had a serious fall from a horse at 11 and turned to guinea pigs as pets instead. At one point she had 300 of them. This prepared her for the mouse work with which she helped her brother.

attraction to genetics was that "early genetics was relatively unmathematical; we talked in terms of dominants and recessives. Chromosomes had not come into their own; the cell mechanism was still obscure"(**Mayr, 1982, p. 303**). J.B. S. on rereading Darbishire's 1904 paper "believed that this revealed linkage in mice. He expanded his views on what appeared to be the first specific gene linkage in vertebrates, [and] gave his paper the pretentious title of 'The Comparative Morphology of the Germ-Plasm'"(**Clark, p.29**). When he showed the paper to R. C. Punnett, his advice was for Haldane to get his own data. Thus, A.D. Sprunt, a fellow undergraduate, and his sister were enlisted in the task of studying mouse genetics. Again, mouse fanciers were a source of mutants. His sister was to remember that "We managed to secure some very odd beasts, including some dwarfs...and at St. Giles Fair, which we always attended if possible, we bought two pink ones who, alas, turned out to be dyed albinos. Still, one had to take a chance." (**Clark, p. 30**). As pointed out in the main text, the results were not published until after Sprunt had been killed in WWI. Although mouse genetics was a very small part of J. B. S.'s brilliant career which influenced almost all aspects of genetics, he remained interested in it. "In 1932 Haldane visited Dunn, in Columbia, and C. C. Little in his private mouse Laboratory at Bar Harbor. When he returned to London on the "Mauritania" he took back three inbred lines of mice... At that time the idea of inbreeding mice was still new, and this was the first time I know of that they had been brought into Great Britain...The mice must have been some of the first occupants of Haldane's new department of eugenics at University College" (NA Mitchison, 1968). He was later to predict that inbreeding would not lead to complete genetic uniformity as confirmed by skin grafts (see Chapter 6).

J.B. S. and his sister were close and shared many activities, She describes the prewar period at Oxford where she met her husband among their friends: "When he [J. B. S.] came up to Oxford...he spread himself in friendships and light and the golden air of prewar years for the upper classes. His friendships were very wide, with young men of completely varied interests. Later, from 1914 onward, most of them were killed. After World War I, he came back, expecting somehow that the world of post-war peace might be the same as that of prewar peace—as did I—but almost all our friends were dead. He never made friends in the same spacious way again. His friends were men and women in his own discipline, work friends. The play world was gone forever." (Naomi Mitchison, 1968).

J. B. S. was in officer training school when WWI broke out. He requested assignment to the Scottish Black Watch regiment. He was a bombardier—head of a small squad that moved from position to position to lobe explosive missiles into the German positions using a mortar. The group was unpopular as they were gone when the returning artillery fire occurred. Nonetheless, he was popular, having his father's ability to get along with all sorts of men. He was recklessly brave and enjoyed killing—he went without taking his boots off for 3 weeks once and could be considered to have been in a "fugue" state.

After 7 months of this, when the Germans first used chlorine gas in April, 1915, he was called behind the lines to help his father work on gas toxicity and prevention. They exposed themselves to the degree of being quite ill since they believed they were better able to tolerate the exposures without fear and to monitor the effects. They developed more effective gas masks. When he returned to the front, he immediately joined a regimental charge during which he was blast shocked and wounded by shrapnel. Eventually recovered in England, he taught bombardiering with reckless bravado but superb leadership—in 9 months not a single casualty occurred which was rare with such training courses.

He was then posted with a second Black Watch group located on the Turkish front. Here there was a different kind of war—primarily enemies sniping at each other. His development as a rifleman ended with a munitions dump explosion and this time he had a more severe wound with shrapnel. His convalescence was in India where he started to develop his love of the people and where he learned Urdu. He also taught bombing and explosives to Indian troops before being finally discharged in 1919. "Now he had a double aim in life. One was to give service to science; the other was to make honorable grocers." (**Clark, p. 54**).

After the War, Haldane returned to New College, Oxford where he taught physiology and did extensive studies of the physiology of breathing, again using himself as a subject to the extent that he was often ill from the experiments. After 3 years, he was recruited to Cambridge by the great biochemist Gowland Hopkins who developed his interest in enzyme kinetics. It was also Hopkins whose defense was so important when he was going to be dismissed from his Readership for adultery—carefully planned, with a private detective hired to observe their liason, so that his future wife (Charlotte Burghes) could get a divorce. He was to stay at Cambridge for 10 years before moving to University College London and it was here that his famous 10 papers on the mathematical theory of evolution were written. He would have liked to return to Cambridge during the Second World War when London was being bombed to the extent that 5,000 people were killed a week. He wrote to RA Fisher, then at Cambridge, in Sept. 1940, saying "The place [University College London] has been heavily bombed. The Great Hall and Physical Laboratory are wiped out. The library has partially burned and partly flooded. We are still carrying on, as we have nowhere to go. Do you know of any possible refuge? We only need electric lights, a wash basin, gas, and a little (very little) artificial heat."(**Bennett, p. 216**).

For the first of the above-mentioned aims, J. B. S.'s important contributions to the Grand Synthesis and early linkage studies are discussed in the text. However, he was probably the most important theoretical biologist of the first half of the 20th Century, making major contributions to genetics, biochemistry and physiology. "Few geneticists have had more influence on the steady course of development of the subject than he during his long career. He contributed by numerous critical analyses of data, often involving novel statistical methods, and by numerous syntheses of the findings in all branches of genetics and of biology in general which frequently suggested fruitful lines of research" (Wright, 1968). His many contributions to biochemical kinetics were summarized in his book "Enzymes" (**Haldane, 1930**). He also could be amazingly wrong. He thought that the age effect in the frequency of Down's syndrome was an environmental effect, not the increased frequency of non-disjunction of chromosome 21(**Haldane, 1985, p. 128**). In 1952, just before the structure of DNA was revealed (1953) and when the fact that the genetic material was DNA had been well established by Avery, he could still speculate that the arginines in the chromatin nucleoprotein might be involved in heredity and that genes were enzymes (**Haldane, 1954, p. 118**).

For the second above mentioned aim, he moved from socialism to communism. He always felt for the common man and was anti-establishment. When the Spanish Civil War broke out, he advised on measures to resist gas attacks and bombing, making 3 trips to Spain to give advice. "The benevolent nonintervention with which the British government greeted the rise of right-wing dictatorship in Europe would probably have driven Haldane into the arms of the Communist party whatever else happened" (**Clark, p. 131**). He joined the party in 1937 and remained a member until 1948, when the Lysenko affair drove him out, even though privately he had many objections to Stalinism, (although he had initially refused to support Vavilov against Lysenko [Vavilov died in a Gulag camp]). However, during the Second World Was he patriotically performed more physiological research on deep sea diving for the Navy to help men escape from submarines (his father was now deceased or he would have been assisting). Such knowledge was imperative after the Thetis disaster in which a submarine sank in relatively shallow water but almost all the crew drowned.

His wife, Charlotte, was one of the first British war correspondents to go to Russia and her disillusionment with Stalinism led her to quit the party, and, eventually, her husband. They were divorced in 1945 and he was then free to marry his graduate student, Helen Spurway—a women with a very strong personality quite similar to his. He was always irascible towards the establishment but warmly supportive of lesser people. He enjoyed losing his temper but soon got over whatever it was that had elicited the outbreak.[3] His brilliance sometimes had negative effects because nobody felt like adding anything further after he had spoken. "I have known a meeting of the Genetical Society to be more or less ruined by Haldane's repeated brilliances. I don't think that anyone doubts that he really was brilliant; and I would think of him myself as a genius who misfired because of his personality deficiencies" (letter from Eliot Slater to C . D. Darlington 13, 1968; Darlington , quoted in **Ruse, p. 319**. He was greatly beloved by his staff and students and spent his own money to support them. When England and France invaded Egypt to secure the Suez Canal, and following 3 earlier visits to the Statistical Institute in Calcutta, he decided to emigrate to India. He said that the reason was "because he did not want to live in a police state, in a criminal state such as the one which had attacked Egypt" (**Clark, p. 229**). His nephew, Av Mitchison, adds that India then was a socialist dream and that the Indians appreciated him much more than the English.[3] He was to die of rectal cancer there in 1966, angry at his physicians for lying about the prognosis.

J. B. S. was, like his sister, a prolific writer. His first two books were *The Inequality of Man* a series of scientific essays for laymen, and *The Causes of Evolution*. He was to write over 300 essays on a variety of subjects, most of them published in the "Daily Worker." His writings included the delightful older children's book *My Friend Mr. Leakey* which was written with immense imagination. It is obvious that he would have been a great father to the children he wished for but never had.

His honors included the Darwin Medal of the Royal Society (he had been elected a fellow in 1932) and the Huxley Memorial Medal of the Royal Institute of anthropology. He received many invitations to give keynote addresses at International Congresses.[4] His politics kept him from receiving Royal Honors.

---

[3] N. Avrion Mitchison, video taped interviews at WEB of Stories.

[4] Salome Gluecksohn-Waelsch provided a vivid description of his presence at such meetings. To paraphrase, "he was a large man and on his entrance to a lecture hall, his doughnut cushion in hand (most chairs were uncomfortable due to pain from his war wounds and/or in his coccyx from the bends suffered during his diving experiments), he would take a front seat making a large noise like a "whoopee" cushion as air escaped from under the doughnut. His questions were always penetrating."

Naomi Mitchison, nee Haldane, 1897–1999

Naomi Mitchison was very close to her brother, J.B.S.—perhaps too close according to both her biographers, Jill Benton (**Benton, p. 44)** and Jenni Calder. They had grown up together in the "golden" (for the well-to-do) prewar Edwardian era. They had romped the fields around Oxford together and shared projects like raising the infamous guinea pigs (all 300 of them). The overlapped at the Dragon School in Oxford and after Jack transferred to Eton, they still shared holidays in the big house at Cherwell and its extensive grounds. Although Naomi was taken out of Dragon School at her menarche (and fully chaperoned while with boys after that), she was then tutored at home. When Jack came back to Oxford to be at New College, his friends became her friends. She directed them in plays, including one she wrote.

NAOMI MITCHISON.

At sixteen she took an exam allowing her to be a day student at St. Anne's College, still very much under her mother's rigid control. She studied biology, chemistry, and physics and passed the first year exams. It was among Jack's friends that she found her husband to be. "Two young men were in love with Naomi Haldane in 1914. Within a year she would know of one of them. He was Mitch, Jack's friend Dick Mitchison. The other was Gervais Huxley, who announced his love fifty years later in his autobiography, *Both Hands* (1978)"(**Benton, p. 24)**. Dick was clearly in love with her while her feelings for him weren't very intense—she didn't even enjoy kissing him. The War, and getting away from her mother were the driving forces towards marriage. At the start of WW I, Dick was sent to train to lead cavalry charges. The initial enthusiasm for the War rapidly died as many of their friends were killed. Naomi passed the nursing exams and worked at St. Thomas Hospital in London as a member of the Voluntary Aid Detachment. Here she saw first hand the horrific results of the war.

As happens with war, the courtship was rapid and she was soon engaged. After their marriage, she returned to live with her parents for the next 3 years. Dick was badly injured in a behind-the-lines accident and she was told, as her parents had been about her brother some years before, that he would be brain injured if he survived. However, as in J. B. S.'s case, this was proven false and he was able to pass his bar exams several months later.

Sensuality was becoming increasingly important to her and on a holiday in Italy in 1918, "she recalled that she could hardly wait to see Dick, for she had sent him a copy of Marie Stopes' *Married Love,* an early sex manual that had taken her breath away. She thought that if she and Dick use just a few simple techniques their love-making might be altogether more satisfying. She reports in her memoirs that it actually did get better" (**Benton, p.35)**.

After 12 years of marriage, both she and Dick found new lovers. "I did not take a lover until such time as Dick too was looking elsewhere; this timing is important and means that there is likelihood of hurting the other partner…Yet looking back on it now, I remember the pain almost more clearly than the delight, whereas with the relationships which did not include love-making—for instance with Lewis Gielgud or Angela Blakeney Booth or Aldous or Zita Baker, the only remembered pain is at their death." (**Mitchison, 1979, p. 72)**. The married couple remained highly committed to each other and she shared his political efforts. Part of their attraction to socialism/communism included the belief that free love would be part of the new world order. The openness was such that Dick and Naomi even went on vacations together with their lovers.

She had 7 children, four males: Geoffrey, Denis, Murdoch, Avrion, and 3 females: Lois, Valentine and Clemency While children were very important to her, she led a very independent life and the nanny took care of them. "Naomi breastfed each baby…her involvement with her children was thereafter not necessarily close. With each baby she became less constrained by the prevailing child-reading [sic] orthodoxies of her time….This relaxation meant more freedom." (**Calder, p. 67)**. This was fairly typical for upper class women and the inter-war period was a time when women's liberation flourished for this class. The sons would go to public boarding schools, again typical for the class and time.[5] Avrion

[5]"Public Schools" in England are private. The appellation comes from the late Renaissance when the emerging middle class could not afford tutors as did the nobility but pooled their resources to educate their (male) children together. Thus, some of the schools have trade guild names such as "Haberdasher's."

(personal communication) felt that "she was a wonderful mother to her sons" but not so much so to her daughters and that her youngest daughter, Valentine, was very negative about her—feeling educationally neglected for having to attend the village school near Carradale rather than a good boarding school.[3] However, when the eldest son, Geoffrey, died of meningitis and she had not been at his bedside all the time, she was racked by guilt which lasted many years. His illness and her role in it was rather transparently depicted in Aldous Huxley's novel *Point Counter Point* and she was greatly hurt by this depiction and did not speak to him for years. It also caused a rift with her brother who blamed her. However, all in all, the children must have had pretty happy childhoods as they were to give her 20 grandchildren, even Val had 5!

Her writing was her major sphere and she was to write nearly a hundred books in her 102 years. Her first novel, *The Conquered*, was very successful. An historical novel, it starts with a brother-sister relationship which is clearly modeled on her relationship to her brother and the novel is also dedicated to him. The book became to be considered a classic and was eventually published in an educational version for schools. A later historical novel, *The Corn King and Spring Queen* is considered her best novel by many. Her first book written based on contemporary events was *We Have Been Warned* and was very difficult to publish because of its contents (especially a rape scene although provoked by the protagonist's visit to the room of a single man late in the evening). It is highly autobiographical and, according to Isobel Murray, it took two female protagonists to cover all her activities and she had trouble keeping her literary distance from them (Murray, 2012, **Mitchison, 1935, pp. viii-ix**). "When the author is passionately involved in world affairs and at rather a loss what if anything to do about them, and just as passionately emotionally involved with love affairs and different men, in birth control and whether it is justifiable to have another child while the world is in such danger, appropriate literary distance may be an unreal quality to look for" (ibid.). The major protagonist and her husband visit Russia where she encourages him to have an affair and she returns first to England while it continues. In England the protagonist becomes rapidly reinvolved in Labor politics for her husband—she and her husband are Communist sympathizers but think only the Labor Party can be successful.

Although Naomi Mitchison was acquainted with, and interacted with, some of the Bloomsbury group, and shared their sexual mores, she and Dick lived in Hammersmith, quite away from Bloomsbury, and she is not viewed as a member of that circle by historians of it. Her views of the most prominent member is reflected in her comment, as told to me by her son, Avrion, "She was always afraid of Virginia Wolfe."[6] She was an influential writer of the inter-Was era. "Any well-read person in Great Britain who grew up during the 1920s and 1930s will have read Naomi Mitchison. Not only was history factually accurate in her historical novels, which pleased the scholarly critics, but she was a good read for fast moving adventure and for complicated, erotic love" (**Benton, p. 51**). Her many books included biographies, science fiction, social treatises, children's books, and travelogues as well as the historical and political novels.

The house in Hammersmith, River Court, was the site of many social gatherings as Naomi was a gregarious hostess. It had also been the place of tragedy. "In 1928 the Thames tide overflowed its banks, crashing with door-tearing force into the River Court basement apartment, drowning two servants in their sleep" (**Benton, p. 55**). This may be why that in 1928, she and Dick bought a large house in Scotland, Carradale, which became their major residence. It had a grouse meadow for outdoor shooting and a salmon river for fishing and her son characterized it as a "castle" with beautiful grounds at the time of purchase.[3] However, the grounds were neglected and further down-graded when she converted them into a farm to grow vegetables for the war effort in WW II. An outbuilding with a study for her and squash court for Dick had been added. The social gatherings continued and were enriched as her children brought home friends from university. Av's older brother, Murdoch, was doing some cellular biophysics at Cambridge and got to know Watson and Crick. Av became friendly with Jim Watson through visits to his brother and took him home to Carradale. His mother instantly recognized Jim's special qualities and "took him up."[3] He was to go on trips to southern France with her and his autobiographical account of the discovery of the structure of DNA, *The Double Helix* is dedicated to her. His description of the Christmas gathering, to which he also invited his sister, in 1951, is very apt. He was impressed by the mix of high intellect and low-brow games. "In the evening there was no way to avoid intellectual games, which gave greatest advantage to a large vocabulary [which he did not have]…Much more agreeable were hours playing 'Murder' in the dark twisting recesses of the upstairs floors." (**Watson, p. 106**). He enjoyed himself immensely: "Almost from the start of my stay I knew that I would depart from Naomi's and Dick's spectrum of the left with the greatest reluctance." (**Watson, p. 106**).

After Dick's death (he had finally won a Labor seat from 1945 to 1964 and was made Baron Mitchison of Carradale in 1964), she met the future chief of Bechuanaland (as Botswana was then called) when she hosted one of many British Council parties. They became acquainted with each other and he invited her to attend his induction as Kgori Linchwe, II. She did and was to make annual trips to Botswana for some years where, among other things, she taught birth control.[3]

---

[6] In answer to my question to Av (N A Mitchison) about his mother's view of Virginia Wolf as we were walking through Bloomsbury, Oct., 2012.

This led to a great interest in the African situation and she was to write 8 books based on her African experiences. She shared her mother's love of the British Empire but with an entirely different view of what the UK's relationship to it should be.[3]

Her life long political activities included the early phase of promoting birth control, a continuous involvement with the Labor Party, service for a long time as a County Council member in Scotland, and anti-nuclear war efforts as she grew older—as well as writer's movements. She was made CBE (Commander of the Order of the British Empire) in 1981. Because of Dick's ennoblement she could have used the title "Lady Mitchison" but did not do so.

Denis Mitchison, 1919-

Although living in London at the time, Naomi returned to Edinburgh for Denis' birth. His initial schooling, as with his sibs, was at Dragon School in Oxford, "boarding" with his grandparents. "Denny had hated the school [Dragon School] but went on to Abbotsholme where he was much happier" (**Calder, p. 131**). While there he had an exchange visit with a German school which had started to "Nazify." "Denny, who in his late teens became a very serious Young Communist, painted red hammers and sickles on various doors and bolted. I met him in Paris (**Mitchison, 1979, p. 195**). He went up to Cambridge to study medicine where he met his future wife. " Denny was in trouble at Cambridge for being found with a WOMAN in his rooms. The woman, Ruth, was very sensible and said they were engaged to be married and got off rather more lightly than Denny, who was sent down [flunked out], but they both were well on their way to being doctors" (**Mitchison, 1979, p. 233-4**). They were married at Carradale in 1941 in a wedding celebration without evidence of wartime hardships. "[Denny's wedding to Ruth] almost exactly a year after the declaration of war, went off well. It was a high-spirited occasion, culminating in a splendid ceilidh [traditional Gaelic social gathering with folk music and dancing] at which Carradale let down its hair....The party ended at 2 a.m. In addition to the unofficial supplies, the company had consumed two dozen bottles of champagne, six of whiskey, tea, lemonade, passion fruit drink, sandwiches and biscuits, 15 pounds of cake, half a wedding cake and ten dozen small cakes" (**Calder, p. 162**).

Denis finished medical school at University College London in 1943 and took his first job in Pathology at Brompton Hospital. His description of how this started his career in tuberculosis research (interview@newtbdrugs. org) follows:

> "Serendipity is the right word. It happened because I always have wanted to do research, but not necessarily in TB. I wanted to do research as much as anything because my family has a number of eminent scientists in it in the past. I was following the family tradition. The Haldanes, my mother's side of the family, are an example. J. B. S. Haldane really pioneered respiratory physiology and did an awful lot of things and was a very eminent person in his own right, and others. That was why I got started in research. It just happened that the first job I took was at the Brompton Hospital in London, which was one of the two first clinical centers in the first streptomycin trial. So it's purely fortuitous."

His initial work was with Phillip D'Aracy Hart at the Medical Research Councils (MRC) Tuberculosis Research Unit. He became an international figure as he set up clinical trials in the UK, Czechoslovakia, Africa, India and East Asia. In 1964, he was appointed director of a new MRC Unit on Drug Resistance in Tuberculosis at the Royal Postgraduate Medical School.

Again, from his interview (interview@newtbdrugs.org) :

> "part of the work was just getting the laboratory basis of these trials going. But the other side, which was in a sense just as, and probably more, important, was to understand the theoretical reasons why drug treatment works or fails. This started initially with understanding what is drug resistance, how do you measure it and what are its implications for treatment. I mean that's a very big subject that took something like 20 years to get all the answers to, and we still don't have all the answers."

Although he retired in 1985, and later transferred to St. George's University of London, he is still active (2015). He was Royally Honored as a Companion of the Order of St. Michael and St. George. He has 3 children but none are scientists.

John Murdoch Mitchison, 1922–2011

Murdoch, as he was known, unlike his older two siblings, was born in London and, at school age, went to live in Oxford with his grandparents to attend Dragon School, which he, unlike his older brother Denis, liked. He did well in all subjects excepting Latin in which he was coached by WH Auden who his mother was championing. He enjoyed the outdoors and fly fishing. He won a scholarship to Winchester College where he excelled at shooting, which he practiced, apparently to avoid cricket. He then received a scholarship to Trinity College Cambridge. According to his mother's biographer (**Calder, p. 160**), "Murdoch had followed Denny to Cambridge also to study medicine, was depressed and unhappy, thinking about leaving university and joining the navy. He was anxious about the future and the family, clearly concerned that no two members were in the same place." Perhaps this was at the time when he became disenchanted with gross anatomy "the conclusive moment came when he had to root around in a bucket full of limbs to find an arm he was supposed to dissect" (Obituary, *The Telegraph*, April 5, 2011). He transferred to zoology where he did double firsts in two years. "It was said that he had a Rolls-Royce brain, though his privileged background meant that he seldom had to drive it on anything other than flat roads, and he himself was always happy to acknowledge his good fortune in being taught by some of the best brains of his generation" (Obituary, *The Guardian*, April 13, 2011). He was drafted in 1941 and initially worked in Army Operational Research. This directed military research was "a period which he acknowledged as being influential in the evolution of his highly focused approach to research. He then undertook officer training before seeing active service in Italy, being demobilized as a major." (ibid.).

Returning to civilian life at Trinity College, he was first a research scholar and then a fellow. He moved to the University of Edinburgh in 1953 where he remained for the rest of his career, eventually as one of the two professors in zoology.[7] Although his initial interest was cell biophysics, his major contribution was developing the yeast *S. pombe* as a model organism. His studies on this organism's cell cycle resulted in his classic book, *The Biology of the Cell Cycle*, published in 1971. The future Nobel Laureate, Paul Nurse, who first used genetics to dissect the cell cycle in yeast and then in human cells, worked with him for 6 years. He described those years "with Mitchison in Edinburgh as 'pivotal' for his entire research career. 'He gave me both complete support and total freedom, spending hours each week talking with me but never instructing me what to do,' Nurse recalled. 'An astonishingly generous supervisor, he never once was a coauthor on any of the papers I produced during my time in Edinburgh. He considered, quite wrongly of course, that he had not made a sufficient contribution to justify inclusion'" (Obituary, *The Telegraph*, April 5, 2011).

He had broad interests, including architecture (he designed his own house), botany, and history (he was working on a history of the WW II Italian campaign, in which he had participated, at the time of his death). "He was not famed for his adherence to conventional manners in the domestic field [took after his parents' sexual mores?]. He was also much given to the creation of scientific devices to further his researches and would display examples of these in his home, almost as if they were sculptures" (Obituary, *The Guardian*, April 13, 2011). He married the social historian Rowy Wrong and they had 3 daughters and one son.

He served the University of Edinburgh as Dean of the Science Faculty for one year and was a member of the university court for several years. He was president of the British Society of Cell Biology in 1973-4 and was elected FRS, London in 1978.

Nicholas Avrion Mitchison, 1928-

N. A. Mitchison was conceived shortly after the death of Naomi and Dick's eldest son Geoffrey (see above). After the trauma of the death and its aftermath, his parents wanted to get away. They (and their lovers) "hired a yacht, the *Avrion*, and explored the [Greek] islands…Naomi was pregnant, a regenerative response to Geoff's death. Nicholas Avrion would be born later that year (**Calder, p. 82**).[8]

As with the other children, he saw more of his nanny than his mother but the nannies were well vetted by his mother. She described one of them, Mrs. Davis, and her interactions with the children: "She was always marvelously calm and

---

[7]Traditionally in Great Britain there was only one Professor per academic department and that person served as Chairperson.

[8]"Nicholas" went by his second name, as did Murdoch, and was "Av" to most. I first met him when he was a visiting professor at Stanford University School of Medicine in1960-61 where he taught an elective on Immunobiology to some of the medical students. Having done my Bachelor's thesis on transplantation immunity in urodeles (Erickson, 1962), I was attracted to the course and enjoyed Av's knowledge. Another student in the course, Harvey Ozer, was working in the Herzenberg laboratory (Leonard and Leanore) and introduced me to it. This was another long term association. I spent a last year of post-doc with Av at Mill Hill in 1969-70 when he allowed me to pursue my own project searching for cell surface antigens determined by *t*-alleles (see Chapter 7). He even supplied a lab technician. He has remained a friend and sometimes collaborator (Erickson and Mitchison, 2014).

sensible in any medical crisis, measles or scarlet fever, not to speak of Val crawling round the table and biting Avrion's leg or Avrion seeing things which were invisible to others" (**Mitchison, 1979, p. 27**). He was very precocious and seemed to charm all. When participating in one of the theatricals his mother arranged for fund-raising, he created his own lines. "Avrion, then three, was the fairy baby who had to give Kate the wand and the cake. He knew this part and it suddenly occurred to him that all the other were allowed to talk—why not him? I said all right, for clearly it would have been useless to say anything else, but explained that it was part of the story. I thought he understood and indeed what he said was quite in character but not very audible. He looked entrancing, all in tinsel and blond curls" (**ibid., p. 35**).

These early experiences created a great sense of confidence[9]. He was "easy going and self-reliant, [and] went off to prep school at seven (**Calder, pp131-2**). This was a Quaker boarding school, the Leighton Park School, and had been proceeded by several years at elementary school in London. In a long interview at *Web of Science*, Av describes some aspects of his childhood. The artist Wyndham Lewis, a friend of his parents, tried to teach him drawing, rather unsuccessfully. At Carradale, he enjoyed collecting in the sea pools and had a microscope to examine specimens—he was particularly interested in hydroids. At age 16, in 1944, he had been accepted at New College, Oxford a year in advance and so went to work for his uncle, J. B. S. Haldane as a "lab boy" at University College London—a job classified as a "servant." This was the beginning of a long association with Univ. College where he would eventually be Professor of Zoology. The situation for J. B. S. and the University were much improved from the early years of the war when J. B. S. had written to Fisher looking for a refuge. Av reports that J. B. S. was a great teacher but that he spent most of his time in his office editing the *Journal of Genetics*.

When Av went up to Oxford, there was no tutor in biology at New College so he was sent to Peter Medawar. At the first interview in Medawar's rooms, he was greeted with a partially operated-on rabbit. Medawar had been working on skin grafting and transplantation as part of the war effort. The association with Medawar was particularly fruitful and Av went on to do his Ph.D. with him. Their separate and collaborative efforts on transplantation immunology were to continue for many years—Av's contributions to mouse immunogenetics will be further discussed in Chapter 7. Immediately after finishing his Ph.D he was a Fellow of Magdalen College, Oxford and his work then continued with a Fullbright Fellowship at the Jackson laboratory in 1952. His initial academic appointment was at the University of Edinburgh where his work showing that graft rejection was cell mediated was performed. He soon moved to University College, London. Again, in his interview, Av recounts time with colleagues Hans Kalmus, who was interested in sex determination and was transplanting gonads, and Michael White, who was a cytogeneticist. His major work in the next decades was on T cell-B cell interactions and the mechanisms of low zone and high zone tolerance. Sir Peter Medawar had become Director of the Medical Research Council's National Institute for Medical Research (NIMR) at Mill Hill, London and invited Av to become head of the Division of Experimental Biology. There was already a star-studded cast of immunologists there with Brigitte Askonas, John Humphrey and David Dresser in the Division of Immunology. Sir Peter continued doing active research while there.

In time, Av returned to University College as Professor of Zoology. He was such a popular mentor that some visiting scientists were assigned a desk in the hall![10] After his retirement (arbitrary at 65 in the UK), he went on to be the founding director of an institute of rheumatology in Berlin—the Deutsches RheumaForschungsZentrum. After his second retirement from this Institute, he returned to University College where he remains a Professor Emeritus (2016).

He married Lorna Martin, a lass from the Isle of Skye, at Carradale in 1957 and they have 5 children: Tim, Matthew, Mary, Hannah, and Ellen. Of these, two are very successful scientists: Tim (see below) is a cell biologist and Hannah is a physician scientist, pediatric geneticist with a special interest in genetic disorders of cilia[11].

Av was elected FRS in 1967 and is a foreign member of the U. S. National Academy of Sciences. His many other honors included the Robert Cock Gold Medal, the Sandoz Prize in Basic Immunology and an honorary Doctorate from the Weizmann Institute.

---

[9]This self-assuredness made him comfortable with his "studied eccentricity." An example related to me by my wife, Sandra Doris Erickson, nee De'Ath, who was a technician in Sir Allan Park's laboratory at the time (and became Av's technician), was during his job interview seminar. He had been invited by the new Director of the MRC's Institute of Medical Research at Mill Hill, Sir Peter Medawar, who had been his thesis adviser, to give a job seminar for the position as Head of the Division of Experimental Biology and told to wear a tie. Meeting the letter, but not the spirit of the request, he appeared with a tie knotted backwards, i.e. showing the lining, and wearing carpet slippers.

[10]According to one such visitor/admirer, Lois Epstein, Av told her that it would be crowded but that she could come if she found her own financial support. She found the desk in the hall a little less than amusing.

[11]They were introduced to science at an early age. My wife recounts coming in on Monday mornings to clean up after they had played "acids and bases"—a game in which they are given one known acid and one known base and in which they are to classify a large number of unknowns by the violent reaction with one of the knowns in the sink. The children are performed some less useful experiments—putting sugar in the gas tank of the head technician's car being one of them.

Timothy John Mitchison, 1961-

Tim, as he is generally known, spent his early childhood in a large house on the campus of the NIMR at Mill Hill, London where his father was working. The large grounds of the campus allowed lots of roaming. There were almost constant scientific visitors at the house and lots of stimulating conversation. He attended The Haberdashers' Aske's Boys' School and then went up to Merton College, Oxford where he majored in biochemistry. He moved on to the University of California, San Francisco (UCSF) to perform his doctorate work with Marc Kirschner. This was a very productive time with important discoveries on the dynamic properties of microtubules in the cytoskeleton and in the mitotic spindle. He returned to the NIMR to do a postdoc with David Trentham on the chemistry of fluorescent probes. The development and use of small molecular probes to explore dynamic cell structure has remained a major interest. He returned to UCSF as an Asst. Professor of Pharmacology but soon moved on to Harvard where he is Professor of Systems Biology.

    He is married to the cell biologist Christine Field and they have 2 children. He was elected FRS in 1997 and to Fellowship in the U.S. National Academy of Sciences in 2014.

While the report of this first linkage of 2 genes in mammals occurred in 1915, by the time of the writing of their classic **The Mechanism of Mendelian Heredity** in this year (**Morgan, et al., 1915**), the authors had assigned 36 genes to 4 linkage groups in *Drosophila.* Of course, with its longer generation time, smaller number of individuals for study, and much larger number of chromosomes (compared to *Drosophila*), it took much longer to establish multiple, large linkage groups in mice. Mapping genes was to become one of the major occupations of mouse geneticists in the first half of the 20th Century. This work would culminate in the construction of a consensus genetic map (by the Mouse Nomenclature committee, see below). During the middle half of the century this was to be primarily an Anglophone effort. As mentioned in Chapter 2, Germans preferred other experimental organisms, mostly insects, for their genetic studies, and France was anti-Mendelian until the 1940s, see Chapter 1. After the first usage of nuclear weapons, studies of radiation effects on living matter (radiobiology) were supported strongly in England (at the Radiobiological Institute at Harwell) and in the United States (at the Oak Ridge National Laboratory)—see Chapter 6. The Jackson Laboratory remained preeminent and, especially, was the center at which the linkage data was assembled.

In the 1920 paper, Dunn followed Haldane's (1919a,b) formulas for probable errors, and observed a higher frequency of recombination in females than in males (Dunn, 1920). He placed 3 genes (*albinism, red-eyed yellow* and *pink-eyed yellow*, the latter 2 being alleles of the now named *pink-eyed dilution*) in one linkage group in rats. He soon pointed out a linkage-length difference between mice and rats for the *pink eye-albinism* distance which was 16% (recombination) in mice and 21% in rats (Dunn, 1921). He noted that this pair of genes is not linked in guinea pigs and suggested that they were on different chromosomes in that species, with its larger number of chromosomes (28). The establishment of linkage groups consisting of more than 2 genes in mice took a little longer.

The absolute linkage of *dilute* to *short-eared*, creating a new linkage group, was reported by Gates in 1927 (Gates, 1927). He and Lord added a third gene, *shaker*, now renamed *shaker*-1, which causes head-shaking, to the *albino-pink eye* linkage group in 1929 (Lord and Gates, 1929). An important advance in the making of linkage maps was to set up linkage stocks to which a new mutant could be crossed in order to determine if it mapped to an established linkage group and, if so, at what position. RA Fisher was one of the first to start doing this. In a letter to GD Snell, Nov. 9, 1943, he stated "I have run a little mouse colony now for more than fifteen years, but it was only when I accepted the Arthur Balfour Chair of Genetics in Cambridge, formerly held by R.C. Punnett, that I decided to put into practice what I had long felt needed doing, namely the creation of permanent inbred lines covering all (or as near as makes no matter) of the genes recognizable in mice. I believe that the advantages offered by segregating inbred lines have never been fully appreciated." (**Bennett, p. 264**). It was about this time that Mary Lyon worked in his lab. Fisher "had a mouse colony in which he was looking for new linkages among the then known mutant genes... At that time the number of known genes was less than 100, but there was already a map consisting of 10 linkage groups, as they were called, each group containing 2-4 gene loci...Although Fisher was very interested in mapping, at that time no systematic method of mapping existed, and he devised his own method. He reasoned that, as there were 20 pairs of chromosomes in the mouse, if he took 21 genes and tested them all against each other he was bound to find a linkage. He seemed to not think about chromosomes over 50 cM [centiMorgan, which equals the cross-over frequency, and 50% is equivalent to non-linkage] long" (Lyon, 2002). Mary Lyon decided that just mapping a gene was not enough to warrant a Ph.D. and so went to Edinburgh where Falconer was chairman of genetics. He and Carter developed the theory for, and started the use of, linkage stocks. They established a formula for calculating accurate linkage distances, taking into account crossing-over suppression, the fact that one cross-over inhibits nearby crossing-over (thought to be due to the physical interference of one chiasma [the visible site of crossing-over] with another). At the time of their work (1950), twelve linkage groups were recognized but it was not certain that some of them might not be on the same chromosome since, as mentioned, two linkage groups whose nearest genes were 50 cM apart would appear to be on separate chromosomes. They deduced that, if their principles were applied, "if suitably arranged, all the known genes that are desirable as markers for testing linkage [fully penetrant, properly spaced] could be contained in five stocks; and that, if only 100 offspring were raised in tests with each of these five stocks, the fraction of the total genetical map that would be covered would be of the order of 0.5-0.6" (Carter and Falconer, 1951). They went on to list 5 stocks of 21 genes in total, 2 with dominant genes and 3 using recessive genes. Of these stocks, three were already constructed as they went to press while 2 were still being developed. Such linkage stocks were also developed at the Jackson Laboratory. George Snell may have been starting them as early as 1939. When Margaret Green moved to the Jackson Laboratory in 1956, and took over the Mouse Mutant Stocks Center, she noted that "some of the stocks in the Center came from George Snell who had also been propagating some of them by inbreeding with forced heterozygosis...We used the same breeding system for maintenance of the linkage testing stocks containing multiple dominant genes" (Green, 1978).

There was a logarithmic rate of increase in the number of mapped genes over the next decades, with 13 loci in 1935, 26 in 1945 and 70 in 1952 (Eicher, 1981). As the number increased, the labor of keeping track of them became more onerous. Margaret Green did more for organizing the mouse linkage map than anyone else. Every year she carefully sifted through newly published data together with recent contributions made in Mouse News Letter, which, as mentioned in Chapter 1, was a very important source of information for the mouse genetics community. She organized all of this information together with the previously known data into what was called 'Marg's Map'" (Eicher, 1981). The most iconic version of this map was one used sometimes at the annual genetic course at the Jackson Laboratory: cages with the mutant mice were placed at the correct positions (distances from each other) on shelves representing the linkage groups. Margaret Green's, and others', efforts were assembled in a seminal book: **Green, 1981**.

Linkage studies went much faster when protein and enzyme markers were added to the mapping effort, since the number of pigment-related and easily observable motor problems was limited. In 1941 Figge and Stong (1941) found a polymorphism for enzyme activity (xanthine dehydrogenase) in different strains of mice. It was 11 years before such differences were shown to be under genetic control—the enzyme studied was beta-glucuronidase (Law, et al, 1952)[1] and differences were even found in sperm (Erickson, 1976). Further advances largely depended on electrophoretic separation of proteins. The original Tiselius (1937) apparatus required large quantities of the protein. The procedure depended on monitoring the movement of the optically detected Schlieren pattern, the wavy pattern visible between two solutions of different viscosity, made by the moving front of each protein in solution as they separated into buffer in a cell with poles of opposite charge at each end. In other words, there was a chamber in which the high density protein solution was over-laid with the low density buffer, creating the initial Schlieren boundary which, as the proteins separated created multiple boundaries. The method was both awkward and wasteful of frequently expensive-to-prepare proteins. Methods were soon developed to resolve on paper strips the 5 fractions of serum proteins that the Tiselius apparatus could separate, and then many other materials were exploited as carriers for electrophoretic protein separations. Hunter and Markert (1957) were the first to do an enzymatic stain to localize electrophoretically separable zones of enzyme activity on cellulose strips. The discovery of many electrophoretic variants of enzymes, both between tissues and individuals, led to an explosive increase in genetic markers. The variants were termed "isozymes" (and should have been termed more appropriately "allozymes"). They became better known when one of them, lactate dehydrogenase (LDH) had variants which were found to be clinically useful for detecting tissue damage (Vesell and Bearn, 1958).

Better resolution of bands was achieved with agar gels (Gordon, et al, 1959) and starch grains (Kunkel and Slater, 1952). A major breakthrough occurred when a gel of starch was produced by heating—this was introduced by Oliver Smithies[2] (1955), a future Nobel Laurate; see Chapters 4 and 9. Even greater resolution of proteins occurred when polyacrylamide gel electrophoresis was introduced (Raymond and Weintraub, 1959).

The use of wild subspecies of mice sped this work. Verne Chapman made the "discovery that wild subspecies of *the genus* Mus, such as *Mus musculus castaneus* and *Mus musculus molossinus*, could provide, at many isozyme encoding loci, the genetic variation missing in laboratory inbred strains (Chapman, 1978) and this was supported by French data (Bonhomme & Selander, 1978). In fact, when one thinks of the *castaneus* or *molossinus* genome as a linkage testing stock, the reason some mouse geneticists have been so successful at finding linkages can be more easily understood" (Eicher, 1981). Chapman had found that 30-40% of the loci studied had useful variants (Chapman, 1978).

The era of gene cloning (see Chapter 9) soon provided many more genetic markers for linkage mapping. The first markers used restriction fragment length polymorphisms (RFLPs). Cloned fragments of mouse DNA (radio-labeled with P^{32}) were hybridized to a blot of genomic DNA that had been cut into manageable fragments with a restriction enzyme (see below) and then separated by size using gel electrophoresis (smaller fragments moving farther than larger fragments). The DNA was denatured (made single-stranded) and transferred to a supporting membrane (a Southern blot;Southern, 1975). The denatured DNA on the membrane was hybridized with a denatured radioactive probe. When the blot was washed and then exposed to film, one or several bands of hybridization were usually detected: one if the hybridizing DNA was flanked by the cutting sites of the restriction enzyme, more if there were also cutting sites within the hybridizing portion of DNA. (Restriction enzymes from bacteria cut at specific sequences of DNA. That is, for each enzyme there is a unique typically palindromic sequence of base pairs, usually from 4 bp to 6 bp, e.g. 5'-GGCC-3', 3'-CCGG-5'. These enzymes evolved in bacteria to protect the host from foreign DNA (which bacteria frequently take up) To protect the host organism's own DNA, other enzymes evolved that specifically alter the host DNA at the same palindromic sequence, usually methylating it, so the host DNA will not be cut. When DNA samples from different inbred strains of mice were compared, there were frequently different banding patterns found, especially when a variety of different restriction enzymes were used, because the different restriction enzymes would detect different single base pair changes in the sequences of the homologous DNA in the mice. The method was popularized in human genetics by Botstein, et al (1980), and was applied to mapping in the mouse and many other species.

Although many markers were found using RFLPs, (Sweet, et al., 1988) the method was rapidly superseded, when the polymerase chain reaction (PCR) was introduced[3], by the use of Simple Sequence Length Polymorphisms (SSLPs). These include short tandem repeat (STR) or microsatellite markers, and variable number tandem repeat (VNTR) or minisatellite markers. The STRs are usually 2-5 bp repeats arranged in a head to tail fashion (e.g., gccgccgccgcc), which, with "slippage" during replication, enlarge or decrease in size rapidly over multiple generations and, thus, frequently have different lengths. VNTRs are similar, but with repeats up to about 25. One DNA probe specific for unique sequences flanking one of these repeats could detect the variation at one site (a probe to the repeat would detect a broad smear since the repeats are so abundant at very many sites[4]. Many such markers were rapidly found to vary between inbred strains (Love, et al, 1990) and among wild mice (Montagutelli, et al, 1991). The use of about 400 markers in humans was responsible for mapping many disease loci segregating in sufficiently large families—the only approach available in humans since controlled matings are not possible as they are in mice (Weber and May, 1989). This was greatly aided by Mendelian Inheritance in Man, a catalog of genetic variants started by Victor McKusick at John Hopkins University. Initially published in ever-increasing-in-size volumes, it became Online Mendelian Inheritance in Man (OMIM).

The assignment of the linkage groups to particular chromosomes heavily involved the use of translocations (see below). In the DNA cloning era, direct hybridization of DNA clones to banded chromosomes sped up the chromosomal assignments of genes. Soon the technique of somatic cell hybridization was developed, and became very useful, especially to assign to particular chromosomes genes for enzymes or other proteins without known strain variation, but with human or hamster/mouse differences. In somatic cell hybridization, human or hamster cultured fibroblasts could be fused with mouse cultured fibroblasts. Such hybrids were first detected as spontaneous events in mixed cultures.

"The occurrence of this phenomenon was discovered in 1960 by a research team working in Paris under the leadership of George Barski., and not by me [Ephrussi] as has been often erroneously stated in reviews of the subject. I hasten to add that one person has not shared in this error: having made in 1965 an important contribution to the field, Henry Harris, whose style sometimes makes one believe that he may be short on modesty, seems to ascribe the discovery of cell hybridization neither to Barski, nor to me, but to himself," **Ephrussi, (1972) pp 5-6**.

Directed cell fusion between human and mouse cells was performed first with viruses and then using polyethylene glycol (PEG; Pontecorvo, 1976) with the resulting hybrid cells preferentially losing mouse chromosomes. The first selection systems were based on specially designed media and on using mutated cell lines which would survive in the media only if fusion and complementation occurred. After several divisions, stable clones with one or a few mouse chromosomes remaining were obtained. A series of different hybrid cell lines could then be typed for which mouse chromosomes they contained and which mouse-specific enzymatic, other protein or DNA markers, the genes for which were thus assigned to these chromosomes. Fournier and Frelinger (1982) prepared a panel of hamster/mouse hybrids using translocation chromosomes which placed different autosomes in association with selectable markers. Eighteen of the 20 autosomes were separately available in cell clones. Such clones only achieved chromosome-specific assignments but, when the hybrids were sub-lethally irradiated to break the chromosomes, only fragments of mouse chromosomes remained and refined localizations could be achieved (Cox, et al, 1990). This was usually performed with somatic cell hybrids containing only 1 mouse chromosome, following which sub-regional colocalizations could be performed.

The great advantage of the radiation hybrids was to allow the construction of "physical maps," independent of the occurrence of crossing over. The efforts of all the many laboratories that were rapidly adding markers were hard to keep track of. In order to solve this problem, Chromosome Committees were established. At their meetings, usually held annually, high resolution/high density maps were made by pooling a large amount of primary data, allowing the establishment of consensus maps and anchoring cloned DNAs, the backbone of the physical maps. As mapping in both mouse and humans progressed, it became apparent that quite long stretches of genetic linkages were the same in both species. This "synteny" indicated a great conservation of mammalian genome organization over many tens of millions of years. Initially, a small number of breaks and rejoins explained many big blocks of synteny between the 2 species (Nadeau & Taylor, 1984; Searle, et al., 1987). These were well displayed in the Oxford grid (Edwards, 1991). However, detailed analysis from whole genome sequencing identify 293 "macroblocks" shared between mice and humans, with 1070 micro-rearrangements (Bourque, et al, 2004). This synteny was very important for the eventual sequencing of the mouse genome in the 21st Century.

Another important technique of mapping, especially for quantitative loci, is the use of Recombinant Inbred (RI) lines, which were introduced in the mouse by Bailey (1971). In this technique, two inbred strains are crossed, ideally in both directions, i.e., with the female of one inbred strain with the male of the other, and vice versa, to balance the number of X chromosomes from each inbred strain, and F2 generations are created. The Y has very few functional genes but, by performing the reciprocal parental cross, both strains' Y chromosomes will also be present.Many sib pairs are mated and their offspring brother-sister mated for about 20 generations. The chromosomes of the resultant inbred strains are then mosaics of parental chromosomes[5]. Since all the individuals of a given recombinant inbred strain are genetically identical, one can then perform quantitative analyses of traits which may have a significant amount of variation. Enough individuals from one line are analyzed to define meaningful parameters for comparison between a number of the RI lines. The genetic content of each RI line is established by the use of strain specific markers, especially STRs, and, thus, the quantitative trait can, if controlled by 1 or a few genes, be mapped. Using this procedure, a variety of genes could be mapped. These included, for examples, a locus affecting beta-adrenergic receptor activity (Markovac and Erickson, 1983), susceptibility to teratogen-induced oral facial clefting (Liu and Erickson, 1986a; Erickson, et al, 2005), glucocorticoid receptor levels (Liu and Erickson, 1986b), and bone shape (Lovell and Erickson, 1986). The method was also very useful for mapping behavioral trait loci, e.g. Gerschenfeld, et al, 1997. In the 21st Century, mose geneticists from around the world collaborated in making hundreds of RI lines from the multiple crosses of 8 inbred strains choosen for diversity, including ones from wild species (Keele, et al., 2019). This was known as the Collaborative Cross (The Complex Trait Consortium, 2004; Threadgill, et al., 2002; Threadgill and Churchill, 2012).

General cytogenetic studies in mice

Elizabeth Cox, working in Painter's lab at the University of Texas, correctly counted the mouse chromosomes (40) in Bouin's fixed, sectioned and stained testicular material (Cox, 1926). Although there was variation in the size of these uniformly stained chromosomes, they all were acrocentric (i.e. apparently possessing centromeres at one end, a very short arm to provide a telomere is also needed). The fact that all chromosomes were acrocentric limited the ability to distinguish between them.

The same kind of material was not adequate to determine the correct number of human chromosomes (46) and Painter's deduced number of 48 was long considered correct (Painter, 1923). The use of cultured cells or rapidly dividing lymphocytes from blood, which, when placed on slides, provided a two-dimensional, instead of a 3-D view, and the use of colchicine to stop cell division at mitosis when chromosomes are condensed, were big advances. A further major improvement occurred when Hughes (Hughes, 1952), working with avian cells, first used hypotonicity to swell nuclei, thus separating the chromosomes for better observation; a technique quickly applied to other species (Hsu & Pomerat, 1953). Use of these technical advances allowed the correct identification of the number of human chromosomes (Tjio & Levan, 1956). These improved techniques soon revealed that one X in XX mouse females remained more condensed, representing late replication (Ohno & Hauschka, 1960). This observation provided an explanation of the origin of the Barr body, a condensed body first seen in neuronal nuclei of only female cats (Barr & Bertram, 1949). It also provided a cytological basis for the inactive X hypothesis Mary Lyon developed (which also became known as the Lyon hypothesis) to account for coat color mosaicism in female mice. She ascribed this variegation to inactivation of one of the two X chromosomes, and therewith of the coat color allele on that chromosome (Lyon, 1961, 1962). More direct cytological confirmation came with Ohno and Cattanach's (1962) studies using an X-autosome translocation. X-inactivation studies will be further considered in Chapter 10.

These new techniques also allowed the rapid clarification of the sex determination mechanism of mouse and man. It had been assumed that sex determination in mammals would be mediated by the ratio of the number of X chromosomes to the number of sets of autosomes, in analogy with *Drosophila*. However, in 1959, it was discovered that an XO sex chromosome karyotype produced a basically female phenotype both in humans (Turner syndrome; Ford, et al, 1959) and in mice (Welshons & Russell, 1959). Cattanach then found that XXY mice are males (1961). Sex determination is the subject of Chapter 11.

As mentioned above, the early assignment of linkage groups to chromosomes depended primarily on the use of chromosomal translocations which provided easily microscopically-detected markers on cytogenetic spreads of chromosomes. Snell was the first to report the production of translocations by X-rays in mice (Snell, 1934). He and Pickens (1935) related the abnormal embryos in the offspring of translocation-carrying mice to the abnormal chromosomal segregation which occurred during meiosis. Snell (1946) and Carter, et al (1955, 1956) described a number of these translocations and Slizynski (1957) was able to assign 10 of the linkage groups to different chromosomes. In addition, reciprocal translocations, in which two chromosomes with their associated linkage groups are broken and two new ones are created, were very useful.

"Quite early in the history of formal mouse genetics, George Snell became interested in determining the centromeric ends of the linkage groups. The date was 1946. The method he used was a very clever one, although a most complicated one, analysis by adjacent-2 segregation (segregation of homologous centromeres during meiosis in a translocation heterozygote). Of historical importance is the fact that he accurately assigned the centromere of Linkage Group V (now Chr 2) to its *Sd* (Danforth short tail) end and the centromere for Linkage Group VIII (now Chr 5) to its *wd* (waddler) end.[Snell, 1946]" (Eicher, 1981). Methods for assigning centromere orientation and linkage groups to particular chromosomes continued to be improved, and Lyon et al (1968) assigned two centromeric locations (for II and IX) while Searle, et al (1971) added 3 more (for V, XIII and XIV).

"Around 1970, a revolution occurred in genetics that permanently affected how the mouse map was displayed: Caspersson and colleagues...discovered that quinacrine mustard bound to chromosomes in a specific and repeatable manner making bands such that metaphase chromosomes were identifiable under the microscope [Caspersson, et al, 1970]. Quickly it was reported by four separate groups, Nesbitt and Francke in California [Francke and Nesbitt, 1971], Miller and coworkers in New York [Miller , et al, 1971], Schnedl in Germany [Schnedl, 1971], and Buckland and colleagues in England [Buckland, et al, 1971], that each mouse chromosome, when stained with quinacrine or treated with trypsin or specific salt solutions, banded in a specific pattern...The next step in the puzzle was to identify which chromosome carried which linkage group." (Eicher, 1981).[6]

The unique banding patterns allowed the identification of the frequently similar-sized chromosomes involved in the many translocations and, thus, the assignment of the linkage groups to particular chromosomes. Robertsonian translocations are the result of the joining of 2 acrocentric chromosomes at their centromeres (after reciprocal translocation). This usually results in a metacentric chromosome, i.e., one with the centromere near the middle, but if the sizes of the acrocentric chromosomes are quite unequal the centromere of the Robertsonian translocation chromosome will be more towards one end. These transloca-tions were particularly useful, as the centromere could be mapped relative to marker loci. Such Robertsonian translocations are found naturally in some wild populations (Adolph and Klein, 1981) and all the chromosomes, excepting the sex chromosomes, can be metacentric in a few populations (Capanna, et al, 1976—see Chapter 5 for more about Robertsonian translocations in wild mice). The banding patterns allowed the assignment of tufted to band 17B (Forejt, et al, 1980).

The cloning era (see Chapter 9) brought other tools for mouse cytogenetics. Cloned frag-ments of DNA could be directly hybridized to spread metaphase chromosomes on slides. Initially, the cloned fragment was radioactively labelled with ^{32}P nucleotides as it was amplified in vitro, and then hybridized, in a dark room, to slides which were than covered with a liquid photo-emulsion. These could later be photographically developed and the resulting grains, indicating radioactive decay, related to the underlying banded chromosomes. This was a very laborious technique which was greatly simplified when, instead, the cloned fragment was labelled with a fluorescent dye. Cytogenetics was revolutionized when the resulting fluorescent probes were used for fluorescent in situ hybridization (FISH), first used in *Drosophila*.

Another advance was the ability to separate some chromosomes by size in the flow cell sorter. Using this device, single cells (or other objects in droplets) were "dropped" past a fine beam of UV or other wavelength of light, and then fell between charged plates, to be separated into different containers per preestablished parameters for absorbance or reflectance. This machine was initially developed in the Herzenberg laboratory at Stanford (Herzenberg, et al, 2002) for immunological studies of separated cells. The small mouse chromosome 19 (Baron, et al, 1990) was successfully separated by this method. Such separated chromosomes allowed the construction of chromosome-specific libraries of frag-ments cloned by non-specific PCR (Telenius, et al, 1992) which could be used for "chromo-some painting" by FISH (Liyanage, et al., 1996). This FISH technique was particularly useful in identifying novel translocations.

In the 1990s, collaborative projects were developed to create very high density chromosome maps using DNA markers (Dietrich, et al, 1994; Rhodes, et al, 1998). The latter group collected 982 DNA samples from the offspring of 2 large (reciprocal) backcrosses of *Mus spretus* and C57/BL6, mapping 3,365 microsatellite markers among them. This work prepared the framework for assigning DNA sequences to particular chromosomal regions, further strengthening the human/mouse syntenic maps. These could be aligned with the complete sequence of the genome of the C57BL/6J strain early in the 21st Century. This was accomplished when massively parallel sequencers, which made 50-100 copies each of many millions of random sequences, became available. This was known as whole-genome sequencing or WGS. Overlapping sequences could be assembled to establish huge lengths, which frequently could be aligned, using the synteny, to human sequences. Thus, the final sequence made relatively little use of the high density/high resolution maps previously established. The human sequence had been established using such a shotgun sequencing strategy aligned to detailed human chromosomal maps, locating many cloned fragments[7]. However, this approach does not allow the sequencing of certain genomic segments such as highly repeated regions, which are still not completely resolved in either species. As more mammalian species were sequenced, the syntenic relationships between many of them sped up the effort. As the maps generated by sequencing were compared to the "classic" linkage maps, it became apparent that recombination events are not random but preferentially occur at "hotspots" which could hundred of kb.s apart (Steinmetz, et al, 1987). There were also chromosomal locations where recombination was suppressed, e.g. by a retropososon insertion (Hsu and Erickson, 2000).

t-allele linkage and cytogenetic studies

The progress of work on mapping the *T/t* complex well illustrates the approaches and methods of gene mapping in mice. Dunn and Gluecksohn-Schoenheimer (1942; later Gluecksohn-Waelsch, see mini-biography and Chapter 4) had thoroughly studied the possibility of cross-overs between *T, t⁰,* and other *tⁿ*s, with negative results. The first studies of linkage in the region concerned 3 genes with similar phenotypic effects on tail morphology. Many linkage studies in a variety of organisms established the general notion that genes affecting similar processes or developmental patterns were not necessarily closely linked. An exception to this was found when Dunn and Caspari (1945) found that *T*, (and *t¹*), *Fused (Fu)* and *Kink (Ki)* were closely linked but separable by crossing-over. *Fu* and *Ki* have phenotypic overlaps with *T*, sharing vertebral abnormalities of the tail. *Fu* is homozygous semi-viable, while *Ki* and *T* are not viable in the homozygous condition. *Fu* variably expresses vertebral abnormalities of the trunk, waltzing behavior, and deafness. *Ki* doesn't show trunkal vertebral abnormalities and is more penetrant than *Fu*. It also shows waltzing behavior and deafness, which are correlated (Dunn & Caspari, 1945). The three genes are not allelic, i.e., they do not interact with each other or *T, t⁰,*or *t¹* to cause taillessness. Dunn and Caspari (1945) found 4% recombination between *T* and *Fu* and between *T* and *Ki*, with a higher recombination frequency in females than males. Although apparently the same physical distance from *T, Ki* and *Fu* showed 2% crossing-over. Dunn and Caspari noted that *Fu* offspring from *Fu* females seldom showed the phenotype, as had been noted in the original description by Sheldon Reed (Reed, 1937). We now know that this is due to imprinting in the region—see Chapter 10. Neither *Fu* or *Ki* showed recombination with *t⁰* or *t¹*, however, and the presence of *t⁰* or *t¹* inhibited *Ki-T* recombination, as had previously been reported between *T* and *ts* (Dunn and Gluecksohn-Schoenheimer, 1942). They concluded "that the segment of chromosome in which these mutations occurred is so constituted that any change is likely to be extensive and to cause abnormalities in early developmental processes, the effects of which converge on axial structures (spinal cord, notochord, etc.) which have 'organizer' relationships with each other and earlier processes" (Dunn and Caspari, 1945). However, this wasn't correct as we will see.

Dunn and Gluecksohn-Schoenheimer (1950; Dunn & Glucksohn-Waelsch, 1953a; Dunn, 1956) found that new t^n alleles arose frequently from T/t^n stocks, which were distinguished by their developmental phenotypes. They interpreted them as new mutations, which led to the studies, described in the previous chapter, in which such "mutations" were sought in wild mice, and the high incidence of t-alleles in wild populations was rediscovered. Gruneberg summarized the state of affairs at about this time succinctly: "The genes T, Fu, and Ki thus show crossing over in the absence of the t-factors; the total length of the chromosome covered is about eight units. In the presence of t-factors, crossing-over throughout the region is blocked. It follows that the t-factors are not single genes in the ordinary sense but they represent some kind of structural change" **(Gruneberg, 1952, p. 343)**.

Dunn's group (Bennett & Dunn, 1958; Bennett, et al, 1959; Dunn & Bennett, 1960) showed that t^n lethals, with many newly acquired from the wild (t^ws), fell into 2 complementation groups with differing times and patterns of death between them. That is, compound heterozygotes in one group were distinguishable from the compound heterozygotes of the other group. These could now be interpreted as representing two different loci separable by crossing-over. In time, many more complementation groups were found, see Chapter 4.

Mary Lyon found a marker, *tufted (tf)*, showing a coat phenotype instead of a tail phenotype, which mapped near T (Lyon, 1956). Lyon and Phillips (1959) showed that the crossover suppression of t^n with T also extended to tf. When new t^n alleles arose, they carried the tf marker, showing that the new t^n alleles did not arise from mutation but from rare cross-over events. Dunn and Glucksohn-Waelsch (1953b) had seen one case in which a t^n (t^3) did not suppress crossing-over. When Dunn's group obtained tf from Mary Lyon[8], they were able to further extend their studies with this marker to measure cross-over suppression. They analyzed many t^n alleles which were maintained as balanced lethal lines (each with T), and which had 46 viable recombinant offspring from 27,751 born (Dunn, et al, 1962)[9]. Twenty-eight were studied. Seven were recombinant with tf, indicating strong cross-over suppression generally. Thus, they postulated that the crossover suppression was due to inversions or deletions,

Lyon and Meredith (1964a,b,c) continued to work on these regional cross-overs and showed that the recombinant chromosome carrying the tf marker still interacted with T to cause taillessness, while the other chromosome carried the t^n homozygous lethality; i.e. recessive t-alleles were separable from T. Interestingly, most alleles no longer showed high segregation distortion, the over 50% transmission by males, but some showed a low transmission ratio, as seen in Table 3.1. They found different percentages of recombination between various t^n alleles and tf, suggesting possible different chromosomal positions, which they tentatively mapped (Lyon & Meredith, 1964a). "Ten alleles...could be divided into three groups on the basis of male segregation ratio, and the low ratio group could be further subdivided, since one allele in this group showed partial crossover-suppression and the others did not–all of these had lost the lethality of t^6, suggesting that the lethality was due to a specific factor lying at a distance from the T-locus. This was confirmed by the finding of the complementary type

Table 3.1 The *T-tf* recombination per cent, viability, and transmission ratio distortion (% offspring mutant) of t^{21} and a number of its derived *t*-alleles found at Harwell.

t-allele	*T-tf* recombination per cent	Viability	Transmission ratio distortion
t^{21}	11.6	Normal	64
th^2	9	Normal	20
th^3	12.5	Normal	47
th^4	8	Normal	53
th^5	1	Normal	14
th^6	10	Normal	10
th^8	5	Normal	23
th^9	7	Normal	21
th^{10}	9.8	Normal	53
th^{11}	7	Normal	54
th^{18}	6.2	Decreased	51

Table 3.2 The transmission ratio distortion, lethality, and fertility in heterozygous *t*-allele combinations of a number of *t*-alleles from the wild or from recombinants compiled from Lyon and Meredith, 1964a; 1964b; 1964c.

t-allele	Source	Transmission ration distortion (% mutant offspring)	Lethality	Fertility	Fertility in combination with other *t*-alleles
t^{w12}	Wild	96	Lethal	Normal	
t^{26}	*tf* recombinant of t^0	57	Homozygous viable	Nearly normal	t^{w12}/t^{26} 0.4% t^0/t^{26} 12%
t^{27}	From t^0	41	Homozygous viable	Nearly normal	t^{w12}/t^{27} 3% t^0/t^{27} 10%
t^{28}	From t^1	Not determined	Homozygous viable	Nearly normal	t^{w12}/t^{28} 0.8%% t^0/t^{28} 1%
t^{w29}	From t^{w18}	48	Homozygous viable	Normal	t^{w12}/t^{w29} sterile t^0/t^{w29} 2%
t^{w31}	From t^{w5}	53	Homozygous viable	Normal	t^{w12}/t^{31} 2.2% t^0/t^{31} normal
t^{24}	From t^0	25	Homozygous viable	Nearly normal	t^0/t^{24} 16%
t^{38}	From t^0	17	Homozygous viable	Nearly normal	t^{w12}/t^{38} 32% t^0/t^{38} normal
t^{w33}	From t^{w1}	41	Homozygous viable	Nearly normal	t^0/t^{w33} 36%
t^{w2}	Wild	95	Homozygous viable	Male sterile	Not tested
t^{w8}	Wild	84	Semilethal	Male sterile	t^{w8}/t^{38} 40%
t^{w1}	Wild	89	Lethal	Nearly normal	t^{w1}/t^{13} 22%
t^{13}	From t^{12}	35	Ciable	Nearly normal	Not tested

of crossover, th^{18}, carrying the lethal factor but lacking the T-modifying effect of t^6." (Lyon & Meredith, 1964a). One derivative t^n allele (th^7) lost the properties of interacting with T to cause taillessness, and was thought to represent a duplication (Lyon & Meredith, 1964b) and it was suggested that recombination occurred more frequently with the duplication. Table two provides information on the transmission ratio distortion, lethality, and fertility of a number of these alleles. In general, a high distorter is more likely to cause sterility than a low distorter when in combination with a viable t-allele.

The addition of H-2, the major histocompatability complex of mice (see Chapter 7), as a marker on this chromosome showed that cross-over suppression of t-haplotypes extended to H-2 (5 centiMorgans past tf) while not as far as *thin fur* (thf), 15.1 cM more distal than H-2 (Hammerberg & Klein, 1975a, b). This work also represents the addition of more non-Anglophone workers, as immunogenetics was a more popular subject than developmental genetics (see Chapter 7). Jan Klein, working in Germany since the early 70s, represents scientists from these middle European communities' developing interests in mouse T/t complex genetics. Forejt in Czechoslovakia was studying hybrid sterility in the overlap zone between the two *Mus* sub-species which crossed his country, and Jan Klein, who fled Czechoslovakia during the 1968 Russian invasion, shared Forejt's (Forejt, 1972, see below) interest in this and in t-alleles. Klein first went to Len Herzenberg's laboratory at Stanford University where he assigned linkage group IX containing H-2 to one of the *Mus poschiavinus* metacentrics (Klein, 1971; Klein and Klein, 1972).

France, also, was starting to develop programs in mouse genetics. Francois Jacob, at the Institut Pasteur in Paris, decided to shift from *C. elegans* (which he started to work on as suggested by Sydney Brenner as a more complex organism) to mice in the late seventies. He hired Jean-Louis Guenet, a veterinarian/Ph.D. with a strong interest in genetics, to develop a mouse facility in 1970 (Guenet, 2014). Jacob sent Hubert Condamine to Beatrice Mintz's laboratory to learn mouse embryology and the making of chimeras, and hired Edwige Jacob, who had worked with Boris Ephrussi, to develop tissue culture facilities for mouse lines and teratomas (Guenet, 2014). Jean-Louis Guenet went on to develop the first mouse colony for genetics at the Institut Pasteur. "Immunology was dominated by work on idiotypes (Oudin, Cazenave, Bordenave). Rabbits were thus dominant. Of course, mice were used as model animals of infectious diseases, but there was no genetics, no modern immunology, no cellular immunology, no research on development, and no immunogenetics until the end of the 1960s" (Guenet, 2014).[10] A formal mouse genetics Division (*l'unite de Genetique des Mammiferes)* was established at the Institute Pasteur in October of 1981[11]. In the South of France, at Montpelier, Francois Bonhomme was studying wild mice. Both units were to become involved in the genetics of the T/t complex, especially Jacob's laboratory.

The question of the possibility that the cross-over suppression of t-alleles was due to inversions was carefully examined by Womack and Roderick (1974), with a negative conclusion. They performed microscopic studies of meiotic divisions in the testes of t^6 /+ males, and did not find an increased frequency of anaphase bridges ("string-like" connections of chromatin between 2 daughter cells, due to chromatin entanglement as is expected when an inversion pairs with the non-inverted segment of the chromosome) as would be expected if there were inversions. "Frequencies of broken bridges at anaphase I and frequencies of fragments appearing alone at anaphase I [potential products of crossing-over in inversions], also were not significantly different" (Womack & Roderick, 1974). A similar result was found by electron microscopy of male meiotic synaptonemal complexes (Tres & Erickson, 1982). However, later data showed that these conclusions were erroneous, and the problem of what caused cross-over suppression was not solved.

This was a period when the *T/t* complex was receiving much international attention among mouse geneticists. Because of this enthusiasm, Eva Eicher and Salome Glueksohn-Waelsch organized a "Mouse *t* Complex Meeting," held at the Jackson Laboratory, in 1976 which had more than 50 attendees including Mary Lyon.[12] This gathering continued at 10 year intervals until 1996 and a number of new players and methods were included (Artzt & Schimenti, 1997)

Mary Lyon summarized her group's, and others', data as of 1980, to suggest that the series of deletions with their varying properties were in a region of intercalary DNA (Lyon, et al. 1980). They presented data for asynchronous synapsis in testes at pachytene, at the region of the T-locus on chromosome 17, even in +/+ but more marked in *T*/+ or +/t6 chromosomes. They confirmed Jiří Forejt's and collaborator's (Forejt, 1972; Forejt & Gregorova, 1977) observations of decreased chiasmata in the presence of t^{12}, using 3 different translocations to identify chromosome 17. They followed Womack and Roderick (1974) in not finding evidence of anaphase bridges or chromosomal fragments in synaptonemal complexes, again arguing against inversions. Thus, they concluded that "although the exact mechanism remains unknown, it seems that *t*-chromatin alters the function of the chromosome at meiosis in such a way that chiasma formation is reduced. Moreover, as previously known, chromosome function in development is altered by *t*-chromatin, leading to embryonic lethality, modification of tail length, and effects on spermatogenesis. In considering the type of change in DNA which could produce such effects, it seems that major structural changes, such as inversions, are ruled out. We postulate that in *t*-chromatin it is the moderately repetitive DNA...that is altered" (Lyon, et al, 1980).

The observation that the variants produced by the rare crossovers from the different *t*-allele complementation groups were less "abnormal" than the original (as expressed in a later time of in utero lethality) suggested that the variants were produced by loss of material in a complex of interacting lethal factors, instead of the loss of independent units (genes) (Bennett, et al, 1976). Lyon and Bechtol (Lyon & Bechtol, 1977; Bechtol & Lyon, 1978) further supported the occurrence of deletions "exposing" *Knobbly* (*Kb*), a new radiation-induced locus in the *T-tf* interval), t^{w5}, and, in one case, *tf* (Lyon & Bechtol, 1977). In another case, an *H-2* allele was exposed (Bechtol & Lyon, 1978) while a deletion in *quaking* (*qk*, mapping proximal to *tf*, i.e. nearer to *T*), was also found (Erickson, et al, 1978). With previous data for a duplication in the case of th^7, (Lyon & Meredith, 1964b), unequal cross-overs creating duplications and deletions were strongly suggested.

Silver and Artzt (1981) demonstrated that recombination occurred between complementing lethal haplotypes (th^{17}/t^{12}). The frequency of recombination of 18.4% was greater than the usual value for the *T-tf* interval. Thus, "the recombination suppression of a complete

t-haplotype cannot be due to an intrinsic factor(s) which suppresses along the length of its own chromosome" (Silver & Artzt, 1981). Using additional markers, this group showed that many of the cross-overs were unequal (Silver, et al, 1980; reviewed in Silver, 1981). This work was extended as this group studied recombination between four t-haplotypes (t^{w12}, t^{w32}, t^{w5}, and t^{w18}) showing that the lethal factors of these 4 "alleles" (haplotypes) map to different locations (Artzt, et al, 1982a). However, the complementation of t-alleles in *trans*, as compared to *cis*, was not full, which led to a variety of speculations (Shin, et al, 1983a). In other crosses, though, the opposite effect was found (better complementation in *trans* than in *cis*), suggesting that different modifying genes, rather than intrinsic properties of the alleles, were involved (Mains, 1986). The group at Columbia University which had been started by L. C. Dunn (Artzt, Bennett) also demonstrated that *H-2* mapped proximal, instead of distal, to *tufted*, suggesting that an inversion was involved (Artzt, et al, 1982b).

The era of molecular cloning (see Chapter 9) brought important new tools for the analysis of the *T*-locus. Initially, clones for the nearby *H-2* complex became available and were used to evaluate t-haplotypes (which are in linkage disequilibrium with *H-2*). It was found that t-haplotypes showed very little variation for restriction fragments detected on Southern blots, even when they differed for their *H-2* serological type (Shin, et al, 1982; Silver, 1982) although they were also not very different serologically (Sturm, et al, 1982). The t-haplotype diversity at the DNA level was only about 10% of that found across the genome among wild mice, which again suggested a relatively recent origin of t-haplotypes with their rapid spread through the subspecies of *Mus* (see Chapter 2). *H-2* DNA probes also suggested that several DNA segments of the complex were lost in t-haplotypes (Rogers and Willison, 1983). The lack of diversity was also found for other new markers in the area of linkage disequilibrium: *Pgk-2* (Rudolph & Vandeberg, 1981) and *Tcp-1* (Alton, et al, 1980).

The continued use of cloned probes from the *H-2* region allowed the first direct demonstration of an inversion in the t-complex. A cloned fragment mapped centromeric to the serologically-defined *H-2* in t-alleles, but with the $E_\alpha 1a$ gene telomeric to the complex, while $E_\alpha 1a$ is centromeric to t in wild type (Shin, et al, 1983b). This inversion could also be found using a serological *H-2* marker (Condamine, et al, 1983). The use of a variety of crosses between viable t-allele pairs (presumably sharing the inversion[s] and, thus, not suppressing recombination) allowed the confirmation of Artzt, et al (1982a), by mapping. The non-complementing t-alleles fell into several groups and, thus, were independent loci (Artzt, 1984; Shin, et al, 1984; Pla & Condamine, 1985; Nadeau, et al, 1985) Fig. 3.1.

The next phase of molecular mapping involved t-haplotype clones. Rohme, et al (1984) micro-dissected DNA from the visualized bands of chromosome 17 on dried slides—a very arduous technique–which were then cloned into bacteria. These probes, mapping to different positions in the t-locus, indicated that t-alleles varied in the amount of t-haplotype DNA (Fox, et al, 1985). Such clones soon led to the discovery of a second, more proximal inversion in the t-haplotypes (Hermann, et al, 1986). Such an inversion well explained the phenotypes of two t-haplotypes (t^{wLub2}, t^{Orl}) which appeared to be reciprocal products of, respectively, a deletion and duplication (Sarvetnick, et al, 1986). Soon genes whose phenotypes were unrelated to classical *T/t* phenotypes were being mapped into the inversions, e.g. *Pim-1* (a T-cell lymphoma

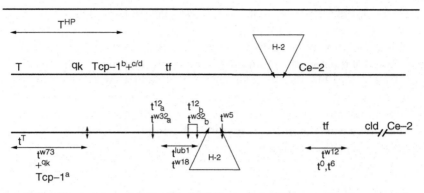

FIG. 3.1 Inversions in the *T/t* complex as understood in 1985. The upper solid line represents a normal chromosome 17 while the lower line represents a *t* chromosome. The upper arrow indicates the *TH*ᵖ deletion while the lower arrows represent approximate positions of various mutants while the thick vertical arrow on the *t* chromosome represents the start of recombination. From Frischauf, A-M (1985) The *T/t* complex of the mouse. Trends in Genet., 1:100-103. Used by permission of Elsevier Press.

oncogene), *Crya-1* (alpha-1-crystallin), and *H-2 I-E* mapped to the distal inversion (Nadeau and Phillips, 1987). Molecular probes also confirmed the presence of duplications (Hermann, et al, 1987) and found a family of repeats (T66, more abundant in *t*-haplotypes than wild-type chromosome 17s) in the *T/t*-complex (Schimenti, et al, 1987). Mary Lyon's group, using probes from other English investigators[13] and in situ hybridization on chromosomal spreads mapped the *t*-complex to chromosome 17, bands A1-A2 to band C (Lyon, et al, 1986) and directly visualized the two larger inversions (Lyon, et al., 1988). These in situ results were confirmed and extended by Silver's group (Mancoll, et al, 1992). Two smaller inversions, one proximal to the most proximal of the first two inversions and one between the original two inversions, were soon found (Hammer, et al, 1989). The finding of two *T/t* complex tail modifying genes was also best explained by an inverted duplicated segment (Nadeau, et al, 1989). Modifiers of the tail phenotypes from the *T/t* complex were also found which did not map near it (Artzt, et al, 1987). Thus, as maps of individual mouse chromosomes were assembled by the chromosomal committees mentioned above, a detailed map of chromosome 17 and its 4 inversions in *t*-haplotypes was ready in 1991 (Comm. Chrom 17, 1991). Much as yet unpublished work was incorporated, and then was soon published (Forejt, et al, 1991; Hammer, et al, 1991; King, et al, 1991; Neufeld, et al, 1991; Sertic, et al, 1992; Uehara, et al, 1991). The multiplicity of the inversions, preventing a single inversion loop, explains the inability to detect "an inversion" in synaptonemal complexes (Womack and Roderick, 1974; Tres and Erickson, 1982).

The last decade of the 20th Century included cloning the gene for *quaking* (an unusual protein with features of an RNA-binding protein and a signal transduction protein; Ebersole, et al, 1996) which was helped by the previous identification of a deletion in *quaking*^viable^ (Ebersole, et al, 1992a). There was a flurry of mapping new genes into the *t*-region (Ebersole, et al, 1992b; Filippov, et al., 1992; Schweifer and Barlow, 1992; Cox, et al, 1993; Burke, et al, 1994) and finer mapping of the region (Himmelbauer and Silver, 1994; Himmelbauer, et al, 1994; Howard, et al,

1990; Kasahara, et al, 1989; Nadeau, et al, 1991; Rossi, et al, 1994; Sutherland, et al, 1995; Vernet and Artzt, 1995; Gregorova, et al., 1996). The T-cell lymphoma oncogene, *Pim-1,* was mapped exceedingly close, and proximal to, to *tufted* (Ark, et al, 1991). The work necessary as a precursor for total sequencing of the region commenced with the development of yeast artificial chromosomes (YACs, Cox, et al, 1994; Barclay, et al 1996) and bacterial artificial chromosomes (BACs, Yoshino, et al, 1998) from the region. Extensive sequencing of *t*-haplotypes took place towards the end of the first decade of the 21st Century, e.g. Wallace and Erhart (2008) who found that only the t^{w32} haplotype was "complete," with others having 100 Kb-2 Mb insertions of wild-type DNA. These were best explained as gene conversion events.

Notes

1. Studies on the genetic regulation of beta-glucuronidase activity by androgen, performed by Ken Paigen's group, have been extensive (Swank, et al, 1973). A cis-regulatory element was found reminiscent of the bacterial beta-galactosidase operon in *E. coli*, the history of which is eloquently told by **Carroll, 2013**.

2. This is a difficult technique—there is a very brief period when the gel is formed and liquid enough to be poured, before it becomes a solid. It has to be de-gassed (a brief application of a vacuum) and then quickly poured into the waiting mold which has a "comb" to create wells into which the samples would later be placed.

3. When first introduced this was quite laborious. Three water baths were required. The first was held at near boiling point to denature the DNA. A second was at a temperature for annealing primers. These were machine-synthesized, about 20 base pair complementary fragments, flanking the desired segment of DNA and from which DNA synthesis was initiated in both directions, 3' to 5' and 5' to 3' on the 2 strands. The third water bath was at the temperature for DNA synthesis to occur (initially with bovine DNA polymerase at 37 C). A test tube rack containing the reagent tubes had to be moved from one water bath to the next in sequence, and more DNA polymerase added to the tubes after the annealing step at each round, for more than 30 rounds. The advent of a heat stable DNA polymerase purified from bacteria living in hot springs, and thermocyclers, made the procedure fast and efficient.

4. These were freely shared—I received them in 1985.

5. A way to visualize the result is to imagine that the C57BL/6J inbred strain had black chromosomes per its skin color, and that it was crossed with the A/J strain (albino) and that it had white chromosomes. The resultant RI lines would have harlequin chromosomes with patches of white and black.

6. The rapid discovery of alternative methods for detecting the banding suggests that they had already been discovered. In the same way that the discovery of hypotonicity to swell nuclei and spread chromosomes may have been the result of a laboratory mistake, cytogeneticists had sometimes seen banded chromosomes (at least Fred Hecht had, as he told me) but considered them lab errors and not grasped the importance (perhaps forgetting that the banded salivary chromosomes of *Drosophila* had been so important).

7. The many years and huge manpower to do this for the human genome was funded by NIH and was publicly available. This allowed Craig Venter and colleagues, with private capital, to buy many sequencers and to complete the sequence of a human genome, assembling it using the public framework, at about the same time as the consortium put together by NIH (which included foreign collaborators, too).

8. Mouse geneticists followed *Drosophila* geneticists in freely sharing mutants and stocks, a tradition which has become much weakened in modern times, although many journals require such sharing as a condition of publishing.

9. An impossible number for a mouse geneticist to obtain nowadays, given the costs of maintaining mice— sometimes over $2/day/cage (such that the cost of maintaining a cage of mice for a year is now more

expensive than shotgun sequencing a mouse genome (2019). This is due to increased regulations and supervisory costs.

10. Personal communication, Gabriel Gachelin, Feb. 23, 2016.

11. Communicated to Gabriel Gachelin by Isabelle Flevirance, November, 2016.

12. I was on the organizing committee for this conference and we were very pleased with the widespread enthusiasm for the meeting.

13. Although probes were cloned in bacterial plasmids and, thus, could be considered to be living forms, they were not shared as readily between labs as were mice or *Drosophila*.

References

Adolph, S., Klein, J., 1981. Robertsonian variation in muc musculus from Central Europe, Spain, and Scotland. J. Hered. 72, 219–221.

Alton, A.K., Silver, L.M., Artzt, K., Bennett, D., 1980. Molecular analysis of the genetic relationship of trans interacting factors at the T/t complex. Nature 288, 368–370.

Ark, B., Gummere, G., Bennett, D., Artzt, K., 1991. Mapping of the Pim-1 oncogene in mouse t-haplotypes and its use to define the relative map position of the tcl loci to(t6) and tw2 and the marker tf(tufted). Genomics 10, 385–389.

Artzt, K., McCormick, P., Bennett, D., 1982a. Gene Mapping within the T/t Complex of the Mouse. I. t-lethal Genes are Nonallelic. Cell 28, 463–470.

Artzt, K., McCormick, P., Bennett, D., 1982b. Gene Mapping within the T/t Complex of the Mouse. II. Anomalous Position of the H-2 Complex in t Haplotypes. Cell 28, 471–476.

Artzt, K., 1984. Gene Mapping within the T/t Complex of the Mouse. III. T-lethal Genes Are Arranged in Three Clusters on Chromosome 17. Cell 39, 565–572.

Artzt, K., Cookingham, J., Bennett, D., 1987. A New Mutation (t-int) Interacts With the Mutations of the Mouse T/t Complex That Affect the Tail. Genetics 116, 601–605.

Artzt, K., Schimenti, J., 1997. Decennial tails. Trends in Genetics 13, 10–11.

Barclay, J., King, T.F., Crossley, P.H., Barnard, R.C., Larin, Z., Lehrach, H., Little, P.F.R., 1996. Physical Analysis of the Region Deleted in the tw18 Allele of the Mouse tcl-4 Complementation Group. Geneomic 36, 39–46.

Bailey, D.W., 1971. Recombinant-inbred strains. An aid to finding identity, linkage, and function of histocompatibility and other genes. Transplantation 11, 325–327.

Baron, B., Metezeau, P., Kiefer-Gachelin, H., Goldberg, M.E., 1990. Construction and characterization of a DNA library from mouse chromosome 19 purified by flow cytometry. Biol. Cell 69, 1–8.

Barr, M.L., Bertram, E.G., 1949. A morphological distinction between the neurons of the male and female, and the behaviour of the nucleolar satellite during accelerated nucleoprotein synthesis. Nature 163, 676–677.

Bateson, W., Punnett, R.C., 1907. Poultry. Rep. Evol. Comm. Roy. Soc. Lond. IV, 18–35.

Bechtol, K.B., Lyon, M.F., 1978. H-2 Typing of Mutants of the t6 Haplotypes in the Mouse. Immunogenet 6, 571–583.

Bennett, D., Dunn, L.C., 1958. Effects on embryonic development of a group of genetically similar lethal alleles derived from different populations of wild houe mice. J. Morphol. 103, 135–158.

Bennett, D., Badenhausen, S., Dunn, L.C., 1959. The embryological effect of four late lethal t alleles in the mouse, which affect the neural tube and skeleton. J. Morphol 105, 105–144.

Bennett, D., Dunn, L.C., Artzt, K., 1976. Genetic change in mutations at the T/t-locus in the mouse. Genetics 83, 361–372.

Bonhomme, F., Selander, R.K., 1978. Estimating total genic diversity in the house mouse. Biochem Genet 16, 287–297.

Bourque, G., Pevzner, P.A., Tesler, G., 2004. Reconstructing the Genomic Architecture of Ancestral Mammals: Lessons from Human, Mouse and Rat Genomes. Genome Res 14, 507–516.

Botstein, D., White, R.L., Skolnick, M., Davis, R.W., 1980. Construction of a genetic linkage map in man using restriction fragment length polymorphisms. Am. J. Hum. Genet. 32, 314–331.

Boveri, T., 1907. Zellenstudien VI: Die Entwicklung dispermer Seeigeleier. Ein Beitrag zur Befruchtungslehre und zur Theorie des Kernes. Jena. Zeitschr. Naturw. 43, 1–292.

Buckland, R.A., Evans, H.J., Sumner, A.T., 1971. Identifying mouse chromosomes with the ASG technique. Exp Cell Res 69, 231–236.

Burke, P.S., Don, J., Wolgemuth, D.J., 1994. Afp-51, a murine zinc finger encoding gene mapping to the t-comple region of Chromosome 17, encodes 19 contiguous zinc fingers and is ubiquitously expressed. Mammal. Genome 5, 387–389.

Capanna, E., Gropp, A., Winking, H., Noack, G., Civitelli, M.-V., 1976. Robertsonian Metacentrics in the Mouse. Chromosoma 58, 341–353.

Carter, T.C., Falconer, D.S., 1951. Stocks for detecting linkage in the mouse, and the theory of their design. J. Genetics 50, 307–323.

Carter, T.C., Lyon, M.F., Phillips, R.J.S., 1955. Gene-tagged chromosome translocations in eleven stocks of mice. J. Genet. 53, 154 -16.

Carter, T.C., Lyon, M.F., Phillips, R.J.S., 1956. Further genetic studies of eleven translocations in the mouse. J. Genet. 54, 462–473.

Caspersson, T., Zech, L., Johansson, C., 1970. Differential binding of alkylating fluorochromes in human chromosomes. Exp Cell Res 60, 315–319.

Cattanach, B.M., 1961. XXY mice. Genet. Res. 2, 156–158.

Chapman, V.M., 1978. Biochemical polymorphisms in wild mice. In: Morse, HC (Ed.), Origin of Inbred MiceIII(ed). Academic Press, New York, pp. 555–568.

Committee for Mouse Chromosome 17, 1991. Maps of mouse chromosome 17: First report. Mammal. Genome 1, 5–29.

Condamine, H., Guenet, J.L., Jacob, F., 1983. Recombination between two mouse t haplotypes (tw12tf and tLub-1). Segregation of lethal factors relative to centromere and tufted (tf) locus. Genet. Res. Camb. 42, 335–344.

Cox, D.R., Burmeister, M., Price, E.R., Kim, S., Myers, R.M., 1990. Radiation hybrid mapping: a somatic cell genetic method for constructing high resolution maps of mammalian chromosomes. Science 250, 245–250.

Cox, R.D., Whittington, J., Shedlovsky, A., Connelly, C.S., Dove, W.F., Goldsworthy, Larin Z., Lehrach, H., 1993. Detailed physical and genetic mapping in the region of the plasminogen, D17Rp17e, and quaking. Mammal. Genome 4, 687–694.

Cox, R.D., Shedlovsky, A., Hamvas, R., Goldsworthy, M., Whittington, J., Connelly, C.S., Dove, W.F., Lehrach, H., 1994. A 1.2-Mb YAC Contig Spans the quaking Region. Genomics 21, 77–84.

Cox, E.K., 1926. The chromosomes of the house mouse. J. Morphol. Physiol. 43, 45–54.

Darbishire, A.D., 1904. On the result of crossing Japanese waltzing with albino mice. Biometrika 3, 1–51.

Dietrich, W.F., Miller, J.C., Steen, R.G., Merchant, M., Damron, D., Nahf, R., et al., 1994. A genetic map of the mouse with 4,006 simple sequence length polymorphisms. Nat. Genet. 2, 220–245.

Dunn, L.C., 1920. Linkage in Mice and Rats. Genetics 5, 325–343.

Dunn, L.C., 1921. Unit character variation in rodents. J. Mammal. 2, 125–140.

Dunn, L.C., Glucksohn-Schoenheimer, S., 1942. Tests for recombination amongst three lethal mutations in the house mouse. Genetics 28, 29–40.

Dunn, L.C., Caspari, E., 1945. A case of neighboring loci with similar effects. Genetics 30, 543–568.

Dunn, L.C., Glucksohn-Schoenheimer, S., 1950. Repeated mutations in one area of a mouse chromosome. Proc. Nat'l Acad. Sci 36, 233–237.

Dunn, L.C., Glucksohn-Waelsch, 1953a. Genetic analysis of seven newly discovered mutant alleles at locus T in the house mouse. Genetics 38, 261–271.

Dunn, L.C., Glucksohn-Waelsch, 1953b. The failure of a t-allele (t3) to suppress crossing-over in the mouse. Genetics 38, 512–517.

Dunn, L.C., 1956. Analysis of a Complex Gene in the House Mouse. Cold Spring Harbor Symp in Quant. Biol 21, 187–194.

Dunn, L.C., Bennett, D., 1960. A comparison of the effects, in compounds, of seven genetically similar lethal T alleles from populations of wild house mice. Genetics 45, 1531–1538.

Dunn, L.C., Bennett, D., Beasley, A.B., 1962. Mutation and recombination in the vicinity of a complex gene. Genetics 47, 285–303 Edwards JH (1991 The Oxford Grid. Ann. Hum. Genet., 55:17-31.

Ebersole, T.A., Chen, Q., Justice, M.J., Artzt, K., 1996. The quaking gene product necessary in embryogenesis and myelination combines features of RNA binding and signal transduction proteins. Nature Genet 12, 260–265.

Ebersole, T., Rho, O., Artzt, K., 1992a. The proximal end of mouse chromosome 17:new molecular markers identify a deletion associated with quakingviable. Genetics 131, 183–190.

Ebersole, T., Lai, F., Artzt, K., 1992b. New molecular markers for the distal end of the t-complex and their relationships to mutations affecting mouse development. Genetics 131, 175–182.

Edwards, J.H., 1991. The Oxford Grid. Ann. Hum. Genet. 55, 17–31.

Edwards, A.W.F., 2012. Reginal Crundall Punnett: First Arthur Balfour Professor of Genetics. Cambridge, p. 1912.

Eicher, E.M., 1981. Foundation for the Future: Formal Genetics of the Mouse. In: Russell, ES (Ed.), Mammalian Genetics and Cancer: The Jackson Laboratory Fiftieth Anniversary Symposium" published as Vol. 45 in the series, Progress in Clinical and Biological Research. New York. Alan R. Liss, Inc., pp. 7–49.

ENCODE Project Consortium, Dunham I., Kundaje A., Aldred S.F., Collins P.J., Davis C.A., Doyle F and several hundred more (2012) An integrated encyclopedia of DNA elements in the human genome. Nature. 489: 57-74.

Erickson, R.P., 1976. Strain variation in spermatozoal beta-glucuronidase in mice. Genet. Res.,Camb. 28, 139–145.

Erickson, R.P., Lewis, S.E., Slusser, K.S., 1978. Deletion mapping of the t-complex of chromosome 17 of the mouse. Nature 274 (5667), 163–164.

Erickson, R.P., Karolyi, I.J., Diehl, S.R., 2005. Correlation of susceptibility to 6-aminonicotinamide and hydrocortisone-induced cleft palate. Life Sci 76 (18), 2071–2078.

Erickson, R.P., Mitchison, A.N., 2014. The low frequency of recessive disease: insights from ENU mutagenesis, severity of disease phenotype, GWAS associations, and demography an analytical review. J. Applied Genetics 55, 319–327.

Erickson, R.P., 1962. Reactions to homografts in Triturus viridescens. Transplant. Bull. 30, 25–28.

Figge, F.H.J., Stong, L.C., 1941. Xanthine oxidase (dehydrogenase) activity in livers of mice of cancer-sensitive and cancer-resistant strains. Cancer Res 1, 779.

Filippov, V.A., Fedorova, E.V., Rogozin, I.B., Kholodilov, N.G., Ruvinsky, A.O., 1992. Molecular organization and evolution of the D17Leh80-like loci in the mouse t complex. Mammal. Genome 3, 11–16.

Ford, C.E., Jones, K.W., Polani, P.E., de Almeida, J.C., Briggs, J.H., 1959. A sex chromosome anomaly in a case of gonadal dysgenesis (Turner's syndrome). Lancet 1, 711–713.

Forejt, J., 1972. Chiasmata and crossing-over in the male mouse (mus musculus). Suppression of recombination and chiasma frequencies in the ninth linkage group. Folia biiol Praja 18, 161–170.

Forejt, J., Gregorova, S., 1977. Meiotic studies of translocations causing male sterility in the mouse. Cytogenet. Cell Genet. 19, 159–179.

Forejt, J., Capkova, J., Gregorova, S., 1980. T(16:17)43H translocation as a tool in analysis of the proximal part of chrormosome 17 (including T-t complex) of the mouse. Genet. Res, Camb. 35, 165–177.

Forejt, J., Vincek, V., Klein, J., Lehrach, H., Loudova-Mickova, M., 1991. Genetic mapping of the t-complex region on mouse chromosome 17 including the Hybrid sterility-1 gene. Mammal. Genome 1, 84–91.

Fournier, R.E.K., Frelinger, J.A., 1982. Construction of Microcell Hybrid Clones Containing Specfic Mouse Chromosomes: Application to Autosomes 8 and 17. Mol. and Cell. Biol. 2, 526–534.

Fox, H.S., Marin, G.R., Lyon, M.F., Herrmann, B., Frischauf, A.-M., Lehrach, H., Silver, L.M., 1985. Molecular Probes Define Different Regions of the Mouse t Complex. Cell 40, 63–69.

Francke, U., Nesbitt, M., 1971. Identification of the mouse chromosomes by quinacrine mustard staining. Cytogenetics 10, 356–366.

Frischauf, A.-M., 1985. The T/t complex of the mouse. Trends in Genet 1, 100–103.

Gates, W.H., 1927. Linkage of Short Ear and Density in the House Mouse. Proc. Natl. Acad. Sci. USA 75, 946–950.

Gershenfeld, H., Neumann, P.E., Mathis, C., Crawley, J.N., Li, X., Paul, S.M., 1997. Mapping Quantitative Trait Loci for Open-Field Behavior in Mice. Behav. Genet. 27, 201–210.

Gordon, A.H., Keil, B., Sebesta, K., 1959. Electrophoresis of proteins in agar jelly. Nature 164, 498–499.

Green, M.C., 1978. Brief Autobiography Related to Work on Inbred Mice. In: Morse, HC III (Ed.), Origins of Inbred Mice. Academic Press, New York, pp. 157–166.

Gregorova, S., Mnukova-Fajdelova, M., Trachtulec, Z., Capkova, J., Loudova, M., Hoglund, M., Hamvas, R., Lehrach, H., Vincek, V., Klein, J., Forejt, J., 1996. Sub-milliMorgan map of the proximal part of mouse Chromosome 17 including the hybrid sterility 1 gene. Mammal. Genome 7, 107–113.

Guenet, J.-L., 2014. Francois Jacob,....an outstanding mentor!. Research in Microbiology 165, 377–379.

Haldane, J.B.S., Sprunt, A.D., Haldane, N.M., 1915. Reduplication in mice. J. Genet. 5, 133–135.

Haldane, J.B.S., 1919a. The combination of linkage values and the calculation of distances between the loci of linked factors. J. Genet. 8, 291–307.

Haldane, J.B.S., 1919b. The probable errors of calculated linkage values, and the most accurate method of determining gametic from certain zygotic series. J. Genet. 8, 299–309.

Hammer, M.F., Schimenti, J., Siolver, L., 1989. Evolution of Mouse Chromosome 17 and the Origins of Inversions Associated With t-Haplotypes. Proc. Natl. Acad. Sci. USA 86, 3261–3265.

Hammer, M.F., Bliss, S., Silver, L.M., 1991. Genetic Exchange Across a Paracentric Inversion of the Mouse t Complex. Genetics 128, 799–812.

Hammerberg, C., Klein, J., 1975a. Linkage disequilibrium between H-2 and t complexes in chromosome 17 of the mouse. Nature 258, 296–299.

Hammerberg, C., Klein, J., 1975b. Linkage relationships of markers on chromosome 17 of the house mouse. Genet. Res., Camb. 26, 203–211.

Hermann, B., Bucan, M., Mains, P.E., Frischauf, A.-M., Silver, L.M., Lehrach, H., 1986. Genetic Analysis of the proximal portion of the Mouse t Complex: Evidence for a Second Inversion within t Haplotypes. Cell 44, 469–476.

Hermann, B.G., Barlow, D.P., Lehrach, H., 1987. A Large Inverted Duplication Allows homologous Recombination between Chromosomes Heterozygous for the Proximal t Complex Inversion. Cell, 48, 813–825.

Herzenberg, L.A., Parks, D., Sahaf, B., Perez, O., Roederer, M., Herzenberg, L.A., 2002. The history and future of the fluorescence activated cell sorter and flow cytometry: a view from Stanford. Clin. Chem. 48, 1819–1827.

Himmelbauer, H., Silver, L.M., 1994. High-Resolution Comparative Mapping of Mouse Chromosome 17. Genomics 17, 110–120.

Himmelbauer, H., Harvey, R.P., Copeland, N.G., Jenkins, N.A., Silver, L.M., 1994. High-resolution genetic anlysis of a deletion on mouse Chromosome 17 extending over the Fused, tufted, and homeobox Nkx2-5 loci. Mammal. Genome 5, 814–816.

Howard, C.A., Gummere, G.R., Lyon, M.F., Bennett, D., Artzt, K., 1990. Genetic and molecular analysis of the proximal region of the mouse t-complex using new molecular probes and partial t-haplotypes. Genetics 126, 1103–1114.

Hsu, T.C., Pomerat, C.M., 1953. Mammalian chromosomes in vitro. II A method for spreading the chromosomes of cells in tissue culture. J. Hered. 44, 23–29.

Hsu, S.J., Erickson, R.P., 2000. Construction of the long-range YAC physical map of the 10 cM region between the markers D18Mit109 and D18Mit68 on mouse proximal chromosome 18. Genome 43, 427–433.

Hughes, A., 1952. Some effects of abnormal tonicity on dividing cells in chick tissue cultures. Quart. J. Microscop. Sci. 93, 207–219.

Hunter, R.L., Markert, C.L., 1957. Histochemical demonstration of enzymes separated by zone electrophoresis in starch. Science 125, 1294–1295.

Janssens, F.A., 1909. La theorie de la chiasmatypie. Nouvelle interpretation des cineses de maturation. La Cell XXV, 389–414.

Kasahara, M., Passmore, H., Klein, J., 1989. A testis-specific gene Tpx-1 maps between Pgk-2 and Mep-1 on mouse chromosome 17. Immunogenetics 29, 61–63.

Keele, G.R., Crouse, W.L., Kelada, S.N.P., William Valdar, W., 2019. Determinants of QTL mapping power in the realized collaborative cross. G3 (Genes, Genomics, Genetics). doi:10.1534/g3.119.400194 doi.org/.

King, T.R., Dove, W.F., Guenet, J.-L., Herrmann, B.G., Shedlovsky, A., 1991. Meiotic mapping of murine chromosome 17: The string of loci around l(17)-2Pas. Mammal. Genome 1, 37–46.

Klein, J., 1971. Cytological Identification of the Chromosome Carrying the IXth Linkage Group (Including H-2) in the House Mouse. Proc. Nat. Acad. Sci., U. S. A. 68, 1594–1597.

Klein, J., Klein, D., 1972. Position of the translocation break T(2;9)138Ca in linkage group !X of the mouse. Genet. Res., Camb. 19, 177–179.

Kunkel, H.G., Slater, R.J., 1952. Zone electrophoresis in a starch supporting medium. Proc. Soc. Exptl. Bio. Med. 80, 42–44.

Law, L.W., Morrow, A.G., Greenspan, E.M., 1952. Inheritance of low liver glucuronidase activity in the mouse. J. Nat'l Cancer Inst. 12, 909–916.

Liu, S.L., Erickson, R.P., 1986a. Genetic differences among the A/J x C57BL/6J recombinant inbred mouse lines and their degree of association with glucocorticoid-induced cleft palate. Genetics 113, 745–754.

Liyanage, M., Coleman, A., du Manoir, S., Veldman, T., McCormack, S., Dickson, R.B., Barlow, C., Wynshaw-Boris, A., Janz, S., Wienberg, J., Ferguson-Smith, M.A., Schröck, E., Ried, T., 1996. Multicolour spectral karyotyping of mouse chromosomes. Nature Genetics 14, 312–315.

Liu, S.L., Erickson, R.P., 1986b. Genetics of glucocorticoid receptor levels in recombinant inbred lines of mice. Genetics 113, 735–744.

Lord, E.M., Gates, W.H., 1929. Shaker, a new mutation of the house mouse (Mus musculus). Am. Nat. 63, 435–442.

Love, J.M., Knight, A.M., McAleer, M.A., Todd, J.A., 1990. Towards construction of a high resolution map of the mouse genome using PCR-analysed microsatellites. Nucleic Acids Res 18, 4123–4130.

Lovell, D.P., Erickson, R.P., 1986. Genetic variation in the shape of the mouse mandible and its relationship to glucocorticoid-Induced cleft palate analyzed by using recombinant inbred lines. Genetics 113, 755–764.

Lyon, M.F., 1956. Hereditary hair loss in the tufted mutant of the house mouse. J. Hered. 47, 101–103.

Lyon, M.F., 1961. Gene action in the X-chromosome of the mouse (Mus musculus L.). Nature 190, 372–373.

Lyon, M.F., 1962. Sex chromatin and gene action in the mammalian X-chromosome. Amer. J. Hum. Genet. 14, 135–148.

Lyon, M.F., Phillips, R.J.S., 1959. Crossing-over in mice heterozygous for t-alleles. Hered 13, 23–32.

Lyon, M.F., Meredith, R., 1964a. Investigations of the nature of t-alleles in the mouse I. Genetic analysis of a series of mutants derived from a lethal allele. Hered 19, 301–312.

Lyon, M.F., Meredith, R., 1964b. Investigations of the nature of t-alleles in the mouse II. Genetic anlysis of an unusual allele and its derivatives. Hered 19, 313–325.

Lyon, M.F., Meredith, R., 1964c. Investigations of the nature of t-alleles in the mouse III. Short tests of some further mutant alleles. Hered 19, 327–330.

Lyon, M.F., Butler, J.M., Kemp, R., 1968. The positions of the centromeres in linkage groups II and IX of the mouse. Genet. Res. 14, 164–166.

Lyon, M.F., Bechtol, K.B., 1977. Derivation of mutant t-haplotypes of the mouse by presumed duplication or deletion. Genet. Res., Camb. 30, 63–76.

Lyon, M.F., Evans, E.P., Jarvis, S.E., Sayers, I., 1980. t-Haplotypes of the mouse may involve a change in intercalary DNA. Nature 279, 38–42.

Lyon, M.F., Zenthon, J., Evans, E.P., Burtenshaw, M.D., Dudley, K., Willison, K.R., 1986. Location of the t complex on mouse chromosome 17 by in situ hybridization with Tcp-1. Immunogenet 24, 125–127.

Lyon, M.F., Zenthon, J., Evans, E.P., Burtenshaw, M.D., Willison, K.R., 1988. Extent of the mouse t complex and its inversions shown by in situ hybridization. Immunorgenet 27, 375–382.

Lyon, M.F., 2002. A Personal History of the mouse Genome. Annu. Rev. Genomics Hum. Genet. 3, 1–16.

Mains, P.E., 1986. The Cis-Trans Test Shows No Evidence for a Functional Relationship Between Two Mouse T Complex lethal Mutations: Implications for the Evolution of T Haplotypes. Genetics 114 (1986), 1225–1237.

Mancoll, R.E., Snyder, L.C., Silver, L.M., 1992. Delineation of the t complex on mouse Chromosome 17 by in situ hybridization. Mammal. Genome, 2201–2205.

Markovac, J., Erickson, R.P., 1983. Genetic variation in b-adrenergic receptors in mice: A magnesium effect determined by a single gene. Genet. Res.,Camb. 42, 159–168.

Miller, O.J., Miller, D.A., Kouri, R.E., Allderdice, P.W., Dev, V.G., Grewal, M.S., Hutton, J.J., 1971. Identification of the mouse karyotype by quinacrine fluorescence, and tentative assignment of seven linkage groups. Proc. Natl. Acad. Sci. USA 68, 1530–1533.

Mitchison, N.A., 1968. Antigens. In: Dronamraju, K.R. (Ed.), Haldane and Modern Biol. John Hopkins Press, Baltimore, pp. 63–72.

Montagutelli, X., Serikawa, T., Guenet, J.L., 1991. PCR-anlyzed microsatellites: data concerning laboratory and wild-derived mouse inbred strains. Mamm. Genome 1, 255–259.

Nadeau, J.H., Phillips, S.J., Egorov, I.K., 1985. Recombination between the t6 complex and linked loci in the house mouse. Genet. Res., Camb. 45, 251–264.

Nadeau, J.H., Phillips, S.J., 1987. The putative oncogene Pim-a in the Mouse: Its Linkage and Variation Among t Haplotypes. Genetics 117, 533–541.

Nadeau, J.H., Varnum, D., Burkart, D., 1989. Genetic Evidence for Two t Complex Tail Interaaction (tct) Loci in t Haplotypes. Genetics 122, 8895–8903.

Nadeau, J.H., Hermann, B., Bucan, M., Burkart, D., Crosby, J.L., 1991. Genetic maps of mouse chromosome 17 including 12 n32 anonymous DNA loci and 25 anchor loci. Genomics 9, 78–89.

Nadeau, J.H., Taylor, B.A., 1984. Lengths of chromosomal segments conserved since divergence of man and Mouse. Proc. Nat'l Acad. Sci., U.S.A. 81, 814–818.

Neufeld, E., Vincek, V., Figueroa, F., Klian, J., 1991. Limits of the distal inversion in the t complex of the house mouse: Evidence from linkage disequilibria. Mammal. Genome 1, 242–248.

Ohno, S., Hauschka, T.S., 1960. Allocycly of the X-chromosome in tumors and normal tissues. Cancer Res 20, 541–545.

Ohno, S., Cattanach, B.M., 1962. Cytological study of an X-autosome translocation in Mus musculus. Cytogenetics 1, 129–140.

Painter, T.S., 1923. Studies in mammalian spermatogenesis II. The spermatogenesis of man. J. Exp. Zool. 37, 465–476.

Pla, M., Condamine, H., 1985. Recombination Between Two Mouse t Haplotypes (tw12tf and tLub-1):.Mapping of the H-2 Complex Relative to Centromere and Tufted (tf). Locu Immunogenet 20, 277–285.

Pontecorvo, G., 1976. Polyethylene glycol (PEG) in the production of mammalizn somatic cell hybrids. Cytogenet. Cell Genet. 16, 399–400.

Raymond, S., Weintraub, L., 1959. Acrylamide gel as a supporting medium for zone electrophoresis. Science 130, p711.

Reed, S.C., 1937. The inheritance and expression of Fused, a new mutation in the house mouse. Genetics 22, 1–13.

Rhodes, M., Straw, R., Fernanado, S., Evans, A., Lacy, T., Dearlove, A., et al., 1998. A high-resolution microsatellite map of the mouse genome. Genome Res 8, 531–542.

Rogers, J.H., Willison, K.R., 1983. A major rearrangement in the H-2 complex of mouse t haplotypes. Nature 304, 549–551.

Rohme, D., Fox, H., Herrmann, B., Frischauf, A.-M., Edstrom, J.-E., Mains, P., Silver, L.M., Lehrach, H., 1984. Molecular Clones of the Mouse t Complex Derived from Microdissected Meaphase Chromosomes. Cell 36, 783–788.

Rossi, J.M., Chen, H., Tilgman, S.M., 1994. Genetic Map of the Fused Locus on Mouse Chromosom 17. Genomics 23, 178–184.

Rudolph, N.S., Vandeberg, J.L., 1980. Linkage Disequilibrium Between the T/t complex and Pgk-2 in wild mice. J. Exp. Biol. 217, 455–458.

Sarvetnick, N., Fox, H.S., Mann, E., Mains, P.E., Elliott, R.W., Silver, L.M., 1986. Nonhomologous Pairing in Mice Heterozygous for a T Haplotype can Produce Recombinant Chromosomes with Dupllications and Deletions. Genetics 113, 723–724.

Schimenti, J., Vold, L., Socolow, D., Silver, L.M., 1987. An Unstable Family of large DNA Elements in the Center of the Mouse t Complex. J. Mol. Biol. 194, 583–594.

Schnedl, W., 1971. The karyotype of the mouse. Chromosoma 35, 111–116.

Schweifer, N., Barlow, D.P., 1992. The mouse plasminogen locu maps to the recombination breakpoints of the tLub2 and TtOrl partial t haplotypes but is not at the tw73 locus. Mammal. Genome 2, 260–268.

Searle, A.G., Ford, C.E., Beechy, C.V., 1971. Meiotic disjunction in mouse translocations and the determination of centromere position. Genet. Res. 18, 215–235.

Searle, A.G., Peters, J., Lyon, M.F., Evans, E.P., Edwards, J.H., Buckle, V.J., 1987. Chromosome maps of man and mouse III. Genomics 1, 13–18.

Sertic, J., Zaleska-Rutczynska, Z., Vincek, V., Nadeau, J.H., Figueroa, F., Klein, J., 1992. Mapping of six DNA markrs on mouse Chromosome 17. Mammal. Genome 2, 138–142.

Shin, J.-S., Stavnezer, J., Artzt, K., Bennett, D., 1982. Genetic structure and origin of t haplotypes of mice, analyzed with H-2 cDNA probes. Cell 29, 969–976.

Shin, H.-S., McCormick, P., Artzt, K., Bennett, D., 1983a. Cis-Trans Test Shows a Functional Relationship between Non-Allelic Lethal Mutations in the T/t Complex. Cell, 925–929.

Shin, H.-S., Flaherty, L., Artzt, K., Bennett, D., Ravetch, J., 1983b. Inversion in the H-2 complex of t-haplotypes in mice. Nature 306, 380–383.

Shin, H.-S., Bennett, D., Artzt, K., 1984. Gene Mapping within the T/t Complex of the Mouse. IV. The Inverted MHC Is Intermingled with Several t-lethal genes. Cell 39, 573–578.

Silver, L.M., White, M., Artzt, K., 1980. Evidence for unequal crossing over within the mouse T/t complex. Proc. Nat. Acad. Sci. USA 77, 6077–6080.

Silver LM 91981) Genetic Organizaion of the Mouse t Complex. Cell 27:239-240.

Silver, L.M., Artzt, K., 1981. Recombination suppression of mouse t-haplotypes due to chromatin mismatching. Nature 290, 68–70.

Silver, L.M., 1982. Genomic analysis of the H-2 complex region associated with mouse t-haplotypes. Cell 29, 961–968.

Slizynski, B.M., 1957. Cytological analysis of translocations in the mouse. J. Genet. 55, 122–130.

Smithies, O., 1955. Zone electrophoresis in starch gels: group variation of the serum proteins of normal human adults. Biochem J 61, 629–641.

Snell, G.D., 1934. The production of translocations and mutations in mice by means of X-rays. Am. Naturalist 68, 178.

Snell, G.D., 1946. An analysis of translocations in the mouse. Genetics 31, 157–180.

Snell, G.D., Pickens, D.I., 1935. Abnormal development in the mouse caused by chromome imbalance. J. Genet. 31, 213–235.

Southern, H.M., 1975. Detection of specific sequences among DNA fragments separated by gel electrophoresis. J. Mol. Biol. 98, 503–517.

Steinmetz, M., Uematsu, Y., Fischer-Lindahl, K., 1987. Hotspots of homologous recombination in mammalian genomes. Trends. Genet 3, 7–10.

Sturm, S., Figueroa, F., Klein, J., 1982. The relationship between t and H-2 complexes in wild mice I. The H-2 haplotypes of 20 t-bearing strains. Genet. Res., Camb. 40, 73–88.

Sutherland, H.F., Pick, E., Francis, F., Lehrach, H., Frischauf, A.-M., 1995. Mapping around the Fused locus on mouse Chromosome 17. Mammal. Genome 6, 449–453.

Sutton, W.S., 1902. On the morphology of the chromosomo group in Brachystola Magna. Biol. Bull. 4, 24–39.

Swank, R.T., Paigen, K., Ganschow, R.E., 1973. Genetic control of glucuronidase induction in Mice. J. Mol. Biol. 81, 225–233.

Sweet, A.M., Cohen, J.A., López, M., Erickson, R.P., 1988. An Eco RI polymorphism for p MetG in inbred mice. Nucleic Acid Res 16, 8745.

Telenius, H., Ponder, B.A.J., Tunnacliffe, A., Pelmear, A.H., Carter, N.P., Ferguson-Smith, M.A., Behmel, A., Nordenskjold, M., Pfragner, R., 1992. Cytogenetic analysis by chromosome painting using dop-pcr amplified flow-sorted chromosomes. Genes, Chrom.s Canc 4, 256–263.

The Complex Trait Consortium, 2004. The Collaborative Cross, a community resource for the genetic analysis of complex traits. Nature Genet. 36, 1133–1137.

Threadgill, D.W., Hunter, K.W., Williams, R.W., 2002. Genetic dissection of complex and quantitative traits: from fantasy to reality via a community effort. Mamm. Genome 13, 175–178.

Threadgill, D.W., Churchill, G.A., 2012. Ten Years of the Collaborative Cross. Genetics 190, 291–294.

Tiselius a (1937) A new apparatus for the separation of colloidal mixtures. Trans. Faraday Soc. 33:524-531.

Tjio, J.-H., Levan, A., 1956. The chromosome number of man. Hereditas 42, 1–6.

Tres, L.L., Erickson, R.P., 1982. Electron microscopy of t-allele synaptonemal complexes discloses no inversions. Nature 299, 752–754.

Uehara, H., Abe, K., Flaherty, L., Bennett, D., Artzt, K., 1991. Molecular organization of the D-Qa region of t-haplotypes suggests that recombination is an important mechanism for generating genetic diversity of the major histocompatibility complex. Mammal. Genome 1, 92–99.

Vernet, C., Artzt, K., 1995. Mapping of 12 markers in the proximal region of mouse Chromosome 17 using recombinant t haplotypes. Mammal. Genome 6, 219–221.

Vesell, E.S., Bearn, A.C., 1958. Observations on the heterogeneity of malic and lactic dehydrogenase in human serum and red blood cells. J. Clin. Invest. 37, 672–679.

Wallace, L.T., Erhart, M.A., 2008. Recombination within mouse t haplotypes has replaced significant segments of t-specific DNA. Mammal. Genome 19, 263–271.

Weber, J.L., May, P.E., 1989. Abundant class of human DNA polymorphisms which can be typed using the polymerase chain reaction. Am J Hum Genet 44, 388–396.

Welshons, W.J., Russell, L.B., 1959. The Y-chromosome as the bearer of male determining factors in the mouse. Proc. Nat'l Acad. Sci., USA 45, 560–566.

Wilson, E.B., 1914. Croonian Lecture: The Bearing of Cytological Research on Heredity. Proc. Roy. Soc. London, Series B 88, 333–352.

Womack, J.E., Roderick, T.H., 1974. T-alleles in the Mouse Are Probably Not Inversions. J. Hered. 65, 308–310.

Wright, S., 1968. I. Contributions to Genetics. In: Dronamraju, K.R. (Ed.), Haldane and Modern Biology. John Hopkins Press, Baltimore Md.

Yoshino, M., Xiao, H., Amadou, C., Jones, E.P., Lindahl, KF, 1998. BAC clones and STS markers near the distal breakpoint of the fourth t-inversion, In(17)4d, in the H2-M region on mouse Chromoome 17. Mammal. Genome 9, 186–192.

4

Developmental genetics and recessive *t*-lethals

The history of the Modern Synthesis of genetics with evolution was dealt with in Chapter 2, another synthesis, that of genetics with embryology, was needed. "If there was a 'Modern Synthesis' between genetics and evolution, there had to have been some '*Unmodern* synthesis' that it replaced. This Unmodern Synthesis was the notion that evolution was caused by changes in *development.* The syntheses of, respectively, E. Haeckel, E. Metchnikov, A. Weismann, W. K. Brooks, and others were that of evolution and *embryology*...by the 1930s, this synthesis had become both racist and scientifically untenable...But in the 1930s and 1940s, embryology had nothing new to substitute for this discredited notion" (Gilbert et al., 1996). As discussed in Chapter 1, the arguments between proponents of nuclear and proponents of cytoplasmic inheritance were strong. This was particularly true in France, while in the United States and Germany, where embryology was more strongly developed, the argument was also heated between the embryologists and the geneticists. For instance, "In 1937, Ross Harrison gave a lecture at the zoological sciences section of the American Association for the Advancement of Science, warning the geneticists to stay on their own turf. He warned of the dire consequences of 'this threatened invasion' caused by the 'wanderlust' of geneticists" (Gilbert, 1991). This conflict has been well reviewed by Burian (1997). The embryologists studied the rapidly changing embryo, and noted marked cytological (cytoplasmic) differences in various tissues and cell types while the nucleus remained unchanged. It was not at all clear how the "fixed" genes could play a role in these processes.

Even though several of the great early geneticists, including Castle and Morgan, had started their scientific careers as embryologists, their role in establishing the new paradigm of genetics led them to turn their backs on many of the embryologists' concerns. While Morgan maintained his interest in, and remained up to date on, embryology, he compartmentalized the two. He wrote a book, *Embryology and Genetics* in 1934 but the two subjects were dealt with separately. About all he could say about the relationship was that "the initial differences in the protoplasmic regions may be supposed to affect the activity of the genes. The genes will then in turn affect the protoplasm, which will start a new series of reciprocal reactions. In this way, we can picture to ourselves the gradual elaboration and differentiation of the various regions of the embryo" (cited in Gilbert, 1988). This notion was broadly true but convincing evidence would not emerge for decades. Morgan, and many geneticists, were content to follow Johannsen's belief that "heredity should be considered as the transmission of genetic traits from one generation to another. The emergence of the phenotype was of secondary importance and belonged to the realm of embryology" (Gilbert, 1988). It was only nearer the end of the Twentieth Century that it was appreciated that changes in the body plans of organisms could be subject to natural selection. This synthesis of evolution and development became known as "Evo-Devo" and has been the topic of extensive essays (e.g., Muller, 2007).

Twentieth Century Mouse Genetics. DOI: https://doi.org/10.1016/B978-0-12-824016-8.00010-6

The great advances of embryology in the 1920s and 1930s gave embryologists confidence in their notions. This was to such an extent that Spemann's summary, *Embryonic Development and Induction* (1938), did not even mention genetics. Many of the outstanding advances, such as Mangold and Spemann's "Organizer," Child's "Gradients," and Witschi's studies on gonad differentiation could be unified by the well accepted, and indeed prescient concept **(Hadorn, p. 302)** of a Morphogenetic Field. "These fields designated areas of embryological information, bound by physiological substrates...The components of these fields created a web of interactions such that any cell was defined by its position within its respective field" (Gilbert et al., 1996). It was defined by Needham in another influential book, *Biochemistry and Morphogenesis* (1942) which includes a long section on genetics (see below): "A morphogenetic field is a system of order such that the positions taken up by unstable entities in one part of a system bear a definite relation to the position taken up by other unstable entities in another part of the system. The field effect is constituted by their several equilibrium positions. A field is bound to a material substratum from which a dynamic pattern arises. It is heteroaxial and heteropolar, has recognizable separate districts, and can, like a magnetic field, maintain its pattern when its mass is either reduced or increased...The morphogenetic *gradient* is a special limited case of the morphogenetic field" (**Needham, p. 684**). But what were these "unstable entities?" Were they waves, as in an electromagnetic field, or gradients of small ions or proteins? The notion was rather like the "Emperor with no clothes." The theory collapsed. John Opitz summarized its demise: "In one of the most outstanding developments in Western scientific history, the gradient-field, or epimorphic field concept, as embodied in normal ontogeny and as studied by experimental embryologists, seems to have simply vanished from the intellectual patrimony of Western biologists" (Opitz, 1985). It was genetics which would eventually lead to the discovery of the molecules involved in positional identification, inductive gradients, and the organizer. These are present at such minute concentrations that the methods of classical biochemistry could not identify them. It was through developmental genetics, which identified the genes responsible, and molecular biology (see Chapter 9), which identified the products of these genes, that these mysteries were solved.

In many ways, development of eye pigments in insects, being studied by Caspari, Ephrussi, and Beadle, could be considered to be studies of development, but they did not deal with the structure of the eye, a cluster of ommatidia, each with a variety of cell types. These studies were relegated to "physiological genetics." Studies in vertebrates with mutants that affected the structure of the adult form, such as the mouse tail in Dobrovalskaia-Zavadskaia's studies in the 1920s, detailed in Chapter 1, and Landauer's studies in chickens (detailed by Oppenheimer, 1981) of the *Creeper* genes (shortened limbs) in the early 30s, clearly showed genetic effects on development. However, Goldschmidt used only invertebrate and protista examples in his book, *Physiological Genetics* (1938), to argue that variable timing of gene expression was what was important in the genetic control of development. Homeotic mutants of *Drosophila,* in which body parts appear at inappropriate place such as a leg at the site of the antenna (*Aristapedia*) were seen as due to inappropriate timing of gene expression. Heat shocks which could change pigment patterns or eye color were also seen as working through altered timing. In addition, Goldschmidt cited Hammerling's nuclear transplants in *Acetabularia,* in which the source of the nucleus determined the complex shape of the differing cap structures, as indicating the key role of genes.

By 1942, in the book mentioned above (**Needham, 1942**), the importance of genetics in development could be more firmly stated. In this nearly 700-page review of embryology, emphasizing quantitative experiments, 90 pages were devoted to the role of genes. Needham, like Goldschmidt, did not see a boundary between physiological and developmental genetics. "It is difficult to consider the action of the genes which cause morphological effects in abstraction from the action of those which govern reaction rates, the expression of which is in terms of the production of coloured substances or the accumulation of chemically defined storage materials" (**Needham, p. 341**). In line with his belief in the morphogenetic field, he felt that genetic lethals were due to defects in organizers, stating that: "...in many of those cases where the mechanism of genetic lethality has been investigated from the embryological point of view, there is ground for the belief that the effect is due to interference with organizer phenomena" (**Needham, p. 365**).[2] His analyses of mammalian genetic lethals focused mainly on *otocephaly* in guinea pigs, *shaker-short* in mice, and, as he called it, *short-tail (Brachyury, T),* for which his analysis depended on Chesley's, rather than Gluecksohn-Schoenheimer's, work (see below). He did not mention Ephrussi's work with cultures of the embryonic material from the *T* mutants, which showed that cell death was not intrinsic, i.e. that cell death did not occur in culture within the time span within which it occurred in vivo (see below), thus supporting his notion of an organizer defect. He thoroughly reviewed Landauer's work, mentioned above, on *creeper,* in chickens.

Scott F Gilbert, in his thoughtful review of the origins of developmental genetics (Gilbert, 1991), sees two individuals as being the most instrumental in creating it: Salome Gluecksohn-Schoenheimer (later Gluecksohn-Waelsch) and Conrad H Waddington. The two were very close friends (see minibiography, Salome Gluecksohn-Waelsch) such that they were certainly influencing each other. Although Paul Chesley had stated the role that morphological analysis could play in analyzing the role of genes in development (see next section), it was Salome Gluecksohn-Schoenheimer who more clearly enunciated the paradigm of using mutants to unravel the roles of genes in development. In her classic paper (Gluecksohn-Schoenheimer, 1938a) she stated: "A mutation that causes a certain malformation as the result of a development disturbance carries out an 'experiment' in the embryo by interfering with the normal development at a certain point. By studying the details of the disturbed development, it may be possible to learn something about the results of the 'experiment' carried out by the gene." Gilbert emphasizes Salome Gluecksohn-Schoenheimer's unique background for creating this synthesis: "In Gluecksohn-Schoenheimer...we see the merging of Spemann's embryological concepts (induction, regulation, etc.) with the particulate gene theory of the Morgan school— the merging of Freiburg and Columbia "(Gilbert, 1991). Virginia Papaioannou, writing in 1999, recognized that Salome Gluecksohn-Schoenheimer's 1938 paper was not a first for its content but found other reasons for considering it record setting: "The article reported early (and beautiful) work from one of several prominent women developmental biologists who were highly influential role models for aspiring women scientists, especially developmental biologists, throughout the following decades" (Papaioannou, 1999).

Salome Gluecksohn-Waelsch, 1907-2007

Salome Gluecksohn was born in Danzig, Germany (now Gdansk, Poland), which was a center of diverse populations (it was an independent, Free City from 1920-1939 and many Jews escaped abroad from it). Although it was a multi-national city, it teemed with anti-Semitism. She said: "I had to fight on the street against the children who ran after me and sang dirty and anti-Semitic rhymes"[1]. Her father was a grain merchant and had been sent to Russia in 1918 "by the German government to help re-establish trade relations between the two countries" (Gluecksohn-Waelsch, 1994). While there, he contracted "Spanish" flu during the great influenza pandemic. Salome

Salome with son and baby granddaughter, 1988, courtesy of Peter Waelsch.

vividly recounted his being brought home from the train station lying down in a farmer's cart, too sick to sit upright[2]. He died at home only a few days later.

The family fortunes were brought low with the massive inflation after WWI. The family had moved to Konigsberg, the former capital of Prussia. As Salome later recalled, "I had completed several semesters studying Zoology and Chemistry in Konigsberg, my home town in East Prussia"[3], but she had started in classics. She said:

"As a young woman, I was an ardent Zionist. I was determined to go to Israel and be a teacher. I went to the university to study classic languages, Greek and Latin. But some friends of mine told me that the last thing they need to be taught in Israel at that time was classic languages. They said I would do very well to include biology among my skills for teaching, so I decided to take some biology lectures"[1].

She continued her education at the University of Berlin, where she was partially supported by the parents of another student, for whom she played the role of chaperone/adviser, as that fellow student was in a relationship with an artist about whom the parents were concerned, and whom the young lady later married[2]. She enjoyed her studies in Berlin:

"In particular, it was Experimental Embryology—now known as Developmental Biology—that stimulated my curiosity. I, therefore, decided to go to the University of Freiburg to find out if Professor Spemann, the leader in this field, would accept me as a doctoral student…Although Spemann accepted me as a student, I was assigned the most boring dissertation project, the purpose of which was that of providing the necessary basic information for the research of a young male student who was Spemann's most favored disciple" (Waelsch, 1994).

She certainly learned amphibian embryology. She described her experience as follows: "Spemann came to our lab every day, stopped at every desk and discussions between him and the rest of us were extremely instructive. Spemann had high standards of perfection, and we had to learn to live up to them. During the "season", i.e. that of Amphibian egg laying, all of us worked day and night and we shared results, interpretations, etc."[3].

Salome said that Viktor Hamburger was her special mentor. "He was Spemann's so-called assistant when I became a graduate student in Freiburg. Spemann turned over to him the close supervision of my dissertation research, and Viktor

[1]Salome Gluecksohn-Waelsch was extensively interviewed for a piece, "On Journeys Well Travelled", An Occasional Series, Yeshiva University, Office of Publicity.

[2]Personal communication, Salome Gluecksohn-Waelsch. I was lucky to have frequent contact with Salome from 1965 until she became incommunicado from Alzheimers Disease in the mid-90s. We had many conversations about her life history and these are the source of such statements. We were good friends and I prefer to use only her first name in this mini-biography.

[3]From a, to my knowledge, unpublished manuscript entitled "The Causal Analysis of Development in the Past Half Century: A Personal History" (Conference on "Embryonic origins and control of neoplasia", Dubrovnik, Yugoslovia, October 13-16, 1986).

saw to it that I remained on a straight path….His approach to science was broad, and he was the only one who provided us students with some introduction to the principles of genetics"[3].

Spemann, despite having been a student of Boveri's, was, as discussed in in this chapter, utterly oblivious to genetics. Thus, she quite enjoyed getting to know Waddington when he visited the lab (see in this chapter):

"I received much encouragement and infinite stimulation in thinking about problems of development from Waddington, who came to Spemann's laboratory in 1931 as a student from Cambridge….We became very close friends. He opened our eyes to the biochemical and molecular problems inherent in inductive interactions between cells. He also fully recognized the involvement of genetic mechanisms in developmental phenomena. Waddington remained one of my closest friends until the time of his death"[3].

Thus, she sought a post-doctoral position in genetics.

"My visit in 1932 to the Kaiser Wilhelm Institute of Genetics in Berlin, where Richard Goldschmidt was the Director [he had visited the Spemann laboratory[3]], turned out to be a total fiasco. I was not received by Goldschmidt: His so-called Assistant, the human geneticist Curt Stern [actually a Drosophila geneticist who was interested in, and wrote the first text book of, human genetics], who screened potential visitors, did not even permit me to enter the inner halls of the institute. When I told him of my desire to receive training in genetics in preparation for a career as a scientist, his response was, to quote his own words, 'You, a woman and a Jew—forget it'" (Waelsch, 1994).

She had met Rudolph Schoenheimer in Freiburg and he became her first husband. The marriage was stormy as he suffered from manic depression. He had graduated in medicine from Freidrich Wilhelm University, and then studied organic chemistry at the University of Leipzig, which stood him in good stead when, later, he introduced stable isotopes into metabolites to create molecular weight variants whose metabolism could be followed in vivo. This was followed by studies of biochemistry at the University of Freiburg, where he rose to become head of Physiological Chemistry. Among his accomplishments in Freiburg was the discovery of glycogen storage in the newly described Von Gierke's disease (Schoenheimer, 1929). He was in America lecturing when he received a telegram firing him, after Hitler's anti-Semitic laws were passed. Schoenheimer was in New York on April 1, 1933.

"That was the day Hitler proclaimed his first anti-Semitic law. All Jewish university professors were dismissed. In Germany, the issue of tenure had been totally different from what it is here. The moment you became a Privatdozent, that is, an assistant professor, in Germany, you had tenure for life. So this was the first time the government had broken old academic rules and traditions. He took the cable from his chief telling him that he was dismissed to the department of biochemistry at Columbia and got a job immediately"[1]

At Columbia he used the newly available stable radioisotopes to study metabolism. This work involved collaboration with David Rittenberg who was in the laboratory of Harold C Urey and Konrad Bloch. The work demonstrated the long half-life of serum gamma-globulin with a more rapid turn-over of serum albumin—these were exciting and unexpected findings. He committed suicide in 1941 but, after his death, the Edward K Denham lectures that he had prepared for this honorific series at Harvard were read for him, and published as a book, *The Dynamic State of Body Constituents*. Salome described him as one of those brilliant people who, no matter how great their accomplishment, are never satisfied.[2]

In New York, Salome again sought to gain training in genetics. As often happens in life, a chance event provided that opportunity.

"A party for Columbia University faculty included Rudolf Schoenheimer and myself, and I found myself sitting at dinner next to L. C. Dunn. I told him of my interests and he talked to me about certain lethal mutations in the mouse that he had just begun to analyze and that seemd [sic] to affect developmental mechanisms in early embryos. He was looking for a developmental biologist to carry out the relevant developmental studies. Even though he had no research support to pay me a salary, we agreed that in return for using my knowledge of experimental embryology in the analysis of the mutational effects on development, he would teach me genetics"[3].

Thus began a collaboration that was fruitful for both of them. She described Dunn as follows:

"DUNN fully realized the potential of this unusual genetic system [the T/t complex] for the analysis of a number of fundamental problems of gene transmission and expression. He did not share the naïve assumption, held at that time by some, of a simple one-to-one relationship between a gene and its phenotype far removed from primary gene action and he was aware of the great complexities of gene actions and interactions. DUNN (1939) [Dunn,

1939] gave credit to GOLDSCHMIDT (1938) for his attempts to deal with developmental effects of genes beyond the level of their transmission or structure. But DUNN felt that, in 1939, developmental genetics had not yet coalesced as a field" (Gluecksohn Waelsch, 1989).

Salome remained at Columbia for 19 years as a Research Associate. However, while still a Research Associate, despite being denied faculty rank because she was a woman, she became independent. After being denied the possibility of faculty rank she

"'called the NIH. And they suggested that I apply for my own grant, even as a research associate. And I did that, and I got the grant.' Buoyed by the NIH funding, Waelsch moved back to the College of Physicians and Surgeons. 'A very enlightened professor of obstetrics there, Howard Taylor, gave me a room in which to work....I was at last independent and totally undisturbed. My husband always said to me if you close the door in that room at P&S and you die in there they wouldn't find your skeleton for fifty years"[1]

After the 19 years as a Research Associate, she became a Full Professor at the newly founded Albert Einstein College of Medicine. At Einstein, she also became Acting Chairman of Genetics when Pontecorvo did not come to be founding Chairman as had been planned. She remained in this position until this anomaly was corrected in her last year before becoming Professor Emeritus. Describing her own personality during the open discussion of her paper at the meeting in which she gave her paper "The Development of Creativity" (Waelsch, 1994), a personality which allowed her to act under adversity at Freiburg and Columbia, she said: "I do not recall that I ever was ready to give up. I also consider myself a rebel, and therefore, when obstacles were put in my way, I tried to overcome them, possibly with unconventional means. I know I was never discouraged. I'm not discouraged now. Now my age is a great factor of adversity, but I'm not discouraged".

Salome's work on the *T/t* complex is described in the text of this chapter (4). In addition, she worked on a variety of other developmental genetic systems: *Sd* (Danforth's short tail), *Kinky,* and lethal albinos. The developmental defects of the posterior skeleton of all 3 loci, with urogenital abnormalities in the first two, provided a n unitary theme (see in this chapter). She saw the medical relevance of this work (Gluecksohn-Waelsh, 1962). She did not continue her work on the *T/t* complex after it became clear, in the late 1970s, that there were no such things as *t*-antigens (see Chapter 7).

Another major line of research towards the end of her career was concerned with the developmental defects found with radiation-induced deletions at the *albino* locus (reviewed in Gluecksohn-Waelsch, 1979). A complex array of biochemical and morphological defects were was found, but not a unitary mechanism. Only in the age of cloning was it discovered that these effects were due to the absence of fumarylacetoacetate hydrolase, the gene for which is located in the deleted region (Grompe, et al, 1993). It is surprising that maternal correction of the metabolic defect did not occur, since there were marked liver abnormalities in the newborn mice.

Salome's second marriage, to Heinrich Waelsch, was very peaceful compared to her first. He had been born in Brno, Czechoslovakia, studied medicine at the German university in Prague and went on to get a Doctorate in Chemistry from this university (Elke, 1967). He had become a Privatdozent at this university, but the family emigrated to the States in the early 30s.[4] At Columbia, he was Professor of Biochemistry and Chief of Psychiatric Research at the New York State Psychiatric Institute, and Director of its Neurochemical Laboratories.

"There was a largesse about Heinrich, which allowed him, with no compromise in regard to the precision, and logic in the laboratory, to roam wide into the implication of his work. He read deeply in neurophysiology and experimental psychology. His friends in the behavioral sciences were many, and his grasp of things behavioral instinctive and direct. He was profoundly interested in, and a shrewd judge of, people; and had a rare understanding for the realities of mental anguish" (Elke, 1967).

Salome, Heine, and their two children, lived in a large apartment in a Columbia University-owned building on Morningside Drive. It had a small kitchen, with a maid's bedroom and bath between it and a large living/dining room area and the 2 bedrooms looking out over the Upper West side and Harlem. On one occasion police entered the apartment abruptly to use the bedrooms to watch an on-going drug deal. The story of how they obtained such a lovely apartment is told by Lee Silver:

---

[4]Salome told the story that his family had had a property dispute with neighbors. When the legal judgment went against the family, they realized that it was only because they were Jewish, and they decided it was time to leave—earlier then many others.

"The apartment was given to her by Dwight Eisenhower, who was president of Columbia before becoming President of the country. Salome told the story that Eisenhower had read an article in the New York Times about one of her freakish mutant mice, and asked her to speak to him about her research. Salome said she had different canned talks about genetics for primary school students, high school student, and undergraduates and graduate students, and she decided that Eisenhower would get the most out of a primary school lesson in genetics, which she proceeded to give to him. He was so impressed that he decided that Salome and her family deserved to live in a better apartment than the tiny tenement to which they had been confined since their arrival in the country" (Silver, 2008).

At one end of the living area, a large portrait of Salome as a beautiful, mature young women was on the wall. A story of how attractive she was when a graduate student is told by Davor Solter:

"In one of the numerous Freiburg boutiques she noticed a beautiful pink cashmere sweater and could not resist the temptation to step in and try it on. Though it fit perfectly and she very much wanted it, she admitted that it was well beyond her means. The owner suggested she could keep it as a gift if she agreed to be photographed in it and that the enlarged photo could be placed in the store window as an advertisement for his wares. She agreed and her picture remained in the window for several months and was much remarked upon. She told this story with unconcealed glee. She was obviously proud of her good looks and even more of her daring: having one's photograph in a boutique window was not the thing for a proper graduate student to do" (Solter, 2008),

She remained a handsome woman and always was well dressed, usually wearing a large, carved jade brooch which Heine had given her.

She and Heine had two children, Naomi and Peter, who were much loved. There were frequent European trips and Salome enjoyed sharing her knowledge of Renaissance art with them. She credited both husbands with supporting her research: "Both my first and second husbands shared with me the excitement I derived from my research activities. They were both biochemists and their knowledge and expertise stimulated and helped my attempts of trying to interpret the experimental results in biochemical and later molecular terms" (Waelsch, 1994). Sadly, Heinrich Waelsch died of a brain tumor in 1966 and she never re-married.[5] She remained in the same Columbia University apartment on the upper West Side in which she and Heine had raised their children, until her death.

Salome was frequently seen as difficult by those people who could not meet her standards[6], and this may be a large part of the reason why only 3 students finished their Ph.D.s with her. She herself, as most mouse geneticists, spent a large amount of time in the mouse room breeding and weaning her stocks. Graduate students and postdocs also worked in the mouse room, helping her and taking care of their own stocks and she was a "stickler" for keeping the breeding book absolutely up-to-date. Despite her very high standards, however, she was very devoted to people who conscientiously performed their tasks, no matter their place on the social scale. Anybody who attended a scientific meeting at which she was present would remember her: she always sat at or near the front and was very attentive to the speakers. Her questions, asked with a slight German accent, were precise and sought the relevance of the presented work to fundamental questions.

She received many honors. She was elected to the United States National Academy of Sciences in 1979 and to the American Academy of Arts and Sciences in 1980. She was presented with the National Medal of Science by President Clinton in 1993. She was elected a Foreign Fellow of the Royal Society in 1995 but was not able to go to London for her induction. She received an honorary D.Sc. from Columbia University and was proffered an Honorary Doctorate from the University of Freiburg, with its Spemann Graduate School of Biology and Medicine, but turned it down with a letter reminding the university authorities of Spemann's anti-Semitism and anti-feminism. "On January 30, 1983, the fiftieth anniversary of Hitler's ascent to power, Freiburg's Dean of Biology appeared before the student body and read them Waelsch's letter"[1].

---

[5]Her brother, a well-known city planner, who had gone to Israel as Salome herself had initially wished to do, and then had emigrated to Australia, also died of the same brain tumor, glioblastoma multiforme.

[6]During a six-week research rotation in her lab in 1967, as part of my pediatric residency at Albert Einstein College of Medicine, I kept a scrupulous lab book, trying to meet her high standards. I later found out that she showed it to rotating graduate students or new technicians as an example of how a lab book should be kept. Sadly, my lab books have not continued to be maintained at that level.

Gluecksohn-Schoenheimer's interpretations of *T* and *t*-allele effects were limited by the mis-understanding of them caused by assigning them as alleles of the same locus (the resolution of which was discussed in chapter 3), so her analyses focused on unitary mechanisms. Thus, by 1949 she could say that "the study of this material makes it very likely that in mammals the notochord plays a role in processes of early organization similar to that of the notochord in amphibians as analyzed with the techniques of experimental embryology" (Gluecksohn-Scholenheimer, 1949a). Her work on the locus *Kinky*, linked to *T*, was also interpreted in terms derived from classical embryology. "Homozygous mutants of *Kinky* were found to have duplications of their dorsal axis, sometimes forming twin embryos....She interpreted the action of the *Kinky* gene as causing constrictions analogous to those done experimentally by Spemann and Holtfreter on salamander eggs which could also lead to double animals [Gluecksohn-Schoenheimer, 1949b]" (Gilbert, 1991). As further discussed (see below and in the minibiography), she was to apply these principles to other non-*T/t*-complex genes as well.

Conrad Waddington also had worked in paleontology and classical embryology (visiting Spemann, see minibiography of Salome Gluecksohn-Waelsch) before coming to genetics. His work in these areas was greatly influenced by philosophical considerations. "Waddington (1975) makes it quite clear that his initial thinking on the topic [development] derived from his reading of Whitehead's books on philosophy, particularly *Process and Reality* (Bard, 2008). His interest in genetics came from his friendship with Gregory Bateson, the son of Bill Bateson.[3] He had worked extensively with Needham (both were at the University of Cambridge) looking for the substance of the organizer. Spemann's lab had found that dead inductive tissue from the blastopore still worked for inducing the new axis, a finding Spemann tended to ignore. Waddington (1933) extended this finding to the chick embryo and found multiple substances that would substitute for the tissue (Waddington and Needham, 1934). Given the "diffuseness" of these results, he turned to genetics. His major work in developmental genetics was a thorough study of a series of mutants which affected development of the *Drosophila* wing (Waddington, 1940). However, his greatest contribution was towards an understanding of how gene expression could be channeled differently in different tissues. His use of "epigenetics" to name this phenomenon and his mathematical formulae of competing biochemical reactions provided a model of how "environmental" differences in different parts of the embryo could allow differential gene expression. This was important, as all genes might be active in all cells: "there are therefore some grounds for thinking that all genes may be active, and producing effects of some kind, in all the cells of the body, even in those in which, owing to the canalization of developmental process, the influences of the mutant alleles do not suffice to produce any divergence from the normal" (Waddington, p. 364). His figure of an epigenetic landscape where a ball rolling down a hill might end up in different valleys is ubiquitous in biology. This was canalization, which "is not unlike the current notion of 'developmental constraints'. Indeed, it was he who suggested [Waddington, 1938] in 1938 that all vertebrates were constrained to have a notochord, since that transitory organ induced the neural plate, and similarly it was he who claimed that birds (and presumably humans) had to have nonfunctional pronephric kidneys since they would give rise to the Wolffian ducts" (Gilbert, 1991).

Early studies of the developmental genetics of the *T/t* complex

As described in Chapter one, Nadine Dobrovalskaia-Zavadskaia had studied the formation of the mouse tail in adults, especially by examination with X-rays, and speculated on the mode of action of the *T* gene. However, she did not study embryonic development. Lang (1914) had studied the development of short tailed mice in utero but, according to Chesley (1935), it seemed probable that heterozygotes were being studied, or a mutation at a different locus, since no resorptions (uterine sites of previous implantations, which would be expected with most early lethals unless the lethality was so early that implantation did not occur) were found. It was Paul Chesley who first studied mouse embryos homozygous for *T* (Chesley, 1935). His opening sentence might be taken as the motto for developmental genetics: "Embryological studies on material bearing known mutations offer an opportunity for a better understanding of the processes of development" (Chesley, 1935). If it were not for his untimely death by suicide, he, and not Salome Glueksohn-Waelsch, might be lauded as the prime mover in mouse developmental genetics (but Nadine Dobrovalskaia-Zavadskaia has a claim, too, see her minibiography).

Chesley, with the help of Dorothy Chesley, performed 66 timed dissections, separated by as little as 4 hours (mouse embryonic development is very rapid), obtaining 615 embryos from *T/+* X *T/+* crosses. These were all prior to 11 days gestation, which was the time he had previously shown that they died (Chesley, 1932). These were studied methodically with external examination and, many, with histology. The expected one-fourth were afflicted with developmental abnormalities. He found that the embryos were normal up to 8 ½ days, when there were 4 somites ("blocks" of mesoderm along the vertebral column; their number is a standard way of staging embryos). At 8 somites, midline blebs and irregularity of the neural tube were grossly visible in the affected embryos. As the embryo flexed (the tail approaches the head for a short time in development), the neural tube became more markedly irregular. Histological sections demonstrated that the notochord was absent and no posterior limb buds appeared, which indicated that there were much greater abnormalities posteriorly than anteriorly. The pericardial cavity was enlarged and heart development was delayed with the presence of abnormal vessels—he thought that poor blood circulation could be the cause of the pyknotic nuclei (condensed and darkened nuclei indicating cell death) which were seen.

Chesley (1935) also studied *T/+* heterozygous embryos which became distinguishable at 11 days by a constriction of the tail. The posterior notochord was reduced and frequently had diverticula, deficiencies consisting of blind-ended sacs, with abnormalities of the over-lying neural tube above these defects. Granules thought to be from pyknotic nuclei and hence indicating dying cells were found in the distal tail which sometimes persisted as a small filament, as had been described by Dobrovolskaia-Zavadskaia. Chesley (1935) concluded, after comparing his results to extirpation studies in amphibians, which indicated that the notochord was essential for neural tube development (Lehman, 1928; Holtfreter, 1933), that: "This is apparently the first instance reported of a genetically controlled condition in which results similar to those obtained by the methods of experimental embryology are encountered. The findings substantiate the experimental results. Moreover, they show that by utilizing changes in the germ-plasm of the organism the inter-relationship of processes can be demonstrated." This is a resounding "call to arms" for developmental genetics.

Boris Ephrussi, who was to make contributions to many aspects of genetics, was fascinated by the problem of embryonic lethality. He asked: "At what level of organization does the action of a lethal gene manifest itself? Is it the cell or tissue level, or the level of organic correlations? Are all cells constituting a lethal embryo inviable or is there a special kind of cell particularly affected by the lethal gene? Have the cells of a lethal embryo normal differentiation potentialities?" (Ephrussi, 1935). He chose to use tissue culture techniques, dissecting embryos aseptically into warm Ringer's solution (a physiological solution of essential salts) and culturing the heart or entire embryo either in Carrel flasks (small transparent "bottles;" Ephrussi, 1933) and/or hanging drops ("hanging" under a cover glass over a chamber of liquid to keep them moist; Ephrussi, 1935). Many of the experiments were performed in Morgan's lab at Cal. Tech. while Ephrussi was a Rockefeller Foundation Fellow there. He found that the hearts would contract for two weeks (and maybe longer; a lab accident had terminated the experiment) while portions of the embryo developed nests of cartilage at ten days. He concluded that there was no inherent lethality of the mutant cells and that "the death of lethal embryos is due exclusively to internal factors" (Ephrussi, 1935) which is what Chesley had suggested. It is interesting that Dorothea Bennett later found that the T/T neural tube could induce cartilage formation in $+/+$ but not T/T somites (Bennett, 1958, see below) which raises questions about which components of Ephrussi's culture media (it included "aged embryo extract") might have been the inducer of cartilage in his experiments.

In the paper discussed above as "foundational" for developmental genetics, Salome Gluecksohn-Schoenheimer extended Chesley's (1935) studies on the development of $T/+$ mice (Glueckson-Schoenheimer, 1938a). She found that there was no notochord in the developing tail while more anterior abnormalities of the notochord were sometimes found (hindgut-notochord connections which do not occur normally, probably explanatory of some $T/+$ newborns found with no anal openings). She compared her results to that of Goldschmidt (1937) with wing mutants of *Drosophila* which develop normally to a certain point and then have degeneration of the margin of the wing, but she doesn't mention Ephrussi's work (above) on T homozygous mutants. She was to replicate Ephrussi's experiments done in tissue culture by studying the development of T homozygous mutants in the extraembryonic coelom of the chicken, which is a benign environment where xenografts (grafts between species) survive for some time (Gluecksohn-Schoenheimer, 1944). These were explanted at day 7, well before any abnormalities of T/T development are visible and 19 out of 81, a good fit to the expected ¼, were abnormal after two days. These showed an abnormal allantois (the mesodermal sac which will become the umbilical cord), underdevelopment of head and trunk, and an absence of somites. The transplanted tissues were only maintained for up to 47 hours so the results did not explore the lethality as had Ephrussi's.

Gluecksohn-Schoenheimer's investigations continued with descriptions of the embryonic defects in homozygous t-alleles. She found that t^1 homozygotes died before implantation while t^0 homozygotes died at 6 days (Gluecksohn-Schoenheimer, 1938b). She provided a detailed histological study of the t^0 homozygotes, finding that they were normal until 5 days, but that growth then ceased, leaving an unorganized inner cell mass (ICM, the few cells that will form the embryo found as a cluster at one inner side of the blastocoele cavity) surrounded by a thick layer of endoderm (Glueckshoen-Schoenheimer, 1940). She interpreted this as a failure of mesodermal formation and concluded that there was a relationship to the abnormalities of T/T with its dorsal notochord (a mesodermal structure) defects. Although t^1/t^0 compound heterozygotes were initially thought to be normal, for example., that t^1 and t^0 were complementary, further study showed that the phenotype of t^1/t^0 embryos varied from death before day 9 (abnormal embryos with head, neural, or heart abnormalities) to viable offspring (Pai and Gluecksohn-Waelsch, 1961). The earliest *t*-allele lethality found was that of t^{12}, homozygotes of which had nucleolar abnormalities and decreased RNA concentrations at the 30-cell stage (Smith, 1956). Selma Selagi (1962) Fig. 4.1 described the partial complementation which could occur between t^x/t^y embryos which resulted in longer-lived, but abnormal, fetuses.

Other early developmental genetic models

Gluecksohn-Schoenheimer's work on T/t complex genes branched out to other mutations with sometimes overlapping phenotype effects. These included a mutation, *urogenital (u)*, found in 2 different T/t balanced lethal stocks (Dunn and Gluecksohn-Schoenheimer, 1947) and Danforth's short tail, *Sd*. The new mutation, *u*, as the name indicates, affected the urogenital system, kidney, genital and urinary ducts, and the bladder. There was great variability of expression which was strongly modified by interaction with *T*. In fact, it probably would not have been discovered if it had not occurred in the balanced *T*-lethal stocks: *u/u* showed nearly full penetrance on the T/t^0 genetic background and only 5% on the normal background—a very strong example of the influence of modifying genes. The authors felt that "It seems to us that the chief interest of the observations...lie in the developmental relationship between the structures affected, which the study of 2 recent mutations in mice [the other being *Sd*] has revealed in so striking a manner" (Dunn and Gluecksohm-Schoenheimer, 1947).

Gluecksohn-Schoenheimer's work on *Sd* focused on development of the kidney. Homozygous *Sd* embryos have no kidneys and die soon after birth. Her early morphological studies showed a failure in formation, growth, and differentiation of the ureteric buds (Gluecksohn-Schoenheimer and Dunn, 1945). Following Ephrussi (and others, see above), she studied the development of kidney rudiments in tissue culture (Gluecksohn-Waelsch and Rota, 1963). In vitro, differentiation of the homozygous *Sd* mutants started later, did not proceed as far, and fewer rudiments developed (50%) than with tissue from wild-type mice (100%). She also performed mixed cultures with normal and mutant fragments of spinal cord (as an inducer) and kidney rudiments, finding even better differentiation of mutant ureteric buds with mutant spinal cord than had been found in the first cultures. This seminal work suggested that the full potential for differentiation was present in the mutant, but that differences in timing of gene expression or amount of "inductive" substance caused the in vivo abnormalities.

The interactions of these mutants, along with *Fused*, sometimes created monsters including sirens (a single leg with urogenital abnormalities), aprosopi (reduction or complete absence of

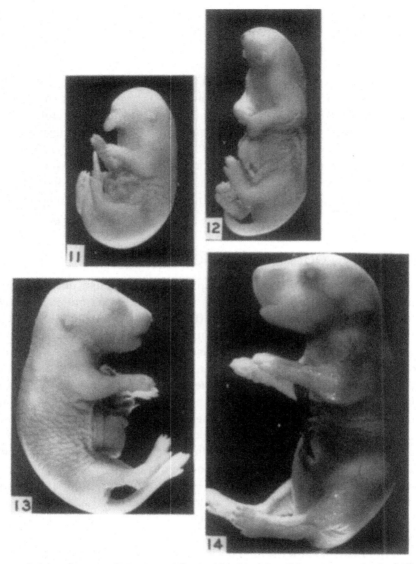

FIG. 4.1 Gross morphology of mutants heterozygous for combinations of t^n alleles. 11, otocephatlic t^1 /t^{12} embryo with absent eyes, mouth and forebrain with proboscis-like outgrowth. 12, more extremely otocephalic embryo t^8 /t^{12} embryo. 13, relatively normal t^1 /t^{12} embryo. 14, tailless T/+ newborn. From Silagi (1962) with permission.

some head structures) and severe intestinal abnormalities (absence of portions of the intestine, all of which could be explained by failure of the normally externalized intestine to be returned to within the body cavity, with subsequent degeneration) (Glueksohn-Schoenheimer and Dunn, 1945).

The general approach of using mouse mutants as embryological experiments gradually became more widespread. Robert Auerbach, for his doctoral studies at Columbia University (acknowledging Salome Gluecksohn-Waelsch for suggesting the problem) studied *Splotch*, a pigmentary mutant with homozygous lethality (Auerbach, 1954). The lethal embryos died at day 13 with defects and overgrowth of neural tissue and tail abnormalities. Given the role of the neural crest in providing precursor pigment cells and in the developing nervous system, he suggested that the mutation caused a primary defect in the neural crest. He followed others who used embryo culture or transplantation as supplements to histology, and found that neural crest tissues transplanted to the chicken coelom or the anterior chamber of the mouse eye never developed pigment (Auerbach, 1954). Green's work on *dominant hemimelia, Dh*, again suggested occurrence of an inductive defect, as the splanchnic epithelium of the *Dh* embryo was absent or abnormal in the 9 ½ day embryo with lack of spleen and hind limbs (Green, 1967).

Dorothea Bennett, also at Columbia and working with Dunn analyzed another pigmentary mutation, *Steel*. She found that Steel displayed pleiotropic effects (diverse effects on phenotype; Bennett, 1956). This did not easily show a "pedigree of causes," which Gruneberg thought could usually be found (Gruneberg, 1943). The heterozygotes had diluted pigment with a white blaze, diminished numbers of germ cells, and a macrocytic anemia. There were no germ cells in homozygotes and the anemia was considered to be lethal. Skin grafts from homozygotes to normal mice never produced pigment. There were many similarities to some alleles of *W*, *dominant spotting*, an unlinked locus studied extensively by Elizabeth (Tibby) Russell (Coulombre and Russell, 1954; Russell et al., 1956). This gene was eventually found to code for the receptor (*kit*) for a ligand coded for by *Steel* (Flanagan and Leder, 1990).

In 1960, Hans Gruneberg could list 42 mouse mutants with early embryonic effects (Gruneberg, 1960). Nine of these were from the *T/t* complex while many other mutants affected the tail or skeleton. He himself published a series of about 30 articles on skeletal mutations in the mouse. However, the period of merely describing the histological features of the abnormal development was coming to an end, as more modern analytical tools and methods were being used. Gluecksohn-Waelsch summarized a number of embryological processes, such as inductive interactions and cell differentiation, which could be interrupted by genetic mutations in mice in a lecture at the Eleventh International Congress of Genetics in 1993 (Gluecksohn-Waelsch, 1994).

New tools for mouse developmental genetics

The use of tissue culture (see above) and embryo manipulations expanded the arsenal of techniques available to the mouse developmental geneticist. A frequently neglected aspect of the history of mammalian developmental genetics is the effort involved in learning to manipulate the pre-implantation mammalian embryo. First, the basic physiology of murine pregnancy needed to be established and various investigators, including Sir Allen Parkes[4] (Parkes, 1926), started these studies in the 3rd decade of the 20th century (Allen, 1922). These advances provided information that allowed investigators to transfer embryos between mice—it required pseudopregnancy, achieved by mating females with vasectomized males. This was first performed for cancer studies (Fekete and Little, 1942) and then for studies of effects of the uterine environment of different inbred strains on development of embryos from other strains (Fekete,

1947). At that time, the embryos were kept in a simple salts solution (Locke's solution) before transfer. Major advances in media composition were needed to allow in vitro culture of mouse embryos. W K Whitten (1956) developed a complex medium which allowed 8-cell embryos to develop to blastocysts, and McLaren and Biggers (1958) showed that such cultured blastocysts could develop into normal mice when transferred to the uteri of pseudo-pregnant females. Further improvements by Whitten (1971) and Whittingham (1971) in the composition of the media with lower osmolality and no glucose (i.e., substrates for anaerobiosis were needed) and the use of 5% oxygen/5% carbon dioxide/ 90% nitrogen instead of 5% carbon dioxide/room air (20% oxygen causes oxygen toxicity for cells/tissues accustomed to 5% oxygen in vivo). The purity of the water used was crucial, and water from Sieur de Monts Spring in Bar Harbor, Maine (the home of the Jackson laboratory, where much of this early work was performed) was particularly good. Culture of fertilized eggs from the 1 cell stage was blocked at 2 cells (the "2 cell block") but this could be overcome by using fertilized ova from F_1 hybrids between 2 different inbred strains (Whitten and Biggers, 1968). [5] Finally, the ability to operate on the embryo, inserting cells for clonal analysis, was especially developed by Richard Gardner. He has described his development of the appropriate tools and methods (Gardner, 1998). Of interest is that his initial attempts were on the much larger rabbit blastocyst, but it did not tolerate these manipulations nearly as readily as did the mouse blastocyst.

The development of teratocarcinomas as models of mouse development was another important advance. The name teratoma refers to the capacity of these tumors to produce "monstrous" growths with differentiated tissue derived from all three germ layers. In humans such tumors occur spontaneously and are clinically important. The ones derived from the female germ line are much more benign that those derived from the male germ line (Erickson and Gondos, 1976). Leroy C. Stevens was the pioneer of their study in mice. He had done his Ph.D. research on the neural crest origin of pigment cells in amphibians under the guidance of Johannes Holtfreter at the Univ. of Rochester (Bagnara, 1993) but continued his developmental studies with mice at the Jackson Laboratory. Spontaneous testicular teratomas occurred much more frequently in the 129/J strain (Stevens and Little, 1954). The malignant potential of these tumors was shown to reside in single cells—so-called embryonal carcinoma cells which are pluripotent, i.e, can give rise to most of the kinds of cells found in mice (Kleinsmith and Pierce, 1964). The differentiated somatic derivatives of these cells, however, are benign (Pierce et al., 1960). When a transplantable teratoma was passed in the ascites form (injected into the peritoneal cavity, where they grow in the peritoneal fluid), embryoid bodies developed from single cells (Jami and Ritz, 1974). The embryoid bodies partially resembled six-day old mouse embryos with embryonal carcinoma cells surrounded by endodermal cells. These tumors could be maintained in culture as hypodiploid (they have lost some chromosomes) teratoma cell lines which frequently produced slower-growing differentiated structures representing many types of tissue (Lehman et al., 1974). An in vivo, one-way differentiation of embryonal carcinoma cells to nontumorigenic cells had occurred (Martin and Evans, 1974). The ability of the normal embryonic environment to control the malignant potential of teratocarcinoma cells was shown by Brinster (1974) and others (Mintz and Illmensee, 1975; Papaioannou et al., 1975), in experiments in which they injected teratoma cells into blastocysts. The adult animals eventually resulting from the injections provided evidence that small colonies of apparently non-malignant cells had been established and were distributed in various tissues. It was also found that mouse eggs transplanted to extrauterine sites revealed properties similar to those of teratoma lines, in that they are able to develop differentiated tissues while still maintaining undifferentiated embryonal carcinoma cells (Stevens, 1968), and about 50% of embryos grafted to extra-uterine sites evoke potentially malignant tumors (Damjanov et al., 1971). These observations strongly argued that mutation does not play a major role in the malignancy of teratomas. The finding that transplanted parthenotes (XX parthenotes have limited viability; Iles et al, 1975) generate a variety of different tissues but no malignant tumors suggested the possible importance of a male gametic contribution (requiring a Y chromosome) for the malignancy of teratomas. Since, in those times, very little was known about the origins of cancer (multiple somatic mutations in cancers had not yet been discovered), this work was strongly supported by research grants from cancer societies and the NIH in the United States.

These experiments, in which teratocarcinoma cells were injected into blastocysts, were creating chimeras—the name given to legendary creatures with parts from several animals. The technique of making chimeras by fusion of two eggs was pioneered by Tarkowski (1961) whose work was soon followed by that of Mintz (1962a, 1962b), while Gardner (1968) developed the microsurgical techniques for injecting single cells into the blastocyst. "The term 'chimera' was used by Tarkowski; Mintz coined an alternative term, "allophenic", to refer to animals derived from embryo aggregation; however, Hadorn had previously used the terms 'allophene' and 'allophenic' in a different sense, referring to any characteristic of a cellular system which derives not from its own genetic constitution but from the genetic constitution of some other cellular system" (**McLaren, 1976, p 6**). 'Chimera' became the generally accepted term. Such chimeras allowed studies of the degree of cell autonomy shown in various characters, and for determining cell lineages. Initially, pigment differences were the major marker for identifying different cell types (only useful for normally pigmented structures such as skin or the eye) but soon viable, balanced chromosomal translocations, identifiable in all dividing cells, were also used as markers (Mystkowska and Tarkowski, 1968) Later, color-producing enzymes (beta-galactosidase, requiring ex vivo staining—Rosa 26, Friedrich and Soriano, 1991) or green fluorescent protein (GFP) only requiring ultraviolet light to visualize (Okabe et al., 1997) allowed less laborious identification of the genetically different cell types in chimeras.

In the cloning era (see Chapter 9), the techniques of transgenesis (Gordon et al., 1980) and embryonic stem cell knockouts (see below) and, later, knockins (replacing the normal copy of the gene with a mutated copy instead of just removing the normal copy), became available as tools for developmental biology. Transgenic mice frequently produce the protein product of an introduced gene, depending on the genetic control regions introduced with it (usually a tissue-specific promoter). Thus, if overproduction of the protein product creates abnormalities of development, transgenics can provide models of abnormal development. For instance, over expression of TGF-alpha caused abnormal pancreatic and mammary gland development (Jhappan et al., 1990). Using tissue specific promoters, expression in abnormal locations can also perturb development, for example Charite et al, (1994). However, these techniques were soon overtaken by embryonic stem cell knockouts. The development of this methodology is described in Chapter 9.

The advent of the cloning era (see Chapter 9) brought tremendous new advances to mouse genetics and mouse developmental genetics. Not surprisingly, pigmentation genes were some of the first to be cloned. Brilliant, et al. (1991) identified the gene for the classic *pink eye dilution* mutation by genome scanning. This turned out to be the gene in which human mutations cause Type II albinism, the most common form. The cloning of murine developmental genes was greatly aided by advances in *Drosophila* developmental genetics. Saturation mutagenesis was performed for developmental genes—this work led to the Nobel Prize in 1995 to Christiane Nusslein-Vollhard and Eric Weischaus, along with Edward Lewis for his proceeding work on homeotic genes (mentioned above, further discussion follows). Their work led to the discovery of many hundreds of developmental genes many of which had homologs in mammals. The historian of science at the Institut Pasteur, Michel Morange, sees these developments as paradigm shifting with the construct of the "developmental gene concept" (Morange, 1996). Saturation mutagenesis for developmental genes was later performed in mice (see Chapter 6).

A particularly instructive group of developmental mutations found were in the homeotic genes which became known, from a shared genetic element, as *Homeobox,* or *Hox* genes. The members of the one set in *Drosophila* provide positional information for cells in the developing embryo. They show partially overlapping expression patterns which are collinear with their order in the complexes, especially in mammals which have 4 linear sets, each with some losses or duplications, while the single set in *Drosophila* is separated into 2 chromosomally distant parts. The history of these genes, with an emphasis on *Drosophila,* is well covered by Gehring (1994); also **Duboule, 1994**. Homologs were discovered in mice by McGinnis et al. (1984) and many other organisms and the degree of functional conservation was amazing—the human homolog could substitute for its *Drosophila* counterpart in vivo (McGinnis et al., 1990).[6] Many of the homeobox genes in mammals have gained functions quite distinct from providing positional information. Another set of genes containing a highly-conserved motif are the *T-box* genes.

One of the advances in the 21st Century was to establish the fact that the mammalian egg has polarity by using single cell marking with GFP (Piotrowska et al., 2001). The negative side of the experiments showing that 1 cell of the 2 cell embryo could develop into a normal embryo was that the other cell could not. This was because only the "animal" pole, and not the "vegetal" pole cell could do so. Such cell marking and tracking experiments provided detailed example of the complex cell lineages in the developing embryo (for a highly readable account of these and other experiments, **Zernicka-Goetz and Highfield, The Dance of Life**). The place of sperm entry highly influenced this asymmetry (Piotrowska and Zernicka-Goetz, 2001) and was related to asymmetric location of molecular markers such as coactivator-associated arginine methyltransferase (Carm1; Hupalowska et al., 2018). This histone modifying enzyme, requiring DNA binding factors, indicated that epigenetic differences were present as early as the 4-cell stage. This pathway of early development was found to continue with variable expression of *Sox2* studied by, among other things, the length of time that Sox2 was bound to DNA (White et al., 2016).

Major advances occurred when the aggregation of embryonic stem cells (ESCs) into embryoid bodies was studied. The self-organizing properties of ESCs in these organoid cultures allowed spontaneous development of a pro-amnionic cavity (van den Brink et al., 2014). Only when the 3 early embryonic cell types (ESCs, trophoblasts, and extraembryonic endoderm cells) were cultured together in the correct numeric combinations did a model of the gastrulating embryo occur (Sozen et al., 2018).

Another major 21st Century development was using the knowledge obtained from studying embryonic stem cells to reverse the differentiation of somatic cells to re-acquire pluripotency, using vectors to elicit expression of a cocktail of early induction factors (Okita, et al. 2007). These induced pluripotent cells (iPSCs) could be prepared from individual patients with a variety of diseases and induced to form a variety of cell types, e.g. neurons or bone marrow stem cells, for study or re-introduction to patients. The combination of human iPSCs with optogenetics (using genes modified by attached chromophores and and light activated or inactivated to watch gene effects in vivo) allowed amazing abilities to analyze rhythms of gene expression, such as the somite forming oscillations in human cell cultures (reviewed in Palla and Blau, 2020). Single cell genomic analyses advanced to such a stage such that the chromosomal organization of embryonic versus adult cells could be studied with "chromosome confirmation capture" establishing the timing of development of local "topical logical-associated domains" (Collombet et al., 2021). These approaches were also important for studying complex human developmental disorders. The us of "Hi-C" (high-throughput chromosomal confirmation capture which involved formaldehyde cross-linking of such DNA associations (TADs, topologically associating domains) on as few as a hundred cells, followed by high-throughput sequencing of captured co-fragments) allowed an understanding of altered gene regulation due to new regulatory interactions in children with complex chromosomal re-arrangements (Melo et al., 2020).

Studies of *T/t* complex developmental genes, continued

Whichever one of the three "pioneers" of mouse developmental genetics (Dobrovalskaia-Zavadskaia, Chesley, or Glueckson-Schonheimer) one wants to choose as the "founder" of mouse developmental genetics, they all chose the same material for some or all of their studies—mutants at the *T/t* complex. Thus, it is fitting to use these mutants to pursue the history of advances in mouse developmental genetics. The mutants also attracted a number of developmental biologists who hadn't previously been involved in *T*-complex genetics. As Virginia Papaioannou described it in the subtitle of a review, "the *T* complex educated a generation of developmental biologists" (1999). The histories of studies of pre-implantation lethals, post-implantation lethals, and *T/T* lethals are separated.

Pre-implantation lethals

The early lethals did not provide much material to study under the light microscope but were explored with biochemical, electron microscopical, and embryo culture techniques. The t^{12} and t^{w32} mutations, when homozygous, led to developmental arrest at the morula stage (when the embryo is a "solid ball;" Smith et al., 1956). These two mutations belong to the same complementation group, i.e., the compound heterozygotes of t^{12}/t^{w32} also arrest at the morula stage and no live-born offspring with that genotypic combination are found. Nonetheless, the two mutations were distinguishable by electron microscopy (Hillman and Hillman, 1975) and, therefore, are not identical. The Hillmans in the Department of Biology at Temple University represent some of the new recruits to the study of the T/t complex mutations, as did Lois Smith at the Jackson Laboratory. Many of the abnormalities visible by electron microscopy, such as aberrations of mitochondrial morphology and lipid inclusions, were compatible with some derangement of intermediary metabolism.

These mutations were studied extensively. Early electron microscopy only demonstrated degenerative changes in these embryos (Calarco and Brown, 1968). Two basic approaches were used to better understand these mutations. In one, the embryos were analyzed before the developmental arrest at the morula stage. Groups of normal embryos, from control crosses, were compared to groups of embryos derived from t^{12} mouse crosses where one-third of the embryos were expected to be abnormal because of transmission ratio distortion—the higher rate of fertilization by t-allele, compared to +-bearing sperm, discussed in Chapter 2. The second approach involved comparing delayed morulae (t^{12}/t^{12}) to littermate early blastocysts (T/t^{12} and +/+).

Utilizing the first approach, both RNA and protein synthesis were found to be quantitatively normal (Hillman and Tasca, 1973; Erickson, et al., 1974). However, ATP levels studied by the first approach were found to be increased in pre-blastocyst embryos from t^{12} litters as compared to normal embryos. ATP levels then dropped in post-blastocyst embryos from t^{12} litters to below normal values (Ginsberg and Hillman, 1975). Using a very sensitive firefly luciferase method which allowed ribonucleotide determinations on single embryos, this drop in ATP levels was only seen when degeneration of the embryo was starting (Spielmann and Erickson, 1983.) The earlier results supported the notion that the mutant embryos had a derangement of intermediary metabolism. Studies done following the second approach of comparing distinguishable embryos demonstrated that compaction, the process in which tight junctions are formed among the outer cells of the embryo in preparation for blastocoel formation, is frequently deficient in t^{w32} homozygotes (Granholm and Johnson, 1978). This also appeared to be the case when t^{12} homozygotes were studied in chimeras with littermate controls: the blastocysts of "mutant" chimeras were less able to control the hydrostatic pressure (formed as water followed ions, pumped by membrane transporters, into the blastocoel cavity which develops in the blastocyst). As a result, the blastocysts ruptured, suggesting poorer connections between the mutant cells and nearby normal cells (Mintz, 1964). The effect of t^{12} beyond the blastocyst stage in these chimeras was unknown, since the embryos were not transferred back to pseudo-pregnant females. There was a suggestion that intermedia filaments were not induced in the t^{12}/t^{12} (Nozaki et al., 1986) embryos but the relationship of this to the early death seemed unexplained.

The t^{w73} mutation was the sole member of a complementation group, i.e., it would interact with all the other t-allele lethals that were tested at the time (another member of this complementation group was found later), to produce some live-born offspring. Although the mutation was frequently characterized as one which failed to implant, this was also true of t^0 homozygotes. A major difference from t^0/t^0 is that the trophectoderm (the outer cells of the embryo destined to form part of the placenta) seemed to have a lesser proliferative capacity in t^{w73} homozygous mutants than in normal embryos (Spielgelman et al., 1976). Despite this, during in vitro culture, the outgrowth of the trophoblast was similar to that of normal embryos (Axelrod et al., 1985).

There were many overlaps in the morphology of t^0 homozygous and t^{w73} homozygous embryos. Such results suggested possible metabolic abnormalities which could be corrected by the culture media and/or conditions. Defective development due merely to a metabolic block was not an interesting finding for developmental geneticists. Luckily, the mutants delayed at later stages revealed problems more specific for unique cellular mechanisms in embryogenesis.

The t^0 and t^6 mutants belonged to the same complementation group and showed very similar histology. True implantation cannot be said to take place; the ectoplacental cone did not proliferate. Embryonic arrest was usually at the early egg cylinder stage (the early implanted embryo with embryonic ectoderm from ICM and extra-embryonic ectoderm from the trophoblast, both surrounded by primitive endoderm) and the ectoderm did not differentiate clearly into embryonic and extra-embryonic portions. At the ultrastructural level of analysis, mitochondrial abnormalities and lipid exclusions, similar to those seen in t^{12} and t^{w32}, were described (Nadijcka and Hillman, 1975). There was a suggestion of a separation between the inner cell mass and the polar trophectoderm (the portion of the trophoblast next to the developing multi-layer embryo) in contrast to the mural trophectoderm which is at the opposite pole) at 4 days of gestation. This could readily explain the lack of proliferation of the polar trophectoderm into an ectoplacental cone, since experiments by Gardner (1972) demonstrated that growth of the ectoplacental cone is dependent on contact between trophectoderm and the inner cells. This was supported by the finding that in vitro cultured t^6/t^6 embryos lacked an inner cell mass (Erickson and Pedersen, 1975). However, the trophoblast was not studied in these embryos (it flattened out as a disc in these cultures, normal embryos having the disc with an inner cell mass in its center; Erickson and Pedersen, 1975). Wudl and Sherman (1978), at the Roche Institute of Molecular Biology, found that the trophoblast cells are also abnormal: they did not undergo polyploidization (multiple replications of the chromosomes without cell division) normally, and apoptosis (programmed cell death) seemed to occur, leading to an absence of the normal expression of the lysosomal enzyme beta-glucuronidase, a marker of viability. Trophoblast giant cells from t^6-homozygous embryos were not saved by being placed under the kidney capsule (Wudl et al., 1977) and embryonic stem cell lines could not be established from the mutant embryos (Martin et al., 1987) as determined by using the *Tcp1* allele associated with the *t*-haplotypes. Thus, these *t*-mutants, which had later developmental delays, did not suggest metabolic abnormalities but rather defects of cellular interactions. Interestingly, pure populations of these t^6-homozygous embryos could be prepared by either delaying overall development by ovariectomy (Nadijcka et al., 1981) or culturing all the embryos in suboptimal media (Pai et al., 1981) since in both cases the normal embryos continued development while the mutants did not. However, sophisticated, e.g. em studies, analyses of these embryos were not performed.

Deletions in the *T* chromosomal region showed a pattern of lethality different from that found with homozygosity for the mutation in *T* (see below). T^{Orl} (Erickson et al., 1978), T^{1Or}, and T^{3Or} (Bennett et al., 1975) show an early lethality in which the cells of a two-layered egg cylinder became pyknotic, eventually disappearing to leave an empty yolk sac cavity. The cloned gene coded for a protein with features of an RNA-binding and signal transduction protein (Ebersole, et al, 1996). These abnormalities present in the *T*-deletions, which were later found by to have loss of neighboring genes, are very different than those that occur with homozygosity for the small deletion in the original *T*. This suggested that multiple loci for which null alleles affect development also map in this region of the *T/t* complex. While the discovery of molecular mechanisms did not emerge from these morphological, embryo culture, and biochemical studies, the results, such as the importance of cell-cell contact, elucidated some critical steps in early development.

Post-implantation lethals

The t^{w5} mutation was the most frequently studied of a complementation group which consisted of a large number of *t*-alleles. In this group, the homozygous mutants formed an egg cylinder with embryonic and extra-embryonic ectoderm and endoderm but no mesoderm. At the time that mesoderm normally formed, the embryonic ectoderm became pyknotic and died (Bennett and Dunn, 1958). When t^{w5} blastocysts were cultured, they developed an inner core of disorganized ectoderm cells with a small pro-amniotic cavity (Hogan et al., 1980) but all the cells died at about 11-14 days gestation equivalent, which was later than the in vivo cell death at 3 days, but compatible with the generally delayed development seen in vitro (Wudl and Sherman, 1976). These results also argued against metabolic abnormalities, as diffusible metabolites from the normal cells did not correct the defects. Chimeras of homozygous mutant cells with normal cells were made, and the two kinds of cells acted autonomously in the resulting blastocysts, with the death of the t^{w5}/t^{w5} cells affecting trophoblast development for the intermixed normal cells (Wudl and Sherman, 1976). Embryonic stem cells could be established from the mutant embryos and teratomas with all 3 germ layers resulted (Martin et al., 1987; Magnuson, et al. 1982). When cloned in the 21st Century, t^{w5} was found to have a frame shift mutation in *Vacuolar protein sorting 52,* a gene encoding a highly conserved protein tethering endosomes to the trans-Golgi apparatus (Sugimoto et al., 2012). Analysis of its role in embryos suggested a role in cell-cell interactions.

The t^9 mutation represented a complementation group whose multiple abnormalities could best be described as defects in neurulation (the development of a neural tube). The two-layered egg cylinder was normal but then there was abnormal thickening of the primitive streak, failure of neural tube closure, and lack of somites followed (Fig. 4.2) (Moser and Gluecksohn-Waelsch, 1967). A defect of mesoderm was usually seen and sometimes there was duplication of the incompletely closed neural tube. Electron microscopy disclosed that primitive streak thickening was due to a failure in the normal transition of primitive streak cells to mesoderm and lack of migration of mesodermal-like cells from the primitive streak (Spiegelman and Bennett, 1974). In another member of this complementation group, t^{w18}, there were deficiencies in intercellular contacts, and mitotic abnormalities were found in the epiblast, the primitive ectoderm where the primitive streak will develop as a groove (Snow and Bennett, 1978). Cell lethality of this mutant was tested by transplantation to extraembryonic sites. When t^{w18}/t^{w18} embryos were transplanted to the testes, malignant embryonic tumors containing ectodermal derivatives but no mesodermal components resulted (Artzt and Bennett, 1972). However, this study relied only on the morphological appearance of the embryos for their classification, and there was no definitive evidence that the particular class of tumors arose from t^{w18} homozygous mutants.

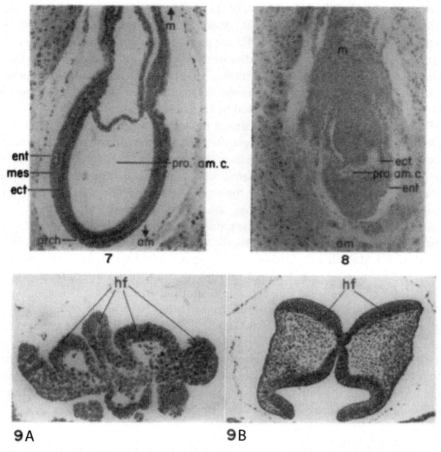

FIG. 4.2 Histological analyses of $t^{4/9}$ heterozygous embryos. 7, normal, 8 mutant embryos at 7 days. The mutant shows a very small proamniotic cavity (Pro am c.) and poorly defined ectoderm (ect) and endoderm (ent). 9A, mutant, and 9B, normal embyos at 9 days with sections through the head folds which are ill-formed in the mutant. From Moser and Gluecksohn-Waelsch (1967), with permission.

Kelly et al. (1979) reinvestigated the growth potential of t^{w18}-homozygous embryos using the metacentric chromosome, Rb(16,17)7Bnr, a cytogenetic marker for identification, containing the wild-type chromosomes, in cell lines obtained from embryos. They found a very limited growth potential from t^{w18}/t^{w18} embryos but, interestingly, found that they could quite easily obtain permanent lines from such embryos when they were transformed by the oncogenic virus, SV40. It was interesting to speculate on what kind of function would be absent in t^{w18}/t^{w18} embryos, preventing their growth, which could be provided after transformation by SV40. One would say that it certainly would not be a "housekeeping" function since the embryos normally survive to a relatively late stage, and it seemed unlikely that SV 40 transformation would provide such a function. It seemed more likely that SV40 transformation altered some cell surface or mitotic regulatory capacity which would allow more rapid growth after the viral transformation.

Some t^{w18}-homozygous cell lines had been obtained from embryos without viral transformation (Wudl et al., 1977; Hammerberg et al., 1980). Embryonic stem cell lines could also be established from these mutant embryos (Martin et al., 1987). When the cloning era came (see Chapter 9), it was discovered that t^{w18} is due to the deletion of a cluster of zinc finger genes (Crossley and Little, 1991; Shannon and Stubbs, 1999). These results with oncogenic viruses and permanent cell lines encouraged the notion that cancer was "development gone awry," and many developmental studies were supported by research funding for cancer.

The t^{w1} complementation group also shows defects classifiable as problems in neurulation. The lethal period for these embryos was variable, and the oldest embryos were small and showed extensive pyknosis of neural tube structures (Bennett et al., 1959). The skeletal defects seen were likely to be secondary to the neural tube defects. When inner cell masses from the mutant embryos were transplanted under the testis capsule, teratomas devoid of cartilage or bone, usually prominent tissues in teratomas, were found (Axelrod et al., 1981).

One unusual t allele, t^{AE5}, had only poor growth, viability and fertility i.e. was semi-lethal. This variant had appeared in Dr. Glueksohn-Waelsch's mouse colony at the new Albert Einstein College of Medicine (Vojtiskova et al., 1976). It's phenotype included accelerated involution of the thymus, necrotic changes during spermiogenesis but normal ovaries. It was suggested that the changes were similar to auto-immune syndromes. These sex specific differences in the reproductive organs were reflected in other sex-specific effects on T/t complex genes. Although t^0/t^{12} embryos are semi-viable with 27% survival, the males are sterile while the females are fertile (Silagi, 1962). The T^{Hp} mutation cannot be transmitted by females, the defective embryos from these females can be rescued in chimeras with normal embryos but still are not able to transmit the mutation, i.e. female transmission is still lethal (Bennett, 1978). Sex also had a great influence on the lethality of t^6/t^{w5} compound heterozygotes: female mutants had 80% viability while males were only 25% viable (Bechtol, 1982).

T/T Homozygous lethality

Homozygosity for T led to a consistent pattern of lethality (see above). The embryos did not form a characteristic notochord. This led to severe kinking of the neural tube, poorly defined somites, and, eventually, a body that terminated just posterior to the anterior limb buds (Chesley, 1935). The embryos were of sufficient size that studies of their cellular properties were possible. The problem attracted the interest of researchers at the Mitsubishi-Kasei Institute of Life Sciences in Tokyo. They found that the mitotic index in the hypertrophied ventral neural tube was not elevated at day 9 while +/+ embryos had an elevated mitotic index in the dorsal neural tube at this time (Yanagisawa and Kitamura, 1975).

In another study, Yanagisawa and Fujimoto (1977) studied the rotation-mediated aggregation of cells dissociated from T/T homozygotes and control embryos. Aggregation was slower and smaller aggregates eventually resulted from the T-homozygous embryos. These deficiencies could be corrected by culturing T/T embryonic cells in the supernatants from the cultured control embryonic cells. Normal cells also made better aggregates in a conditioned medium. Surprisingly, conditioned medium from the T/T-embryos was perfectly adequate for the aggregation of wild-type embryonic cells (Yanagisawa and Fujimoto, 1978). This suggested that there were likely to be quantitative differences in materials secreted into the media—sufficient for normal embryos but not for T/T embryos—or that the T/T embryos were deficient in some receptor. Cell lines obtained from the T homozygotes also formed smaller aggregates than wild type cells (Yanagisawa et al., 1980). Since the mitotic index was not abnormal in the T/T embryos, the altered mesodermal/ectodermal ratio was considered to be due to migrational defects (Yanagisawa et al., 1981). There was evidence that the extracellular matrix of the T/T embryo would be abnormal since cell surface galactosyltransferase was 2 to 6-fold increased, which would increase glycosylation of the matrix proteins (Shur, 1982). Scanning electron microscopy visualized defects of matrix with these embryos (Jacobs-Cohen et al., 1983a). The defects were further confirmed when mesodermal cells from the T homozygous embryos were cultured on an extra-cellular matrix, where their migration was significantly reduced compared to that of the normal mesodermal cells (Hashimoto, et al., 1987). An early indication of the later notochord defect was found by scanning electron microscopy of these embryos, which found fewer pre-notochord cells in the archenteron area (Fujimoto and Yanagisawa, 1983). Although initial studies concluded that isolated somites could not form cartilage in vitro (Fig. 4.3) (Bennett, 1958), later experiments were successful (Jacobs-Cohen et al., 1983b) and the previous failure was blamed on toxicity of horse serum, which was replaced in the second experiment by fetal calf serum. It is also possible that growth factors present in fetal calf serum were responsible.

The nature of T homozygous lethality was also explored by analyses in chimeras, at ectopic sites, and in ovarian teratocarcinomas. In England, the "home" of stem cell research[5], Rashbass, et al (1991) derived embryonic stem cell lines from T homozygous and T heterozygous embryos and made chimeras by injecting the stem cells into blastocysts. The stem cells can fuse with the inner cell mass and contribute a variable portion of cells to the developing embryo. The authors found that embryos derived from T/T injected blastocysts showed the typical

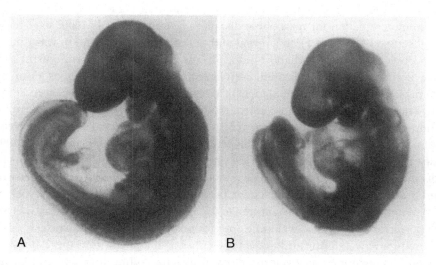

FIG. 4.3 Comparison of normal and mutant embryos, 30X magnification. A, normal, B T/T where somites are not distinct. From Yanagisawa and Kitamura (1975), with permission.

phenotype of the mutant even though normal cells were also present. The $T/+$ chimeras were normal. They concluded that the T gene product was required to maintain the notochord and its precursors in the primitive streak, and would have a role in an upstream pathway necessary to specify other mesodermal precursors. This was an extension of Yanagisawa and Fujimoto's (1978) work which also involved ectopic embryo culture (under the testis capsule;Fujimoto and Yanagisawa, 1979). Since the presence of normal cells in the chimera did not rescue the phenotype, they thought there must be a threshold for the normal gene product, which was diluted by its absence in the T homozygous cells. A further analysis of more embryos (Wilson et al., 1993) suggested that the T/T cells accumulated near the primitive streak, an outcome which could be due to the migratory defect described above. A correct placental connection was not made because of the presence of the T homozygous cells in the allantois. They found some chimeras which developed normally, except for a very abnormal tail when the T/T cells were predominately in the tail. The transplantation of T/T embryos to ectopic sites led to teratomas involving all 3 germ layers, also suggesting a lack of cell autonomous lethality (Fujimoto and Yanagisawa, 1979). Spontaneous ovarian teratocarcinomas from T or t^n homozygous embryos were studied in the LT/Sv inbred strain, which has a high spontaneous incidence of these tumors (Artzt et al., 1987). The tumors are due to parthenogenetic activation of oocytes, and the breeding scheme was such that homozygous t^n/t^n or T/T eggs were present. (Tumors for which the mother is heterozygous will be homozygous unless crossing over between the marker and the centromere has occurred. They found a distance of about 5 centiMorgans between the T/t complex and the centromere.) The T homozygous, parthenogenetic embryos made ovarian teratocarcinomas at a normal rate, but only one t^n teratocarcinoma was found, indicating the greater lethality of the latter (Artzt et al., 1987).

T was cloned using positional cloning This involved a laborious but exciting combination of probes micro-dissected from chromosome 17 identified in metaphase spreads, genetic studies to map these probes, and "chromosome jumping" which allows directional cloning from one probe in a known direction for reasonable distances (Hermann et al., 1990; see Chapter 9). As previously discussed (Chapter 1), this revealed that the original T mutation was due to a small deletion and was likely to be radiation-induced. The analysis of the gene's expression in the early mouse embryo showed localization both in epithelial progenitors of mesoderm and in the early stage mesoderm (Wilkinson et al., 1990). With time, T's expression became limited to the notochord. These findings helped to confirm the role of the T-product in mesoderm formation, a role its homologues in many species seem also to perform (Technau, 2001). The role of T as a transcription factor was soon established (Kispert and Hermann, 1993; Kispert et al., 1995). In the 21st Century, its role in the development of the primitive streak was elucidated. The in vitro development of embryoid bodies to develop a primitive streak described above depended on Wnt signalling and Brachury was one of the first gene products expressed in these "embryos" (ten Berge et al., 2008). When blasatocysts were cultured to the egg cylinder stage, Brachyury was the earliest marker for posterior cells in the developing anterior-posterior gradient (Morris et al., 2012).

The cloned gene was found to have a very conserved sub-portion, the T-box, which was found in many genes for transcription factors of developmental importance (Agulnik, et al. 1995). In this regard, it is the founding member of a gene family similar to that of the *Hox* genes discussed above. The patterns of expression of many members of the family suggest roles in inductive processes (Chapman et al., 1996). The T-box genes will be more fully discussed in Chapter 9 in which the cloning of *T* is used as an example for the "cloning era."

In conclusion, studies of *t*-complex developmental mutations led to many advances in understanding the underlying cellular and molecular processes in developmental biology, especially in the last quarter of the 20th Century. Other advances in the first 1½ decades of the 21st Century are reviewed by Sugimoto (2014). Although there have been claims of related genetic mutations affecting human development and mapping to HLA, the human region analogous to the mouse H-2 complex (e.g. Morrison,1996), these remain unsubstantiated. These attempts ignored the cross-over suppression which "locked in" a number of genes in the *T/t* complex. For example, the *tctex-1* locus of the mouse complex maps to the human X (Roux et al., 1994). The subject has been reviewed (Erickson, 1988).

Notes

1. Short overviews of the history of mouse developmental genetics are those of Artzt (2012) and Glueksohn-Waelsch (1983).
2. Joseph Needham was ahead of his time in attempting to apply biochemical techniques to embryology. He was inspired philosophically to do so. "Needham's scientific agenda in biochemical embryology, deemed unusual by most of his colleagues in both biochemistry and embryology, was both enabled and constrained by his philosophical interests in reconciling *mechanism,* which Needham considered the best methodological approach to biology, with *organicism,* which Needham came to regard as a metaphysically superior approach " (Abir-Am, 1991), Sadly, the tools of biochemistry at the time were not nearly sensitive enough to separate and identify the molecules (which work in nanoMolar concentrations) involved in induction and other developmental process. It would take developmental genetics and the tools of molecular biology to identify them. Needham, having married a Chinese post-doc, went on to become the major Western historian of Chinese science—perhaps inspired by his earlier work on the history of embryology which included Indian but not Chinese items (Needham, 1934).
3. Conrad Waddington has had several minibiographies: Robertson, 1977; Hall, 1992; Slack, 2002). His life long friendship with Gregory Bateson, who became a well know anthropologist and one-time husband of Margaret Mead, is documented and interpreted in Erik Peterson's doctoral thesis for the University of Notre Dame: "Finding Mind, Form, Organism, and Person in a Reductionist Age: The challenge of Gregory Bateson and C.H. Waddington to Biological and Anthropological Orthodoxy, 1924-1980" **(Peterson, 2010)**. It is interesting to note that they were so engaged despite living on different continents. "Much of their communication passed through anthropologist Margaret Mead, Bateson's former spouse and partner in ethnographic fieldwork, and also a frequent host to Waddington in his travels to the United States. Through their relationship, Bateson and Waddington found personal and intellectual support for continued work on their organismic evolutionary theory in the life and social sciences" **(Peterson, p. 5)**.

Although Bateson and Mead only stayed married for 15 years (1935-1950), They worked well together and spent 2 years together in Bali where "They were tackling a new set of problems at the intersection of psychology and anthropology: mainly in the area of the cultural determinants of mental health" **(King, p. 291)**. Peterson sees the relationship of Bateson and Waddington as encouraging the latter to see organicism as "an alternative to the genetic reductionism at the base of contemporary neo-Darwinism and sociobiology" **(Peterson, p. 5)**.

4. Sir Allen Parkes was the head of the Division of Experimental Biology at the National Institute of Medical Research prior to N. A. Mitchison (see Haldane-Mitchison clan biography). He was very interested in comparative reproductive biology and his studies included the elephant—he had an elephant ovary in a keg of formalin in the lab (verbal communication from Sandra D Erickson).

5. John D. Biggers, 1923-2018, trained as a veterinarian at the Royal College of Veterinary Medicine. He was a peripatetic scholar with many academic positions, including one in Australia and Cambridge University but ending up at Harvard where he was a Professor of Physiology. A very friendly man, he stayed at our home in Tucson in the 90's after I had gotten to know him at Gordon Conferences on early mammalian development (one of which I co-chaired in 1982). He was full of good stories and we joked about obscure facts of sex determination in a variety of species. These included eels with sex inversion and lower chordates in which H-Y antigen had been described (which already was becoming a laughing matter in the 80s). He described the difficulties of convincing embryologists that the early mammalian embryo used anaerobic metabolism.

6. When this author learned that the mouse had homologs of *Drosophila* homeobox genes, he thought that they must be doing something quite different in mammals than in *Drosophila,* since development is so different in the two species. (The adult fly develops from small patches of cells set aside in the pupa [imaginal discs] which grow and meet as they differentiate, dissolving the internal organs of the pupa for energy as they do so). Boy, was I wrong! The *Hox* genes must have been present in the earliest segmented organisms and almost certainly in the 300 million year history of the trilobites (Richard Fortey's "Trilobites" fascinatingly documents their history and distribution). I was later to play a part in the discovery that mutations in *HOXA1* were the cause of Athabaskan Brainstem Dysgenesis (Tischfield et al., 2005), a rare autosomal recessive disorder found among the Athabaskans of the Southwest states (Holve et al., 2003). (Figs. 4.1, 4.2 and 4.3).

References

Abir-Am, P.G., 1991. The Philosophical Background of Needham's Work in Chemical Embryology. In: Gilbert, S.F. (Ed.), A Conceptual History of Modern Embryology. Vol. 7 in Developmental Biology: A Comprehensive Synthesis. Plenum Press, NY, pp. 159–180.

Agulnik, S.I., Bollag, R.J., Silver, L.M., 1995. Conservation of the T-box gene family from Mus musculus to Caenorhabditis elegans. Genomics 25, 214–219.

Allen, A., 1922. The estrous cycle in the mouse. Am. J. Anat. 30, 297–348.

Artzt, K., 2012. Mammalian Developmental Genetics in the Twentieth Century. Genetics, 192, 1151–1163.

Artzt, K., Bennett, D., 1972. A genetically caused embryonal ectodermal tumor in the mouse. J. Nat'l Canc. Inst. 48, 141–158.

Artzt, K., Calo, C., Pinheiro, E.N., DiMeo-Talent, A., Tyson, F.L., 1987. Ovarian Teratocarcinomas in LT/Sv Mice Carrying t-mutations. Develop. Genet. 8, 1–9.

Auerbach, R., 1954. Analysis of the developmental effects of a lethal mutation in the house mouse. J. Exp. Zool. 127, 305–330.

Axelrod H., R., 1985. Altered trophoblast functions in implantation-defective mouse embryos. Dev. Biol. 108, 185–190.

Axelrod H., R., Artzt, K., Bennett, D., 1981. Rescue of embryonic cells homozygous for a lethal haplotype of the T/t complex: t^{w12}. Dev. Biol. 86, 419–425.

Bagnara, J.T., 1993. In Memoriam: Johannes F.C. Holtfreter (1901-1992). Devel. Biol. 158, 1–8.

Bard, J.B.L., 2008. Waddington's Legacy to Developmental and theoretical. Biology. Biol. Theory 3, 188–197.

Bechtol, K.B., 1982. Lethality of heterozygotes between *t*-haplotype complementation groups of mouse: sex-related effect on lethality of *t6/tw5* heterozygotes. Genet. Res. 39, 79–84.

Bennett, D., 1956. Developmental analysis of a mutation with pleiotropic effects in the mouse. J. Morphol. 98, 199–233.

Bennett, D., 1958. *In vitro* study of cartilage induction in *T/T* mice. Nature 181, 1286.

Bennett, D., 1978. Rescue of a lethal *T/t* locus genotype by chimaerism with normal embryos. Nature 272, 539.

Bennett, D., Dunn, L.C., 1958. Effects on embryonic development of a group of genetically similar lethal alleles derived from different populations of wild house mice. J. Morphol. 103, 135–157.

Bennett, D., Badenhausen, S., Dunn, L.C., 1959. The embryological effects of four late-lethal *t*-alleles in the mouse, which affect the neural tube and skeleton. J. Morph. 105, 105–144.

Bennett, D., Dunn, L.D., Spielgelman, M., Artzt, K., Cookingham, J., Schermerhorn, E., 1975. Observations on a set of radiation-induced dominant T-like mutations in the mouse. Genet. Res. (Camb.) 26, 95–108.

Brilliant, M.H., Gondo, Y., Eicher, E.M., 1991. Direct molecular identification of the mouse pink eye unstable mutation by genome scanning. Science 252, 566–569.

Brinster, R.L., 1974. The effects of cells transferred into the mouse blastocyst on subsequent development. J. Exp. Med. 140, 1049–1056.

Burian, R.M., 1997. On conflicts between genetic and developmental viewpoints-and their attempted resolution in molecular biology. Structures and Norms in Science, vol. 260 of the series Synthese Library, 243–264.

Calarco, P.G., Brown, E.H., 1968. Cytological and Ultrastructural Comparisons of t^{12}/t^{12} and Normal Mouse morulae. J. Exp. Zool 168, 169–186.

Chapman, D.L., Garvey, N., Hancock, S., Alexiou, M., Agulnik, S.I., Gibson-Brown, J.J., Cebra-Thomas, J., Bollag, R.J., Silver, L.M., Papaioannou, V.E., 1996. Expression of the T-box Family Genes, *Tbx1-Txs5*, During Early Mouse Development. Develop Dynamics 206, 379–390.

Charite, J., de Graff, W., Shen, S., Deschamps, J., 1994. Ectopic expression of *Hoxb-8* causes duplication of the ZPA in the forelimb and homeotic transformation of axial structures. Cell 78, 589–601.

Chesley, P., 1932. Lethal action in the short-tailed mutation in the house mouse. Proc. Soc. Exp. Biol. Med. 29, 437–438.

Chesley, P., 1935. Development of the Short-tailed mutant in the House Mouse. J. Exp. Zool. 70, 429–459.

Collombet, S., Ranisavljevic, N., Nagano, T., Varnai, C., 2021. Shisode T and 8 more (2020) Parental-to-embryo switch of chromosome organization in early embryogenesis. Nature 580, 142–146.

Coulumbre, J.L., Russell, E.S., 1954. Analysis of the pleiotropism at the *W*-locus in the mouse: The effects of *W* and *Wv*substitution upon postnatal development of germ cells. J. Exp. Zool. 126, 277–295.

Crossley, P.H., Little, P.F.R., 1991. A cluster of related zinc finger protein genes is deleted in the mouse embryonic lethal mutation *t^{w18}*. Proc. Nat'l Acad. Sci. 88, 7923–7927.

Damjanov, I., Solter, D., Belicza, M., Skreb, N., 1971. Teratomas derived from extrauterine growth of seven-day mouse embryos. J. Nat'l Canc. Inst. 46, 471–480.

Dunn, L.C., Glueksohn-Schoenheimer, S., 1947. A new complex of hereditary abnormalities in the house mouse. J. Exp. Zool. 104, 25–51.

Ebersole, T.A., Chen, Q., Justice, M.J., Artzt, K., 1996. The *quaking* gene product necessary in embryogenesis and myelination combines features of RNA binding and signal transduction proteins. Nature Genet 12, 260–265.

Elke, J., 1967. Heinrich Waelsch, 1905-1966. Psychopharm.(Berl.) 10, 285–288.

Ephrussi, B., 1933. Sur le facteur lethal des souris brachyures. Compt. Rend. Acad. Sci 197, 96–98.

Ephrussi, B., 1935. The behavior in vitro of tissues from lethal embryos. J. Exp. Zool 70, 197–204.

Erickson, R.P., 1988. Minireview: The 6's and 17's of Developmental Mutants near the Major Histocompatabilitiy Complex:The Mouse *t*-Complex Does Not Have a Human Equivalent. Am. J. Hum. Genet. 43, 115–118.

Erickson, R.P., Betlach, C.J., Epstein, C.J., 1974. Ribonucleic acid and protein metabolism of *t12/t12*embryos and *T/t12* spermatozoa. Differentiation 2, 203–209.

Erickson, R.P., Pedersen, R.A., 1975. *In vitro* development of t^6/t^6 embryos. J. Exp. Zool. 194, 377–384.

Erickson, R.P., Gondos, B., 1976. Alternative explanations of the differing behavior of ovarian and testicular teratomas. Lancet 1976, 407–410.

Erickson, R.P., Lewis, S.E., Slusser, K.S., 1978. Deletion mapping of the *t* complex of chromosome 17 of the mouse. Nature 174, 163–164.

Fekete, E., Little, C.C., 1942. Observations on the mammary tumor incidence in mice born from transferred ova. Canc. Res. 2, 525–530.

Fekete, E., 1947. Differences in the effect of uterine environment upon development in the DBA and C57 Black strains of mice. Anat. Rec. 98, 409–415.

Flanagan, J.G., Leder, P., 1990. The *kit* ligand: A cell surface molecule altered in steel mutant fibroblasts. Cell 63, 185–194.

Friedrich, G.A., Soriano, P., 1991. Promoter traps in embryonic stem cells: a genetic screen to identify and mutate developmental genes in mice. Genes Devel 5, 1513–1523.

Fujimoto, H., K.O., Yanagisawa, 1979. Effect of the T-mutation on histogenesis of the mouse embryo under the testis capsule. J. Embryol. Exp. Morphol. 50, 21–30.

Fujimoto, H., K.O., Yanagisawa, 1983. Defects in the archenteron of mouse embryos homozygous for the T-mutation. Differentiation 25, 44–47.

Gardner, R.L., 1968. Mouse chimaeras obtained by the injection of cells into the blastocyst. Nature 220, 596–597.

Gardner, R.L., 1972. An investigation of inner cell mass and trophoblast tissues following their isolation from the mouse blastocyst. J. Embryol. Exp. Morphol 28, 279–312.

Gardner, R.L., 1998. Contribiutions of blasatocyst micromanipulation to the study of mammalian development. BioEssays 20, 168–180.

Gehring, W.J., 1994. A history of the homeobox. In: Duboule, D. (Ed.), Guidebook to Homeobox Genes. Sambrook and Tooze at Oxford University Press, Oxford.

Gilbert, S.F., 1988. Cellular Politics: Ernst Everett Just, Richard B. Goldschmidt, and the Attempt to Reconcile Embrylogy and Genetics. In: Rainger, R., Benson, K.R., Maienschein, J. (Eds.), The American Development of Biology. University of Pennsylvania Press, Philadelphia.

Gilbert, S.F., 1991. Induction and the Origins of Developmental Genetics. In: Gilbert, SF (Ed.), A Conceptual History of Modern Embryology. Plenum Press, New York.

Gilbert, S.F., Opitz, J.M., Raff, R.A., 1996. Resynthesizing Evolutionary and Developmental Biology. Develop. Biol. 173, 357–372.

Ginsberg, L., Hillman, N., 1975. ATP metabolism in t^n/t^n mouse embryos. J. Embryo. Exp. Morph. 33, 715–723.

Glucksohn-Schoenheimer, S., 1938a. The development of two tailless mutants in the mouse. Genetics 23, 573–584.

Glucksohn-Schoenheimer, S., 1938b. Time of death of lethal homozygotes in the *T* (Brachury) series of the mouse. Proc. Soc. Exp. Biol. Med. 39, 267–268.

Glucksohn-Schoenheimer, S., 1940. The effect of an early lethal (t^0) in the house mouse. Genetics 25, 391–400.

Gluecksohn-Schoenheimer, S., 1944. The development of normal and homozygous brachy (*T/T*) mouse embryos in the extraembryonic coelom of the chick. Proc. Nat'l. Acad. Sci., USA 30, 134–140.

Gluecksohn-Schoenheimer, S., Dunn, L.C., 1945. Sirens, aprosopi, and intestinal abnormalities in the house mouse. Anat. Rec. 92, 201–213.

Gluecksohn-Schoenheimer, S., 1949a. Causal analysis of mouse development by the study of mutational effects. Growth Symp 9, 164–176.

Gluecksohn-Schoenheimer, S., 1949b. The effects of a lethal mutation responsible for duplications and twinning in mouse embryos. J. Ex.;. Zool. 110, 47–76.

Gluecksohn Waelsch, S., 1962. Mammalian Genetics in Medicine. Chapt. 8 in Progress in Medical Genetics 2, 295–330.

Gluecksohn-Waelsch, S., 1994. Genetic control of mammalian differentiation. Genetics Today, Proc. Of the XI Int. Cong of Genet.The Hague1993. Pergamon Press, Oxford, The Netherlands Sept.

Gluecksohn-Waelsch, S., Rota, T.R., 1963. Development in Organ Tissue Culture of Kidney Rudiments from Mutant Mouse Embryos. Develop. Biol. 7, 432–444.

Gluecksohn-Waelsch, S., 1979. Genetic control of morphogenetic and biochemical differentiation: lethal albino deletions in the mouse. Cell 16, 226–237.

Gluecksohn-Waelsch, S., 1983. Fifty years of developmental genetics. Trans. N. Y. Acad. Sci. 41, 243–251.

Gluecksohn-Waelsch, S., 1989. In Praise of Complexity. Genetics 122, 721–725.

Goldschmidt, R., 1937. Gene and Character IV. Univ41. California Publ. Zool, pp. 1–277.

Gordon, J.W., Scangos, G.A., Barbosa, J.A., D.J., Plotkin, 2021. Ruddle FH (1980) genetic transformation of mouse embryos by microinjection of purified DNA. Proc. Nat'l Acad. Sci. 77, 7380–7384.

Granholm, N.H., Johnson, P.M., 1978. Identification of eight-cell *tw32* homozygous lethal mutants by abnormal compaction. J. Exp. Zool. 203, 81–88.

Green, M.C., 1967. A defect of the splanchnic mesoderm caused by the mutant gene dominant hemimelia in the mouse. Dev. Biol. 15, 62–89.

Grompe, M., al-Dhalimy, M., Finegold, M., Ou, C.N., Burlingame, T., Kennaway, N.G., Soroiano, P., 1993. Loss of fumarylacetoacetate hydrolase is responsible for the neonatal hepatic dysfunction phenotype of lethal albino mice. Genes Develop. 7, 2298–2307.

Gruneberg, H., 1943. Congenital hydrocephalus in the mouse, a case of spurious pleiotropism. J. Genet. 45, 1–21.

Gruneberg, H., 1960. Developmental genetics in the mouse, 1960. J. Cellular Comp. Physiol. 56 (S1), 49–60.

Hall, BK, 1992. Waddington's legacy in development and evolution. Amer. Zool. 32, 113–122.

Hammerberg, C., Wudl, L.R., Sherman, M.I., 1980. H-2 analyses of blastocyst-derived cell lines from Mice bearing *T* and t^{w18} mutations. Transplant 29, 484–486.

Hapalowska, A., Jedrusik, A., Zhu, M., Bedford, M.T., Glover, D.M., Zernicka-Goetz, M., 2018. CARM1 and paraspeckles regulate pre-implantation mouse embryo development. Cell 175, 1902–1916.

Hashimoto, K., Fujimoto, H., Nakatsuji, N., 1987. An ECM substratum allows mouse mesodermal cells isolated from the primitive streak to exhibit motility similar to that inside the embryo and reveals a deficiency in the *T/T* mutant cells. Develop 100, 587–598.

Herrmann, B.G., Labeit, S., Poustka, A., King, T.R., Lehrach, H, 1990. Cloning of the T gene required in mesoderm formation in the mouse. Nature 343, 617–622.

Hillman, N., Hillman, R., 1975. Ultrastructural studies of t^{w32}/t^{w32} mouse embryos. J. Embryo. Expt'l Morph. 33, 685–695.

Hillman, N., Tasca, R.J., 1973. Synthesis of RNA in t^{12}/t^{12} embryos. J. Reprod. Fert. 33, 501–506.

Hogan, B., Spiegelman, M., Bennett, D., 1980. *In vitro* development of inner cell masses isolated from *to/to* and t^{w5}/t^{w5} mouse embryos. J. Embryol. Expt. Morphol. 60, 419–428.

Holtfreter, J., 1933. Der Einflus von Wirtsalter und verschiedenen Organbezirken auf die Differenzierung von angelagertem Gastrulaektoderm. Arch. Fur Entwmech. Die Organ 127, 619–775.

Holve, S, Freidman, B, Hoyme, HE, Tarby, TJ, Johnstone, SJ, Erickson, RP, Clericuzio, CL, Cunniff, C, 2003. Athabascan brainstem dysgenesis syndrome. Am. J. Med. Genet. 120A, 169–173.

Iles, S.A., McBurney, M.S., Bramwell, S.R., Deussen, Z.A., Graham, C.F., 1975. Development of parthenogenic and fertilized mouse embryos in the uterus and extra-uterine sites. J. Embryol. Expt'l Morph. 34, 387–405.

Jacobs-Cohen, R.J., Spiegelman, M., Bennett, D., 1983a. Abnormalities of cells and extracellular matrix of *T/T* embryos. Differentiation 25, 48–55.

Jacobs-Cohen, R.J., Spiegelman, M., Bennett, D., 1983b. *T/T* somite mesoderm is able to differentiate into cartilage in vitro. Cell Differen 12, 219–223.

Jami, J., Ritz, E., 1974. Multipotentiality of single cells of transplantable teratomas derived from mouse embryo grafts. J. Nat'l Canc. Inst. 52, 1547–1552.

Jhappan, C., Stahle, C., Harkins, R.N., Fausto, N., Smith, G.H., Merlino, G.T., 1990. TGFalpha overexpression in transgenic mice induces liver neoplasia and abnormal development of the mammary gland and the pancreas. Cell 61, 1137–1146.

Kelly, F., Guenet, J.-L., Condamine, H., 1979. Karyological identified homozygous *tw18* embryos: extrauterine growth properties and transformation by SV40. Cell 16, 919–927.

Kisbert, A., Herrmann, B.G., 1993. The *Brachury* gene encodes a novel DNA binding protein. EMBO J 12, 3211–3220.

Kisbert, A., Koschorz, B., Herrmann, B.G., 1995. The T protein encoded by *Brachury* is a tissue-specific transcription factor. EMBO J 14, 4763–4772.

Kleinsmith, L.J., Pierce, G.B., 1964. Multipotentiality of single embryonal stem cells. Cancer Res 24, 1544–1551.

Lang A. (1914) Die experimentelle Verebungslehre in der Zoologie seit 1900. Jena.

Lehman, F.E., 1928. Die Bedeutung der Unterlagerung fur die Entwicklung der Medullarplatte von Triton. Arc. Fur Entwmec. Die organ 113, 1123–1171.

Lehman, J.M., Speers, W.C., Swartzendruber, D.E., Pierce, G.B., 1974. Neoplastic differentiation: characteristics of a cell line derived from a murine teratocarcinoma. J. Cell Physiol. 84, 13–27.

Magnuson, T., Epstein C., J., Silver L., M., Martin G., R., 1982. Pluripotent embryonic stem cell lines can be derived from t^{w5}/t^{w5} blastocysts. Nature 298, 750–753.

Martin, G.R., Evans, M.J., 1974. The morphology and growth of a pluripotent teratocarcinoma cell line and its derivatives in tissue culture. Cell 2, 163–172.

Martin, G.R., Silver, L.M., Fox, H.S., Joyner, A.L., 1987. Establishment of embryonic stem cell lines from preimplantation mouse embryos homozygous for lethal mutations in the *t*-complex. Dev. Biol. 121, 20–28.

McGinnis, W., Hart, C.P., Gehring, W.J., Ruddle, F.H., 1984. Molecular cloning and chromosomal mapping of a mouse DNA sequence homologous to homeotic genes of Drosophila. Cell 38, 675–680.

McGinnis, N., Kuziora, M.A., McGinnis, W., 1990. Human *HOX 4.2* and Drosophila *Deformed* encode similar regulatory specificities in Drosophila embryos and larva. Cell 63, 969–976.

McLaren, A., Biggers, J.D., 1958. Successful development and birth of mice cultivated in vitro as early embryos. Nature 182, 877–878.

Melo, U.S., Schopflin, R., Acunna-Hidalgo, R., Mensah, M.A., 2020. Fischer—Zimsak B and 17 more (2020) Hi-C identifies complex genomic rearrangements and TAD -shuffling in developmental diseases. Am. J. Hum. Genet. 106, 872–884.

Mintz, B., 1962a. Experimental recombination of cells in the developing mouse egg: notmal and lethal mutant genotypes. Amer. Zool 2, 541–542.

Mintz, B., 1962b. Experimental study of the developing mammalian egg: removal of the zona pellucida. Science 138, 594–595.

Mintz, B., 1964. Formation of Genetically Mosaic Mouse Embryos, and Early Development of "Lethal (*t12/t12*)-Normal" Mosaics. J. Exp. Zool. 157, 273–292.

Mintz, B., Illmensee, K., 1975. Normal genetically mosaic mice produced from malignant teratocarcinoma cells. Proc. Nat'l Acad. Sci. 72, 3585–3589.

Morange, M., 1996. Construction of the developmental gene concept, the crucial years: 1960-1980. Bull. Zent.bl 115, 132–138.

Morris, S.A., Grewal, S., Barrios, F., Patankar, S.N., Strauss, B., Buttery, L., Alexander, K.M., Shakesheff, M., Zernicka-Goetz, M., 2012. Dynamics of anterior–posterior axis formation in the developing mouse embryo. Nature Comm 3, doi:10.1038/ncomms1671 doi.org/.

Morrison, K., Papapetrou, C., Attwood, J., Hol, F., Lynch, S.A., et al., 1996. Genetic Mapping of the Human Homologue (*T*) of Mouse *T* (*Brachyury*) and a Search for Allele Association between Human *T* and Spina Bifida. Hum. Mole. Genet. 5, 669–674.

Moser, G.C., Glucksohn-Waelsch, S., 1967. Developmental genetics of a recessive allele at the complex T-locus in the mouse. Develop. Biol. 16, 564–576.

Muller, G.B., 2007. Evo-devo: extending the evolutionary synthesis. Nature Rev Genet 8, 943–949.

Mystkowska, E.T., Tarkowski, A.K., 1968. Observations on CBA-p/CBA-T6T6 mouse chimeras. Develop 20, 53–71.

Nadijcka, M., Hillman, N., 1975. Studies of t^6/t^6 mouse embryos. J. Embryol. Exp. Morph. 33, 697–713.

Nadijcka, M., Morris, M., Hillman, N., 1981. The effect of delay on the expression of the t^6/t^6 genotype. J. Embryol. Exp. Morph. 63, 267–283.

Nozaki, M., Iwakura, Y., Matsushiro, A., 1986. Studies of Developmental Abnormalities at the molecular Level of Mouse Embryos Homozygous for the t^{12} Lethal Mutation. Devel. Biol. 115, 17–28.

Okaabe, M., Ikawa, M., Kominami, K., Nakanishi, T., Nishimune, Y., 1997. Green mice as a source of ubiquitous green cells. FEBS Lett 407, 313–319.

Okita, K., Ichisaka, T., Yamanaka, S., 2007. Generation of germline-competent induced pluripotent stem cells. Nature 448, 313–317.

Opitz, J.M., 1985. Editorial Comment: The Developmental Field Concept. Am J. Med. Genet. 21, 1–11.

Oppenheimer, J.M., 1981. Walter Landauer and Developmental Genetics. In: Subtelny, S., Abbott, U.K. (Eds.), Levels of Genetic Control in Development. Alan R liss, Inc., New York.

Pai, A., Glueicksohn-Waelsch, S., 1961. Interaction of *t* alleles at the *T* locus in the house mouse. Experientia 17, 372–374.

Pai, A.D., Wudl, L.R., Sherman, M.I., 1981. *In vitro* studies of mouse embryos bearing mutations in the T complex: effects of culture in suboptimal medium upon t^6/t^6 and mormal embryos. J. Embryol. Exp. Morph. 63, 99–110.

Palla, A., Blau, H., 2020. The clock that controls spine development in a dish. Nature 580, 32–34.

Papaioannou, V.E., McBurney, M.W., Gardner, R.L., Evans, M.J., 1975. Fate of teratocarcinoma cells injected into early mouse embryos. Nature 258, 70–73.

Papaioannou, V.E., 1999. The Ascendency of Developmental Genetics, or How the T Complex Educated a Generation of Developmental Biologists. Genetics 151, 421–425.

Parkes, A.S., 1926. Observation on the estrous cycle of the albino mouse. Proc. Roy. Soc., B 100 (151) . doi:doi. org/10.1098/rspb.1926.0040.

Pierce, G.B., Dixon, F.J., Verney, E.L., 1960. Teratocarcinogenic and tissue-forming capabilities of the cell types compromising neoplastic embryoid bodies. Lab. Invest. 9, 583–602.

Piotrowska, K., Zernicka-Goetz, M., 2001. Role for sperm in spatial patterning or the early mouse embryo. Nature 409, 517–521.

Piotrowska, K., Wianny, F., Pedersen, R.A., Zernicka-Goetz, M., 2001. Blastomeres arising from the first cleavage division have a distinguishable fates in normal mouse development. Devel 128, 3739–3748.

Rashbass, P., Cooke, L.A., Herrmann, B.G., Beddington, R.S.P., 1991. A cell autonomous function of *Brachyury* in *T/T* embryonic stem cell chimaeras. Nature 353, 348–350.

Robertson, A., 1977. Conrad Hal Waddington 8 November 1905—26 September 1975. Biograph. Memoirs Fellows Roy. Soc. 23, 575–622.

Roux, A.-F., Rommens, J., McDowell, C., Anson-Cartwright, L., Bell, S., Schappert, K., Fishman, G.A., Musarella, M., 1994. Identification of a gene from Xp21 with similarity to the tctex-1 gene of the murine t complex. Hum. Mole. Genet. 3, 257–263.

Russell, E.S., Murray, L.M., Small, E.M., Silvers, W.K., 1956. Development of embryonic mouse gonads transferred to the spleen: effects of transplantation combined with genotypic autonomy. J. Embryol. Exp. Morph. 4, 347–357.

Schoenheimer, R., 1929. Uber eine eigenartage Storung des Kohlehydrat-Stoffwechsels. Hoppe-Seyler's. Zeitschrift fur Physiologische Chemie 182, 148–150.

Shannon, M., Stubbs, L., 1999. Molecular characterization of *Zfp54*, a zinc-finger containing gene that is deleted in the embryonic lethal mutation *t*w18. Mammal. Genome 10, 739–743.

Shur, B.D., 1982. Cell surface glycosyltransferase activities during normal and mutant (*T/T*) mesenchyme migration. Develop. Biol 91, 149–162.

Silagi, S., 1962. A genetical and embryological study of partial complementation between lethal alleles at the T locus of the house mouse. Devel. Biol. 5, 35–67.

Silver, L., 2008. Salome Gluecksohn-Waelsch 1907-2007. Nature Genet 40, 376.

Slack, J.M.W., 2002. Waddington's legacy in development and evolution. Nature Rev. Genet. 3, 889–895.

Smith, L.J., 1956. A morphological and histochemical investigation of a pre-implantation lethal (t^{12}) in the house mouse. J. Exp. Zool. 132, 51–84.

Snow, M.H.L., Bennett, D., 1978. Gastrulation in the mouse: assessment of cell population in the epiblast of *tw18/tw18* embryos. J. Embryol. Exp. Morph. 47, 39–52.

Solter, D., 2008. Memoriam: Salome Gluecksohn-Waelsch (1907-2007) Develop. Cell 14, 22–24.

Sozen, B., Amadei, G., Cos, A., Wang, R., Na, E., 2018. Self-assembly of embryonic and two extra-embryonic stem cell types into gastrulating embryo-like structures. Nature Cell Biol 20, 979–989.

Spielgelman, M., Bennett, D., 1974. Fine structural study of cell migration in the early mesoderm of normal and mutant mouse embryos (*T*-locus: t^9/t^9). J. Embryol. Exp. Morph. 32, 723–738.

Spielgelman, M., Artzt, K., Bennett, D., 1976. Embryological study of a *T/t* locus mutation (t^{w73}) affecting trophoectoderm development. J. Embryol. Exp. Morph. 32, 723–738.

Spielmann, H., Erickson, R.P., 1983. Normal adenylate ribonucleotide content in mouse embryos homozygous for the t^{12} mutation. J. Embryol. Exptl. Morph. 18, 43–51.

Stevens, L.C., Little, C.C., 1954. Spontaneous testicular teratomas in an inbred strain of mice. Proc. Nat'l Acad. Sci. 40, 1080–1087.

Stevens, L.C., 1968. The Development of Transplantable Teratocarcinomas from Intratesticular Grafts of Pre- and Postimplantationn Mouse Embryos. Develop. Biol. 21, 364–382.

Sugimoto, M., Kondo, M., Hirose, M., Suzuki, M., Mekada, K., Abe, T., Kiyonari, H., Ogura, A., Takagi, N., Artzt, K., et al., 2012. Molecular identification of t^{w5}: V^{ps52} promotes pluripotential cell differentiation through cell-cell interactions. Cell Rep 2, 1363–1374.

Sugimoto, M., 2014. Developmental genetics of the mouse *t*-complex. Genes Genet. Syst. 89, 109–120.

Tarkowski, A.K., 1961. Mouse chimeras developed from fused eggs. Nature 190, 857–860.

Technau, U., 2001. *Brachury,* the blastopore and the evolution of the mesoderm. BioIEssays 23, 778–794.

ten Berge, D., Koole, W., Fuerer, C., Fish, M., Eroglu, E., Nusse, R., 2008. Wnt Signaling Mediates Self-Organization and Axis Formation in Embryoid Bodies. Cell Stem Cell 3, 508–513.

Tischfield, M., Bosley, T.M., Salih, M.A.M., Alorainy, I.A., Sener, E.C., Nester, M.J., Oystreck, D.T., Chan, W.-M., Andrews, C., Erickson, R.P., Engle, E.C., 2005. Homozygous HOXA1 mutations disrupt human brainstem, inner ear, cardiovascular and cognitive development. Nature Genet 37, 1035–1037.

Van den Brink, S.C., Baillie-Johnson, P., Balayo, T., Hadjantonakis, A.-K., Nowotschin, S., Turner, D.A., Marinez Arias, A., 2014. Symmetry breaking, germ layer specification and axial organization in aggregates of mouse embryonic stem cells. Develop. 141 4321-4242.

Vojtiskova, M., Viklicky, V., Voracova, B., Lewis, S.E., Glueckschn-Waelsch, S., 1976. The effects of a *t*-allele (*tAE5*) in the mouse on the lymphoid system and reproduction. J. Embryol. Exp. Morph. 36, 443–451.

Waddington, C.H., 1933. Induction by coagulated organizers in the chick embryo. Nature 131, 275.

Waddington, C.H., Needham, D.M., 1934. Physico-chemical experiments on the amphibian organizer. Proc. Royal Soc. London 114, 393.

Waddington, C.H., 1938. The morphogenetic function of a vestigial organ in the chick. J. Exp. Biol 15, 371–384.

Waddington, C.H., 1940. The genetical control of wing development in Drosophila. J. Genet. 41, 75.

Waelsch, S.G., 1994. The Development of Creativity. Creativity Research Journal 7, 249–264.

White, M.D., Angiolini, J.F., Alvarez, Y.D., Kaur, G., Zhao, Z.W., Mokos, E., Bruno, L., Bissiere, S., Levi, V., Plachta, N., 2016. Long-lived binding of Sox2 to DNA predicts cell fate in the four-cell mouse embryo. Cell 165, 75–87.

Whitten, W.K., 1956. Culture of tubal mouse ova. Nature 177, 96.

Whitten, W.K., 1971. Nutritional requirements for the culture of preimplantation embryos in vitro. Adv. Biosci. 6, 129–139.

Whitten, W.K., Biggers, J.D., 1968. Complete development in vitro of the pre-implantation stages of the mouse in a simple chemically defined media. J. Reprod. Develop. 17, 399–401.

Whittingham, D.G., 1971. Culture of mouse ova. J. Reprod. Devel. Suppl. 1 (14), 7–21.

Wilkinson, D.G., Bhatt, S., Herrmann, B.G., 1990. Expression pattern of the mouse *T* gene and its role in mesoderm formation. Nature 343, 657–659.

Wilson, V., Rashbass, P., Beddington, R.S.P., 1993. Chimeric analysis of *T (Brachyury)* gene function. Develop 117, 1321–1331.

Wudl, L.R., Sherman, M.I., 1976. *In vitro* studies of mouse embryos bearing mutations in the T locus: t^{w5} and t^{12}. Cell 9, 523–531.

Wudl, L.R., Sherman, M.I., 1978. *In vitro* studies of mouse embryos bearing mutations in the T complex: t^6. J. Embryol. Exp. Morph. 48, 127–151.

Wudl, L.R., Sherman, M.I., Hillman, N., 1977. Nature of lethality of *t* mutations in embryos. Nature 270, 137–140.

Yanagisawa, K.O., Kitamura, K., 1975. Effects of the Brachury (*T*) Mutation on Mitotic Activity in the Neural Tube. Develop. Biol. 47, 433–438.

Yanagisawa, K., Fujimoto, H., 1977. Differences in rotation-mediated aggregation between wild-type and homozygous brachyury (T) cells. J. Embryool. Exp. Morph. 40, 277–283.

Yanagisawa, K., Fujimoto, H., 1978. Aggregation of homozygous brachyury (T) cells in the culture supernatant of wild-type or mutant embryos. Exp. Cell Res. 115, 431–435.

Yanagisawa, K.O., Urushihara, H., Fujimoto, H., Shiroishi, T., Moriwaki, K., 1980. Establishmentt and Characterization of Cell Lines from Homozygous Brachyury (*T/T*) Embryos of the Mouse. Differentiation 16, 185–188.

Yanagisawa, K.O., Fujimoto, H., Urshihara, H., 1981. Effects of the Brachyury (*T*) Mutation on Morphogenetic Movement in the Mouse Embryo. Develop. Biol. 87, 242–248.

5

More population genetics with particular attention to the effects of *t*-allele mediated transmission ratio distortion

Population genetics was introduced in Chapter 2 as part of the Modern Synthesis of genetics and evolution where it was seen that mouse genetics did not contribute much to the development of the Synthesis but contributed greatly to the later increase in understanding of genetic barriers between species. In this chapter, selection and variation in numbers of individuals in a species are the main foci. Variation in wild house mouse populations has been much studied and is discussed at length in several reviews (Berry, 1977, 1986; Boursot et al., 1993; Sage et al., 1993). These variations lead to great fluctuations in the frequency of *t*-alleles in wild populations.

Early studies on island populations (1 such study was described in Chapter 2) emphasized unique features of the mice in isolated island populations and considered the possibility that they were distinct subspecies, e.g. Clarke (1904). There were also speculations about the persistence of such island populations from interglacial times (Beirne, 1952) but, as previously discussed (Chapter 1), more recent dispersals accompanying humans are well documented. A systematic study of rodent numbers and dispersal was a WWII effort to learn to better control "pests" so as to increase agricultural production. Thus, a division of the Ministry of Agriculture and Food[1] was the Rodent Research Branch–much of this research was performed at the Oxford University Branch of Animal Population. Publications such as "The house mouse (*Mus musculus*) in corn ricks (piles of unharvested corn stalks stored over winter in piles; Southern and Laurie, 1946) resulted. These efforts were summarized in 3 volumes, the third of which was on mice (Southern, 1952). Its scope was "devoted to the house mouse and commences with reports on its environment and feeding habits. The remainder of the book deals with the various poisons that can be used to. control mice, and the best ways of handling poisons for effective eradication" (Southern, 1952, p 1). The studies found that *Mus musculus* lived in hedgerows and fields in the summer but over-wintered in burrows near buildings and corn ricks (Laurie, 1946). In another such study, the mice were only found in hedgerows from September to December and the following May (Rowe et al., 1959). Once a population achieved maximum density in small isolates there seemed to be little movement between them (Anderson, 1965) but many exceptions of migration were found.

Most longitudinal studies of mouse populations, especially when the environment was stressful, have found a fair amount of movement between the small demes. Berry and Jacobson (1971) performed a 7 year release-capture study on a small Welsh island. While most mice remained local, fully one-fourth bred at a site different from the one at which they had been born.

Twentieth Century Mouse Genetics. DOI: https://doi.org/10.1016/B978-0-12-824016-8.00012-X

The maximum number of animals in a single group seemed to be about six, but the composition of all groups changed constantly because of the relatively high mortality at all ages (life expectation at birth was c. 100 days, compared with over two years in commensal populations: Bellamy et al., 1973; Takada, 1985).

There was great interest as to what degree natural selection was on-going in *Mus musculus* wild populations. Skeletal variants were much studied, e.g. Deol (1958) but these were greatly affected by environment (McLaren and Michie, 1958). Many kinds of genetic variants were used as biochemical genetics provided new tools for such studies. Electrophoretic variations in serum esterases and hemoglobin were used by Selander (1970) to show that genotypes were segregating randomly among dense mouse populations living in chicken barns while Petras (1967) found heterozygote deficiencies, suggesting homozygous advantage, for similar variants in farm barns near Ann Arbor, Michigan (home of the University of Michigan). In a later study, Petras and Topping (1983) found fitness values of 0.4 and 0.2 for Hbb^s/Hbb^s and Hbb^d/Hbb^d respectively, resulted in a frequency approaching 0.70 for the common allele. These high selection coefficients, along with assumptions of migration rates of 5-10%, allowed a model fitting the gene frequencies fairly well. Many other studies confirmed the presence of strong selective forces on the hemoglobin locus. Berry (1977) found that the frequencies of heterozygotes for Hbb^s/Hbb^d increased during the breeding season and decreased in the summer among mice on the island of Skokholm off the coast of Wales. These selection coefficients differed between the summer and winter. Berry and Peters (1975), in a study on genetically isolated mouse populations on a sub-Antarctic island, found changes in the frequency of the Hbb^d allele between age groups: the frequency of the Hbb^d allele decreased from 71% in younger mice (< 3 mo.s) to 54% in older ones, especially in the animals trapped further away from any of the buildings. However, temperature was not the only selective force. Myers (1974) studied mice in the more temperate climate of California and found that the Hbb^d allele was favored in a more or less permanent population whereas the Hbb^s allele was favored in in a new colony. Studying 36 protein allozyme loci among mice trapped at the edge of sugar cane fields in the even milder climate of Hawaii, Berry et al. (1980) found no evidence for selection with high heterozygous rates (16%) and no change with age. The basis for the apparent selection among hemoglobin alleles, which differ in oxygen affinity (Newton and Peters, 1983) is, thus, not clear and could be the result of selection on closely linked genes rather than on hemoglobin itself.

In addition to the environmental forces and olfactory discrimination, discussed in regards to species isolation in Chapter 2, other modes of selection were considered. Louis Levine at the City College of New York, with frequent advice from L.C. Dunn who was nearby at Columbia, studied the possibility of sexual selection in mice which we have previously mentioned in Chapter 2 where odorants and the hybrid zone were discussed. Darwin had considered sexual selection to be distinct from natural selection since traits maladaptive for survival (e.g. the gigantic antlers of the extinct Irish elk, also known as giant deer). He (Levine, 1958) confirmed multiple inseminations in one pregnancy (see Chapter 8) but his conclusions as to male preference are marred by the *H-2* allelic vaiation and many other differences between the two inbred strains (chosen for distinguishable coat color), used in the experiments. His further studies (Levine and Lascher, 1965; Levine and Krupa, 1965) were similarly subject to multiple interpretations. He also performed deliberate double inseminations in which female mice were mated with a male of 1 strain and after 10 min. the copulatory plug was removed and a male of the second strain introduced (Levine, 1967). In these experiments, preferential fertilization of the female's eggs by sperm of the same inbred strain as the mother, rather than the other inbred strain, occurred independent of the order of the matings. Since the sex of the resulting offspring was not determined, these results could have been the result of sex chromosome drive (see Chapter 8) or sperm competition for other reasons.

As more studies were made of isolated populations, it became clear that there were very wide fluctuations in the size of the population over the year's seasons. Berry and associates (summarized in Berry, 1977) studied the population of wild house mice on the 1 km² island of Skokhom, located 3 km. of the coast of Wales, for 7 years. Mark-release-recapture experiments using 40 trap sites allowed an accurate estimate of the number of mice. The numbers varied up to ten-fold (Fig. 5.1). The lowest numbers were in the spring and were as low as 75 breeding pairs representing as few as 100 individuals. These low numbers provided important support for Sewall Wright's theory of evolution, rather than R.A. Fisher's theory of evolution of dominance (see **Provine, 1986, pp. 243-276**) for the estimates of population sizes used in their calculations. Fisher even thought that "for the relevant purpose, I believe N [population number] must usually be the total population on the planet, enumerated at sexual maturity" **(letter to Wright, cited by Provine, 1986, p. 255)**. Studies in the field mouse species *Microtus pensylvanicus* used fencing to limit spread and found that dispersal was essential for their large population size fluctuations (Krebs et al., 1973).

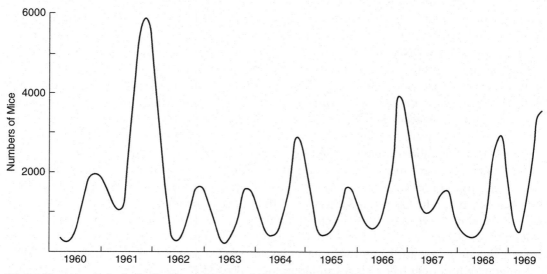

FIG. 5.1 Numbers of mice in the Skokhom Island population from 1960–1970, from Berry and Jacobson (1971), with permission.

These small populations and their fluctuation created the conditions for marked shifts in some allelic frequencies, such as that for *Dip-1^c*, a electophoretic distinguishable dipeptidase. In addition to the protein electrophoretic polymorphisms, which included hemoglobin, a major source of variation studied in wild mice were histocompatability antigens. Jan Klein's group in Tubingen was a leader in these efforts which were also extensively contributed to by Joseph Nadeau and others. The *H-2* types of 5 mouse populations varying in karyotype were correlated with the variations in these karyotypes (Figueroa et al., 1983). In one major study, using mice from Texas, Scotland, the Federal Republic of Germany, Denmark, Spain, Greece, Israel, Egypt, and Chile, 16 new H-2 haplotypes were isolated (Zaleska-Rutczynska et al., 1983). Many of these were recombinants of previously known haplotypes. It was found that more than 95% of wild mice are heterozygous at their *H-2* loci (Duncan et al., 1979). This suggested marked selection against homozygosity for these antigens. However, the degree of variability at loci for alloenzymes linked to *H-2* was even greater (Nadeau, 1983) which argued against selection occurring. The results were re-interpreted by Sved (1983) as being compatible with selection if an associative over-dominance model was assumed. Nadeau and Collins (1983) counter-argued that the condition for associative overdominance were not met since there was no evidence for disequilibrium for most populations and most alleles (Nadeau et al., 1981). However, different degrees of selection might be occurring in the isolated island populations than in more pan-mictic, continental populations. In this regard, several island populations were found to have only one or two haplotypes (Helgoland had two while the Isle of May and Fara Island only had one; Figueroa et al., 1986). These results were likely to be due to founder effects and strongly suggested that great variability, as found in continental populations, was not necessary for disease resistance.

One of the selection forces likely to be contributing to the continental H-2 diversity was maternal/fetal interactions. The are a number of cases of fetal maternal incompatibility in which offspring of the same genotype as their mother are less frequent than offspring of differing genotype (Palm, 1970). Clarke and Kirby (1966) detailed the potential contribution to maintaining *H-2* diversity.

Unique sequence DNA probes from the *H-2* region of chromosome 17 also contributed to the understanding of the *M. m. musculus*/*M. m. domesticus* separation (Figueroa et al., 1987). Only 1 restriction enzyme digestion pattern was found with one of the probes in *musculus* while 8 were found in *domesticus*. Surprisingly, many of these polymorphisms were not found in inbred strains suggesting that inbred strains derived from other populations than those sampled for the study.

Other histocompatiability loci studied in wild mice include the Mta, maternally transmitted, weak antigen (Fischer Lindahl and Burki, 1982). As might be expected, it is mitochondrial in origin and variation correlates with unique restriction fragment patterns of mitochondrial DNA (Ferris et al., 1983). Its expression depends on the nuclear-encoded *Hmt* gene which maps to the distal end of the *H-2* complex (Fischer Lindahl et al., 1983). This may code for a class I H-2 molecule (see Chapter 7) and it, β-2 microglobulin, and a mitochondrial protein or fragment thereof constitute the antigen for skin rejection after transplantation (Fischer Lindahl, 1986). β-2 microgloulbin only has 2 alleles in laboratory mice (Michaelson, 1983) but rare variants have been found in wild mice (Robinson et al., 1984).

Another diverse polymorphism occurs in the karyotype. As discussed in Chapter 3, mice normally have pairs of 19 autosomal acrocentric chromosomes (and the sex chromosome pair). Some relatively isolated populations differ greatly by having fused the acrocentric chromosomes at their very short p arms to make metacentrics, chromosomes, i.e., have Robertsonian translocations. *Mus mucsculus poschiavinus*, a *M. m. domesticus* sub-species, has 7 such metacentics leading to a total chromosome number of 26 instead of 40 (Gropp and Winking, 1981). The occurrence of such variants is limited to the *M. m. domesticus* range and the hybrid zone with *M. m. musculus* but is not found North of this zone (Winking, 1986; see Chapter 2, Fig. 5.1). Eighty-six of the 162 possible combinations of arms have been found in distinct populations (ibid.). Only the sex chromosomes and the smallest chromosome, number 19, have not been involved in some one of these. Although heterozygosity for different metacentrics or a metacentric and normal chromosomes decreases fertility because of nondisjunction and lethal chromosomal combinations, it is possible to deliberately mate desired combinations to generate particular trisomies as detailed by **Epstein (1986, pp. 208-210)**. Thus, since mouse chromosome 16 has a significant amount of synteny to human chromosome 21, trisomics for 16 have been studied as models of Down's syndrome (Epstein et al., 1985). Although mice with this trisomy did not survive to birth, aggregation chimeras were useful for studying the effect of trisomy 16 on cardiac development (Cox et al., 1984), and radiation chimeras rescued by injected hematopoietic stem cells were useful for analyzing the activities of trisomy 16 cells in immunological responses (Herbst et al., 1982).

These Robertsonian translocations are of recent derivation—less than the 10,000 years (Britton-Davidian et al., 1989). Despite the reduced fertility of heterozygotes for these metacentrics (and, of course, the first occurrence would be in a heterozygote) there must be selective forces favoring these Robertsonian translocations. All the mice trapped in the Northwestern Scottish county of Caithness, isolated from the rest of Scotland by moors occupied by field mice instead of house mice, had at least 2 homozygous pairs of fusion chromosomes (Brooker, 1982). On the nearby Orkney Islands, the mice on 8 of the 11 islands had differing Robertsonian fusions (Nash et al., 1983). Despite, this chromosomal variation, these populations seem very closely related on the basis of mandible morphometrics and allozymic analysis of 23 polymorphisms (Nash et al., 1983). RJ Berry, a major figure in studies of wild mice concludes "The distribution of fused chromosomes in Caithness and Orkney seems inexplicable on the basis of random spread: either natural selection or some 'drive mechanism must be operating'" (Berry, 1986). On the basis of studying 38 electrophoretic variations in 28 populations of "metacentric" mice, Britton-Davidian, et al (1989) also found very little divergence from the "non-metacentric" populations. These authors came up with 2 models for the results, one with severe and the other with lesser, transient bottlenecks—both compatible with the population fluctuations described above. The evolutionary geneticist, Allan Wilson, and collaborators have suggested that the social behavior of mice creates the conditions for the occurrence of the "metacentric" populations (Larson et al., 1984) while Ganem (1993) emphasized the ecological importance of commensal habitats for the phenomenon. Nachman and Searle reviewed all the data available in 1995 and concluded that, as well as genetic drift, meiotic drive, although not found in laboratory studies might still be important (Nachman and Searle, 1995).

In the 20th Century, new tools allowed a better understanding of selective forces causing hybrid sterility. Moriwaki and colleagues (Oka et al., 2007) used inbred strains from predominately *domesticus* mated to one of predominately *molossinus* to study hybrid sterility since the X chromosome of the latter on the former's background, shows male sterility. They performed a genome wide linkage analysis and subsequent QTL analysis using the sperm shape anomaly which caused the sterility and found two chromosomal locations causing the interaction. Nachman and collaborators (Good et al., 2007; Teeter et al., 2008), using similar interspecies crosses, found an influence of the X chromosome and that many autosomal markers exhibit asymmetrically broad clines, usually with high frequencies of domesticus alleles on the musculus side of the hybrid zone. They concluded that these alleles are spreading from one species to the other. The X influence fitted Haldane's law (see Chapter 2 and Laurie's [1997] review). There also was a small influence of the Y chromosome (Campbell et al., 2012). Dureje et al., (2012) found that mitochondrial DNA introgressed in both directions across the boundary to a great distance while Y chromosomal markers introgressed from the *M. musculus* side into the *M. domesticus* side but not vice-versa. When the underlying genes were found they were sometimes unsurprising, e.g. dynein involved in the sperm tail (see below), and sometimes surprising, e.g. a histone H3 methyltransferase (Mihola et al., 2009). Using microsatellite loci at various distances from the metacentric centromere, Britton-Davidian et al. (2017) could demonstrate that there was variation in recombination patterns as shown by the rate of divergence of centromere-proximal to centromere-distal loci. Unlike Oka, et al. (2007) their analysis of the clines of variation in the hybrid zone suggested that the difference in selection did not involve linked genetic incompatibilities.

Also, an awareness of the greater diversity in mice than in humans developed. As the full power of modern genomics was applied to diverse populations, a greater understanding of this variation was discovered. By sequencing 17 inbred strains and using these to correlate with 718 quantitative trait loci, variation in expression was found to show frequent allelic imbalance (Keane et al., 2011). For instance, studying gene expression in 3 inbred strains derived from 3 different subspecies which had been sequenced, Crowley, et al (2015) could conclude that greater than 80% of mouse genes have *cis* regulatory variation. Surprisingly, at least one in every thousand SNPs created a *cis* regulatory effect. There were many examples of imprinting and an overall allelic imbalance favoring the paternal allele.

Another advance in the 20th Century was the realization that selection could not only work on the nuclear and mitochondrial genome of the mice, but also on genetic variation in the microbiome which was especially studied in the gut. The parental genome and diet influenced the composition of the biome which in turn influenced the metabolome with, of course, great potential for selective forces, for example Snijders et al. (2016).

Population genetics of t-alleles

The t alleles have existed for at least 3 million years. They have grown steadily in size since their origin to become more than 1% of the genome. Their spread has generated adaptation on both sides, improvements in t allele action, as well as evolution of countermeasures by the rest of the genome. The t alleles have been studied for the past 75 years and a great deal is known about them, although unfortunately on several key points the information is contradictory

Burt and Trivers, p. 19

Those 75 years of research involved a great variety of approaches and individuals. Variation in frequency of *t*-alleles in natural populations was an early focus of population studies in mice. As seen in Chapter 2, they were readily found in wild mice.. Theorectical studies focused on the effects of male transmission ratio distortion (trd, see Chapter 8) and several theoreticians calculated the balance needed between trd and homozygous lethality to maintain them in the population. One of the early derivations was that by Bruck, who was at Columbia and aware of the problem from LC Dunn, who communicated his paper to the Proceedings of the National Academy of Science. (Bruck, 1957). His calculations showed that an equilibrium frequency of the *t*-allele would occur after about 10 generations but the calculations were based on the assumption of a large inter-breeding population. These calculations were applied to, and modified for smaller populations by Dunn and Levene (1961). The persistence of a *t*-allele in feral mice, which were maintained in a laboratory as a random-bred small population at the Rockefeller Institute had been demonstrated by Dunn and Bennett (1967). When applied to this small, mixed *t*-allele containing population, the predictions were compatible with the percentage detected but the range for the predictions was very high. The average values predicted were higher than those found (but within the range predicted). Lewontin and Dunn (1960). Richard Lewontin[2], who became the Alexander Agassiz Professor of Zoology at Harvard) refined these calculations coming up with results highly compatible with the findings using the assumption that mice bred in small endogamous family units Lewontin (1962) as had been described (above; Anderson, 1964). Young (1967) used the newly available high speed (for the time) computer to calculate a series of curves for population levels of *t*-alleles assuming different levels of fitness for the heterozygote and homozygote. Levin, et al (1969), using similar assumptions, still found the frequency of *t*-alleles to be lower than expected on the hypothesis of genetic drift between demes that had lost *t*-alleles and those that had not. These calculations all focused only on the male effects through sperm trd while Nigel Bateman at the University of Edinburgh had shown some effects of the egg's *T/t*-complex genotype on the trd (Bateman, 1960). This was in also in a "closed laboratory stock" originally derived from "a number of inbred lines and heterozygous populations", similar to the Rockefeller population the Columbia group studied. However, the sexual influences on *T/t*-complex phenotypes described in Chapter 11 were not considered. In addition, the calculations had assumed that, although sterile, *t/t* homozygotes were fully viable and Lewontin added this correction to the calculations in 1968 (Lewontin, 1968). However, Johnston and Brown's (1969) data showing decreased viability of t^{w2} heterozygotes was pretty much ignored.

Also, in this period before the size of the chromosomal "block" maintained by these inversions was known, the importance of many linked genes was not considered. There would also be negative effects of such inversions. As summarized in **Trivers and Burt**, "selfish genetic elements almost invariably set in place forces that cause their own deterioration. This can come in a variety of ways, the *t* haplotype illustrates one: increasing linkage (e.g., via inversions) provides a way for the selfish element to grow but robs the now-linked DNA of the benefits of recombination" **(p. 25)**.

Charlesworth and Hartl (1978) calculated the rate of recombination between "killer" (the t allele responder loci) and resistor (the Tcr, the t-complex response element, see Chapter 8) had to be less than 1/3 for the killer to spread. Other killers achieved this by being located near centromeres.

One approach to understand t-alleles in nature was to introduce them to island populations in which they had not been found. An early study followed the frequency of a t^w (presumably the t^{w2} found in the laboratory-maintained stock previously studied; Dunn and Bennett, 1967) after its release in 1957 on Great Gull Island in Long Island sound near New York City. By 1964, the allele had reached high frequencies in the release zone but "was spreading very slowly" (Anderson et al., 1964). However, by 1967, it could be reported that it had spread to the other end of the island (Bennett et al., 1967). However, the gene was subsequently lost in the population (personal communication by L.C. Dunn to Myers in 1973; Myers, 1974). A somewhat similar result was found when the t^{w2}-allele was introduced into a large, enclosed populations (Pennycuik et al., 1978). Only female carriers successfully bred but the allele could not be detected after 2 years. In contrast, the release of 77 mice marked by allozymes, mitochondrial DNA markers, Y chromosomal DNA markers, and Robertsonian translocations onto the island of Eday in the Orkneys resulted in rapid spread and hybridization of the introduced mice with the resident mice (Berry et al., 1990). These results emphasized the probably stochastic nature of the t-allele selection in natural populations.

Robert C. Lacy, (1978) reviewing the published data to the date of his summary emphasized the unanswered questions which needed to be addressed before the evolutionary role of t-alleles could be answered: "more needs to be known about the causes of variability in the ratio... The nature and evolutionary role of Mus deme structure should also be clearly defined... We must learn how important coadapted gene complexes have been in the evolution of Mus...". These questions were slowly addressed.

By 1988, Lenington, et al (1988) could review 11 published studies of the frequency of t-alleles in wild populations as well as 1 unpublished data set and their own collection data from 13 sites. Not surprisingly, they could conclude that the frequency of semi-viable t-alleles was higher than the frequency of lethal t-alleles. The frequency found was higher in males than females but did not change between trapped young mice and trapped older mice. There was also little variation in the frequency over many years in the majority of populations. They also confirmed Johnson and Brown's (1969) evidence for reduced litter size in the offspring of t-heterozygote males, at least in the first litter (Lenington et al., 1994).

Gradually the study of t-alleles in populations from other parts of the world were again performed. Dunn and Bennett (1971) studied 7 populations from Denmark including $M.$ $m.$ $musculus$ from the Northern part of Jutland and $M.$ $m.$ $domesticus$ from the Southern part. Although the sample was small, 17% of the males carried t-alleles which, by complementation tests, fell into allelic classes already known in America. Soon after, the linkage of H-2 to the T/t complex was realized and each lethal t-complementation groups was found to have a different, unique H-2 haplotype (Hammerberg and Klein, 1975; Levinson and McDevitt, 1976). As antisera were made to the lymphocytes of the different t-allelic lines of mice, many new antigens were discovered and the uniqueness of each t-allele associated haplotype, despite coming from widely different geographic regions was confirmed (Hammerberg and Klein 1976).

The findings led to the conclusion that "It suggests more than a casual relationship, at least at the population level, between the *t* and *H-2* loci" (ibid.). A rare isozyme of the sperm-specific phosphoglycerate kinase 2 (see Chapter 8) was found that was also linked to *t* and it was speculated that only because of this linkage and trd that it persisted in the population (Rudolph and Vandeberg, 1981). On the other hand, Joseph Nadeau, at the Jackson laboratory, did not find linkage disequilibrium for this allele of *Pgk-2* or 3 other allozyme-encoding loci although his samples were small (Nadeau et al., 1981). Jan Klein and his associates derived an ancestral tree of the *H-2* loci and were able to infer an ancient origin of the *T/t* complex on the basis of the antigenic variation and single nucleotide substitutions in the introns of the Class 2 loci (Klein et al., 1986). This predicted an origin predating the *M. m. musculus* and *M.m. domesticus* separation. A similar evolutionary scheme was devised by Schimenti and Silver (1986) on the basis of analyses with a clone for repetitive DNA (T66) which had amplified in the *T/t* complex. Restriction fragment analyses of the DNA from many mouse species to create a phylogenetic analysis of the alpha-globin pseudogene, located in one of the inversions, suggested that the first inversion arose about 3 million years before present (Hammer and Silver, 1993). Although the subspecies of *Mus* diverged nearly a million years ago, the *t*-complexes they share are so similar that a variant must have swept through them, despite the species barriers discussed in Chapter 2, as recently as 100,000 years ago (Morita et al., 1992).

As the descriptive data on *t* alleles and linked loci in natural populations accumulated, theoretical studies continued. Levin's et al studies showing the lack of previous theory to account for *t* allele frequencies in natural populations (above) found better fit when a degree of migration between demes was included. The theory was further developed and applied to natural populations in Ontario, Canada by Petras and Topping (1983). They found that migration rates of between 0.05 and 0.1 adequately accounted for the frequencies of *t*-alleles in these populations. As pointed out by Haig and Bergstrom (1995), another aspect of mouse reproduction which favors trd is the existence of multiple inseminations. In the case that one of these is by a *t*-allele carrying male, his *t*-sperm will not only have an advantage over his wild type sperm but over those of another male as well. However, there is some evidence that the opposite effect, i.e. decreasing the trd, occurs (Ardlie and Silver, 1998).

Another advance was, with the realization that about 1% of the genome was involved in the *T/t* complex inversions, other effects than lethality and trd needed to be considered. Olfactory cues were found in female preference of males differing in their *t*-allele status (Lenington, 1983). Of course, this might well be due to the linked H-2 effects on olfactory discrimination (see Chapter 2)—as Potts, et al. (1991) found in seminatural populations. Lenington and Heisler (1991) continued these studies on male preference and found that the frequency of trd was higher with females caught outside the local area, suggesting that migration was indeed necessary for the establishment of *t*-alleles in the wild. Further studies showed that *t*-allele heterozygous mice were more aggressive than wild type males and that they had higher survival rates as well when tested in four large outdoor enclosures (Lenington et al., 1996). Others sought to determine genetic factors affecting trd. Gummere, et al. (1986; Bennett's lab) made F_1 and F_2 crosses between two stocks differing in transmission ratio for the t^{12} allele. These crosses allowed them to distinguish between the effects of chromosomal

17 and other genetic modifiers allowing them to conclude that both chromosome 17 and other genetic factors had profound influences on trd. However, studies of trd in newly caught wild mice did not find evidence for modifiers of drive in natural populations (Ardlie and Silver, 1996). Thus, genetic drift may have occurred in these laboratory stocks which had long been maintained by brother-sister mating.

Since this 1% of the genome also contained hybrid sterility loci, e.g. *Hybrid sterility 6, (Hst-6)* (Pilder et al., 1993), this was another factor which would influence trd. This locus had an effect on flagellar assembly and movement (ibid.) and was later found to code for an axonemal dynein (Fossella et al., 2000).

Although earlier studies had emphasized the frequent presence of *t*-alleles in the wild, other studies did not find such high frequencies. Myer (1974) did not detect any on Grizzly Island, an 8,000 acre island in the California central valley where the delta of the Sacramento and San Juaquin rivers creates many swamps and sloughs (and where much of America's rice is grown). The island was only separated from the mainland by a slough and connected to it by a bridge, conditions thought to be compatible with genetic exchange with the "mainland" populations. Similarly, low levels were found in Siberia (Ruvinsky et al., 1991). As the century ended, a very comprehensive study was performed by Ardlie and Silver (1998). They reported *t*-allele frequencies found among 3263 mice trapped at 63 locations across the world (and, with repeats across time at some sites, 80 samples). The much larger numbers were possible because DNA analyses, not breeding, were used for the determinations. For some populations, marking-release-recapture analyses were performed to estimate total population size. They found an overall low frequency of *t*-alleles of 6.2% with only 46% of populations containing *t*-alleles (Ardlie and Silver (1998)). They postulated that one reason for this lower frequency than had generally been reported was due to the non-inclusion of populations found to have none (on the assumption that in small samples, they had simply been missed) in previous reports. Interestingly, if only the first sampling at the multiple-tested sites was included, the percentage rose to 10.7% suggesting a decline with time. This may have been due to poisoning, reducing the population size, at some sites. Overall, there was a very marked decrease in frequency, the larger the population size (Fig. 5.2). This might have been due to decreased migration rates between demes in large, stable populations. They considered the results to be compatible with their previous conclusion that "there is no stable, low-level *t*-polymorphism. Rather wild populations are in one of two stable states characterized by extinction of the *t*-haplotype and a high *t*-haplotype frequency, respectively, or in transition between the two (Durand et al., 1997). They did not find the male/female difference in frequencies previously reported by Lennington, et al. (1988) and found higher frequencies in populations where homozygous *t/t* mice were also found (Ardlie and Silver, 1998).

Studies of *t*-allele segregation distortion, both theoretical and experimental, continued into the 21st Century. Durand, et al's (1997) calculations predicting no stable levels of them was extended by van Boven and Weissing (1999) at the very end of the 20th Century. They concluded that high ratio distorters could reach high levels in single demes but would be underrepresented in the migrant pool and might lead to deme extinction. A growing appreciation of the importance of inbreeding in large populations, with

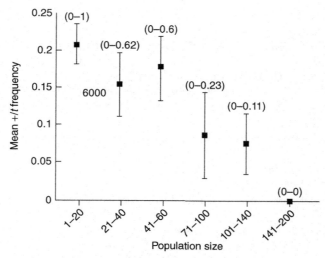

FIG. 5.2 The frequency of T/+ mice as a function of population size. From Ardlie (1998), with permission.

less migration, as a negative factor for *t*-allele frequencies also developed (**Burt and Trivers, 2006, p. 38**). Further evidence for lower success of *t*-alleles in multiple insemi-nations was provided (Baker, 2008). Game theory was added to enhance the models of segregation distortion (Traulsen and Reed, 2012; Sarkar, 2016). The importance of female selection (which requires energy expenditure to find the appropriate male and, thus, is a negative selection factor) against males carrying distorters, perhaps identified by olfactory cues discussed above, was further developed (Manser et al., 2017). Evidence for increased migration by juvenile mice carrying the *t* haplotype within a long-term free-living house mouse population was provided (Runge and Lindhom, 2018).

Notes

1. Many of the developments of this ministry helped the British civilian population to be healthier than ever be-fore during the war. A single type of bread, the National Loaf, was produced which was fortified with vitamins and minerals. The nutritionally highly valuable yeast left over from beer production (well maintained for mor-al of troops and home) was utilized to make a paste for bread—Marmite (a similar product is called Vegemite in Australia), a product still loved by some. Sugar and candy were heavily rationed, a restriction which was not lifted until 1972 (the post-war available sugar and chocolate was being shunted to export candy production to help the national economy). The many Victory Gardens helped supply vegetables and fruit for a healthy diet.

2. Lewontin played an important role in the renewed discussion of the synthesis of genetics and evolution as much of genetic variation was found to be neutral, a position much championed by T. Nomura. In a series of lectures he focused on the need to find the kinds of genetic variation on which selection acted and whether it was single genes or more complex genetic entities, such as chromosomal regions, which indeed, later genome sequencing suggested. He said in his Jessup lectures "For I now realize that the question was not simply, How much genetic variation is there? nor even, How much genetic variation in fitness is there? but rather, How much genetic variation is there that can be the basis of adaptive evolu-tion?" (**Lewontin, 1974, p.x**).

References

Anderson, P.K., 1964. Lethal *t* alleles in *Mus musculus:* local distribution and evidence for isolation of demes. Science 145, 177–178.

Anderson, P.K., 1965. The role of breeding structure in evolutionary processes of *Mus musculus* populations, In Proceedings of a symposium on the mutation process, Prague, Czech. Acad. Sci., Aug. 9-11, 1965, pp. 17–21.

Anderson, P.K., Dunn, L.C., Beasley, A.B., 1964. Introduction of a Lethal Allele into a Feral House Mouse Population. Amer. Natural. 98, 57–64.

Ardlie, K.G., Silver, L.M., 1996. Low frequency of mouse *t* haplotypes in wild populations is not explained by modifiers of meiotic drive. Genetics 144, 1787–1792.

Ardlie, K.G., Silver, L.M., 1998. Low frequency of mouse *t* haplotypes in natural populations of house mice (*Mus musculus domesticus)*. Evol 52, 1558–1646.

Ardlie, K.G., 1998. Putting the brake on drive: meiotic drive of *t* haplotypes in natural populations of mice. Trends in Genet. 41, 189–193.

Baker, A.E.M., 2008. Mendelian inheritance of *t* haplotypes in house mouse (*Mus musculus domesticus*) field populations. Genet. Res. 90, 331–339.

Bateman, N., 1960. High frequency of a lethal gene (*to*) in a laboratory stock of mice. Genet. Res. 1, 214–225.

Beirne, B.P., 1952. The Origin and History of the British Fauna. Methuen, London.

Bellamy, D., Berry, R.J., Jakobson, M., Lidicker, W.Z., Morgan, J., Murphy, H.M., 1973. Ageing in an island population of the house mouse. Age Ageing 2, 235–250.

Bennett, D., Bruck, R., Dunn, L.C., Klyde, B., Shutsky, F., Smith, L.J., 1967. Maintenance of gene frequency of a male sterile, semi-lethal *T*-allele in a confined population of wild mice. Amer. Natural. 101, 535–553.

Berry, R.J., 1977. The population genetics of the house mouse. Sci Prog., Oxford 64, 3341–3370.

Berry, R.J., 1986. Genetical Processes in Wild Mouse Populations. Past myth and Present Knowledge. Curr. Top. Microbiol. Immunol. 127, 86–94.

Berry, R.J., Jacobson, M.E., 1971. Life and death in an island population of the house mouse. Exptl. Gerontol. 6, 187–197.

Berry R., J., Peters J., J., 1975. Macquarie Island house mouse: a genetical isolate on a sub-Antarctic island. J. Zool. 176, 375–389.

Berry, R.J., Sage, R.D., Lidicker, W.Z., Jackson, W.B., 1980. Genetical variation in three Pacific house mouse populations. J. Zool. Lond. 193, 391–404.

Berry, R.J., Triggs, G.S., Bauchau, V., Jones, C.S., Scriven, P., 1990. Gene flow and hybridization following introduction of *Mus domesticus* into an island population. Biol. J. Linnean Soc. 41, 279–283.

Boursot, P., J., Aujfrai J.-C., Britton-Davidian, J., Bonhomme, F., 1993. The evolution of house mice. Ann. Rev. Ecol. Syst. 24, 119–152.

Britton-Davidian, J., Nadeau, J.H., Croset, H., Thaler, L., 1989. Genic differentiation and origin of Robertsonian populations of the house mouse (*Mus musculus domesticus* Rutty). Genet. Res. 53, 29–44.

Britton-Davidian, J., Caminade, P., Davidian, E., Pagès, M., 2017. Does chromosomal change restrict gene flow between house mouse populations (*Mus musculus domesticus*)? Evidence from microsatellite polymorphisms. Biol. J. Linn. Soc. 122, 224–240.

Brooker, P.C., 1982. Robertsonian translocations in *Mus musculus* from N.E. Scotland and Orkney. Hered., 48, 305–309.

Bruck, D., 1957. Male segregation ratio advantage as a factor in maintaining lethal alleles in wild populations of house mice. Proc. Nat'l Acad. Sci., U.S.A. 43, 152–158.

Campbell, P., Good, J.M., Dean, M.D., Nachman, M.W., 2012. The contribution of the Y chromosome to hybrid male sterility in house mice. Genetics 191, 1271–1281.

Charlesworth, B., Hartl, D.L., 1978. Population dynamics of the segregation distorter polymorphism of *Drosophila melanogaster*. Genetics 89, 171–192.

Clarke, W.E., 1904. On some forms of *Mus musculus*, Linn., with descriptions of a new subspecies from the Faeroe Islands. Proc. R. Phys. Soc., Edinburgh. 15, 160–167.

Clarke, B., Kirby, D.R.S., 1966. Maintenance of histocompatibility polymorphisms. Nature 211, 999–1000.

Cox, D.R., Smith, S.A., Epstein, L.B., Epstein, C.J., 1984. Mouse trisomy 16 as an animal model of human trisomy 21 (Down syndrome): Production of viable trisomy 16↔ diploid mouse chimeras. Develop. Biol. 101, 416–424.

Crowley, J., Zhabotynsky, V., Sun, W., Huang, S., Pakatci, I.K., et al., 2015. Analyses of allele-specific gene expression in highly divergent mouse crosses identifies pervasive allelic imbalance. Nat Genet 47, 353–360.

Deol, M.S., 1958. Genetical studies on the skeleton of the mouse: XXIV. Further data on skeletal variation in wild populations. Develop 6, 569–574.

Duncan, E.R.W., Wakeland, K., Klein, J., 1979. Heterozygosity of H-2 loci in wild mice. Nature 281, 603–605.

Dunn, L.C., Bennett, D., 1967. Persistence of an introduced lethal in a feral house mouse population. Amer. Natural. 101, 538–539.

Dunn, L.C., Bennett, D., 1971. Lethal alleles near locus *T* in house mouse populations on the jutland peninsula. Denmark. Evol. 25, 451–453.

Dunn, L.C., Levene, H., 1961. Population dynamics of a variant *t*-allele in a confined population of wild House mice. Evolution 15, 385–393.

Durand, D., Ardlie, K., Buttel, L., Levin, S.A., Silver, L.M., 1997. Impact of migration and fitness on the stability of lethal *t*-haplotype polymorphism n *Mus musculus*: a computer study. Genetics 145, 1093–1108.

Dureje, L., Macholan, M., Baird, S.J.E., Pailek, J., 2012. The mouse hybrid zone in Central Europe: from morphology to molecules. J. Vertebrate Biol. 61, 304–318.

Epstein, C.J., Cox, D.R., Epstein, L.B., 1985. Mouse trisomy 16: an animal model of human trisomy 21 (Down syndrome). Ann. N. Y. Acad. Sci 450, 157–168.

Ferris, S.D., Ritte, U., Fischer Lindahl, K., Prager, E.M., Wilson, A.D., 1983. Unusual type of mitochondrial DNA in mice lacking a maternally transmitted antigen. Nucleic Acids Res 11, 2917–2926.

Figueroa, F., Zaleska-Rutczynska, Z., Adolph, S., Nadeau, J.H., Klein, J., 1983. Genetic variation of wild mouse populations in southern Germany. II. Serological study. Genet. Res. 41, 135–144.

Figueroa, F., Tichy, H., Berry, R.J., Klein, J., 1986. MHC Polymorphism in Island Population of Mice. Curr, Top. Microbiol. Immunol. 127, 100–105.

Figueroa, F., Kasahara, M., Tichy, H., Neufeld, E., Ritte, U., Klein, J., 1987. Polymorphism of unique noncoding DNA sequences in wild and laboratory mice. Genetics 117, 101–108.

Fischer Lindahl, K., 1986. Genetic Variants of Histocompatibility Antigens from Wild Mice. Curr. Top. Microbiol. Immunol. 127, 272–278.

Fischer Lindahl, K., Burki, K., 1982. Mta, a maternally inherited cell surface antigen of the mouse, is transmitted in the egg. Proc. Nat'l Acad. Sci., U.S.A. 79, 5362–5366.

Fischer Lindahl, K., Hausmann, B., Chapman, V.M., 1983. A new *H-2*-linked class I gene whose expression depends on a maternally inherited factor. Nature 306, 383–385.

Fossella, J., Samant, S.A., Silver, L.M., King, S.M., Vaughan, K.T., Olds-Clarke, P., Johnson, K.A., Mikami, A., Vallee, R.B., Pilder, S.H., 2000. An axonemal dynein at the *Hybrid sterilitiy 6* locus: implications for *t* haplotype-specific male sterility and the evolution of species barriers. Mammal. Genome 11, 8–15.

Ganem, G., 1993. Ecological characteristics of Robertsonian populations of the house mouse: Is their habitat relevant to their evolution? Mammalia 57, 349–357.

Good, J.M., Handel, M.A., Nachman, M.W., 2007. Asymmetry and polymorphism of hybrid male sterility during the early stages of speciation in house mice. Evol 62, 50–65.

Gropp, A., Winking, H., 1981. Robertsonian translocations: cytology, meiosis, segregation patterns and biological consequences of heterozygosity. Zool. Soc. London Symp. 47, 141–181.

Gummere, G.R., McCormick, P.J., Bennett, D., 1986. The influence of genetic background and the homologous chromosome *17* on *t*-allele transmission ratio distortion in mice. Genetics 114, 235–245.

Haig, D., Bergstrom, C.R., 1995. Multiple mating, sperm competition and meiotic drive. J. Evol. Biol. 8, 265–282.

Hammer, M.F., Silver, L.M., 1993. Phylogenetic analysis of the alpha-globin pseudogene-4 (Hba-ps4) locus in the house mouse species complex reveals a step-wise evolution of *t*-haplotypes. Mol. Biol. Evol. 10, 971–1001.

Hammerberg, C., Klein, J., 1975. Linkage disequilibrium between *H-2* and *t* complexes in chromosome 17 of the mouse. Nature 258, 296–299.

Hammerberg, C., Klein, J., 1976. Histocompatibility-2 system in wild mice II. *H-2* haplotypes of *t*-bearing mice. Transplant 21, 199–212.

Herbst, E.W., Gropp, A., Nielsen, K., Hoppe, H., Freymann, P.D.H., 1982. Reduced ability of mouse trisomy 16 stem cells to restore hemopoiesis in lethally irradiated animals. In: Baum, S.J., Ledney, G.D., Thierfelder, S., Karger, S. (Eds.), Experimental hematology today. Karger, Basel, pp. 119–126.

Johnston, P.G., Brown, G.H., 1969. A comparison of the relative fitness of genotypes segregating for the t^{w2} allele in laboratory stock and its possible effect on gene frequency in mouse populations. Amer. Natural. 103, 5–21.

Keane, T., Goodstadt, L., Danecek, P., White, M.A., Wong, K., et al., 2011. Mouse genomic variation and its effect on phenotype and gene regulation. Nature 477, 289–294.

Klein, J., Golubic, M., Budimir, O., Schopfer, R., Kasahara, M., Figueroa, F., 1986. On the origin of *t* chromosomes. Curr. Top. Microbiol. Immunol. 127, 239–246.

Krebs, C.J., Gaines, M.S., Keller, B.L., Myers, J.H., Tamarin, R.H., 1973. Population Cycles in Small Rodents Demographic and genetic events are closely coupled in fluctuating populations of field mice. Science 179, 35–41.

Lacy, RC, 1978. Dynamics of t-alleles in *Mus musculus* populations: review and speculation. The Biologist 60, 41–67.

Larson, A., Prager, E.M., Wilson, A.C., 1984. Chromosomal evolution, speciation and morphological change in vertebrates: the role of social behaviour. In: Bennett, M.D., Gropp, A., Wolf, U. (Eds.), Chromosomes Today. Springer, Dordrecht.

Laurie, C.C., 1997. The weaker sex is heterogametic: 75 years of Haldane's rule. Genetics 147, 937–951.

Laurie, E.M.O., 1946. The reproduction of the house-mouse (*Mus musculus*) living in different environments. Proc. Rou. Soc B 133, 248–281.

Lenington, S., 1983. Social preferences for partners carry 'good genes' in wild house mice. Anim. Behav. 31, 325–333.

Lenington, S., Franks, P., Williams, J., 1988. Distribution of *T*-haplotypes in natural populations of wild house mice. J. Mammol. 69, 489–499.

Lenington, S., Heisler, I.L., 1991. Behavioral reduction in the transmission of deleterious t haplotypes by wild house mice. Amer. Natural. 137, 366–378.

Lenington, S., Cooopersmith, C.B., Erhart, M., 1994. Female Preference and Variability Among t-Haplotypes in Wild House Mice. Am. Natural. 143, 766–784.

Lenington, S., Drickamer, L.D., Robinson, A.S., Erhart, M., 1996. Genetic basis for male aggression and survivorship in wild house mice (*Mus domesticus*). Aggressive Behav. 22, 135–145.

Levin, B.R., Petras, M.L., Rasmussen, D.L., 1969. The effect of migration on the maintenance of a lethal polymorphism in the house mouse. Amer. Natural. 103, 647–666.

Levine, L., 1958. Studies on sexual selection in mice I. Reproductive Competition between Albino and Black-Agouti Males. Amer. Natural. 92, 21–26.

Levine, L., Lascher, B., 1965. Studies on sexual selection in mice II. Reproductive competition between black and brown males. Amer. Natural. 97, 67–72.

Levine, L., Krupa, P.L., 1965. Studies on sexual selection in mice. III. Effect of the gene for albinism. Amer. Natural. 100, 227–234.

Levine, L., 1967. Sexual selection in mice. IV. Experimental demonstration of selective fertilization. Amer. Natural 101, 289–294.

Levinson, J.R., McDevitt, H.O., 1976. Murine *t* factors: an association between alleles at *t* and *H-2*. J. Exp. Med. 144, 834–839.

Lewontin, R.C., Dunn, L.C., 1960. The evolutionary dynamics of a polymorphism in the house mouse. Genetics 45, 705–722.

Lewontin, R.C., 1962. Interdeme selection controlling a polymorphism in the house mouse. Amer. Natur. 96, 65–78.

Lewontin, R.C., 1968. The effect of differential viability on the population dynamics of *t*-alleles in the house mouse. Evolution 22, 262–273.

McLaren, A., Michie, D., 1958. An effect of the uterine environment upon skeletal morphology in the mouse. Nature 181, 1147–1148.

Manser, A., Lindhom, A.K., Weissing, F.J., 2017. The evolution of costly mate choice against segregation distorters. Evol 27, 2817–2818.

Michaelson, J., 1983. Genetics of beta-2-microglobulin in the mouse. Immunogenet 17, 219–259.

Mihola, O., Trachtulec, Z., Vicek, C., Schimentia, J.C., Forejt, J., 2009. A mouse speciation gene encodes a meiotic histone H3 methyltransferase. Science 323, 373–375.

Myers, J.H., 1974. The absence of *t*-alleles in feral populations of house mice. Evol 27, 702–704.

Morita, T., Kubota, H., Murata, K., Nozaki, M., Delarbre, C., Willison, K., Satta, Y., Sakaizumi, M., Takahata, N., Gachelin, G., Matsushiro, A., 1992. Evolution of the mouse *t* haplotype: recent and worldwide introgression to *Mus musculus*. Proc. Nat'l Acad. Sci., U.S.A. 89, 6351–6355.

Nachman, M.W., Searle, J.B., 1995. Why is the house mouse karyotype so variable? Trends Ecol. Evol. 10, 397–402.

Nadeau, J.H., 1983. Absence of detectable genetic disequilibrium between the *t*-complex and linked allozyme-encoding loci in house mice. Genet. Res. 42, 323–333.

Nadeau, J.H., Wakeland, E.K., Gotze D, D., Klein, J., 1981. The population genetics of the *H-2* polymorphism in European and North African populations of the house mouse (*Mus musculus* L.). Genet. Res. 37, 17–31.

Nadeau, J.H., Collins, R.L., 1983. Does associative overdominance account for the extensive polymorphism of *H-2*-linked loci. Genetics 105, 2441–2444.

Nash, H.R., Brooker, P.C., Davis, J.M., 1983. The Robertsonian Translocation House-mouse Populations of North East Scotland: A Study of Their Origin and Evolution. Hered 50, 303–310.

Newton, M.F., Peters, J., 1983. Physiological variation of mouse haemoglobins. Proc. Roy. Soc., Lond. B 218, 443–453.

Oka, A., Aoto, T., Totsuka, Y., Takahashi, R., Ueda, M., Mita, A., et al., 2007. Disruption of Genetic Interaction Between Two Autosomal Regions and the X Chromosome Causes Reproductive Isolation Between Mouse Strains Derived From Different Subspecies. Genetics 175, 185–197.

Palm, J., 1970. Maternal and fetal interactions and histocompatibility antigen polymorphisms. Transplant. Proc. 2, 162–173.

Pennycuik, P.R., Johnston, P.G., Lidicker, W.Z., Westwood, N.H., 1978. Introduction of a male sterile allele (tw2) into a population of house mice housed in a large outdoor enclosure. Aust. I. Zool. 26, 69–81.

Petras, M.L, 1967. Studies of natural populations of Mus. I. Biochemical populations and their bearing on breeding structure. Evolution 21, 259–274.

Petras, M.L., Topping, J.C., 1983. The maintenance of polymorphisms at two loci in house mouse (Mus musculus) populations. Canad. J. Genet. Cytol. 25, 190–201.

Pilder, S.H., Olds-Clarke, P., Phillips, D.M., Silver, L.M., 1993. *Hybrid sterility 6:* A mouse *t* complex locus controlling sperm flagellar assembly and movement. Develop. Biol. 159, 631–642.

Potts, W.K., Manning C.J., C.I., Wakeland, E.K., 1991. Mating patterns in seminatural populations of mice influenced by MHC genotype. Nature 352, 619–621.

Robinson, P.J., Steinmetz, M., Moriwaki, M., Fischer-Lindahl, K., 1984. Beta-2-microglobulin types in mice of wild origin. Immunogenet 20, 655–665.

Rowe, F.P., Taylor, E.J., Chudley, A.H.J., 1959. The numbers and movements of house-mice in the vicinity of four corn ricks. J. Anim. Ecol. 32, 87–97.

Rudolph, N.S, Vandeberg, J.L., 1981. Linkage disequilibrium between the *T/t* complex and *Pgk-2* in wild mice. J. Exp. Zool 217, 455–458.

Runge, J.-N., Lindhom, A.K., 2018. Carrying a selfish genetic element predicts increased migration propensity in free-living wild house mice. Proc. Roy. Soc. London, B 285. doi:10.1098/rspb.2018.1333 doi.org/.

Ruvinsky, A., Polyako, A., Agulnik, A., Tichy, H., Figueroa, F., Klein, J., 1991. Low diversity of *t* haplotypes in the eastern form of the house mouse. Mus musculus L. Genetics 127, 161–168.

Sage, R.D., Atchley, W.R., Capanna, E., 1993. House mice as models in systematic biology. System Biol 42, 523–561.

Sarkar, B., 2016. Random and non-random mating populations: Evolutionary dynamics in meiotic drive. Marh. Biosci. 271, 29–41.

Schimenti, J., Silver, L.M., 1986. Amplification and rearrangement of DNA sequences during the evolutionary divergence of *t* haplotypes and wild-type froms of mouse chromosome 17. Curr. Top. Microbiol. Immunol. 127, 247–252.

Selander, R.K., 1970. Behavior and genetic variation in natural populations. An. Zool. 10, 53–66.

Snijders, A.M., Langley, S.A., Kim, Y.-M., Brislawn, C.J., Noecker, C., 2016. Influence of early life exposure, host genetics and diet on the mouse gut microbiome and metabolome. Nature Microbiol. 2, 1622. doi:10.1038/nmicrobiol.2016.221.

Southern, H.N., 1952. *Control of Rats and Mice. Vol. III House Mice.* Clarendon Press, Oxford (London: Geoffrey Cumberlege, Oxford University Press).

Southern, H.N., Laurie, E.M.N., 1946. The house mouse (*Mus musculus*) in corn ricks. J. Anim. Ecol. 15, 134–149.

Sved, J.A., 1983. Does natural selection increase or decrease variability at linked loci? Genetics 105, 239–240.

Takada, Y., 1985. Demography in island and mainland populations of the feral house mouse, Mus musculus molossinus. J. Mammal. Soc. Japan 10, 179–191.

Teeter, K.C., Payseur, B.A., Harris, L.W., Bakewell, M.A., Thibodeau, L.M., et al., 2008. Genome-wide patterns of gene flow across a house mouse hybrid zone. Genome Res 18, 67–76.

Traulsen, A., Reed, F.A., 2012. From genes to games: Cooperation and cyclic dominance in meiotic drive. J. Theor. Biol. 299, 120–125.

Van Boven, M., Weissing, F.J., 1999. Segregation distortion in a deme-structured population: opposing demands of gene, individual and group selection. J. Evol. Biol. 12, 80–93.

Winking, H., 1986. Some aspects of Robertsonian karyotype variation in European wild mice. Curr. Topics in Microbiol. And Immunol. 127, 68–73.

Young, S.S.Y., 1967. A Proposition on the Population Dynamics of the Sterile t Alleles in the House Mouse. Evol. 23, 190–193.

Zaleska-Rutczynska, Z., Firgueroa, F., Klein, J., 1983. Sixteen new H-2 haplotypes derived from wild mice. Immunogenet 18, 189–203.

6

Mutation studies including those targeting the *T/t*-complex—a century long project

At the end of the Second World War, nuclear forces were released onto the world with a stunning display of power. The physical destruction was obvious but little was known about the effects of all the radiation released. There were hints of the danger from the cancers of exposed parts of the body in the early X-ray pioneers, the radiation sickness and necrosis of the jaw in the watch dial painters who used radium-containing paint, and the known mutagenic effects in *Drosophila* but what were the effects of whole body radiation in humans? Some sought quick answers with dosing uninformed patients with radioisotopes[1] or exposing troops to fall out from A-bomb tests. Others sought more humane scientific answers.

Direct human research also was performed on the genetic consequences of the atomic bombing of Hiroshima and Nagasaki. Survivors and their children were studied by the Atomic Bomb Casualty Commission from 1946 to 1974 and the Radiation Effects Research Foundation since 1975. These were joint effort with many Japanese and American, physicians and scientists and were operated by the U. S. National Academy of Sciences, the Japanese Institute of Health (to 1974), and the Japanese Ministry of Health and Welfare (since 1975). The results of these human studies were summarized in **The Children of Atomic Bomb Survivors: A Genetic Study, ed. By James V. Neel and William J. Schull, 1991.** Briefly, these studies could show no statistically significant, harmful effects on pregnancy outcomes or the resultant children's life expectancies and incidence of cancer in over 60,000 exposed pregnancies. Obviously, the huge number of variables in such human studies might have hidden significant effects and, thus, mouse model studies became all the more critical. Both Jim Neel (**Physician to the Gene Pool: Genetic Lessons and Other Stories, 1994**) and Bill Schull (**Song Among the Ruins, 1990**) have written highly readable accounts of their lives and roles in the Atomic Bomb Casualty Commission.

Used as models for human exposures, mutagenesis became extensively studied in mice. As discussed in chapter 1, the requirement to assess radiobiological effects led to the establishment of two major mouse research institutions shortly after the war: the Harwell Laboratory (Medical Research Council [MRC] Radiobiology Unit—now known as the MRC Mammalian Genetics Unit and the U.K. Mouse Genome Centre) in England and the Oak Ridge National laboratory in the U.S.A. Germany was to add a big mouse germ-cell mutagenesis project in Neuherberg in the midsixties. Early attempts with X-rays in mice have been mentioned in regards to the origin of Dobrovalskaia-Zavadskaia's *T* mutation (Chapter 1) and include ones by Little and Bagg (Little & Bagg,1924; Bagg & Little, 1924) when there was no direct quantitation of the degree of exposure (gap and amperage of the spark of the machine, distance from the mouse, duration of

Twentieth Century Mouse Genetics. DOI: https://doi.org/10.1016/B978-0-12-824016-8.00004-0

exposure and whether or not filters were used were recorded). Little and Bagg were as defensive as to whether the X-rays had caused the mutants as Dobrovalskaia-Zavadskaia had been. Acceptance that X-rays really caused mutations in mice occurred with the work of Snell (1933, 1935) and Hertwig (1935) with findings that X-irradiation of post-meiotic germ cell stages in male mice could result in dominant lethal mutations and sterility.

The history of the MRC Radiobiology Unit at Harwell is central to Chadarevian's (2006) article titled "Mice and The Reactor: The 'Genetics Experiment' in 1950s Britain". Knowing that the Manhattan Project had included radiation-induced mutation studies in mice, and expecting the data to be shared (from Rochester, see Ernst Caspari minibiography) and Oakridge, the Harwell group initially planned to study mutation rates in rabbits (at the suggestion of J.B.S. Haldane (see Haldane-Mitchison Clan minibiography, Chapter 3). However, setting up to do this was to take considerable time so "work on the genetic effects on mice under chronic gamma irradiation started at the Institute of Animal Genetics in Edinburgh under the supervision of the distinguished embryologist and geneticist Conrad Waddington, the institution's new head…. The project was funded by the MRC. Loutit, the director of the Harwell Radiobiological Research Unit, was involved as a consultant" (Chadarevian, 2006).

Part of the reason for doing it in Edinburgh was that H.J. Muller had been visiting there just before the war and left a suitable X-ray machine behind when he returned to the United States, and a radium source was found for the gamma-ray experiments. They also used the Specific Locus Test design (see below). However, it soon became clear that the mutation rates were so low that more than a hundred thousand mice would need to be treated to find statistically significant results and this was beyond the capacity at Edinburgh.

Thus, this group was moved to Harwell in 1953 which was to build larger facilities and where there were also better gamma-ray sources. At this time, they also renewed their mouse stocks, receiving mice from the Oakridge Laboratory (see below) so that they would be using the same genetic material and so that results could be compared. Tobias Carter, who had moved with the unit (Carter, 1957, 1959), initially called the Genetics Unit, left in the late 1950s, possibly in part because of the difficulties communicating with the Oak Ridge staff (See Chapter 1 and its endnote 5). "A personal statement drafted by Carter and submitted as confidential document for discussion to the MRC Genetics Research Committee in 1958 revealed the extent to which Anglo-American relations dominated, and threatened to undermine, the mice irradiation experiments at Harwell…The document went on to list in detail the many occasions in which cooperation had been hampered and trust had been misused" (Chadarevian, 2009). The emphasis on these radiation-induced mutation experiments were diminished after the Nuclear Test Ban Treaty of 1963, but radiation experiments in mice were continued until the 1990s at Harwell by John Searle and Mary Lyon (reviewed in Searle, 1974). The unit was down-sized in 1968 and the cytogeneticist, Charles Ford (see Chapters 3 and 11), moved to Oxford. In the mid1990s there was further re-organization into a Radiation and Genome Stability Unit and a Mammalian Genetics Unit in which Mary Lyon played a major role. The latter was further enlarged when the U.K. M.R.C. Mouse Genome Unit was added. This unit provided sperm and embryo banking for all of the U.K. and Harwell continues to play an important role in mouse genetic studies.

Lianne B. Russell (2013) has written a history of the Oak Ridge Laboratory describing its many efforts in areas other than just mutation research. It was founded in 1946 by Dr. Alexander Hollander who recruited the Russell's (Lianne and Bill) from the Jackson laboratory to develop the mouse program. Accumulating mice was a problem because the Jackson laboratory had just had its disastrous fire (Chapter 1) and it, too, was seeking to replace its stocks. A mouse fancier helped them out: "we discovered an unusual source of supply—a Florida pharmacist whose hobby was mouse fancy. In a spotlessly clean converted garage in his backyard, Mr. Holman was breeding the coat-color mutants that were needed as markers in the specific-locus test, as well as numerous other variants" (Russell, 2013).

After extensive discussion, Bill Russell decided to develop the Specific Locus Test: the treated mouse was crossed to a stock homozygous for 7 visible recessive mutations, mostly affecting coat color, loci (Russell,1989). This avoided the usual problem with detecting recessive mutations—the possibly mutated progeny would have to be mated back to the parent, usually now sterile, so the progeny would need to be mated and daughters mated back to their G1 fathers, adding two generations to the screen. Only in males would recessive X-linked mutations be detectable in the first generation. As William Dove (1987) has pointed out, this approach had a long historical tradition with Muller (Muller and Altenburg, 1919) using it in *Drosophila* in 1919, Snell using it in mice in 1935, and Carter, et al 1958, (1957), applying it to mice for ionizing radiation in 1956. Any F1s showing the trait had to have a new mutation. While only assaying for limited kinds of mutations, it had the advantage of being rapidly evaluated—a technician could screen 2000 loci in an hour (Russell, 2013). It was very useful for comparing doses, stage of germinal cell development, etc. The sheer size of their facility was amazing—at its peak in the midfifties, they had 36,000 cages (Russell, 2013)![2] The first experiments were started in 1949 and were with X-rays. A gamma-ray source was soon added. The experiments conducted included a range of exposures over a range of times and a range of stages, from the embryonic to specific stages of male gametogenesis (measured by knowing the rate at which sperm from specific stages of meiosis reached the vas deferens for ejaculation, see Chapter 8 for a detailed discussion of the stages of spermatogenesis). These studies were performed on a massive scale and continued into the 1990s. The Oakridge Laboratory, like its Harwell counterpart, also started freezing embryos in the mid1980s, rather than maintaining their more than 1,000 mutant stocks. The experiments finally came to an end in 2009 when the DOEs funding became limited to energy research and the lab was closed. The huge collection of frozen embryos and frozen sperm was then taken over by the Jackson laboratory.

Mutagenesis by ionizing radiation

The ionizing radiations studies were primarily X-rays (lower wavelength than UV [higher energy]) and gamma-rays (even lower wavelength [even higher energy]). Amounts were originally expressed as rads but the use of the Gray as a unit soon became standard (1 Gray = 1 Joule/kg = 100 rad). The measure of linear energy transfer (LET), since heavy particles, such as alpha-particles, have a high LET and dissipate their energy over a short distance while passing through living matter, was also relevant. Both of these are affected by differing densities of tissues but this difference is relatively small. There were a few attempts to orally dose mice with alpha-emitters or beta-emitters but the dosimetry for the important tissues (the gonads) were problematic and, therefore, studies of particle emitters were not much pursued. Since there was great uncertainty about the kinds and amounts of radioactivity released at Hiroshima and Nagasaki, the controlled experiments in mice were considered to be very important.

The multiple, specific-loci test was widely used in a number of laboratories. As reviewed by Neel and Lewis (1990), on the basis of 2,346,687 loci tested, the mutation rate for acute single or multiple, spermatogonial doses of radiation was 2.2×10^{-7} per rad (0.01 Gy) while the control rate was 9.0×10^{-8} in 6,417,082 loci. The doubling dose was thus 41 rad. A lower rate was found in females but the number of loci tested was dramatically lower since the females quickly became sterile. There was great variation between the different loci of the 7 locus test. This is relevant to the difference found between mostly the Oakridge data and the Harwell data which used 6, mostly non-overlapping loci. Their mutation rate, but only on 149,004 loci, was 7.6×10^{-7} per rad. Other assays for mutations found a number of different rates.

Because of the need for additional crosses, very few studies of mutation rates for other recessive genes were made. Schlager and Dickie (1966) reported the results of years of experience at the Jackson Lab. Mice from 18 inbred strains maintained by brother-sister matings, and six F1 hybrids produced by crossing various pairs of these strains, were examined two to four times between birth and weaning for alterations in coat color. The variation detected had to be transmissible to be scored. Of these strains, 15 were homozygous for 5 specific coat-color alleles. They scored 5.2 million loci in 1.5 million mice, finding a forward rate of 11.1×10^{-6} and a reverse rate of 2.7×10^{-6} per rad with a great variability among loci. Similar results were found by the Russells (Russell & Russell, 1996) with the seven, mostly overlapping between the 2 labs, loci of the specific locus test—a rate of 11×10^{-6} per rad.

Dominant mutations grossly visible in the offspring; especially skeletal, neurological, cataracts, fur and skin, including coat color; of radiated males gave an incidence of 4.7×10^{-7}/ gamete vs. a spontaneous rate of less than 1×10^{-7} (although one small subset had shown a spontaneous rate of 0.8×10^{-5}/gamete) which led to a doubling rate of 17 rad (Searle, 1974; Searle & Beachey, 1986). Many of the coat color mutations appeared as patches and were found to be due to somatic mutations. Most of these had occurred during pre-meiotic DNA synthesis and the spontaneous rate was as high as the induced rate (Russell, 1999). When slit lamp exams were added for the cataracts and cadaver clearing and skeletal staining added for the skeletal defects, little was added because these laborious techniques led to the study of fewer animals and almost no control mutations were found for base rates.

A test for mitotic, somatic mutations in coat color genes was also developed (Russell & Major, 1957). Irradiation of the pregnant mother at 9.5 days resulted in frequent patches of coat color mutations in the offspring. It was estimated that about 150-200 prospective pigment cells were being targeted and the resultant mice could be rapidly screened (after a few weeks of age since even darkly pigmented adults are only "dusky" at birth). The conclusion was that the frequency of these mitotic mutations was equal to that found in spermatogonial stages was important.

The addition of protein electrophoresis (see Chapter 3) to detect mutational events brought a new dimension to mutation studies. Both shifts in electrophoretic movement or absence of an electrophoretic band could be scored. As summarized by Neel and Lewis (1990), the spontaneous rate was 0.7×10^{-6}/locus/generation while the increase per rad was 6.3×10^{-8}. This was extended to two-dimensional electrophoresis where, among hundreds of spots, indicating specific proteins, and different in different tissues, could be studied (Baier, et al, 1984).

There was also an attempt to use gain or loss of mutations in the histocompatibility system (Bailey & Usuma, 1960) to study the mutation rate. A circular exchange of grafts (round robin) among a group of offspring of irradiated males of an inbred strain could detect gain of antigens or loss (the deficient mouse would react against one of its strain's antigens since it no longer would be tolerant; see Chapter 7). However, the control rate detected was so low that the estimate of the radiation-induced rate had very wide confidence intervals (Bailey & Kohn, 1965; Kohn et al.,1976). The approach would detect mutations in any of the 15-17 histocompatibility genes thought to be present (see Chapter 7; now it is known that there are many more) but high spontaneous rates in *H-2* (Yeom, et al) and sex differences other than the classic H-Y were confounding variables. Donald W. Bailey has provided an historical account of these attempts and their eventual success in finding useful *H-2* mutations (Bailey, 1988).

A number of important conclusions came from the multiple studies. One was that the mutation rate per rad was significantly higher in mice than in *Drosophila* (Russell, 1956). A firm conclusion on a doubling rate for man or mouse was impossible. The specific locus test was probably too sensitive—coat color mutations have higher spontaneous frequencies than do many other types of mutations and this was the most extensively used test. Neel and Lewis (1990) concluded that the doubling dose was about 1.34 Gy in mice but they thought that this should be multiplied by 3 since these were acute doses of high radiation and not chronic low doses. Over all they felt "that the genetic risk of low-dose rate, low-LET [linear energy transfer] radiation for humans is less than has generally been assumed during the past 3 years" (Neel & Lewis, 1990).

Another major conclusion was that most of the radiation-induced mutations were due to deletions (see Chapter 1 and the discovery of the *T*-locus; Russell, 1987). These radiation-induced alleles were also instrumental in the identification of several of the coat color genes used in the specific-loci test, e.g., the *pink-eyed dilution* locus from which over 100 radiation-induced alleles were recovered (Lyon, et al 1992). In addition, many translocations were found (Generoso et al.,1984; see Chapter 3) and increases in sex chromosomal aneuploidy (Griffin & Tease, 1988).

Chemical mutagenesis

Although the rate of radiation-induce mutagenesis was many times the spontaneous rate, it was far too low to be used in any screening program for mutations. Also, the high frequency of deletions did not provide many single gene mutations. Thus, attention was turned to chemical agents. N-ethyl-N-nitrosourea (ENU), a potent alkylating agent, was found to be a powerful mutagen (Russell et al., 1979). Its chemical reaction is to add an ethyl group to nitrogen or oxygen radicals in the DNA bases, thus inducing mispairing. It can induce mutation rates as high as one in one thousand loci in an appropriately dose-treated male (Russell et al., 1982). The

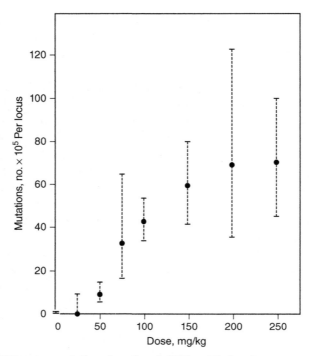

FIG. 6.1 Dosage curve of ENU mutagenesis (from Russell et al.,1982), public domain.

dosage rate was amazingly linear (Fig. 6.1 Russell et al., 1982). These are mostly point mutations (Popp et al., 1983) since its target is premeiotic spermatogonia (Russell et al., 1990).

Chlorambucil, another alkylating agent but one working at post-meiotic germ cells, unlike ENUs action in pre-meiotic stages where DNA synthesis is occurring, is also a powerful chemical mutagen (Russell LB et al., 1989) but it, like radiation, mostly produced deletions (Rinchik & Russell, 1990). However, they are so frequent that deletion mapping could be considered.

ENU was the first mutagen with a high enough mutation frequency such that it could be used for targeted genetic screens, especially to identify models of human genetic diseases, e.g. phenylketonuria (McDonald et al., 1990). The screen used to detect mutations was the Guthrie test, which was used in human newborn screening and is based on elevated phenylalanine levels in blood. The relatively common human mutation is in phenylalanine oxidase but the first mutation found turned out to be the result of a deficiency in GTP-cyclohydrolase instead and this has a different biochemical phenotype from phenylketonuria in humans (where the GTP-hydrolase deficiency is very rare). Increased levels of phenylalanine similar to those in human phenylketonuria patients occur during the first 10 days of life in this mutant but normal levels are found after 20 days of age unless a phenylalanine load is given (while they remain greatly elevated for life, unless treated, with the phenylalanine oxidase deficiency in humans). Mutations in X-linked genes could be directly detected in the first generation as was found for Erb-Duchenne muscular dystrophy (Chapman et al., 1989).

Screening methods were particularly fruitful when electrophoretic differences for a target enzyme exist between two parental strains. In this case, one strain is subjected to mutagenesis and mated to the other. The appearance of a uniparental electrophoretic pattern in the F1, instead of the hybrid electrophoretic pattern, would indicate the loss of the gene product in the mutagenized parent, for example, a null mutation. Breeding can then be performed to maintain the mutation and to create homozygotes. When used with triethylenemelamine, this approach has resulted in a mouse model of alpha-thalassemia (Whitney and Russell, 1980) while an ENU-induced electrophoretic hemoglobin mutant provides a precise model of hemoglobin Rainier with polycythemia (Peters et al., 1985) or alpha-globin, thalessemia (Popp, et al 1985). An example of the successful application of this procedure to another known human genetic disease concerns carbonic anhydrase II deficiency as a model for human osteopetrosis with renal tubular acidosis which is associated with a deficiency of carbonic anhydrase II, at least in erythrocytes (Sly et al., 1983). During an ENU mutagenesis experiment, a carbonic anhydrase II null was found and the homozygotes had many features of the human disorder (Lewis et al., 1988).

The availability of the radiation- and chlorambucil-induced deletions led to a number of newly possible studies. For instance, gene dosage effects for a mitochondrial enzyme (malic dehydrogenase, Mod-2, which mapped near the *c* locus and was deleted in some of the radiation-induced mutants) was found (Diamond & Erickson, 1974). Similar dosage effects were found for a number of cytoplasmic enzymes in which human mutations had been found. In fact, such dosage phenomena were used for mutational screens in the children of atomic bomb survivors by using highly automated enzyme assays (the Miniature Centrifugal Fast Analyzer; Neel et al., 1988). The size of the deletions was such that multiple polypeptides might be absent on 2-dimensional protein gels (Baier et al., 1984) and attempts were made to use 2-dimensional gels to study mutations in the children of atomic bomb survivors (Neel et al.,1989). When one had a marker near a known gene, but not a marker for the gene itself, it was also possible to search for deletions which deleted both the marker and the gene of interest.

Insertional mutagenesis

Although most spontaneous mutations were found to be due to nonsense (base changes) or missense (base changes or indels leading to stop mutations [frequently found further in the reading frame]) when the cloning era came (see Chapter 9), some were due to insertions of ectopic viruses or mobile elements. The A^{vy} allele of the non-agouti locus, which is much studied for transmissible epigenetic effects (see Chapter 10), is due to the insertion of an intracisternal A-particle, a retroposon (Duhl et al., 1994). Other examples include the insertion of a 7.1-kb. LINE (long interspersed element) into the gene encoding the beta unit of the glycine receptor with a phenotype of spasticity (Mulhardt et al., 1994) or the *hairless* mutation which was found to be due to the insertion of a murine leukemia virus (Stoye, et al., 1988). Somatic mutations, a relatively common cause of human non-inherited but genetic disease (Erickson, 2016), also occurred in mice due to intracisternal A insertion (Wu et al., 1997).

Thus, multiple approaches were used to deliberately induce mutations by insertional mutagenesis. Infection with retroviruses was an attractive approach for exogenously induced mutations with "tagged" markers (Soriano et al., 1987) but the yield of useful genes was much lower than with ENU (Jenkins & Copeland, 1985). In addition, there were marked position effects with retroviral insertions influencing the activity of other, usually nearby, genes (Breindl et al., 1984).

These are also examples of mutations involving autonomous transposons which encode their own transposase. Non-autonomous transposons require a transposase to be supplied by another transposon which can be provided from a distant location in the genome by another element. A group of transposons which are widespread in many species and which mostly don't have functional copies are the Tc1/mariner elements (Plasterk et al., 1999). By sequencing a large number of inactive ones from fish, Ivics, et al. (1997) deduced an active sequence which was inserted into a cloned copy of an inactive Tc1/mariner element (by targeted mutagenesis in bacterial plasmids) and then cloned to create a "sleeping beauty" element which could be inserted into the mouse germ line by transgenesis (see Chapter 9). When such a mouse was crossed with a mouse transgenic for inactive Tc1/mariner transposons, activation and multiple insertions occurred, frequently causing mutations (Izvack et al., 2000). The advantage of such mutations was that they were "attached" to a "foreign" DNA marker which provided for rapid cloning of the mutated DNA and identification of the gene involved in the screened phenotypic variation. Rick Woychik started a program at Oakridge using random transgenesis with such an element as a general approach but found only a low percent of interesting mutations. These included a recessive mutation causing polycystic kidney disease (Moyer et al., 1994) and one causing deafness and waltzing (Algraham et al., 2001). Although this approach was also amenable for embryonic stem cells, targeted replacement with a cloned mutation or deletion using mitotic recombination was more widely used (see Chapters 4 and 9). A number of methods, collectively called "gene traps" were developed to capture a variety of types of mutations in embryonic stem cells which could then be screened for relevance in the stem cells before the stem cells were used in embryo injection (reviewed in Zambrowicz and Friedrich, 1998). These included promoter and enhancer traps to identify active genes, gene traps to identify transcriptionally active gene by providing splice site acceptors, and PolyA addition traps which would also capture non-expressed genes since the marker encoded in the vector would require a poly-adenylation signal acquired by landing in such a sequence. Several private companies were founded which screened many insertional mutations and provided the mutant mice to the interested investigator. For instance, interest in the embryonic role of thioredoxin, an important regulator of radical oxygen species, was studied with such a commercially available mutation with the finding of early lethality (Nonn et al., 2003). As less expensive sequencing became available, pooled replicas could be sequenced to identify mutations of interest which then could be recovered from stored stocks. The disadvantages of all these methods was the requirement to then develop the mouse line by embryo injection. Targeting particular genes initially involved providing selectable markers in cis but a sequential pooling strategy (see Fig. 6.2) allowed elimination of such exogenous DNA.

Mutational screens using insertional mutational approaches were further developed in the 21st Century but were largely superseded by ENU-mutagenesis combined with massively parallel DNA sequencing. It was possible to treat a male with ENU and cross him with females of the same strain to generate large numbers of offspring whose DNA could be sequenced, finding thousands of potential mutations per offspring of which nearly a hundred will be in coding or splicing sequences and,

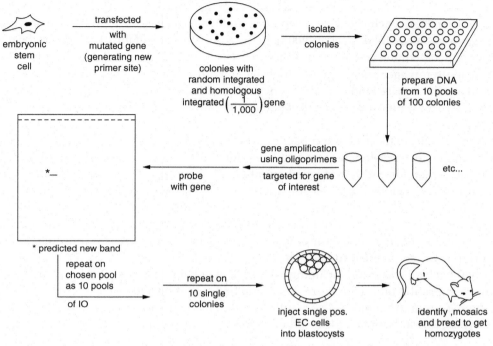

FIG. 6.2 A generalized schemata for creating directed mutations in mice by using homologous recombination in embryonic stem cells (from Erickson, 1988, with permission).

therefore, more easily understood. The sperm would have been frozen from the treated male and artificial insemination could be used to recover the mutation (Gondo et al., 2010). For instance, the method generated a mutaton in N-acetyl transferase, an important drug modifier (see Chapter 12, Erickson, et al 2008).

A form of targeted mutagenesis which entered the arena with the cloning era was by the addition of a mutated gene, the mutation being performed in bacterial plasmids, with subsequent incorporation into the mouse genome by transgenesis. Since the normal gene would also be present, this was primarily useful with dominant negatives but a great excess of the mutated copy, since multiple copies of the transgene were usually incorporated, could create a hypomorph, i.e. the majority of the protein would be abnormal. A good example of the use of this strategy involves studies related to collagenopathies such as osteogenesis imperfecta. Stacey, et al. (1988) performed in vitro mutagenesis on alpha-I collagen, introducing mutations similar to those causing osteogenesis in humans. When introduced into transgenic mice, a marked similarity in pathology to that found in perinatal lethal osteogenesis imperfecta type II (OI II) was found. These experiments allowed the exploration of quantitative relationships between the abnormal collagen and the disease state—an amount as low as 10% of the abnormal collagen caused a severe disorder despite the continued presence of normal amounts of the normal collagen (due to the disruption of the folding of the triple helix of collagen).

Transgenesis for antisense constructs could also inactivate genes (Erickson et al., 1993) but this approach did not achieve wide acceptance while transgenesis for short hairpin, interfering RNAs was less successful (Cao et al., 2005). In the second decade of the 21st Century, the development of gene editing (reviewed in Delerue & Ottmer, 2015) made all the other insertional approaches to mutagenesis pretty much obsolete. With this advance, not only the gene to be mutated but the precise kind of mutation desired could be delivered. The method became the gold standard for testing the relevance of potentially pathological DNA changes in the causation of human disease.

The availability of these new approaches to mutagenesis allowed screens for genes affecting almost any stage of development, as had been done in *Drosophila* (see Chapter 4). A phenotype-based screen for embryonic lethal mutations in the mouse (Kasarkis, et al., 1998) using ethylnitrosourea identified five new mutant lines, of which 3 affected the development of the neural tube. Two of the new mutations mapped to chromosomal regions for which no developmental mutations had yet been found. As methods for targeted mutagenesis became more efficient, an international consortium was developed (the International Mouse Phenotyping Consortium) which planned to knockout more than 5,000 genes. As reviewed in 2016, 1,751 genes had been knocked out, including 410 prenatal lethal mutations (Dickinson et al., 2016). Of course, the embryonic analyses of these mutations required a huge collaborative effort and this summary report included 84 co-authors, one of which was to the International Mouse Phenotyping Consortium to indicate other members of the group.

Mutagenic studies of the *T/t* complex

Initially there was no systematic search for *T/t* complex mutations but a number were found from the radiation studies at Oakridge (Bennett et al., 1975). Paul Selby had found 7 transmissible, dominant short-tailed mutations form 186,000 offspring which he shared with the Columbia University group gathered around L. C. Dunn (ibid.). Six out of the 7 interacted with the viable *t*-allele, t^{38} and, thus, were allelic to the classic *T*. Embryonic studies of these new *T* alleles (numbered T^{Or1}- T^{Or6} after their Oak Ridge origin) found that that they had earlier lethality than the classic *T* which might suggest that they were larger deletions.

With the advent of ENU-mutagenesis which made targeted screens possible, Vernon Bode made a specific search for *T/t* complex mutations (Bode, 1984). He used a Specific Locus Test design in that he looked for new mutations at two recessive (*quaking* and *tufted*), and an interaction with a *T* dominant (*T*) gene. The treated males of the test stock had the chromosome consisting of $qk^+T^+ tf^+$ and were mated to $qk + tf$ females. Targeted mutations in *T* would result in tailless mice (*t-ints*, *t-interactings*) while mutations in *qk* would result in quaking mice and mutations in *tf* would results in *tufted* mice. There was one *t-int* mutation and two *qk* mutations among 2200 offspring and 3 *tf* mutations among 4500 offspring, as well as other visible mutations. In an extension of this work, Justice and Bode (1986) used t^{w5} + heterozygotes as the test offspring. t^{w5} is considered to have a mutation in a gene located in the inversion-chromosome that interacts with *T* and was now named *tail interacting* with the symbol *tct* (previously *t-int*). In this expanded screen which now provided a *tct* allele, 7 *T* mutations were found (numbered t^{kt1} to T^{kt7} but possibly only 3 were independent) as well as 5 *qk* (again, possibly only 3 were independent) and 3 *tf* (possibly only 2 were independent). Since transmission ratio distortion (trd) occurred with the new T^{kt} s, they must have occurred on the t^{w5} chromosome. The characteristics of the new *tf* and *qk* alleles also suggested that they had occurred on the t^{w5} chromosome.

The ENU approach was extended by groups at the University of Wisconsin and Institute Pasteur (Shedlovsky et al., 1986) to look for new recessive *t* alleles. This required further crossed of the offspring of the mutagenized males. The t^{w73} allele was used as a marker for the untreated chromosome 17 with + *tf*/ + *tf* homozygous males being treated with ENU. Offspring females from the cross with the marker stock *T tf*/ t^{w73} (males could be used, too, but transmission ratio distortion would mean that a lot more offspring would have to be screened) were back crossed to the original marker stock and the absence of normal tailed, *tufted* offspring would indicate that a new lethal *t* mutation had occurred. The new lethals, *l*(17-1) to *l*(17-4), were tested for distance from *T* with two being at less than 1% recombination while the other 2 were more distant at 4 and 7%. They recovered 7 more recessive lethals in these crosses which were not characterized and concluded (from the known frequency of ENU hits at the dose used [1/2000] and the number of lethals found for the number of gametes) that the region has between 50 and 100 single-copy essential genes. They continued this work with a slightly different cross and found 4 more definitely, and 2 possibly, more independent mutant lethals as well as a new lethal mutation at *qk* and 2 perinatal lethals (Shedlovsky et al., 1988). They mapped 12 new lethals, using a new spontaneous mutant to *T*, T^{Wis}, and *H-2* as markers. They found one lethal that was centromeric to *T*, one that was telomeric to *H-2*, while the rest mapped to the interval between *T* and *H-2*. However, they had insufficient resolution to determine overlapping intervals, even with the distal markers. Importantly, they performed complementation testing of 11 of the lethals and all complemented each other, therefore different genes that there was potential for a very large number of lethal genes in the region.

The new mutations were quickly put to use. This occured when the field realized that the *T/t* complex consisted of a number of inversions (see Chapter 3). Justice and Bode (1988) were able to use their new *quaking*, *T* and *tufted* mutations against other complete *t* haplotypes (theirs had been performed on the t^{w5} haplotype) and *H-2*. Their results indicated that there were two inversions in *t* chromatin, one involving *T* and *qk* and one involving *tf* and the *H-2* complex. A decreased frequency of recombination between *T* and *qk* suggested further complexities, possibly another inversion, as was soon found with molecular markers (see Chapter 3).

With the advent of the cloning era and the development of transgenesis as a way to modify the mouse genome (see Chapter 9), the random insertions made by transgenes sometimes caused interesting mutations, including of *T* region genes. Tilgman's group (Perry III et al., 1995) were performing transgenesis with a human epsilon hemoglobin gene and found an insertional mutation of the *Fused* locus. The mutation did not complement the known *Fused* gene, i.e. had the same gene product, and its homozygous embryonic phenotype was very similar. The insertion of known DNA allowed cloning an adjacent gene whose 3.9 kb RNA transcript was disrupted by the transgenic insertion, making the coding DNA a very strong candidate for the *Fused* gene. The cloned DNA provided probes to study the other known *Fu* mutations (Fu^{Ki} and $Fu^{Knobbly}$) which were found to be due to transposon insertions (Vasicek et al., 1997). In a similar fashion, a transgene for an *H-2* allele caused an insertional mutation which did not complement, or alter the DNA surrounding *T,* (which had been cloned and for which DNA probes were available; Rennebeck, et al 1995). It was assigned the name *T2* and mapped within 3 cM of *tf.* It was homozygous lethal with a slightly later in utero mortality than the classic *T,* achieving an allantoic connection which does not occur with *T/T* (see Chapter 4). Cloning the disrupted fragment was difficult because the inserted fragment was close to the location of *T* and *H-2*, i.e. other *H-2* genes. Nonetheless, a specific probe adjacent to the insertion was obtained which was found to be deleted in the known 200 kb deletion of *T*. Thus, although the classic *T* was not disrupted, the new *T2* must map close to it. A transgene for the rat *c-myc* gene driven by the human metallothionein promoter also created a null allele of *Brachyury* (Abe et al., 2000). The transgene inserted into the 7th intron and no transcripts of the normal gene were found.

In the cloning era, targeted mutagenesis using embryonic stem cells to create mutants in the *T/t* was also started. For instance, Chapman and Papaioannou, 1998) created a mutation in the T-box gene, *Tbx6,* which maps to mouse chromosome 7 (see Chapter 9 re: *T-box* genes). Homozygosity for the mutation caused the embryos to develop 3 neural tubes and the pattern of gene expression was very similar to that of *Brachyury*. Thus, analyses of the *T/t* complex were aided greatly by the new mutagenic approaches. Another targeted mutation was used to show that disruption of a candidate *T/t* complex transmission ratio distortion gene had no effect on transmission ratio distortion (see Chapter 8).

Notes

1. For a full account of some of these experiments, see **The plutonium files: America's secret medical experiments in the cold war** by Welsome (1999) and **Against their will: The secret history of medical experimentation on children in cold war America** by Hornblum, Newman and Dober (2013).

2. Thirty-six thousand cages might hold over 150,000 mice. I thought I had a very large colony with only 400 cages at University of Michigan in the early 80s! With current mouse room costs, the Oak Ridge colony would cost $25,000 per day or over 9 million dollars per year. It is not surprising that the unit had eventually to be closed.

References

Abe, K., Yamamura, K.-I., Suzuki, M., 2000. Molecular and embryological characterization of a new transgene-induced null allele of mouse Brachyury locus. Mammalian Genome 11, 238–240.

Alagraham, K.N., Murcia, C.I., Kwon, H.Y., Pawlowski, K.S., Wright, C.C., Woychik, R.P., 2001. The mouse *Ames waltzer* hearing-loss mutant is caused by mutation in *Pcdh25,* a novel protocadherin gene. Nat. Genet. 27, 99–102.

Bagg, H.J., Little, C.C., 1924. Hereditary Structural Defects in The Descendants of Mice Exposed to Roentgen Ray Exposure. Amer. J. Anat. 33, 119–145.

Baier, L.J., Hanash, S.M., Erickson, R.P., 1984. Mice homozygous for chromosomal deletions at the albino locus region lack specific polypeptides in two-dimensional gels. Proc. Natl. Acad. Sci. USA 81, 2132–2136.

Bailey, D.W., 1988. *H-2,* A Serendipitously Encountered obstacle. Intern. Rev. Immunol. 3, 305–312.

Bailey, D.W., Usuma, B., 1960. A rapid method of grafting skin on tails of mice. Transplant. Bull 7, 424–425.

Bailey, D.W., Kohn, H.I., 1965. Inherited histocompatibility changes in the progeny of irradiated and unirradiated inbred mice. Genet. Res. 6, 330–340.

Bennett, D., Dunn, L.C., Spiegelman, M., Artzt, K., 1975. Observations on a set of radiation-induced dominant *T*-like mutations in the mouse Genet. Res., 26, 95–108.

Bode, V.C., 1984. Ethylnitrosourea mutagenesis and the isolation of mutant alleles or specific genes located in the *T* region of mouse chromosome *17*. Genetics 108,: 457–470.

Breindl, M., Harbers, K., Jaenisch, R., 1984. Retrovirus-induced lethal mutation in collagen 1 gene of mice is associated with an altered chromatin structure. Cell 38, 9–16.

Cao, W., Hunter, R., Strnatka, D., McQueen, C.A., Erickson, R.P., 2005. DNA constructs designed to produce short hairpin, interfering RNAs in transgenic mice sometimes show early lethality and an interferon response. J. Appl. Genet. 46, 217–225.

Carter, T.C., 1957. Recessive lethal mutation induced in the mouse by chronic y-irradiation. Proc. R. Soc. B 147, 402–411.

Carter T.C., Lyon M.F., Phillips R.J.S., 1958, Genetic hazards of ionizing radiation. Nature, 182:409.

Carter, T.C., 1959. A pilot experiment with mice, using Haldane's method for detecting induced autosomal recessive lethal genes. J. Genet. 56, 353–362.

Chadarevian, S., 2006. Mice and The Reactor: The "Genetics Experiment" in 1950s Britain. J. Hist. Biol. 39, 707–735.

Chapman, D.L., Papaioannou, V.E., 1998. Three neural tubes in mouse embryos with mutations in the T-box gene *Tbx6*. Nature 391, 695–697.

Chapman, V.M., Miller, D.R., Armstrong, D., Caskey, C.T., 1989. Recovery of induced mutations for X chromosome-linked muscular dystrophy in mice. Proc. Natl. Acad. Sci., U.S.A. 86, 1292–1296.

Delerue, F., Ottmer, L.N., 2015. Genome editing in mice using CRISPR/Cas9: achievements and prospects. Con Transgen 4 (2). doi:10.4172/2168-9849.1000135.

Diamond, R.P., Erickson, R.P., 1974. Gene dosage in a deletion for a nuclear-coded, mitochondrial enzyme. Nature 248, 418–419.

Dickinson, M.E., Flenniken, A.M., Xiao, J., Teboul, L., Wong, M.D., White, J.K., et al., 2016. High-throughput discovery of novel developmental phenotypes. Nature 537, 508–514.

Dove, W.F., 1987. Molecular Genetics of *Mus musculus:* point mutagenesis and millimorgans. Genetics 116, 5–8.

Duhl, D.M.J., Vrieling, H., Miller, K.A., G.l., Wolff, Barsh, G.S., 1994. Neomorphic *agouti* mutations in obese yellow mice. Nature Genet 8, 59–65.

Erickson, R.P., McQueen, C.A., Chau, B., Gokhale, V., Uchiyama, M., Toyoda, A., Ejima, F., Maho, N., Sakaki, Y., Gondo, Y., 2008. An N-ethyl-N-nitrosourea-induced mutation in *N*-acetyltransferase 1 in mice. Biochem. Biophys. Res. Comm. 370, 285–288.

Erickson, R.P., 2016. The Importance of De Novo Mutations for Pediatric Neurological Disease—It's not all In Utero or Birth Trauma. Reviews in Mutation Research. 767, 42–58.

Erickson, R.P., 1988. Creating animal models of genetic disease. Amer. J. Hum. Genet. 582–586.

Erickson, R.P., Lai, L.-W., Grimes, J., 1993. Creating a conditional mutation of *Wnt*-1 by antisense transgenesis provides evidence that *Wnt*-1 is not essential for spermatogenesis. Dev. Genet. 14, 274–281.

Generoso, W.M., 1984. Dominant-lethal mutations and heritable translocations in mice. In: Chu, E.H., Generoso (Eds.), "Mutation, Cancer, and Malformation". Plenum Press, New York, London, pp. 369–388.

Gondo, Y., Fukumura, R., Murata, T., Makino, S., 2010. ENU-based gene-driven mutagenesis in the mouse: a next generation gene-targeting system. Exp. Anim. 59, 537–548.

Griffin, C.S., Tease, C., 1988. Gamma—ray induced numerical and structural chromosome anomalies in mouse immature oocytes. Mut. Res. 202, 209–213.

Hertwig, P., 1935. Sterilatatserscheingungen bei rontgenebestrahlten Mause. Zeit. Fur Induk. Abstam. Und Vererbungs. 70, 517–523 (cited by Searle AG, 1974).

Ivics, Z., Hackett, P.B., Plasterk, R.H., Izsvak, Z., 1997. Molecular reconstruction of Sleeping Beauty, a Tc1-like transposon from fish, and its transposition in human cells. Cell 91, 501–510.

Izsvak, Z., Ivics, Z., Plasterk, R.H., 2000. Sleeping Beauty, a wide host-range transposon vector for genetic transformation in vertebrates. J. Mol. Biol. 302, 93–102.

Jenkins, N.A., Copeland, N.G., 1985. High frequency germ- line acquisition of ecotropic MuLV proviruses in SWR/J-RF/J hybrid mice. Cell 43, 811–819.

Justice, M.J., Bode, V.C., 1986. Induction of new mutations in a mouse t-haplotype using ethylnitrosourea mutagenesis. Genet. Res. Camb. 47, 187–192.

Justice, M.H., Bode, V.C., 1988. Genetic analysis of mouse t haplotypes using mutations induced by ethylnitrosourea mutagenesis: The order of T and qk is inverted in t mutants. Genetics 120, 533–543.

Kasarkis, A., Manova, K., Anderson, K.V., 1998. A phenotype-based screen for embryonic lethal mutations in the mouse. Proc. Nat'l Acad. Sci. U.S.A. 95, 7485–7490.

Kohn, H.I., Melvold, R.W., Dunn, G.R., 1976. Failure of X-rays to mutate Class II histocompatibility loci in BALB/c mouse spermatogonia. Mut. Res. 37, 237–244.

Lewis, S.E., Erickson, R.P., Barnett, L.B., Venta, P.J., R.E., Tashian, 1988. Ethylnitrosourea-induced null mutation at the mouse Car-2 locus: an animal model for human carbonic anhydrase II deficiency. Proc. Natl. Acad. Sci. U.S.A 85, 1962–1966.

Little, C.C., Bagg, H.J., 1924. The occurrence of four inheritable morphological variations in mice and their possible relation to treatment with X-rays. J. Exp. Zool. 41, 45–91.

Lyon, M.F., King, T.R., Gondo, Y., Gardner, J.M., Nakatsu, Y., Eicher, E.M., Brilliant, M.H., 1992. Genetic and molecular analysis of recessive alleles at the pink-eyed dilution (p) locus of the mouse. Proc. Nat'l Acad. Sci., U.S.A. 89, 6968–6972.

McDonald, J.D., Shedlovsky, A., Dove, W.F., 1990. Investigating inborn errors of phenylalanine metabolism by chemical mutagenesis by efficient mutagenesis in the mouse germ line. In: Allen, J.W., Bridges, B.A., Lyon, M.F., Moses, M.J., Russell, L.B. (Eds.), "Biology of Mammalian Germ Cell Mutagenesis". Cold Spring Harbor Press, Cold Spring Harbor, N.Y., pp. 259–270.

Moyer, J.M., Lee-Tischler, M.J., Kwon, H.Y., Schrick, E.D., Avner, W.E., Sweeney, V.I., Godfrey, N.I.A., Cacheiro, J.E., Wilkinson, R.P., Woychik, R.P., 1994. Candidate gene associated with a mutation causing recessive polycystic kidney disease in mice. Sci. 264, 1329–1333.

Mulhardt, C., Fischer, M., Gass, P., Simon-Chazottes, D., Guenet, J.L., Kuhse, J., Betz, H., Becker, C.M., 1994. The spastic mouse: aberrant splicing of glycine receptor beta subunit mRNA caused by intronic insertion of L1 element. Neuron 13, 1003–1015.

Muller, H.J., Altenburg, E., 1919. The rate of change of hereditary factors in Drosophila. Proc. Soc. Exp. Biol. Med. 17, 10–14.

Neel, J.V., Lewis, S.E., 1990. The comparative Radiation Genetics of Humans and Mice. Ann. Rev. Genet. 24, 327–362.

Neel, J.V., Satoh, C., Goriki, K., Asakawa, J.-I., Fujita, M., et al., 1988. Search for mutations altering protein charge and/or function in children of atomic bomb survivors. Final report. Am. J. Hum. Genet. 42, 663–776.

Neel, J.V., Strahler, J.R., Hanash, S.M., Kuick, R., Chu, E.H.Y., 1989. 2-D PAGE and genetic monitoring: progress, prospects and problems. In: Endler, A.T., Hanash, S.M. (Eds.)," Two-dimensional electrophoresis." VCH, Weinheim, Federal Repblic of German, pp. 79–89.

Nonn, L., Williams, R.R., Erickson, R.P., Powis, G., 2003. The absence of mitochondrial thioredoxin 2 causes massive apoptosis, exencephaly, and early embryonic lethality in homozygous mice. Mol. Cell. Biol. 23, 916–922.

Perry III, W.L., Vasicek, T.J., Lee, J.J., Rossi, J.M., Zeng, L., Zhang, T., Tilghman, S.M., Costanti, F., 1995. Phenotypic and molecular analysis of a transgenic insertional allele of the mouse Fused locus. Genetics 141, 321–332.

Peters, J., Andrews, S.J., Loutit, J.F., Clegg, J.B., 1985. A mouse beta-globin mutant that is an exact model of hemoglobin Rainer in man. Genetics 110, 709–721.

Plasterk, R.H., Izsvak, Z., Ivics, Z., 1999. Resident aliens: the Tc1/mariner superfamily of transposable elements. Trends Genet 15, 326–332.

Popp, R.A., Bailiff, E.G., Skow, L.C., Johnson, F.M., Lewis, S.E., 1985. Analysis of a mouse alpha-globin gene mutation induced by ethylnitrosourea. Genet 105, 157–167.

Rennebeck, G.M., Lader, E., Chen, Q., Bohm, R.A., Cai, Z.S., Faust, C., Magnuson, T., Pease, L.R., Artzt, K., 1995. Is there a *Brachury* the second? Analysis of a transgenic mutation involved in notochord maintenance in mice. Develop. Biol. 172, 206–217.

Rinchik, E.M., Russell, L.B., 1990. Germ-line deletion mutations in the mouse: Tools for intensive functional and physical mapping of regions of the mammalian genome. In: Davis, K.E., Tilgman, S.M. (Eds.), "Genetic and physical Mapping, Genome Analysis 1". Cold Spring Harbor laboratory Press, Cold Spring Harbor, N.Y., pp. 121–158.

Russell, L.B., Major, M.H., 1957. Radiation-induced presumed somatic mutations in the house mouse. Genetics 42, 161–175.

Russell, L.B., 1987. Definitions of functional units in a small chromosomal segment of the mouse and its use in interpreting the nature of radiation-induced mutations. Mut. Res. 11, 107–1123.

Russell, L.B., Hunsicker, P.R., Cachiero, N.L.A., Bangham, J.W., Russell, W.L., Shelby, M.D., 1989. Chlorambucil effectively induces deletion mutations in mouse germ cells. Proc. Nat'l Acad. Sci. U.S.A. 86, 3704–3708.

Russell, L.B., Russell, W.L., Rinchik, E.M., Hunsicker, P.R., 1990. Factors affecting the nature of induced mutations. In: Allen, J.W., Bridges, B.A., Lyon, M.F., Moses, M.J., Russell, L.B. (Eds.), "Biology of Mammalian Germ Cell Mutagenesis". Cold Spring Harbor Press, Cold Spring Harbor, N.Y., pp. 271–289.

Russell, L.B., Russell, W.L., 1996. Spontaneous mutations recovered as mosaics in the mouse specific locus test. Proc. Nat'l Acad. Sci. U.S.A. 93, 13072–13077.

Russell, L.B., 1999. Significance of the perigametic interval as a major source of spontaneous mutations that result in mosaics. Environ. Mol. Mutagen 34, 15–23.

Russell, L.B., 2013. The mouse house: a brief history of the ORNL mouse-genetics program from 1947–2009. Mutation Res., Reviews in Mutation Res. 753, 69–90.

Russell, W.L., 1956. Comparison of X-ray-induced mutation rates in Drosophila and mice. Amer. Natural. 90, 69–80.

Russell, W.L., Kelly, P.R., Hunsicker, P.R., Bangham, W.J., Maddux, S.C., Phipps, E.L., 1979. Specific-locus test shows ethylnitrosourea to be the most potent mutagen in the mouse. Proc. Nat'l Acad. Sci. U.S.A. 76, 5918–5922.

Russell, W.L., Hunsicker, P.R., Raymer, G.D., Steele, M.H., Stelzner, K.F., Thompson, H.M., 1982. Dose response curve for ethylnitrosourea specific-locus mutations in mouse spermatogonia. Proc. Nat'l Acad. Sci. U.S.A. 79, 3589–3591.

Russell, W.L., 1989. Reminiscences of a mouse specific locus test addict. Environ. Mol. Mutagen. 14 (Suppl. 15), 16–22.

Schlager, G., Dickie, M.M., 1966. Spontaneous Mutation Rates of Five Coat-Color Loci in Mice. Sci. 151, 205–206.

Searle, A.G., 1974. Mutation Induction in Mice. Adv. Rad Biol. 131–207.

Searle, A.G., Beachey, C., 1986. The role of dominant visibles in mutagenicity testing. In: Ramel, C., Lambert, B., Magnussen, J. (Eds.), Genetic Toxicology of Environmental Chemicals Part B Genetic Effect and Applied Mutagenesis. Liss, New York, pp. 511–518.

Shedlovsky, A., Guenet, J.-L., Johnson, L.L., Dove, W.F., 1986. Induction of recessive lethal mutations in the *T/t-H-2* region of the mouse genome by a point mutagen. Genet. Res. Camb. 47, 135–142.

Shedlovsky, A., King, T.R., Dove, W.F., 1988. Saturation germ line mutagenesis of the murine *t* region including a lethal allele at the quaking locus. Proc. Nat'l Acad. Sci. U.S.A. 85, 180–184.

Sly, W.S., Hewett-Emmett, D., Whyte, M.D., Yu, Y.L., Tashian, R.E., 1983. Carbonic anhydrase II deficiency identified as the primary defect in the autosomal recessive syndrome of osteopetrosis with renal tubular acidosis and cerebral calcification. Proc. Natl. Acad. Sci. U.S.A. 80, 2752–2756.

Snell, G.D., 1933. X-ray sterility in the male house mouse. J. Exp. Zool. 65, 421–441.

Snell, G.D., 1935. The induction by X-rays of hereditary changes in mice. Genetics 20, 545–567.

Soriano, P., Gridley, T., Jaenisch, R., 1987. Retroviruses and insertional mutagenesis in mice: proviral integration at the Mov 34 locus leads to early embryonic death. Genes and Devel 1, 366–375.

Stacey, A., Bateman, J., Choi, T., Mascara, T., Cole, W., Jaenisch, R., 1988. Perinatal lethal osteogenesis imperfectas in transgenic mice bearing an engineered pro-alpha$_1$ (I) collagen gene. Nature 332, 131–136.

Stoye, J.P., Fenner, S., Greenoak, G.E., Moran, C., Coffin, J.M., 1988. Role of endogenous retroviruses as mutagens: the hairless mutation of mice. Cell 54, 383–391.

Vasicek, T.J., Zeng, L., Guan, X.-J., Zhang, T., Constantini, F., Tilghman, S.M., 1997. Two dominant mutations in the mouse *Fused* gene are the result of transposon insertions. Genetics 147, 777–786.

Whitney J. B. III, Russell, E.S., 1980. Linkage of genes for adult alpha-globin and embryonic X-like globin chains. Proc. Natl. Acad. Sci. U.S.A. 77, 1087–1090.

Wu, M., Rinchik, E.J., Wilkinson, D.K., Johnson, D.K., 1997. Inherited somatic mosaicism caused by an intracisternal A particle insertion in the moue tyrosinase gene. Proc. Natl. Acad. Sci. U.S.A. 94, 890–894.

Yeom, Y.I., Abe, K., Artzt, K., 1992. Evolution of the Mouse H-2K Region: a hot Spot of Mutation Associated With Genes Transcribed in Embryos and/or Germ Cells. Genetics, 130:629-638.

Zambrowica, B.R., Friedrich, G.A., 1998. Comprehensive mammalian genetics: history and future prospects of gene trapping in the mouse. Int. J. Dev. Biol. 42, 1025–1036.

Immunogenetics and the failed search for *t*-antigens–from mid-century and continuing to it's end

Blood groups and the major histocompatability complex

While studied in multiple organisms, most of experimental immunogenetics was, and is, performed in mice. Because of the importance for transplantation, humans are also extensively studied but, of course, cannot be submitted to the genetic manipulations possible in mice. There is a vast literature of mouse immunogenetics and there have been a number of historical summaries. Leslie Brent's **1996** monograph," A History of Transplantation Immunobiology", is the most thorough and contains many "mini-biographies" of the major players. Jan Klein's two monographs on the major histocompatibility complex of mice (**1975, 1986**) contain brief, but illuminating histories. Irwin (1974) has commented on the early history of immunogenetics and Owen (1989) has amplified Irwin's role. Shreffler (1988) has written a short history of studies of the major histocompatibility complex while Amos (1981) provides a broader short history. Of course, immunogenetics is a part of the more general, brief histories of mouse genetics of Mary Lyon (2002), Ken Paigen (2003a, 2003b), and "Tibby" Russell (1985).

The history of immunology predates that of the history of genetics. Many would consider Edward Jenner's introduction of vaccination with cowpox against smallpox at the turn of the 18th century or Louis Pasteur's (1885) vaccination for rabies as the start of immunology while others might consider the development of antitoxins in the sera of animals in the 1890's (Silverstein, 1985) as the appropriate start, at least of humoral immunology.[1] "Von Behring received the lion's share of credit for introducing 'serum therapy', whereby antibodies generated in animals could be used in treating human disease [1890]. Ehrlich developed the scientific model in his famous 'lock and key' theory of the nature of the immune response" (Bynum, 2006). Metchnikoff's (1893) studies of phagocytosis could be seen as the start of cellular immunology. The animals of choice were larger mammals such as rabbits (and even horses) for production of antibodies while guinea pigs were chosen for studies of anaphylaxis. Mice were to come into their own for studies of immunogenetics in the 20th century.

The start of immunogenetics in mice had its origins in cancer biology. Spontaneously arising tumors were found to be transplantable to other mice where they would grow and again become of lethal size. These could be serially transplanted. However, it was found that the recipient mice had to be closely related to the donor mice. Jensen (1903), working in England transplanted a tumor arising in the available, presumably random-bred, laboratory mice, found that the tumors were not transplantable to wild mice, which he caught in his building

Twentieth Century Mouse Genetics. DOI: https://doi.org/10.1016/B978-0-12-824016-8.00017-9

while the famous physiologist, Loeb (1908), found that tumors arising in his Waltzing mice in Philadelphia could not be transplanted into wild mice also caught in his building (such vermin infestation is, of course, not tolerated in today's laboratories). Tyzzer followed up these observations by performing genetic experiments, studying the results of transplanting the tumors to F1 and F2 mice between his Waltzing mice and his "common" mice (again assumed to random-bred). He found that the tumors grew in all the F1s but none of the F2s (Tyzzer, 1909). He was working at Harvard and showed the results to Castle (see Chapter 1) who then communicated them to Little. Little proposed that the result could be explained if several genes were responsible for the intolerance. "These points...are readily understandable on the Mendelian hypothesis of multiple factors. We have supposed that...a number of factors are found in the germ cells of one race, which are either or not present or are replaced by allelomorphic factors in the other race. We have further supposed that the successful growth of a transplanted tumor depends on the simultaneous presence in the zygote, at least in a single representation, of a considerable numberof factors of the susceptible race." (Little and Tyzzer, 1916). Little came up with the formula $N^{3/4} = \%$ susceptible and calculated that the number of loci was about 15 for the observed frequencies of about 1% tumor acceptance in the F2s of several crosses. This was a very prescient prediction but it would take several decades, and work in other species, before the differences were identified as transplantation antigens.

Karl Landsteiner (1901) is credited with the discovery of the human ABO blood groups system, initially termed the ABC group. These were detected with naturally occurring antibodies in the sera of donors with O donors having antibodies to A and B, B donors (who could be BB or BO) with antibodies to A, etc. This is unlike the situation with other blood groups where prior immunization with red blood cells must occur, such as Rh negative mothers becoming immunized after the birth of the first Rh positive child (due to red blood cells released from the placenta to the maternal circulation at the time of delivery) or patients who have received transfusions of other blood types. That the ABO blood types were inherited was not established until 1910 (von Dungern and Hirschfeld, 1910). The continued development of immunogenetics, and the creation of the name, was with studies of blood groups in many species. M R Irwin, working in the Dept. of Genetics at the University of Wisconsin was the leader of these developments and the coiner of the name "immunogenetics" (Irwin, 1974).

Irwin received his Ph.D in genetics at Iowa State and did post-docs at the Bussey Institution (see Chapter 1) and the Rockefeller Institute where he met Landsteiner and Alexander Wiener (the two discovered the Rh system) and Philip Levine who was also to work on the Rh system (found in Rhesus monkeys but soon studied in man). In 1930, Irwin became the 4th member of the Dept. of Genetics at the University of Wisconsin where the Chairman, Leon Cole, was heavily involved in studying the genetics of doves, a species in which hybrids are fertile. Irwin found that "it was very easy to obtain a reagent, which would react specifically with the erythrocytes of any species [of doves] in comparison with the cells of another [species], following absorption of an antiserum [made in rabbits] " (Irwin, 1974, p. 67). Cole had backcrossed one species of doves against another creating a series of backcross hybrids, which could be studied by absorption of the original antiserum to one of the parent species' cells (Owen, 1989). Studies of the distribution of these reactivities among the many backcross lines established that there were eleven distinctive gene-antigen units distinguishing the two original species (Irwin, 1939). These studies on doves (Irwin, 1932, 1939) were followed by studies of blood groups in many domesticated species including pigs (Andresen and Irwin, 1959), cattle (with 2 outstanding graduate students of Irwin's: Clyde Stormont and Ray Owen [co-discoverer of tolerance]; Stormont et al., 1951) and others.

The two fields of tumor transplantation and blood group genetics were to come together from the inspiration of JBS Haldane (see The Haldane-Mitchison Scientific Clan) and the work of Peter A Gorer.

As mentioned in the mini-biography, JBS Haldane visited Little's lab in 1932 and brought back 3 inbred strains of mice. Haldane "suggested that the tumor resistance factors were akin to blood group antigens and that the failure of the transplants to grow was not unlike the destruction of incompatible blood cells following transfusion...He encouraged one of his students, Peter A. Gorer, who had just completed his medical studies to take up the project."(**Klein, 1986, p. 8²**).

Gorer, following the notion that the ABO blood groups are detected with normal human sera, tested the sera of 4 horses, 3 fowl, 6 rabbits and his own sera (he was group A so an anti-B sera), which was the only one that would agglutinate the red blood cells of one of the inbred strains of mice and not another strain, with the third strain intermediate. Some random bred mice also reacted with the sera but a wild mouse did not. The reaction only occurred at 37° C (instead of at room temperature) and the sera could be absorbed with the target cells (Gorer, 1936a). He then immunized rabbits to obtain specific sera. Normal rabbit sera would agglutinate the red blood cells of all strains but only at low dilutions, for example , 1 to 20 or 30 v.s. 1 to thousands, immune sera showed specificity with strong reactions with the donor strain's red blood cells when greatly diluted (Gorer, 1936b). After absorption with the donor's red blood cells, these sera again could distinguish other inbred strains' cells and he could establish that the factors were inherited in a dominant fashion. He went on to study the relationship of these antigens to tumor susceptibility per Haldane's suggestion (he was no longer with Haldane but at the Lister Institute for Preventive Medicine) using a tumor that had spontaneously occurred in the A strain. The tumor was, of course, accepted by the A strain and rejected by the C57BL strain. Luckily, the number accepting the tumor in the F2 suggested that only 2 tumor susceptibility loci were involved (Gorer, 1937, 1938). Using his rabbit sera, he demonstrated that one half of the mice positive for his "antigen II" accepted the tumor indicating that his antigen II was one of the 2 susceptibility loci in the cross. Sera of the mice, which rejected the tumor detected this antigen (II). Thus, at least one tumor susceptibility locus was equivalent to one coding a red blood cell antigen detected by the rabbit sera. (Gorer, 1937, 1938). This was to become the H-2 antigen and, later, H-2 complex. Because his work always involved antibodies, he believed that tumor rejection was humoral, not cellular. "From Gorer's work the notion became firmly established that 'histocompatibility genes' ... the genes determining whether mice were susceptible to allogenic tumor -were identical to those present on red blood cells and on other tissues of normal individuals" **(Brent, p. 132)**. It should be noted that the antisera made in rabbits were not equivalent in the specificities they were detecting compared to later antisera made in mice. In large part because the second World War intervened, his work was not appreciated for some time.

However, the second World War also stimulated studies of skin grafting, which became the major way to study histocompatibility reactions (in many species, even including salamanders [Erickson, 1972]). It was Peter Medawar, stimulated by hopes of treating burn patients during the war, who first applied the principles of tumor immunity to normal tissues using skin grafts. First with human (Gibson and Medawar, 1943) and then with rabbits (Medawar, 1944). He established that the rejection of a second transplant involved significant cellular infiltration. This became known as "second set". "Medawar considered these three characteristics of the rejection process—accelerated response to the second grafts, specificity [not found with a different skin donor], and systemic character [the transplant could be to skin anywhere on the body] of sensitization—as evidence for the immunological nature of the *allograft reaction,* as the rejection process is referred to. Presumably, the antigens eliciting the allograft reaction were those that Gorer identified in his experiments, and the blood group antigen II was one of them" (**Klein, 1986, p.9**). Like Gorer, he believed at this time that antibodies were the cause (Medawar, 1945). He realized that multiple loci were involved in the antigenic differences. However, he had some notion that leukocytes, perhaps because of the cellular infiltration he had seen histologically in the grafts as they were rejected, might be involved in rejection. In a note describing his experiments, which showed that leucocytes shared antigens with skin (Medawar, 1946a), the paper concludes that "there is no evidence that leukocytes play any...part in the reactions of **blood transfusions** [my emphasis] but the possibility...is worth keeping in mind".[3] Medawar had just shown that rabbit red blood cells were not antigenic for inducing skin graft rejections (Medawar, 1946b), so it is unlikely that he was thinking of Gorer's antigen II system on red blood cells (mice are unusual in having MHC antigens on red blood cells). Perhaps it was a prescient notion of the Human Leukocyte Antigen system, which can elicit transfusion reactions and that was partially discovered because of antisera to them found in multiparous women immunized by their babies with paternal antigens (see below).

At the end of the war, Gorer was able to spend a year at the Jackson laboratory, which was then the mecca for mouse genetics. Snell had found a tumor resistance gene near *fused,* which, as we have seen (Chapter 3), is very near the *t*-complex. The two soon established that the resistance factor was a consequence of Gorer's antigen II and this was linked to *fused* (Gorer et al., 1948). This event was to have major consequences: it involved Snell in studies of the *H-2*-complex, studies, which would lead him to receive the Nobel prized many years later, and it eventually led to thoughts about the immunogenetics of the *t*-complex.

Snell introduced the use of congenic lines to study the *H-2* complex. By crossing an inbred strain with a different *H-2* complex to different inbred strains, and then again backcrossing these to the parental inbred strains and continuing this process for usually, at least 12 generations (N12), new inbred strains differing for their *H-2* complex were obtained (Snell, 1948). This was frequently done in reciprocal directions. Usually an intercross of the backcross mice intervened in order to find homozygotes for the H-2 complex being introduced from one inbred strain to the other, making it 24 instead of 12 generations. These strains allowed the demonstration that the *H-2* complex determined rapid rejection of skin grafts and so the *H-2* complex became designated as the Major Histocompatibility Complex or MHC (Counce et al., 1956). With the discovery of Human Leuckocyte Antigens, HLA, in man (Dausset, 1958; Payne and

Rolfs, 1958), it became apparent that they defined the equivalent locus and the MHC became the focus of major studies in many species (**Snell, Dausset, and Nathenson, 1976**).

These congenic strains also allowed the study of non-MHC or minor histocompatibility) loci. When 2 congenic strains sharing the same *H-2* complex were used for genetic studies of skin graft rejection, a number of unlinked genes controlling multiple minor histocompatibility loci were identified. "Snell and co-workers isolated and identified *H-1* and *H-3* to *H-13*, Bailey and Kohn (1965) and Bailey (1975) identified *H-15* to *H-30* and *H-34* to *H-38* by crossing *H*-locus alleles from BALB/cBy into C57BL/6By, using tail skin grafts for identification. A few more minor autosomal histocompatibility loci have been identified, bringing the total to at least 40." (Russell, 1985).

Gorer (Gorer and Mikulska, 1959) continued to study the serologically-detected specificities of the *H-2* complex. The effort was soon joined by others (Amos et al., 1955) and genetic complementation showing non-overlapping regions was demonstrated (Snell, 1951). Soon recombinants were found separating some of the specificities (Allen, 1955). Similar discoveries were being made for the HLA complex and, with increasing numbers of scientists involved, workshops where all the data could be compared, as had been also done for sharing linkage data (Chapter 3) were established (Workshop on Histocompatibility Testing. Part B. Edited by D. B. Amos. Nat'l Acad. Sci. Res. Council. Washington).

> *"The first workshop which finished in triumph, started as a near disaster. Each participant was asked to bring sera to a central location. The sera would then be distributed to others for testing on a common panel of cells. The variables to be tested were those introduced by technology because each participant used common sera and targets but different procedures or cell preparation and testing… On the eve of the workshop not 1 serum was offered for joint use. Chaos was averted only because the host laboratory [Duke University] had previously been given small samples of sera from many of the participating laboratories. A quick check in the freezer revealed there was just enough of a few sera for everyone to share. The participants agreed to use these reagents and the workshop was begun. Within 2 hr the atmosphere had completely cleared. In the free times while blood was centrifuging or sedimenting or cell suspensions were being prepared, investigators used the opportunity to observe their neighbors and exclamations such as 'Oh, that's how you do it' and 'You didn't mention that step in your paper' were heard." (Amos, 1981).*

The workshops thereafter provided a central repository for sharing reagents. "The first two meetings were attended by only a handful of researchers, but since the 1970 Los Angeles meeting, the workshops attract over 500 participants each" (**Klein, 1986, p. 16**).

These workshops reflected a growing international effort. Much work emanated from Prague where several institutes of the Czechoslovakia Academy of Science were making major contributions to mouse immunogenetics. Alena Lengorova was a major influence as reflected in her organization of a symposia on the "Immunogenetics of the H-2 System," which met near Prague in 1970 (Lengerova and Vojtiskova, 1971).[4] She was a seminal thinker and pointed out that congenic lines were such a powerful genetic tool that very minor effects, as might be found with a modifier gene carried along with the target transfer gene in the development of the congenic line, could show a great statistical significance and be over-valued. Jan Klein, also of the Czech. Acad. of Sci, escaped from Czechoslovakia at the time of the Russian invasion (1968) and, after a brief sojourn at the Herzenberg lab at Stanford University, moved to the University of Michigan where he stayed until 1975, followed by the University of Texas Southwestern Medical School. In 1977 he was made Director of the Max-Planck-Institute of Biology in Tubingen, Germany. As already multiply cited for the DNA analyses of the *H-2* complex in Chapter 3, the Mammalian Genetics Unit of the institute Pasteur was also very active in immunogenetic studies that reflected the interest in the possibility of *t*-antigens.

It gradually became clear that there are both class I and class II histocompatibility antigens in the complex. The class I antigens are defined as consisting of 45,000 molecular weight glycoproteins non-covalently attached to beta-2-microglobulin (which maps to chromosome 2, Michaelson, 1981) in the plasma membrane. Class II histocompatibility antigens were found to have chains of 28,000 and 32,000 molecular weight that are not associated with beta-2-microglobulin. The detailed genetic analyses revealed that the genes for class I histocompatibility antigens flank a region of genes for class II, one of many subgroups of class I, histocompatibility antigens and, using the recombinant mice, it was possible to make antisera to a particular class I antigen, for example a D or K or L antigen or particular class II antigens, for example A region, B region, etc. Thus, syngeneic antisera prepared between mice differing in one subportion of the major histocompatibility complex more accurately defined specific antigens. Gradually a vast number of antigenic specificities were accumulated by many workers and tables containing up to 15 or 20 rows and 15 to 20 columns showing the varying specificities in different *H-2* complexes were developed.[5] It also became apparent that more than antigenic specificities were encoded in the complex. The serum substance, Ss, controlling the amount of complement (Demant, 1973) and the Ss-related sex-limited protein were mapped between *H-2K* and *H-2D* specificities (Shreffler and Owen, 1963; Shreffler, 1964).

Many great advances in the understanding the biology of transplantation and the role of the MHC were being made. "It was N. A. Mitchison [Mitchison, 1953][6] who in the course of his Ph.D. studies at Oxford [with Peter Medawar—see The Haldane-Mitchison scientific clan] found...that the passive cellular transfer of immunity—this time to allogeneic tumors—was far more easily accomplished when the cell donors and recipients belonged to the same inbred strains" (**Brent, p. 12**). This was due to what became known as MHC restriction—the need for MHC compatibility between antigen-presenting and responding cells. This concept could only be developed after the role of the MHC in immune responses and the distinction between T (thymus-derived) and B (bone marrow-derived) cells were developed. The distinction between T and B cells depended on the discovery of the Thy-1 antigen. Found by Reif and Allen (1964) on thymocytes, Martin Raff (1969) found that it was only present on immunoglobulin negative cells, which must be T cells. The role of the MHC in immune response can trace its origins to the work of Lilly et al. (1964) on the response to leukemia viruses. Differences in immune responses to synthetic peptides (the *Ir-1* locus) were mapped to the *H-2* complex (McDevitt and Chinitz, 1969). This followed work by McDevitt and Humphrey, which showed strain differences in rabbits that, because they were not inbred, didn't allow a demonstration of a monogenic response (McDevitt, 2000). "Numerous experiments... finally established beyond any doubt that the I region[immune response region] contains several genes encoding class II MHC molecules (also called Ia antigens) that are responsible for the observed genetic control of specific immune responses and for stimulation in the mixed lymphocyte culture reaction, an in vitro test of a cellular immune reaction which was rapid and easily quantified (McDevitt, 2000). It soon became apparent that T and B cells had to be of the same *H-2* specificities for a number of immune responses (Kindred and Shreffler, 1972; Katz et al., 1973; Zinkernagel and Doherty, 1979). "The concept of MHC restriction provided the first clear indication that histocompatability antigens served a profoundly important biologic function, both in the induction and execution of certain immunologic responses" (**Brent, p. 32**).

H-2 congenic lines allowed rapid mapping of genes nearby, such as, variation in the expression of gangliosides (Hashimoto et al., (1983). The availability of *H-2* congenic lines also allowed studies of the potential role of the complex in many other physiological pathways: such as, hepatic cyclic cAMP levels (Meruelo and Edidin, 1975), the incidence of steroid-induced cleft palate (Bonner and Slavkin, 1975), sex-related traits such as testis and body weight (Ivanyi, 1978) or ovulation rate (Spearow et al., 1991) and glucagon receptor function (Due et al., 1986)[7], These many associations are not surprising given that the H-2 glycoproteins are major cell surface molecules with the possibility of influencing many other cell surface molecules, even transmembrane methyl transferase (Markovac and Erickson, 1985) or actin attachments (Koch and Smith, 1978).

Tolerance

Another important advance was the development and demonstration of tolerance. It is pretty obvious that, in general (auto-immune diseases being the exception), we do not make antibodies or reject our own tissues. Before it was understood how antibodies are made, it was possible to think that antibodies were only made to foreign (for example introduced) substances. Thus, in the induced fit theory of antibody formation, newly synthesized antibodies were somehow formed on the surface of the foreign substance and maintained a configuration that could combine with it. When the understanding that protein-folding was determined by the sequence of the amino acids (coded by DNA and mediated by RNA), the natural selection theory of antibody formation developed. Following the realization that antibodies were selected from the continuous production of a nearly infinite number of antibody specificities, the lack of self-reactivity (tolerance) require an explanation.

Owen's finding of chimerism for genetically distinct red blood cells in bovine twins (Owen, 1945), which he correctly surmised occurred because of shared circulation between the developing calves, indicated that there was a mechanism preventing the development of antibodies to the other twin's blood group. It was Medawar's group, which produced tolerance by injecting newborn mice with "foreign" cells (Billingham et al., 1953). Burnet and Fenner (1949) were the first to suggest that tolerance resulted from deletion of clones reactive to the foreign antigen and it was Medawar and Burnet who shared the Nobel Prize in 1960.[8]

Immunological tolerance was usually experimentally produced by injections of spleen, lymph node, and bone marrow cells within a few hours of birth (Billingham et al., 1953). If injected later, graft-versus-host disease could occur. As discussed in Chapter 4, chimeras made for the study of embryonic development (Chapter 4) were employed, since they obviously had developed tolerance when they were between different strains of mice, although the degree of tolerance was found to be variable when the chimeras were grafted with parental strain skin. Unlike the results with MHC differences, it was found that tolerance to minor histocompatibility loci could be developed with injections a few days after birth. As Jan Klein reports (**Klein, 1986, p. 407**): "Investigators [took} advantage of the entire arsenal of congenic lines, Mhc recombinants, and Mhc mutants to assess the tolerance potential of the individual loci and the combinations of loci. This striving for refinement...made the mouse, which has the largest arsenal, the popular choice for investigations".

It gradually became apparent that the mechanism of tolerance was due to suppression by either antibodies or T cells. It can sometimes be transferred with antibodies (so-called "anti-idiotypes"; Kaliss et al., 1953; Bansal et al., 1973) which are thought to be targeting T-cell receptors (Snell et al., 1960). These are the hugely variable, cell-limited equivalents of the B-cell antibodies and which use recombination and mutation, as do B-cells, to achieve their huge repertoire (reviewed in Davis and Bjorkma, 1988). In some cases T-repressor cells were also involved.

A well-studied minor histocompatibility antigen: H-Y

It was 1955 when Eichwald and Silmser (1955) reported that females of some inbred strains would reject skin grafts from the males of that strain. The interpretation of this as defining a transplantation antigen was made by Hauschka (1955) and Billingham and Silver (1960) suggested calling this locus and antigen H-Y. This observation attracted much attention for two major reasons: 1) this was a system in which a minor histocompatibility antigen could be studied without having to make congenic strains and 2) because of a hypothetical role in sex determination. That the result was due to "standard" transplantation rejection was further supported by the finding that a second graft to the female was rejected more quickly (Eichwald et al., 1957). A major alternative explanation was that this was an autosomally-linked antigen turned on by androgens or turned off by estrogens. Vojtiskova and Polackova (1966) showed that skin grafts from males castrated at birth were still rejected, albeit at a slower rate. The lack of dependence on testosterone was confirmed by making skin grafts from C57BL/6J males to F1 males of C57BL/6J X A/J, for example there must be more than one form of the antigen as well (Hildemann et al., 1974).

The story became much murkier with the finding and use of a supposedly equivalent serological reagent made in mice (Goldberg et al., 1971), which was called Male Specific Antigen (MSA). The sera cross-reacted with cells from other mammals, including man (Wachtel et al., 1974) as well as amphibia and birds, where the female heterogametic sex has the antigen (Wachtel et al., 1975a) and it was claimed to be the sex-determining factor (Wachtel et al.,1975b). It was much less clear that this serological specificity was not androgen-mediated (Erickson, 1977). It was found to be determined by a carbohydrate moiety (Shapiro and Erickson, 1981) instead of a protein, as was the transplantation antigen, and it seemed to map to chromosome 17, the location of *H-2* (Shapiro and Erickson, 1984). MSA tuned out to be a "red herring" although its putative role as the sex-determining factor even reached book status (Wachtel, 1983).

The route to the proper understanding of the *H-Y* antigen(s) lay through the findings that responses were heavily dependent on immune response genes. Using congenics for H-2^b from many strains, Bailey and Hoste (1971) showed that H-2^b females were responders, while many other H-2 haplotypes were non-responders. Using recombinants in the complex, Bailey further showed that the control of the response mapped to the K end of the complex (Bailey, 1971). There seemed to be no heterogeneity of the antigen, even among other mouse species (Simpson et al., 1979), which paralleled findings for other minor histocompatibility antigens in wild species (Roopenian et al., 1993). Using immunoselection against H-Y histocompatibility in cultured cells showed, however, that there were multiple loci involved—at least 2 and possibly up to five (King et al., 1994). Elizabeth Simpson and collaborators (Scott, et al, 1995) unraveled the complexity using the many recombinants of the H-2 complex available in congenic strains and in vitro tests using the combination of lymphocytes "primed" by sensitization to H-Y tested in vitro on radio-labelled male cells "identified the MHC Ir genes as those encoding the class I and class II molecules able to present H-Y epitopes [the part of an antigen combining with the antibody or T-cell receptor], the portions presented to the B cell antibody or the T-cell receptor are usually peptides digested from the larger protein] to cytotoxic (CD8$^+$) and helper (CD4$^+$) T cells" (Simpson et al., 1997). The cloning and sequencing of the Y chromosome eventually allowed, using transfection of cDNAs clones into cells, which would express the encoded proteins from various segments of the Y, and from X/Y sequence comparisons, the identification of 6 different H-Y peptide epitopes (Greenfield, et al, 1996; Simpson et al., 1997). Davis and Roopenian (1990) had found similar findings of a complex of genes at other minor histocompatibility loci. The cloning work also allowed the precise location of these determinants between *Uby* and *Smcy* on the short are of the Y chromosome, well away from the sex determination gene, *Sry*. The history of the possible role of these variants in male fertility will come up in Chapter 11, "Sex Determination".

Association of the *T/t*-complex with histocompatibility antigens

There were many findings, which tentatively associated the *T/t*-complex with antigens. Hammerberg and Klein (1975) and Hammerberg et al. (1976) found new H-2 antigens among strains of mice bearing *t*-alleles and demonstrated that the genetic determinants were in linkage disequilibrium with the *T/t*-complex (see Chapter 3). Both the Klein laboratory (then at the Univ of Texas Southwestern Medical Center; Hauptfeld et al., 1976) and the McDevitt lab (at Stanford university School of Medicine; Levinson and McDevitt, 1976) used the mixed lymphocyte reaction (see above, which shows particular specificity for Ia antigens) to find unique Ia antigens for each *t*-allele complementation group. Studies of the new H-2 antigenic specificities allowed comparisons suggesting evolutionary relationships among the many *t*-alleles (Nizetic et al., 1984). When 33 different strains of mice carrying *t*-alleles were compared, their ancestral chromosome was predicted to predate the separation of the *Mus musculus* and *domesticus* species. Mary Lyon's group used cDNA clones of the *H-2* complex to perform similar studies. They found duplications, deletions, and reshuffling of many portions of the *t*-allele DNA but concluded that "the differences between t and standard mouse strains are similar, in nature and in degree, to those between different standard strains" (Rogers et al., 1985). Klein's group, also using *H-2* cDNA clones, interpreted their results differently as they were amazed at the small amount of diversity among the *t*-allele bearing wild mouse strains (Figueroa et al., 1985*)*. They confirmed the evolutionary conclusions they had made using serological methods with the then new Southern blots. Further complexities of *t*-allele *H-2* genes were found by Paul Mains (1986) of the Cold Spring Harbor Laboratory and Jiri Forejt's group in Prague (Mosinger, Jr. et al., 1988).

Immunological complexity of the region was suggested by the discovery of 4 minor histocompatibility loci mapping near the *H-2* and *T/t* complexes. Flaherty found 2 new histocompatibility loci mapping to the distal end of *H-2*, *H-31* and *H-32*, and one, *H-33* mapping to proximal (Flaherty and Wachtel, 1975; Flaherty, 1975). Artzt et al. (1977) found *H-39* mapping to this region also.

The failed search for *t*-complex antigens

There were many reasons why one would search for antigens determined by the *T/t* complex. There was much interest in molecular changes in the cell surface of the embryo, which might be related to changes in its development. "The abnormalities of development in embryos homozygous for lethal mutations at the *T*-locus involve defects in processes, which could include cell-to-cell recognition and morphogenetic movements" (Glucksohn-Waelsch and Erickson, 1971). It was also thought that haploid expression of such antigens could allow differing properties of sperm, which would be necessary for segregation distortion, "A type of recognition phenomenon would have to be assumed, which protects sperm of identical *t*-genotypes from affecting each other" ibid. Immunological approaches were showing great progress in delimiting various kinds of cells involved in the immune response, such as Thy-1 (see above). However, as might be anticipated by all the antigenic specificities already found in the region, the search would be fraught with difficulties. The idea appealed to many investigators, such as Dorothea Bennett (1975). "Based on the known effects of the *t* mutants on sperm antigens..., Bennett proposed that normal development involves a fixed sequence of changing cell surface properties, each essential for the next developmental transition. The role of the *t*-complex would be to determine this programmed sequence of changing cell surface properties" (**Wilkins, 1986, pp399-400**).

It was assumed that any antigens controlled by the *T/t*-complex would be present on both sperm and embryos. The presence of antigens encoded by the H-2 region in mouse spermatozoa was amply demonstrated using multiple techniques: by absorption of anti-H-2 sera (Vojtiskova, 1969; Vojtiskova and Pokorna, 1972), by presensitization of mice with sperm to subsequent skin grafts (Vojtiskova et al., 1969), by direct cytotoxicity. (Goldberg et al., 1970; Johnson and Edidin, 1972), by an antiglobulin technique utilizing [125] I-labelled rabbit-anti-mouse gammaglobulin (Erickson, 1972) by immunofluorescence (Vojtiskova et al., 1969; Erickson, 1972), and by immune-electron microscopy Vojtiskova et al., 1974), Fig. 7.1. Quantitative studies utilizing radioisotope-labelled reagents suggested that the amount of H-2 antigen present on the surface of spermatozoa was about one-tenth that present on the surface of spleen cells (on a per cell

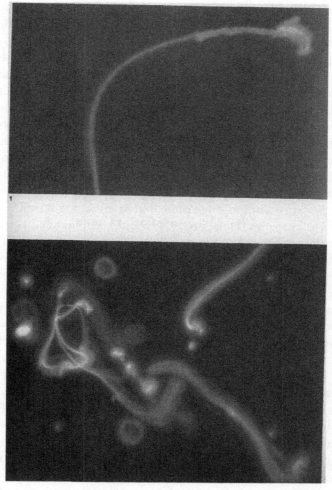

FIG. 7.1 Detection of H-2 antigens on sperm by immunofluorescence. Above, CBA sperm with anti-H-2k followed by fluorescein-labelled rabbit-anti-mouse-g-globulin; below, BALB/cJ sperm with anti-H-2d similarly treated. From Erickson, 1972, copy right holder.

basis; Erickson, 1972, Tsuzuku et al., 1972). The demonstration of Ia antigens on spermatozoa raised the question of whether only these Class II antigens were present since the reagents used in the above studies would have detected both classes. However, the absorption studies of Vojtiskova and co-workers (1974) utilized erythrocytes as the target cells (hemagglutination was being tested). Erythrocytes do not contain Ia antigens so the decrease in hemagglutination reactivity of anti-H-2 sera with spermatozoa suggested that both classes of antigen were present. Very specific antisera to the separate H-2 regions were readily absorbed by target spermatozoa but not by spermatozoa isogenic to the antibody-producing strain or by erythrocytes corresponding to the maximal number contaminating the spermatozoal preparations (Erickson et al., 1977). It seemed unlikely that these antigens were functionally important: Lyon et al. (1972) had shown that spermatozoa bearing a chromosome complement deficient in the H-2 region successfully fertilize eggs.

Because of segregation distortion, to be discussed more fully in the next chapter, it was of great interest as to whether there could be haploid expression of these H-2 antigens on sperm. The weight of the evidence was against haploid expression of H-2 antigens on murine spermatozoa (Goldberg et al., 1970 ; Erickson, 1972; Johnson and Edidin, 1972). Another type of information related to the possibility of haploid expression of histocompatibility antigens concerned the time of expression during spermatogenesis: if they first appeared after meiosis there would be a greater likelihood of haploid expression (although RNA, or even the protein, could have been synthesized some time earlier). Vojtiskova and Pokorna (1971; 1972) demonstrated by absorption that H-2 antigens are present on spermatocytes, while their immunofluorescence studies suggested that spermatogonia and testicular somatic cells did not bear H-2 antigens. However, this would be influenced by the turn over time of H-2 antigens in the membrane (6 hours on splenocytes; Schwarz and Nathenson, 1971). If, however H-2 antigens had a low turnover rate on spermatids, their presence on spermatozoa could represent persistence from premeiotic stage. Fellous et al. (1976) studied the time of appearance of Ia antigens on sperm utilizing absorptions of antisera with testicular cells from immature mice (which do not have the later stages of spermatogenesis until after weaning) and homozygous *Weaver* mice, which have almost no germ cells. They found that Ia antigens are not detectable until the primary spermatocyte stage of spermatogenesis but, of course, this is premeiotic and the antigens would expected to remain on all the products of meiosis, for example, not be different between the 4 sperm resulting from the meiotic division of one spermatocyte.

There were also many studies of H-2 antigen expression on embryos. There had been an earlier period of studying antigens on embryos exemplified by Salome Gluecksohn-Waelsch's study of the effects of immunizing maternal mice with brain extracts and adjuvant, finding that 9% of the resulting embryos showed nervous system abnormalities (Gluecksohn-Waelsch, 1957). The complexities of antigen expression on preimplantation embryos were caused largely by the gradually increasing knowledge of narrower specificities as the evolution of knowledge about H-2 described above occurred. In the first studies, it was assumed that the same array of antigens would be expressed on embryonic and adult tissues. Since there are multiple copies of class I loci that are expressed on them, and many more copies in the genome that can be detected by hybridization of *H-2* cDNA clones (Steinmetz et al., 1981), it was possible that the preimplantation embryos would express class I antigens from alleles not normally expressed in adults. Before this possibility was known, a number of studies were performed with antisera that might have detected minor histocompatibility antigens. The latter were clearly detectable on preimplantation embryos and allowed paternal expression to be detected as early as the eight-cell stage (Muggleton-Harris and Johnson, 1976). When due account was taken of such contaminating specificities, many authors were unable to find class I H-2 antigens on preimplantation embryos (Heyner et al., 1969; Palm et al., 1971; Billington et al., 1977). Searle et al. (1974) could not detect H-2 antigens on the preimplantation embryo by cell-mediated responses but thought they could with antibody staining which is a more subjective approach (and must be read blind). Cytotoxicity provides a less subjective way of estimating antigen expression on cells since the all-or-none phenomenon of life or death for the cell is more easily scored. Krco and Goldberg (1977) used cytotoxicity to detect paternal H-2 antigens as early as the 8-cell stage. Techniques using radioisotopes, in which the reactivity with the embryo is counted by a radioactivity detector, also provided a more objective technique. Hakansson et al. (1975) could detect paternal H-2 antigens on mouse blastocysts during experimental delay of implantation (the delay, called diapause, occurs when ovariectomy is performed to prevent estrogen and progesterone stimuli needed for implantation) but the H-2 antigens disappeared after the onset of implantation. Webb et al. (1977) performed immunoprecipitation of radiolabeled antigens from embryos grown in media containing isotopically labelled amino acid precursors. They did not find H-2 antigens until the late blastocyst stage when they could precipitate them from the inner cell mass. Interestingly, beta$_2$microglobulin, which is normally associated with class I H-2 antigens is expressed on preimplantation embryos (Sawicki et al., 1981). They performed immunoprecipitation and two-dimensional electrophoresis to detect beta$_2$ microglobulin in preimplantation embryos. Using electrophoretic variants, they were able to demonstrate that the paternal form was expressed as early as the two-cell embryo. However, it is not clear why it is expressed, as its role with H-2 antigens doesn't seem relevant.

The class II antigens were also sought on the preimplantation embryo. The first studies found them as far back as day 11 but not earlier (Delovitch et al., 1978). Immunofluorescent studies of preimplantation embryos using the anti-class II reagents were also negative (Heyner and Hunziker, 1981). Although the overall weight of the evidence was against the expression of either class I or class II H-2 antigens on the preimplantation embryo, other than during delayed implantation or, perhaps on the inner cell mass later in development, it was clear that genetic variation in the H-2 region affects preimplantation development. Blastocysts bearing the H-2^b haplotype, in several different congenic combinations, have more cells than do embryos with the H-2^k haplotype (Goldbard et al., 1982). These differences could well be due to some *H-2*-linked gene not expressed as an antigen on the surface of embryos.

Thus, given the conflicting evidence for H-2 antigens on sperm and embryos, the search for *T/t*-complex antigens was, not surprisingly, full of controversy. The earliest attempt seems to be that of Caspari and Dalton which were not published but were part of their annual report to the Carnegie Institution in 1949 (Annual Report of the Director of the Dept. of Genetics, Carnegie Institution of Washington, Year Book #48 for the year 1948-1949, pp 188-201). In a section entitled "serological studies with the Brachury mouse", they reported that they prepared rabbit antisera to *T*-mouse testicular or splenic extracts, absorbed the sera with control extracts from non *T*-mice and tested them by precipitation and complement fixation. They found some evidence of specificity concluding that "serological differences between *T*/+ and +/+ animals are therefore clearly indicated". However, the differences depicted in one figure are small and they did not follow up the work. As already mentioned in regards to Gorer's work on H-2, his rabbit antisera were probably not detecting the H-2 specificities later found with mouse sera and Caspari and Dalton's antisera probably detected nothing related to the *T/t*-complex.

Ernst Caspari, 1909–1988

Ernst Caspari had an important influence on the development of mouse genetics[1]. Although his *ouvre* was not large, he contributed greatly through teaching and mentoring. Ulrich Grossbach has provided a short biography in two parts, the first emphasizing his contributions to biochemical genetics (1996) while the second discusses more fully the effects of Nazi anti-Semitism on his, and other investigators in Gottingen, careers. His doctoral student, Eva Eicher, has provided a fuller account of his life (1987).

Ernst Wolfgang Caspari was born in Berlin to a professional family. His father was a physician and chairman of the Dept. of Animal Physiology at the Royal School of Agriculture. "The family was Jewish and had lived in Berlin since the 18th century, but both parents, as well as Ernst, his brother Fred, and his sister Irene were protestants" (Grossbach, 2009). Ernst developed a strong interest in biology, in part from visiting his

Ernst Caspari in a dedicatory photo to Dr. Esther Lederberg. From Esther M. Zimmer Lederberg Memorial Website, http://www.esthermlederberg.com/.

father's lab. The family moved to Frankfurt when his father was appointed head of the Cancer Department of the Institute of Experimental Therapy there. He attended gymnasiums in both Berlin and Frankfurt which provided the classical Greek and Latin curricula. Visits to his father's laboratory and, no doubt, discussions with him, kindled an interest in biology. "When Caspari told his father that he wanted to be a geneticist, his father was not enthusiastic because there was not much money to be earned in science, and he would have to support his son for a long time. However, his father finally agreed that Caspari could go to the university to pursue a career in zoology or botany. Caspari choose zoology" (Eicher, 1987). He started at the University of Freiberg but his interest in genetics was so strong that he left after one semester when he discovered that the Professor, Spemann, had no interest in genetics (see mini-biography of Salome Gluecksohn-Waelsch). He then spent a semester at the University of Berlin (German students were quite free to move among universities as there was no worry about transferring credits—you passed exams when you were ready). However, the famous geneticist, Goldschmidt, did not take graduate students and the well known geneticist Curt Stern (who, although a *Drosophila* geneticist, was to author a highly influential textbook of human genetics [Principles of Human Genetics, 1949, 1960, Freeman & Co., San Francisco]) was about to leave for the United States. Thus, he moved on to Gottingen and Alfred Kuhn's laboratory. At the University of Gottingen in the mid-1920s, the professor of botany, Fritz von Wettstein and the professor of zoology, Alfred Kuhn, were unusual in collaborating in teaching biology and studying developmental genetics. "The common interest of both groups, gene action in development, induced them to look at plant and animal development from the same point of view. This approach became very successful and led into the new field of biochemical genetics" (Grossbach, 1996). This is the department that Caspari joined as a graduate student. The laboratory and the experimental approaches are well described by Rheinberger (2000) who details the development of the experimental model and the initial studies of it. Color mutants in the moth *Ephestia* had been found by Kuhn. Caspari used tissue transplantation from dark mutant developing tissues to larvae of light mutants which then became dark indicating that a diffusible substance was controlled by the product of the mutant gene. These results "had revealed the first case for a 'one gene-one enzyme' relation several years before the beautiful studies of Beadle and Ephrussi (1937) on the development of eye color in *Drosophila* and Beatle and Tatum (1941) on biochemical effects of genes in *Neurospora* led to the one gene-one enzyme hypothesis' (Grossbach, 1996). However, Caspari saw alternative explanations to his work. He was to write Curt Stern in 1947 that I "have become convincd that Beadle's on gene-one enzyme theory is certainly oversimplified. It is not so much my own research, which brought me

---

[1] I never met Ernst Caspari (although I was at the University of Rochester as an undergraduate in 1956-1958 but he had not yet arrived). However, Eva Eicher's devotion was communicated in many encounters with her. I spent part of the summer of 1972 with Eva at the Jackson Laboratory. We were mixing *t*-allele-bearing and normal sperm in equal numbers for artificial insemination in an attempt to learn more about segregation distortion but artificial insemination wasn't working that summer. Of course, as mentioned in Chapter 5, he was the first investigator to look for *t*-antigens.

to this conclusion but a thorough study of the papers by Beadle's collaborators. The number of cases in which he must make additional assumptions is certainly very striking"[2].

After this exciting work, Kuhn would have liked to hire Caspari as a researcher but after 1933, the Nazi's forbid it. As briefly mentioned in Chapt 1, he was offered a job in Istanbul which he took. He initially applied to a zoologist but the microbiologist Braun replied first and so it was that he worked in microbiology. The studies involved malaria in birds through which he learned bacteriological and immunological techniques. "The most important thing that happened in Istanbul, however, was personal: he met and fell in love with Hermine (Hansi) Abraham. They were married on August 16, 1938" (Eicher, 1987). He was soon to be able to go to the United States. Beverly Kunkel of Lafayette College in Easton, Pennsylvania had asked LC Dunn for recommendations for scientists to take fellowships at the College. Dunn replied "My own impression of Caspari was gained from two days spent with him in Gottingen in the summer of 1936. I thought him the best man at Kuhn's institute and resolved then to take any opportunity to recommend him in the United States... Personally, Caspari is an attractive fellow of very broad culture and I think he would make his way in any university group".[3] He initially went on his own with Hansi to join him later.

He first worked for Paul David, a biologist at the college. David assigned him the study of a mouse mutant sent to David by a breeder in Florida. Caspari established that the mutant, *Kinky,* was autosomal dominant and homozygous lethal. This led to his heavy involvement in mouse genetics and with LC Dunn (see Chapter 3). There is a large correspondence between Dunn and Caspari concerning developmental mutants, especially *Kinky,* for the years 1938-1945 archived at the American Philosophical Society. As mentioned in Chapter 3, he collaborated with Dunn to find that *T,* (and *t*[1]), *Fused (Fu)* and *Kink (Ki)* were closely linked but separable by crossing-over. Caspari showed a critical attitude to this work, writing to Dunn, before it was published "Finally, it is a question whether it should not be mentioned, that there might be a case where the developmental unit called "gene" is larger than one crossover unit. That would be more or less Goldschmidt's hypothesis. I feel very strongly that such an hypothesis, contradictory to any current assumptions, should not appear in print before all other more conservative explanations have found to be unsatisfactory....Actually, I think, that our case [at this time, apparent lack of crossovers between *Fu* and *Ki*] is the best evidence for Goldschmidt's theory hitherto brought forth, and better than his own experiments."[4] It was good that they waited since a rare recombination was soon found.

Caspari desperately tried to get his family out of Germany "In 1941, he successfully obtained a Cuban visa and funds to get his parents out of Germany, but the visa arrived at the American Consulate in Frankfurt two days after his parents had been deported to a concentration camp in Lodz, Poland. His father served at Lodz as a physician treating the sick until he died of starvation" (Eicher, 1987). His mother's death is unrecorded.

Caspari continued his research on *Ephestia* as well as doing mouse genetics. He was made an Assistant Professor in 1941 and taught comparative anatomy as well as continuing his research. With the US entry into the WWII in 1941, he started teaching German to the prospective soldiers. In a letter of Dec. 12, 1942 to Dunn, he reports that he has lost his money for mouse experiments and is concerned about taking money from the science honorary society, Sigma Xi, for an incubator for his *Ephestia* experiments because he might be drafted. Soon Lafeyette College was impecunious (it was an all men's school and the student body plummeted) and Caspari needed a new job. It was a chance encounter with a University of Rochester plant physiologist following a job interview for industry, to whom he described his genetic studies, that led to a recommendation to Curt Stern who hired him to work on mutation studies for the Manhattan Project. He was also given the opportunity to continue his work on *Ephestia.*

His work on mutations caused by gamma-rays in *Drosophila* at Rochester were to cause some controversy. There was initially controversy in the lab as well. Curt Stern wrote to Caspari with concerns about the controls: "The radiation data continue to be puzzling....Muller informs me that his data give an aged control similar to yours. Thus, my first idea that your results could be 'explained away' by assuming that your control value happened to be unusually high, seems unlikely"[5]. In a very thorough paper (which had to be declassified before it could be published) comparing doses of radiation spread out over time compared to earlier dosage work with single doses carried out by Spencer and Stern, concluded that "mutation rates of irradiated...sperm is, with a high degree of probability, smaller if the irradiation is spread over 21 days than if the same dose is given at once" (Caspari and Stern, 1948). As discussed by Calabrese (2011), Herman Muller believed there was no lower limit for radiation-induced mutations which these results invalidated. Despite

[2]Letter of Jan 18, 1947, to Stern, Curt Stern papers, American Philosophical Society archives.

[3]Letter dated May 24, 1938 to Professor Kunkel., L C Dunn, papers, American Philosophical Society archives.

[4]Undated letter from Caspari to Dunn, after Jan. 5, 1942 and before Feb. 2, 1942, L C Dunn, papers, American Philosophical Society archives.

[5]Letter of Feb. 4, 1947 from Stern to Caspari, Curt Stern papers, American philosophical Society archives.

the communication of these results to Muller one month before his 1948 Nobel Prize lecture, Muller continued to declare that there was "no [lower] threshold" of mutation rate by dose conclusion.

Although his work for the Manhattan Project mostly consisted of studies of radiation-induced mutations in *Drosophila*, as he reported to Dunn, he and Curt Stern discussed the *Ki/Fu/T* linkage studies in detail, indicating his continued interest in the *T/t* complex. This continuing interest led to his attempt to find *T/t* complex antigens reported to the Carnegie Foundation in 1948 which are described in Chapter 7. He also was interested in cytoplasmic inheritance, writing a 66 page review of the evidence in favor of it (it seems surprising that it was needed given all the evidence that led many to favor it over Mendelian inheritance; Caspari, 1948). His interest in mutation included somatic cell mutation which he studied some time later with H J Pohley, using purine analogues as the agent and wing scales of *Ephestia* as the target (Caspari et al., 1965).

A friendship with a fish geneticist from Weslayan University, Hubert Goodrich, led to a recommendation for a position there. During the job interview with the President, "they discussed Aristotle and Aristotelian teleology. Caspari's ability to discuss philosophical subjects must have impressed [the President] who had a Ph.D. in philosophy, because he offered Caspari a position at Wesleyan at the level of Associate Professor" (Eicher, 1987). This was in 1946 but his very heavy teaching role led him to take an early leave of absence in 1947 at the Carnegie Institution of Washington at Cold Spring Harbor. Here he unsuccessfully tried to purify the enzyme responsible for pigment production in *Ephestia* and studied the causes of albinism in *axolotl*. He found that the pigmentary defect was in the migration of melanophores. This work was only reported, as were the *T/t* antigen studies, in the annual reports to the Carnegie Institution (Annual Report of the Director of the Dept. of Genetics, Carnegie Institution of Washington, Year Book #47 for the year 1947-1948, p. 183). Although he could have stayed on at Cold Spring Harbor, his wife was bored and wanted to study psychology while Caspari was discouraged about his slow rate of scientific progress. He returned to Wesleyan as a full professor (1949) and his wife obtained her M.A. there which led to a successful career in clinical psychology (Eicher, 1987).

Caspari was named the Daniel Ayres Professor of Biology in 1956 and took on many important administrative responsibilities at Wesleyan, including investigative committees on scientific training during the post-Sputnik era. He continued to visit the Cold Spring Harbor laboratories during the summer where he had important interactions with Ken Paigen, a future Director of the Jackson Laboratory. It was on one of these visits in 1958 that Demerec asked him to take over the editorship of *Advances in Genetics* which he accepted. He developed an interest in behavior genetics and published several highly cited reviews (Caspari, 1963, 1968) and co-edited a book on the subject (Ehrman et al., 1972). He was twice a visiting fellow (1956-1957, 1965-1966) at the Center for Advanced Studies in the Behavioral Studies at Stanford University and received the Thomas Dobzhansky Award for Research in Behavior Genetics. He moved to the University of Rochester in 1960 to be Chairman of Biology. His lectures were much enjoyed and biology became a popular major. However, the administration was not sympathetic to departmental needs and he stepped down in 1966. He became President of the Genetics Society of America in 1966. He accepted the Editorship of their journal, *Genetics*, in 1968 and hired an assistant to help find reviewers to speed up publication. He published, with his Rochester botanist colleague Arnold W. Ravin, a collection of articles in genetics titled "Genetic Organization: A Comprehensive treatise, Vol. 1" (1969) (Caspari and Ravin, 1969).

Ernst Caspari was not highly "focused", a modern requirement for competitive researchers. He used many organisms to seek an understanding of basic biological questions. He allowed graduate students and fellows a wide latitude in choosing research topics and organisms. Eva Eicher (1987) lists *Galleria melonella,* a moth that feeds on bee honey; *Ephestia; Habrobracon,* a wasp parasitic on *Ephestia;* lambda phage; and mice as organisms being studied during her doctoral studies time. She herself worked on X-inactivation of mice, a topic tangentially suggested by Caspari when he recommended Mary Lyon's recently published *Nature* paper on the subject. "Caspari ran his laboratory in the way he approached science: ask a question and proceed from there, utilizing as many organisms and as many techniques as necessary. This approach did not result in hundreds of papers directed at answering a few specific questions." (Eicher, 1987), Eva Eicher later added : "Ernst was different. He did not steal his student's ideas. He never put his name on publications that resulted from my PhD research. He considered these were my ideas and my work. He was an exception".[6]

He became professor emeritus at the University of Rochester in 1975 and he and his wife decided to re-visit Germany. He was a visiting professor in the Dept. of Genetics at Justus-Liebig University in Giessen in 1976 and, after his wife's death in 1979, again in 1982. He was awarded an honorary doctorate degree by Justus-Liebig University in 1983 and also received the Golden Ph.D. honorary degree from the University of Gottingen that year. He passed away in 1988.

---

[6]Personal communication, Eva Eicher, Nov., 2018.

Most of the ensuing work on this problem utilized syngeneic immunizations between inbred strains of mice with *t*-allelic tissues as sources of antigens. Since most of the *t*-alleles used for these experiments suppress crossing-over, the *t*-alleles have unique *H-2* haplotypes maintained in *cis,* as discussed above. Absorption of the resulting antisera with other tissues is not an adequate precaution—for instance, there are loci in the *T*-region determining antigens on t^{w5} lymphocytes used for absorptions. The work claiming the detection of *t*-allele specific antigens can be divided into 2 major topics: 1) putative evidence for *t*-allele determined antigens on sperm, testicular cells, and embryos and 2) evidence suggesting that the structural gene for an antigen from teratocarcinoma cells, the F9 antigen, is the t^{12} allelic product. Positive results detecting *t*-allele determined antigens on sperm and embryos were reported only by the international collaborating groups of Dorotha Bennett at Columbia University and Francois Jacob at the Institut Pasteur (Bennett et al., 1972; Yanagisawa et al., 1974; Kemler et al, 1976; Artzt and Bennett, 1977; Marticorena et al, 1978; Cheng and Bennett, 1980).[9] Their papers and others were sometimes in conflict. For instance, Vitetta et al. (1975) reported that the F9 antigen was a protein with structural similarities to H-2 antigens and that it was associated with beta2 microglobulin, while Dubois et al. (1976) found no association of beta2 microglobulin with F9 antigen and Muramatsu et al. (1979) found it to be predominantly polysaccharide. In addition, Kemler et al. (1976) reported the reactivity of antisera to t^{w32}, t^{0}, t^{w18} and t^{w5} sperm with morula in an immunofluorescence assay. However, Marticorena et al. (1978) failed to detect the presence of t^{w5} sperm antigens on four-cell and morula embryos using a complement-mediated cytotoxicity test.[10]

Negative evidence for the presence of *t*-allele antigens on sperm was provided by workers from Hugh McDevitt's groups. Goodfellow et al. (1979) used an antiglobulin technique to detect the binding of immunoglobulins from immunized mice with target sperm and could find no evidence for such antigens despite a large number of different sera and different target sperm. More importantly, Gable et al. (1979) used a direct cytotoxicity test on sperm, a test directly comparable to the tests that had been claimed to demonstrate such antigens, and found completely negative results, although the number of antisera and target sperm tested were smaller than in the first paper. Hammerberg (1982) used a cell-mediated lymphocytotoxicity test to demonstrate an antigen shared by all the *t*-haplotypes tested, for example, not one comparable to the ones previously claimed to be different with the different *t*-alleles.

The evidence for a determination of the F9 teratocarcinoma antigen by the t^{12} allele was of two sorts: 1) Absorption studies with sperm showed that it took twice as many sperm from a t^{12} -heterozygous animal (t^{12} /+ or t^{12} /*T*) to absorb anti-F9 activity as it took of spermatozoa from a variety of other mice, including other *t*-alleles (Artzt and Jacob, 1974; Marticorena et al., 1978); 2) The apparent segregation of the F9 antigen on embryos of +/ t^{12} X +/ t^{12} cross but also in +/t^{w5} X +/t^{w5} cross (Kemler et al., 1976). The latter result indicated that the F9 antigen was found with multiple *t*-alleles while the former result could be the result of a general, *T*-linked effect on levels of surface antigens as was soon found for H-Y (Kralova and Lengerova, 1979). Erickson and Lewis (1980) using 2 different anti-F9 antisera and several different immunofluorescent secondary antibodies, studied over 1000 embryos from crosses where t^{12} segregated (all read blind) and found no evidence for the absence of F9 antigen in t^{12} homozygous embryos (Fig. 7.2).[11] Although there were no retractions, soon a new "t^{12} -modified", but not coded by t^{12}, antigen was

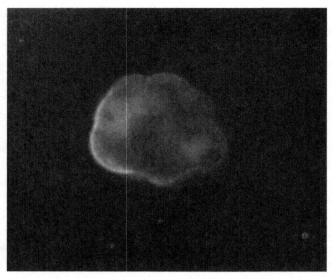

FIG. 7.2 Delayed morulae from +/t12 × +/t12 on day 3 1/2 showing positive fluorescence with anti- F9 antiserum–the faint fluorescence indicates why such images need to be read "blind" (from Erickson and Lewis,1980 with permission).

being studied by the Bennett group (Cheng et al., 1983). Although there were continuing studies of the "t[12]" antigen, such as the finding of a Yin/Yang exchange in expression of the F9 antigen and H-2 in somatic cell hybrids (Artzt et al., 1981), studies instead turned to more general gly-colipid and polysaccharide antigens on embryos (Willison et al., 1982; Sato et al., 1984; Iwakura and Nozaki, 1988), cultured cells (Campbell and Stanley, 1983; Iwakura et al., 1983) and sperm (Scully and Shur, 1988). As the excellent synthesizer of genetic development, Adam Wilkins, puts it: "Solutions to the puzzle of how particular *t* lethals affect critical early transitions in the mouse embryo may prove highly informative about these transitions, but the concept that the complex consists of a set of switch genes is probably incorrect (**Wilkins, 1986, p. 402**).[12]" This was the concept to which Francois Jacob was quite devoted.

The historian of science, Michael Morange (2000) has speculated about the reasons why Francois Jacob was so ready to accept these fallacious results (the French version was originally titled "*Le complexe T de la souris: un mirage riche de'enseignemts*" [The T complex of mice: a mirage rich in lessons]). He relates that Jacob greatly admired Sydney Brenner and his choice of the nematode as a model organism. Jacob "was looking for an organism whose genetics had been extensively studied and for which there were *in vitro* systems allowing the study of development with biochemical tools. In addition, when Jacob finally decided at the end of the sixties to stay at the Pasteur Institute, the organism he choose had to be of some interest for the study of human pathologies" (Morange, 2000). Jacob planned to study genetic control in this model and, with the hiring of Jean-Louis Guenet to head the mouse genetics unit, a strong group was soon built. The *T/t* complex had a special attraction with its apparent cell surface markers and the linkage of the genes involved since molecular biologists were hopeful that mammals would have linked regulatory complexes as had been found in bacteria. In addition, the tera-tocarcinoma lines with their apparently related antigens (such as, F9) provided the important in vitro systems. Thus, Jacob was prepared to embrace the immunological approaches Karen Artzt brought to Paris. However, as Morange emphasizes, the regular members of the laboratory did not have expertise about immunological techniques "and therefore [the results] could not be seriously criticized by the members of the lab who had not been trained to perform these experiments" (Morange, 2000). This is especially true for studies with antibodies where a sec-ondary antibody uses fluorescein as the detection system. There is always some background fluorescence and it is easy to over read or under read the results. As Laura Snyder points out, paraphrasing William James: "sometimes we convince ourselves to believe what we chose to believe, even without rational evidence, so too, it can be said, we sometimes will ourselves to see what we want to see, or what we are accustomed to seeing" (**Snyder**, p. 121). Perhaps his disappointing results with studies of the *t*-antigens is the reason that his book, **The Possible and the Actual, 1982** is much less optimistic about solving the problems of development than his earlier book, **The Logic of Life, 1973**.[13]

Notes

1. The history of immunology has its own large literature. Arthur Silverstein's "A History of Immunology" (2nd Ed., 2009, Elsevier, London) is particularly comprehensive.

2. Leslie Brent (**1996**) pp. 176-177 provides an excellent mini-biography of Peter Gorer. Jan Klein (**1986**) pp. 3-4, tells a delightful story of visiting Gorer' study as the guest of Gorer's widow.

3. I thank Walter Bodmer for pointing this sentence out to me.

4. Although I was a co-author of a paper presented at this symposia (Gluecksohn-Waelsch and Erickson, 1971), I was not in attendance but I was able to meet with this group in Prague in 1972. The walls of many buildings downtown were heavily pockmarked by bullets—living conditions were austere and spirits down but the on-going science excellent. Socialist planning wasn't very good for arranging laboratory supplies or reagents. My friend, Ladja Moravek of the Institute of Protein Chemistry, reported that one year they had no labora-tory grade urea, heavily used for denaturing proteins in experiments. The institute basically closed down for several weeks to purify and crystallize urea using fertilizer as the starting material.

5. These tables could be very confusing and might be made fun of. Frank Lilly, Salome Gluecksohn-Waelsch's successor as Chairman of Genetics at Albert Einstein College of Medicine, provided an example (as told to me by Salome). The various genetic departments in NYC would meet of an evening about monthly to pres-ent each department's work to the members of the other departments. Frank was presenting one of these complex slides and pretended to become confused and, eventually, was in tears (he was also a sometimes actor in off-Broadway theatricals). Soon the audience was in hysterics but all were not entertained (Kurt Hirschhorn for one).

6. As well as Av Mitchison, whom I know well, having spent a last year of postdoc with him in 1970-71 at the National Institute of Medical Research of the Medical Research Council at Mill Hill, outside London (where he gave me free rein and provided a lab technician to start my search for *t*-antigens), I met "Sir Peter" there many times. Although confined to an electrical wheel chair (a monstrous entity in those days) due to a severe stroke, he was a stimulating thinker and a charming conversationalist.

7. The finding of an influence on cAMP levels and on susceptibility to glucocorticoid-induce cleft palate particularly intrigued as cAMP changes had been implicated in palatal formation (Pratt and Martin, 1975). We measured cAMP levels in palate shelves and tongues of fetuses from steroid- treated and untreated mice of the parental and reciprocal *H-2* congenic inbred strains between A/J and C57B/10J. We found evidence

for an epistatic interaction of unidentified genes in the genetic background with *H-2* alleles to determine the level of palatal shelf cAMP (Erickson et al., 1979). The epistatic effect persisted through days 14 and 15, an important period of palatal shelf development. The data suggested that a portion of the *H-2* controlled component of susceptibility to steroid-induced cleft palate is mediated through alterations in the metabolic effects of cAMP. We also studied the *brachymorphic* mouse *(bm*/bm) which has greatly increased suscep- tibility to glucocorticoid-induced cleft palate (Pratt et al., 1980). The results suggested that both A/J and *brachymorphic* mice have a delay in palatal shelf rotation and elevated levels of cAMP, increasing susceptibility to glucocorticoid-induced cleft palate. Methylmercury also increases the incidence of cleft palate but neither the *H-2* allele nor the residual genetic background played a role in the effect of methymercury on cAMP concentrations in fetal tongues (Harper et al., 1981). We pursued the role of the MHC in determining cAMP levels in humans, but in readily-available lymphocytes, not liver. We found that about 19% of the variability in isoproterenol-stimulated cAMP was determined by the HLA-B18 antigen while 38% of the variability of was attributable to factors that aggregate in families independent of HLA-B18 (Erickson et al., 1985). Since the glucocorticoid-inhibition of phytohemagglutinin stimulation had been shown to be influenced by *H-2* alleles (Pla et al., 1976), we looked for a similar association with HLA. HLA-B7 enhanced the glucocorticoid-inhibition of phytohemagglutinin stimulation while HLA-A10 decreased it (Erickson et al., 1985). Perhaps not surprisingly, we did not find an association of human, sporadic cleft palate with HLA antigens (Pairitz et al., 1985). Another glucocorticoid-induced cleft palate-related claim was that the high- and low-incidence strains differed in the amount and affinity of the glucocorticoid receptor and that this difference mapped to the *H-2* complex (Salomon and Pratt, 1976; Goldman et al., 1977). In a series of papers, and with the collaboration of the glucocorticoid receptor expert, William Pratt, we demonstrated that this was incorrect (Butley et al., 1978; Leach et al., 1982; Harper and Erickson, 1983; Liu et al., 1984; Liu and Erickson, 1986a, 1986b).

8. Len Herzenberg, who received his Ph.D. in 1955 working with Ray Owen at Cal Tech., reported to me (he was my scientific mentor in the Dept. of Genetics at Stanford University School of Medicine in 1961-1963) that once Ray's curriculum vitae was requested by the Nobel Comm. during the 50's but he did not send it. Did this deprive him of a chance to share this Nobel Prize (up to 3 can share it for a single discovery)?

9. This work was started at the Columbia University laboratory of Dorothea Bennett and continued with the Jacob laboratory at the Institute Pasteur after Karen Artzt was a postdoctoral fellow there, presumably taking antisera from Columbia with her.

10. The work I had started in Av Mitchison's lab at Mill Hill was continued during my first academic appointment at University of California, San Francisco School of Medicine where I was lucky enough to obtain an NIH Research Career Development Award for this research. I performed many immunizations of mice with various preparations of spermatozoa, with and without adjuvants, and used multiple tests to detect the results, including cytotoxicity and immunofluorescence. To limit the target, immunizations included ones between siblings where the donor was t^n and the recipient was *T*. I was aware of the unique H-2 specificities that would be linked to these t^n-alleles and did things such as absorbing the sera with red blood cells to prevent detecting such reactions. My results were negative. Thus, I was surprised when Karen Artzt and Dorothea Bennett published papers claiming the existence of *t*-antigens. I had been sent preprints by Salome Gluecksohn-Waelsch (with a letter dated Oct. 25, 1973) after Karen Artzt had given a seminar at the Dept. of Genetics, Albert Einstein College of Medicine. I was due for a sabbatical and Karen Artzt had taken this technology to Francois Jacob's lab. Thus, I felt that a less contentious way to "look in" on the Artzt-Bennett work was to have a sabbatical year in Francois' laboratory for which I was able to obtain a John Simon Guggenheim Jr. Memorial Fellowship. While I did not directly work on the *t*-antigens during the year (instead working on spermatogenesis-related matters such as the time of appearance of Ia antigens on sperm [Fellous et al., 1976]), I watched what was going on very closely. I was surprised to see that all the immunological typing of embryos was unblinded. Thus, it was very easy to see what one wanted to see. Because the observations seemed so unfounded, I spent a fair amount of time arguing against the notion of *t*-antigens. Many of these discussions were with Gabriel Gachelin, one of the associate investigators in the large laboratory and the individual I worked with most closely, and I think I only talked to Francois several times about my negative views. Nonetheless, during the year the mirage of *t*-antigens continued. As reported to me by

Gabriel Gachelin, Francois gave up the notion when a former postdoc., Thiery Boon, who he had sent to the States for further training, became involved. As told to me, Thiery Boon wanted some of the antisera for his studies. When he obtained the antisera and set up various control studies, he apparently couldn't confirm any of the reported observations. Supposedly, a phone call to Francois convinced Francois of the irreproducibility of the results and no further work emanated, or at least was published, from Jacob's group on *t*-antigens. Many of the papers had been published in the Proceedings of the National Academy of Sciences, U.S.A, of which Jacob was a foreign member, but none of these or the other papers was retracted.

When asked more recently about the discrepancies of the data published from the Jacob lab on the subject of *T/t* complex antigens, Gabriel Gachelin (Dec. 17, 2018, duplicated with permission) added: "since we discuss about T-locus and Jacob's lab, let me tell you that things have been much more complex and difficult that you can imagine. If we come back to Kemler's experiments, the truth is that he has been the only one to "read" the labelled eggs under the IF [immunofluorsescent] microscope, largely because the signal faded out very fast. Actually we were all excited about Jacob's /Atzt' hypothesis and we relied on his objectivity. The psychological pressure exerted by Jacob was enormous and that probably explains Kemler's results... Concerning the link between F9, H-2 and t-antigens, most of us understood that something was going wrong when, after Vitetta's paper is published and faced to our inability to reproducing her experiments, Hubert [Condamine], Charles [Babinet] and myself went back to the copy of Artzt's note book kept in the lab. The first shock was the discovery that, out of the three absorption tests ran on tw32 spermatozoa, only one showed a slight bias; the two others superimposed with absorption on +/+ cells. Results were by no means conclusive, and the paper should never have been published. Concerning the homology of structure of F9 and H-2 antigens, Ph. Dubois and myself ran a number of immunoprecipitation experiments. All attempts to immunoprecipitate F9 antigen using amino-acid labelled EC cells failed. By contrast, we got results with EC cells labelled with sugars but the precipitate migrated in 2D-thin layer chromatography as a large glycolipid. I was myself convinced that the main differences at the cell surfaces lied in the different structure of the carbohydrates of EC cells. That's a work developed with Nathan Sharon, Takashi Muramatsu and others; the conclusion was that the epitopes of the F9 antigen were located within HMW [high molecular weight] carbohydrates. All that has been published since 1974, as well as the absence of b2m [beta-2 microglobulin] at the surface of EC cells. The final conclusion was that here was no link between F9 and transplantation antigens not to speak of the hypothetical link between F9 and t^{32}, that had become rather fuzzy or wrong: we have been studying embryo-specific blood-group-like substances[.] I presented an overview of these results at a meeting in Japan in 1980. Artzt attended it and she was exceedingly aggressive. That was a month before Lake Placid meeting, remember. Much later, I met with Ellen Vitetta in 1993 in Adelaide, Australia and she could not explain her 1974 paper. In the meantime, between 1978 and 1980, we nearly all had to quit Jacob's lab, I presume because we have been witnesses of all that story (and others)."

11. Despite the extensive data, because it was a negative observation going against a famous name, it could not be published in a "top" journal but was eventually published in the new journal, Journal of Reproductive Immunology. This was also the fate of others with negative results (Goodfellow et al., 1979).

12. Wilkins did not seem to know, or choose to ignore, the already published negative papers on the existence of such antigens but rejected the notion on internal inconsistencies in the data.

13. Francois Jacob was a brilliant scientist who stimulated and guided many molecular biologists. This is well illustrated by the meeting he organized on gene regulation in May, 1990 at the Fondation Les Treilles in Provence, which gathered many of the workers at the forefront of the field (Brenner et al., 1990). After his death, an issue of Research in Microbiology was devoted to eulogies and remembrances, which are potent reminders of his wide influence. One of these eulogies, by Mark Ptashne, a major figure in studies of gene regulation, commented on Jacob's work on mammals as follows: "Jacob's mastery failed him when he dropped bacteria to work on developmental gene regulation in mice (teratomas). The principles he invented, properly framed we now know, go a long way towards bringing to light all forms of gene regulation. But the experimental manipulations in mice, their time frames, and so on, are so different from those he had mastered with bacteria that he had no feel for them. This work was quickly forgotten" (Ptashne, 2014).

References

Allen, S.L., 1955. Linkage relations of the genes histocompatibility 2 and fused tail, brachyury and kinky tail in the mouse, as determined by tumor transplantation. Genetics 40, 627–650.

Amos, D.B., 1981. Presidential Address to the American Association of Immunologists in Atlanta, Georgia, April 15, 181: The Era of the Immunogeneticist. J. Immunol. 127, 1727–1734.

Amos, D.B., Gorer, P.A., Mikulska, Z.B., 1955. An analysis of an antigenic system in the mouse (the *H-2* system). Proc. Roy. Soc. Lond B 144, 369–380.

Andresen, E., Irwin, M.R., 1959. The K blood groups system of the pig. Acta Agric. Scand. 9, 253–260.

Artzt, K., Bennet, D., Jacob, F., 1974. Primitive teratocarcinoma cells express a differentiation antigen specified by a gene at the T-locus in the mouse. Proc. Nat'l Acad. Sci., U.S.A. 71, 811–814.

Artzt, K., Bennett, D., 1977. Serological analysis of sperm of antigenically cross-reacting *T/t*-haplotypes and their recombinants. Immunogenet 5, 97–107.

Artzt, K., Hamburger, L., Flaherty, L., 1977. *H-39,* a histocompatibility locus closely linked to the *t/t* complex. Immunogenet 5, 477–480.

Artzt, K., Jacobs-Cohen, R.J., DiMeo, A., Alton, A.K., Darlington, G., 1981. Reexpression of a T/t-complex antigen (t^{12}) in thymocyte x embryonal carcinoma cell hybrids. Somatic Cell Gene 7, 423–434.

Bailey, D.W., 1971. Allelic forms of a gene controlling the female immune response to the male antigen in mice. Transplantation 11, 426–428.

Bailey, D.W., 1975. Genetics of histocompatibility in mice. I. New loci and congenic lines. Immunogenetics 2, 249–256.

Bailey, D.W., Kohn, H.I., 1965. Inherited histocompatibility changes in progeny of irradiated and unirradiated inbred mice. Genet. Res. 6, 330–340.

Bailey, D.W., Hoste, J., 1971. A gene governing the female immune response to the male antigen in mice. Transplantation 11, 404–407.

Bansal, S.C., Hellstrom, K.E., Hellstrom, I., Sjogren, H.O., 1973. Cell-mediated immunity and blocking serum activity to tolerated allografts in rats. J. Exp. Med. 137, 590–602.

Bennett, D., Goldberg, E., Dunn, L.C., Boyse, E.A., 1972. Serological detection of a cell-surface antigen specified by the T(Brachyury) mutant gene in the house mouse. Proc. Nat'l Acad. U.S.A. 69, 2076–2080.

Bennett, D., 1975. *T*-locus mutants: suggestions for the control of early embryonic organization through cell surface components. In: Balls, M., Wild, A.E. (Eds.), The Early Development of Mammals, The Second Symposium of the British Society for Developmental Biology. Cambridge University Press, Cambridge, London, New York, Melbourne.

Billingham, R.E., Brent, L., Medawar, P.B., 1953. "Actively acquired tolerance," of foreign cells. Nature 172, 603–806.

Billlingham, R.E., Silvers, W.K., 1960. Studies on Tolerance of the Y Chromosome Antigen in Mice. J. Immunol. 85, 14–26.

Billington, W.D., Jenkinson, E.J., Searle, R.F., Sellens, M.H., 1977. Alloantigen expression during early embryogenesis and placental ontogeny in the mouse: immunoperoxidase and mixed hemadsorption studies. Transplant. Proc. 9, 1371–1377.

Bonner, J.J., Slavkin, H.C., 1975. Cleft palate susceptibility linked to histocompatibility 2 (*H-2*) in the mouse. Immunogenetics 2, 213–218.

Brenner, S., Dove, W., Herskowitz, I., Thomas, R., 1990. Genes and Development: Molecular and Logical Themes. Genetics 126, 479–486.

Burnet, F.M., Fenner, F., 1949. The production of antibodies. Monographs of the Walter and Eliza Hall Institute, Melbourne.

Butley, M.S., Erickson, R.P., Pratt, W.B., 1978. Hepatic glucocorticoid receptors and the H-2 locus. Nature 275, 136–138.

Bynum, W.F., 2006. The rise of science in medicine. 1850-1913. In: Bynum, WF et al. (Ed.), The Western Medical Tradition 1800 to 2000. Cambridge University Press, Cambridge, New York, Melbourne, Madrid, Cape Town, Singapore, Sao Paolo, Delhi, Mexico City, p. 134.

Calabrese, E.J., 2011. Muller's Nobel lecture on dose-response for ionizing radiation: ideology or science? Arch. Toxicol. 85, 1495–1498.

Campbell, C., Stanley, P., 1983. Regulatory mutations in CHO cells induce expression of the mouse embryonic antigen SSEA-1. Cell 293–304.

Caspari, E.W., 1948. Cytoplasmic inheritance. Adv. Genet. 2, 1–66.

Caspari, E.W., Stern, C., 1948. The influence of chronic irradiation with gamma-rays of low intensity dosages on the mutation rate in *Drosophila*. Genetics 33, 75–95.

Caspari, E.W., 1963. Selective forces in the evolution of man. Amer. Naturalist 97, 5–14.

Caspari, E.W., 1968. Genetic Endowment and Environment in the Determination of Human Behavior: Biological Viewpoint. Amer. Res. Ed. J. 5, 43. doi.org/10.3102/00028312005001043.

Caspari, E.W., Ravin, A.W., 1969. Genetic organization: A comprehensive treatise vol. 1. Academic Press, New York and London.

Caspari, E.W., Muth, W., Pohley, H.J., 1965. Effects of DNA Base Analogues on the Scales of the Wing of Ephestia. Genetics 52, 771–794.

Cheng, C.C., Bennett, D., 1980. Nature of the antigenic determinants of *T* locus antigens. Cell 9, 537–543.

Cheng, C.C., Sege, K., Alton, A.K., Bennett, D., Artzt, K., 1983. Characterization of an antigen present on testicular cells and preimplantation embryos whose expression is modified by the *t12* haplotype. J. Immunogenet. 10, 465–485.

Counce, S., Smith, P., Barth, R., Snell, G.D., 1956. Strong and weak histocompatibility gene differences in mice and their role in the rejection of homografts of tumors and skin. Ann. Surg. 144, 198–204.

Dausset, J., Nenna, A., Brecy, H., 1958. Leukoagglutinins: V. Leuko-agglutinins in chronic idiopathic or symptomatic pancytopenia and in paroxysmal nocturnal hemoglobinuria. Blood 1954 (9), 696–720.

Davis, A.P., Roopenian, D.C., 1990. Complexity of the mouse minor histocompatibility locus H-4. Immunogenetics 31, 7–12.

Davis, M.M., Bjorkman, P.J., 1988. T cell antigen receptor genes and T cell recognition. Nature 334, 395–402.

Delovitch, T.L., Press, J.L., McDevitt, H.O., 1978. Expression of murine Ia antigens during embryonic development. J. Immunol. 120, 818–824.

Demant, P., 1973. H-2 gene complex and its role in alloimmune reactions. Transplant. Rev. 15, 162–200.

Dubois, P., Fellous, M., Gachelin, G., Kemler, R., Jacob, F., 1976. Absence of a serologically detectable association of murine beta2-microglobulin with the embryonic F9 antigen. Transplantation 22, 467–473.

Due, C., Simonsen, M., Olsson, L., 1986. The major histocompatibility complex class I heavy chain as a structural subunit of the human cell membrane insulin receptor: implications for the range of biological functions of histocompatibility antigens. Proc. Nat'l Acad. Sci. U.S.A. 83, 6007–6011.

Ehrman, L., Omenn, G.S., Caspari, E., 1972. Genetics, environment, and behavior: implications for educational policy. Academic Press, New York, London.

Eicher, E., 1987. Ernst W Caspari: Geneticist, Teacher and Mentor. Adv. Genet. 2, xv–xxxi.

Eichwald, E.J., Silmser, C.R., 1955. Communication, Transpl Bull, 2, 148–149.

Eichwald E.J., Silmser C.R., Weissman I.L. 1958. Sex-linked rejection of normal and neoplastic tissue. I. Distribution and specificity. J. Natl. Cancer Inst. 20, 563–575.

Erickson, R.P., 1972. Alternative modes of detection of *H-2* antigens on mouse spermatozoa. In: Beatty, R.A., Gluecksohn-Waelsch, S. (Eds.), Proceedings, International Symposium on the Genetics of the Spermatozoon. University of Edinburgh Press, Edinburgh, New York, pp. 191–202.

Erickson, R.P., 1977. Gene expression of a region of chromosome 17 during murine spermatogenesis. J. Immunogenet. 4, 353-362.

Erickson, R.P., Gachelin, G., Fellous, M., Jacob, F., 1977. Absorption analysis of H-2D and K antigens on spermatozoa. J. Immunogenet. 4, 47-51.

Erickson, R.P, Butley, M.S., Sing, C.F., 1979. H-2 and non-H-2 determined strain variation in palatal shelf and tongue adenosine 3′:5′ cyclic monophosphate: a possible role in the etiology of steroid-induced cleft palate. J. Immunogenet. 6, 253–262.

Erickson, R.P., Pairitz, G.L., Karolyi, J.M., Kapur, J.J., Odenheimer, D.J, Schultz, J.S., Sing, C.F., 1985. HLA-B18 is associated with decreased levels of isoproterenol-stimulated cAMP in lymphocytes. Am. J. Hum. Genet. 37, 124–132.

Erickson, R.P., Heidel, L., Kapur, J.J., Karolyi, J.M., Odenheimer, D.J., Pairitz, G.L., Schultz, J.S., Sing, C.F., 1985. HLA antigens, phytohemagglutinin stimulation, and corticosteroid response. Am. J. Hum. Genet. 37, 761–770.

Erickson, R.P., Lewis, S.E., 1980. Cell surfaces and embryos: Expression of the F9 teratocarcinoma antigen in *T*-region lethal, other lethal, and normal pre-implantation embryos. J. Reprod. Immunol. 2, 293-304.

Fellous, M., Erickson, R.P., Gachelin, A., Jacob, F., 1976. The time of appearance of Ia antigens during spermatogenesis in the mouse. Transplantation 22, 440-444.

Figueroa, F., Golubic, M., Nizetic, D., Klein, J., 1985. Evolution of mouse major histocompatibility complex genes borne by *t* chromosomes. Proc. Nat'l. Acad. Sci., U.S.A. 82, 2819-2823.

Flaherty, L., 1975. *H-33*—a histocompatibility locus to the left of the *H-2* complex. Immunogenetics 2, 325–329.

Flaherty, L., Wachtel, S.S., 1975. H (*Tla*) system: Identification of two new loci, *H-31* and *H-32*, and alleles. Immunogenet. 2, 81–85.

Gable, R.J., Levinson, Jr, McDevitt, H.O., Goodfellow, P.N., 1979. Assay for antibody mediated cytotoxicitiy of mouse spermatozoa by 86rubidium release. Tissue Antigens 13, 177–185.

Gibson, T., Medawar, P.B., 1943. The fate of skin homografts in man. J. Anat. 77, 299-310.

Gluecksohn-Waelsch, S., 1957. The effect of maternal immunization again organ tissues on embryonic differentiation in the mouse. Development 5, 83–92.

Gluecksohn-Waelsch, S., Erickson, R.P., 1971. Cellular membranes: A possible link between H-2 and T-locus effects. In: Lengerova, A., Vojtiskova, M. (Eds.), Proceedings Symposium Immunogenetics of the H-2 System. Liblice-Prague. Karger, Basel, pp. 120–122.

Goldbard, S.B., Verbanac, K.M., Warner, C.M., 1982. Genetic analysis of H-2 gene(s) affecting early mouse embryo development. J. Immunogenet. 9, 77–82.

Goldberg, E.H., Aoki, T., Boyse, E.A., Bennet, D., 1970. Detection of H-2 antigens on mouse spermatozoa by the cytotoxicity test. Nature 228, 570–572.

Goldberg, E.H., Boyse, E.A., Bennett, D., Scheid, M., Carswell, E.A., 1971. Serological demonstration of H-Y (male). Nature 232, 478–480.

Goldman, A.S., Katsumata, M., Yaffe, S.J., Gasser, D.L., 1977. Palatal cytosol cortisol-binding protein associated with cleft palate susceptibility and H-2 genotype. Nature 265, 643–645.

Goodfellow, P.N., Levinson, J.R., Gable, R.J, McDevitt, H.O., 1979. Analysis of anti-sperm sera for *T/t* locus-specific antibody J. Reprod. Immunol. 1, 11–21.

Gorer, P.A., 1936a. The detection of a hereditary antigenic difference in the blood of mice by means of human group a serum. J. Genet. 32, 17.

Gorer, P.A., 1936b. The Detection of Antigenic Differences in Mouse Erythrocytes by the Employment of Immune Sera Br. J. Exp. Pathol. 17, 42–50.

Gorer, P.A., 1937. The genetic and antigen basis of tumour transplantation. J. Path. Bacteriol. 44, 691.

Gorer, P.A., 1938. The antigenic basis of tumour transplantation. J. Path. Bacteriol. 47. doi.org/10.1002/path.1700470204.

Gorer, P.A., Lyman, S., Snell, G.D., 1948. Studies on the genetic and antigenic basis of tumour transplantation: linkage between a histocompatibility gene and 'fused' in mice. Proc. Roy. Soc. Lond. B 135, 499–505.

Gorer, P.A., Mikulska, Z.B., 1959. Some further data on the H-2 system of antigens. Ibid. 151, 57.

Greenfield, A., Scott, D., Pennisi, D., Ehrmann, I., Ellis, P., Cooper, L., Simpson, E., Koopman, P., 1996. An H-YDb epitope is encoded by a novel Y chromosome gene. Nature Genet. 14, 474–478.

Grossbach, U., 1996. Genes and development: an early chapter in German developmental biology. Int. J. Dev. Biol. 40, 83–87.

Grossbach, U., 2009. Seventy-five years of developmental genetics: Ernst Caspari's early experiments on insect pigmentation, performed in an academic environment of political suppression. Genetics 181, 1175–1182.

Hakansson, S., Heyner, S., Sundqvist, K.-G., Bergstrom, S., 1975. The presence of paternal H-2 antigens on hybrid mouse blastocysts during experimental delay of implantation and the disappearance of these antigens after onset of implantation. Int. J. Fertil. 20, 137–140.

Hammerberg, C., 1982. Detection of a *t*-complex antigen by secondary cell-mediated lymphocytotoxicity J. Immunogenet. 9, 179–184.

Hammerberg, C., Klein, J., 1975. Linkage disequilibrium between H-2 and t complexes in chromosome 17 of the mouse. Nature 258, 296–299.

Hammerberg, C., Klein, J., Artzt, K., Bennett, D., 1976. Histocompatibility-2 system in wild mice II. H-2 haplotypes of t-bearing mice. Transplantation 21, 199–212.

Harper, K., Burns, R., Erickson, R.P., 1981. Genetic aspects of the effects of methylmercury in mice: the incidence of cleft palate and concentrations of adenosine 3':5' cyclic monophosphate in tongue and palatal shelf. Teratology 23, 397–401.

Harper, K., Erickson, R.P., 1983. Ionic effects on strain differences in hepatic cytosolic glucocorticoid receptor levels in mice. Teratology 27, 43–49.

Hashimoto, Y., Suzuki, A., Yamakawa, T., Miyashita, N., Moriwaki, K., 1983. Expression of GM1 and GD1a in mouse liver is linked to the H-2 complex on chromosome 17. J. Biochem 94, 2043–2048.

Hauptfeld, V., Hammerberg, C., Klein, J., 1976. Histocompatabilitiy-2 System in Wild Mice III. Mixed lymphocyte reaction and cell-mediated lymphocytotoxicity with *t*-bearing mice. Imunogenetics 3, 489–497.

Hauschka, T.S., 1955. Probable Y-linkage of a histocompatibility gene (discussion of Eichwald and Silmser skin graft data). Transplant. Bull. 2, 154.

Heyner, S., Brinster, R.L., Palm, J., 1969. Effect of iso-antibody on preimplantation mouse embryos. Nature 222, 783–784.

Heyner, S., Hunziker, R.D., 1981. Oocytes react with antibody directed against *H-2* but not Ia antigens. J. Immunogenet. 8, 523–528.

Hildemann, W.H., Mullen, Y., Inai, M., 1974. Anergy to dual HX and HY antigens occurring In the same skin allografts between reciprocal F1 hybrid mice. Immunogenet 1, 297–303.

Irwin, M.R., 1932. Dissimilarities between antigenic properties of red blood cells of dove hybrid and parental genera. Proc. Soc. Exp. Biol. Med. 29, 850–851.

Irwin, M.R., 1939. A genetic analysis of species differences in Columbidae. Genetics 24, 709–721.

Irwin, M.R., 1974. Comments on the early history of immunogenetics. Anim. Blood Grps biochem Genet. 5, 65–84.

Ivanyi, P., 1978. Some aspects of the H-2 system, the major histocompatibility system in the mouse. Proc. Roy. Soc. B 202, 117–158.

Iwakura, Y., McCormick, P., Artzt, K., Bennett, D., 1983. A class of large polysaccharides contains the antigenic determinants for the cytotoxic antibodies in a conventional syngeneic anti-F9 serum as well as a monoclonal antibody prepared against F9 cells. Cell Differ. 13, 41–48.

Iwakura, Y., Nozaki, M., 1988. Synthesis and distribution of carbohydrate chains in cleavage-stage mouse embryos carrying the t12 lethal mutation. Dev. Biol. 128, 474–476.

Jensen, C.O., 1903. Experimentelle untersuchungen über Krebs bei Mäusen. Zentralblat. Bacteriol. Parasitenk. Infektionskrankh. 34 (28-34), 122–143.

Johnson, M.H., Edidin, M., 1972. H-2 antigens on mouse spermatozoa. Transplantation 14, 781–786.

Kaliss, N., Molomut, N., Harriss, J.L., Gault, S.D., 1953. Effect of previously injected immune serum and tissue on the survival of tumor grafts in mice. J. Nat'l Cancer Inst. 13, 847–850.

Katz, D.H., Hamaoka, T., Dorf, M.E., Maurer, P.H., Benacerraf, B., 1973. Cell interactions between histoincompatible T and B lymphocytes. IV. Involvement of the immune response (Ir) gene in the control of lymphocyte interactions in responses controlled by the gene. J. Exp. Med. 138, 734–739.

Kemler, R., Babinet, C., Condamine, H., Gachelin, G., Guenet, J.L., Jacob, F., 1976. Embryonal carcinoma antigen and the T/t locus of the mouse. Proc. Nat'l Acad. Sci. USA 73, 4080–4084.

Kindred, B., Shreffler, D.C., 1972. H-2 dependence of cooperation between T and B cells in vivo. J. Immunol. 109, 940–943.

King, T.R., Christianson, G.J., Mitchell, M.J., Bishop, C.E., Scott, D., Ehrmann, I., Simpson, E., Eicher, E.M., Roopenian, D.R., 1994. Deletion mapping by Immunoselections against the H-Y Histocompatibility Antigen Further Resolves the Sxra Region of the Mouse Y Chromosome and Reveals Complexity of the Hya locus. Genomics 24, 159–168.

Koch, G.L.E., Smith, M.J., 1978. An association between actin and the major histocompatibility antigen H-2. Nature 273, 274.

Kralova, J., Lengerova, A., 1979. H-Y antigen: genetic control of the expression as detected by host-versus- graft popliteal lymph node enlargement assay maps between the T and H-2 complexes. J. Immunogenet. 6, 429–438.

Krco, C.J., Goldberg, E.H., 1977. Major histocompatibility antigens on preimplantation mouse embryos. Transplant. Proc. 9, 1367–1370.

Landsteiner, K., 1901. Uber Agglutinationserscheinungen normalen menschlichen Blutes. Wien Klin Wschr 14, 1132–1134.

Leach, K.L., Erickson, R.P., Pratt, W.B., 1982. The endogenous heat-stable glucocorticoid receptor stabilizing factor and the H-2 locus. J. Steroid Biochem. 17, 121–123.

Lengerova, A., Vojtiskova, M. (Eds.), 1971. Immunogenetics of the H-2 System. In: Proceedings Symposium Immunogenetics of the H-2 System. Karger, Liblice-Prague, Basel, pp. 120–122.

Levinson, J.R., McDevitt, H.O., 1976. Murine t factors: an association between alleles at t and H-2. J. Exp. Med. 144, 834–839.

Little, C.C., Tyzzer, E.E., 1916. Further experimental studies on the inheritance of susceptibility to a Transplantable tumor, Carcinoma (J. W. A.) of the Japanese waltzing mouse. J. Med. Res. 33, 393–453.

Lilly, F., Boyse, E.A., Old, L.J., 1964. Genetic basis of susceptibility to viral leukaemogenesis. Lancet ii, 1207–1209.

Liu, S.L., Grippo, J.F., Erickson, R.P., Pratt, W.B., 1984. Murine glucocorticoid receptors and the H-2 locus—a reappraisal. J. Steroid Biochem. 21, 633–637.

Liu, S.L., Erickson, R.P., 1986a. Genetic differences among the A/J x C57BL/6J recombinant inbred mouse lines and their degree of association with glucocorticoid-induced cleft palate. Genetics 113, 745–754.

Liu, S.L., Erickson, R.P., 1986b. Genetics of glucocorticoid receptor levels in recombinant inbred lines of mice. Genetics 113, 735–744.

Loeb, L., 1908. Ueber Entstehung eines Sarkoms nach Transplantation eines Adenosarcomas einer Japanischen Maus. Z. Krebsforsch. 7, 80–110.

Lyon, M.F., 2002. A Personal History of the mouse Genome. Annu. Rev. Genomics Hum. Genet. 3, 1–16.

Lyon, M.F., Glenister, P.H., Hawker, S.G., 1972. Do the H-2 and T-loci of the mouse have a function in the haploid phase of sperm? Nature 240, 152–153.

Mains, P.E., 1986. A region flanking H-2K is duplicated to a distant site in most mouse t haplotypes. Immunogenet 23, 357–363.

Markovac, J., Erickson, R.P., 1985. A component of genetic variation among mice in activity of transmembrane methyltransferase I determined by the H-2 region. Biochem. Pharmacol. 34, 3421–3425.

Marticorena, P., Artzt, K., Bennett, D., 1978. Relationship of F9 antigen and genes of the T/t complex. Immunogenet 7, 337–347.

McDevitt, H.O., Chintz, A., 1969. Genetic control of the antibody response: relationship between immune response and histocompatibility (H-2). Science 163, 1207–1208.

McDevitt, H.O., 2000. Discovering the role of the major histocompatibility complex in the immune response. Ann. Rev. Immunol. 18, 1–17.

Medawar, P.B., 1944. The behaviour and fate of skin autografts and skin homografts in rabbits: A report to the War Wounds Committee of the Medical Research Council. J. Anat. 78, 176–199.

Medawar, P.B., 1945. A second study of the behaviour and fate of skin homografts in rabbits: A report to the war wounds committee of the Medical Research Council. J. Anat. 79, 157–176.

Medawar, P.B., 1946a. Relationship between the antigens of blood and skin. Nature 157, 161.

Medawar, P.B., 1946b. Immunity to homologous grafted skin. ii; the relationship between the antigens of blood and skin. Brit. J. Exp. Path. 27, 15–24.

Meruelo, D., Edidin, M., 1975. Association of mouse liver adenosine 3':5' cyclic monophosphate (cyclic AMP) levels with Histocompatability-2 genotype. Proc. Nat'l Acad. Sci. USA 72, 2644–2648.

Metchnikoff, E., 1893. "Lectures on the Comparative Pathology of Inflammation." Kegan, Paul, Trench, Trubner and Co, Ltd, London 1893.

Michaelson, J., 1981. Genetic polymorphism of beta2-microglobullin (Beta2m) maps to the H-3 region of chromosome 2. Immunogenetics 13, 167–171.

Mitchison, N.A., 1953. Passive Transfer of transplantation immunity. Nature 171, 267–268.

Morange, M., 2000. François Jacob's lab in the seventies: the T-complex and the mouse developmental genetic program. Hist. Philos. Life Sci 22, 397–411.

Mosinger, B., Kralova, J., Forejt, J., 1988. A cloned H-2 class I gene from a tw32 -derived recombinant t haplotype identified as functional H-2Kq gene. Immunogenetics 28, 283–285.

Muggleton-Harris, A.L., Johnson, M.H., 1976. The nature and distribution of serologically detectable alloantigens on the preimplantation mouse embryo. J. Embryol. Exp. Morph. 35, 59–72.

Muramatsu, T., Gachelin, G., Damonneville, M., Delarbre, C., Jacob, F., 1979. Cell surface carbohydrates of embryonal carcinoma cells: polysaccharidic side chains of F9 antigens and of receptors to two lectins. FBP PNA. Cell 18, 183–191.

Nizetic, D., Figueroa, F., Klein, J., 1984. Evolutionary relationships between the t and H-2 haplotypes in the house mouse. Immunogenetics 19, 311–320.

Owen, R.D., 1945. Immunogenetic consequences of vascular anastomoses between bovine twins. Sciences 102, 400–401.

Owen, R., 1989. M. R. Irwin and the Beginnings of Immunogenetics. Genetics. 123, 1–4.

Paigen, K., 2003a. One hundred years of mouse genetics: an intellectual history. I. The classical period (1902-1980). Genetics 163, 1–7.

Paigen, K., 2003b. One hundred years of mouse genetics: an intellectual history II. The molecular revolution (1981-2002). Genetics 163, 1227–1235.

Pairitz, G., Erickson, R.P., Schultz, J., Sing, C.F., 1985. Failure to detect association of isolated cleft palate with HLA antigens. J. Immunogenet. 12, 259–262.

Palm, J, Heyner, S., Brinster, R.L., 1971. Differential immunofluorescence of fertilized mouse eggs with H-2 and non-H-2 antibody. J. Exp. Med. 133, 1282–1293.

Pasteur, L., 1885. Methode pour prevenir la rage après morsure. Comptes rendus de l'Academie des Sciences 101, 765–774.

Payne, R., Rolfs, M.R., 1958. Fetomaternal leukocyte incompatibility. J. Clin. Invest. 37, 1756–1763.

Pla, M., Zakany, J., Fachet, J., 1976. H-2 influence on corticosteroid effects on thymus cells. Folia Biol. 21, 49–50.

Pratt, R.M., Martin, G.M., 1975. Epithelial cell death and cyclic AMP increase during palatal development. Proc. Natl Acad. Sci. USA. 72, 874–877.

Pratt, R.M., Salomon, D.S., Diewert, V.M., Erickson, R.P., Burns, R., Brown, K.S., 1980. Cortisone-induced cleft palate in the brachymorphic mouse. Teratog., Carcinog. Mutagen. 1, 15–23.

Ptashne, M., 2014. Francois Jacob 1920-2013. Res. In Microbiol. 165, 396–398.

Raff, M., 1969. Theta isoantigen as a marker of thymus-derived lymphocytes in mice. Nature 224, 378–379.

Reif, A.E., Allen, J.M.V., 1964. The AKR thymic antigen and its distribution in leukemias and nervous tissues. J. Exp. Med. 120, 413–433.

Rheinberger, H.J., 2000. Ephestia: the experimental design of Alfred Kühn's physiological developmental genetics. J. Hist. Biol. 33, 535–576.

Rogers, J.H., Lyon, M.F., Willison, K.R., 1985. The arrangement of H-2 class I genes in mouse t haplotypes. J. Immunogenet. 12, 151–165.

Roopenian, D.C., Christianson, G.J., Davis, A.P., Zuberi, A.R., Mobraaten, L.E., 1993. The genetic origin of minor histocompatabilitiy antigens. Immunogenetics 38, 131–140.

Russell E., S., 1985. A history of mouse genetics. Annu. Rev. Genet. 19, 1–28.

Salomon, D.S., Pratt, R.M., 1976. Glucocorticoid receptors in murine embryonic facial mesenchyme cells. Nature 264, 174–177.

Sato, M., Muramatsu, T., Berger, E.G., 1984. Immunological detection of cell surface galactosyltransferase in preimplantation mouse embryos. Dev. Biol. 102, 514–518.

Sawicki, J.A., Magnuson, T., Epstein, C.J., 1981. Evidence for the expression of the paternal genome in the two-cell mouse embryo. Nature 294, 450–451.

Schwartz, B.D., Nathenson, S.G., 1971. Regeneration of transplantation antigens on mouse cells. Transplant. Proc. 3, 180–182.

Scott, D.M., Ehrmann, I.E., Ellis, P.S., Bishop, C.E., Agulnik, A.I., Simpson, E., Mitchell, M.J., 1995. Identification of a mouse male-specific transplantation antigen, H-Y. Nature 376, 695–698.

Scully, N.F., Shur, B.D., 1988. Stage-specific increase in cell surface galactosyltransferase activity during spermatogenesis in mice bearing t alleles. Dev. Biol. 125, 195–199.

Searle, R.F.l., Johnson, M.H., Billington, W.D., Elson, J., Clutterbuch-Jackson, S., 1974. Investigation of H-2 and non-H-2 antigens on the mouse blastocyst. Transplantation 18, 136–141.

Shapiro, M., Erickson, R.P., 1981. Evidence that the serologic determinant of H-Y antigen is carbohydrate. Nature 290, 503–505.

Shapiro, M., Erickson, R.P, 1984. Genetic effects on quantitative variation in serologically detected H-Y antigen. J. Reprod. Immunol. 6, 197–210.

Shreffler, D.C., 1964. A serologically detected variant in mouse serum: further evidence for genetic control by the histocompatibility-2 locus. Genetics 49, 973–978.

Shreffler, D.C., 1988. Seventy-five years of immunology: the view from the MHC. J. Immunol. 141, 1791–1798.

Shreffler, D.C., Owen, R.D., 1963. A Serologically Detected Variant in Mouse Serum: Inheritance and Association with the Histocompatibility-2 Locus. Genetics 48, 9–25.

Silverstein, A.M., 1985. A history of theories of antibody formation. Cell. Immunol. 91, 263–283.

Simpson, E., Brunner, C., Hetherington, C., Chandler, P., Brenan, M., Dagg, M., Bailey, D., 1979. H-Y antigen: no evidence for alleles in wild strains of mice. Immunogenetics 8, 213–219.

Simpson, E., Scott, D., Chandler, P., 1997. The male-specific histocompatibility antigen H-Y: a history of transplantation, immune response genes, sex determination, and expression cloning. Annu. Rev. Immunol. 15, 39–61.

Snell, G.D., 1948. Methods for the study of histocompatibility genes. J. Genet. 49, 87–108.

Snell, G.D., 1951. A fifth allele at the histocompatibility-2 locus of the mouse as determined by tumor transplantation. J. Nat'l Cancer Inst. 11, 1299–1305.

Snell, G.D., Winn, H.J., Stimpfling, J.H., Parker, S.J., 1960. Depression by antibody of the immune response to homografts and its role in immunological enhancement. J. Exp. Med. 112, 293–314.

Spearow, J.L., Erickson, R.P., Edwards, T., Herbon, L., 1991. The effect of *H-2* region and genetic background on hormone-induced ovulation rate, puberty, and follicular number in mice. Genet. Res. 57, 41–49.

Steinmetz, M., Frelinger, J.G., Fisher, D., Hunkapiller, T., Pereira, D., Weissman, S.M., Uehara, H., Nathenson, S., Hood, L., 1981. Three cDNA Clones Encoding Mouse Transplantation Antigens: Homology to Immunoglobulin Genes. Cell 24, 125–134.

Stormont, C., Owen, R.D., Irwin, M.R., 1951. The B and C systems of bovine blood groups. Genetics 36, 134–161.

Tsuzuku, O., Nakamuro, K., Saji, F., Ogawa, M., Wakso, T., 1972. Quantitative estimation of histocompatibility antigen on the surface of mouse spermatozoa. Acta Obstet. Gynaecol. Jpn. 19, 257–265.

Tyzzer, E.E., 1909. A Series of spontaneous tumors in mice with observations on the influence of heredity on the frequency of their occurrence. J. Med. Res. 21, 479–518.

Vitetta, E.S., Artzt, K., Bennett, D., Boyse, E.A., Jacob, F., 1975. Structural similarities between a product of the T/t-locus isolated from sperm and teratoma cells, and H-2 antigens isolated from splenocytes. Proc. Nat'l Acad. Sci. USA 72, 3215–3219.

Vojtiskova, M., Polackova, M., 1966. An experimental model of the epigenetic mechanism of autotolerance using the H-Y antigen in mice. Folia Biol. (Praha) 12, 137–140.

Vojtiskova, M., 1969. H-2 antigens on mouse spermatozoa. Nature 222, 1293–1294.

Vojtiskova, M., Pokorna, Z., August,1971. H-2 antigens on diploid spermatogenic cells. In: Beatty, R.A., Gluecksohn-Waelsch, S. (Eds.), Proceedings, International Symposium on the Genetics of the Spermatozoon. Edinburgh, 261–266.

Vojtiskova, M., Pokorna, Z., 1972. Developmental expression of H-2 antigens in the spermatogenic cell series: possible bearing on haploid gene action (haemagglutination-inhibition-immunofluorescence-juvenile testes-irradiated testes). Folia Biol. (Praha) 18, 1–9.

Vojtiskova, M., Pokorna, Z., Viklicky, V., Boubelik, M., Hattikudur, N.S., 1974. The expression of H-2 and differentiation antigens on mouse spermatozoa. Folia Biol. (Praha) 20, 321–324.

Von Dungern, E., Hirschfeld, L., 1910. Über Vererbung gruppenspezifischer des blutes. Strukturen des blutes. Z. Immun. Forsch. 6, 284–292.

Wachtel, S.S., Koo, G.C., Zuckerman, E.E., Hammerling, U., Scheid, M.P., Boyse, E.A., 1974. Serological crossreactivity between H-Y (Male) antigens of mouse and man, Proc. Nat'l Acad. Sci. USA, 71, 1215–1218.

Wachtel, S.S., Koo, G.C., Boyse, E.A., 1975a. Evolutionary conservation of H-Y ('male') antigen. Nature 254, 270–272.

Wachtel, S.S., Ohno, S., Koo, G.C., Boyse, E.A., 1975b. Possible role for H-Y antigen in the primary determination of sex. Nature 257, 235–236.

Wachtel, S.S., 1983. H-Y Antigen and the Biology of Sex Determination. Grune and Stratton, New York.

Webb, C.G., Gall, W.E., Edelman, G.M., 1977. Synthesis and distribution of H-2 antigens in preimplantation embryos. J. Exp. Med. 146, 923–932.

Willison, K.R., Karol, R.A., Suzuki, A., Kundu, S.K., Marcus, D.M., 1982. Neutral glycolipid antigens as developmental markers of mouse teratocarcinoma and early embryos: an immunologic and chemical analysis. J. Immunol. 129, 603–609.

Yanagisawa, K., Bennett, D., Boyse, E.A., Dunn, L.C., Dimeo, A., 1974. Serological identification of sperm antigens specified by lethal *t*-alleles in the mouse. Immunogenet 1, 57–67.

Zinkernagel, R.M., Doherty, P.C., 1979. MHC-restricted cytotoxic T-cells: studies on the biological role of polymorphic major transplantation antigens determining T-cell restriction-specificity, function, and responsiveness. Adv. Immunol. 27, 51–177.

8

Gametogenesis and the genetics of gametes, including *t*-haplotype segregation distortion

The study of gametogenesis is one of developmental biology aspects which have been lightly touched on in Chapter 4, as well as of genetics. Speculation about gametes, of course, was of interest to the "ancients". Little was known of the mammalian egg but that of many species was visible and contemplated with it well known that the embryo could develop in it. The mammalian "seed" was more visible in the ejaculate but little was known about it until the advent of the microscope. The great authority, through the Middle Ages, Aristotle, did not even think the testes was essential because his dissections did not find it in fishes or snakes (where it is usually elongate, sometimes one anterior to the other, sometimes attached to the kidney). He thought "semen is hyper-refined blood….testes store semen… their function is to regulated its flow" (**Leroi, p. 185**). It could be said that the source of gametes from the blood persisted in Darwin's theory of gemmules carrying information from formed organs to the ovary or testes (see Chapter 2). The advent of better optical microscopy and histological techniques allowed the correct theory to develop, especially by Weissman (**The Germ-Plasm**) but errors, such as Nicolaas Hartsoeker's sighting of the homonculus in the sperm head, a frequently reproduced figure, persisted for a while.

The subject of gametogenesis has become of great interest in the 21st Century due to the many advances in cloning and re-programming of differentiated cells. The "de-differentiation" of a variety of cells to a pluripotent state, assumed the state of the gametes, and close to, but not identical with the fertilized oocyte, has led to a mostly bogus at the present (2019), industry of cell therapies for a variety of disorders. The advances in molecular biology (the development of the CRISPR techniques) which have led to the potential for gene editing of the embryo (tragically, already prematurely done in China while the data is just coming out that CRISPR-edited mice seem to have shorter life spans). Thus, the somewhat longer chapter deals with many aspects of this subject and is a history of biological developments as well as genetics.

Oogenesis

The primordial germ cells (PGCs) in females, as well as males, arise in the extraembryonic mesoderm at E 7.5 in mice (Chiquoine, 1954). PGC's were observed among the endoderm cells of the mid- and hindgut and the yolk stalk by about 8 days and they migrated as the embryo elongated to the dorsal mesentery at 10-11 days and are in the dorsal coelomic lining, and in the rudimentary genital ridge such that, by 13 days, the gonad was abundantly populated with germ cells (Spiegelman and Bennett, 1973). This migration and the start of their development in the gonads was dependent on a number of growth factors including the Stem Cell Growth factor (which binds to the c-kit receptor), deficient in *Steel* mutant mice which were sterile due to the lack of this growth factor (McCoshen and McCallion, 1975). The PGCs shared many gap junctions between each other and adjacent stromal cells, which allowed small molecules to pass between the cells, and this suggested that these other follicular cells influenced oocyte development.

It was only at the end of the Century, and the beginning of the next, that the mechanisms leading to male v.s. female specificity of these germ cells were found. Prostaglandin D_2 was found to masculinize female germ cells (Adam and McLaren, 2002) while retinoid signaling was essential for female germ cells (Bowles et al., 2006).

The oocytes then remained in meiotic prophase arrest until about 2 weeks of age with female fertility taking 4 more weeks. XX cells entered meiosis then, as also would ovarian XY cells (as might occur in chimeras, see Chapter 4). At 15 days the ovary has thousands of immature oocytes but, if cultured, these will continue into meiosis while younger oocytes won't (Sorenson and Wasserman, 1976). The release into culture isolated the cells from the follicular, inhibitory cells which, however, play an important role in oocyte development in vivo (Ohno and Smith, 1964) The progression of meiosis was also dependent on follicle stimulating hormone (FSH) and luteinizing hormone (LH; Baker and Neal, 1973). It was eventually found that the granulosa cells produced cyclic GMP which diffuses into the oocyte through the gap junctions. In the oocyte, cyclic GMP prevented the breakdown of cyclic AMP by the phosphodiesterase PDE3, and thus maintained meiotic arrest (Norris et al., 2009). The cyclic AMP was generated in the oocyte by adenylyl cyclase in the oocyte membrane (Eppig, 1991). During normal maturation in vivo, LH would cause dissolution of the gap junctions and allow maturation to proceed.

The oocytes could be prepared quite cleanly and studied biochemically (Mangia and Epstein, 1975; Mangia et al., 1976). It took many years, during which cell cycle proteins were discovered (first in yeast), to find that the pre-15 day lack of entry into meiosis with release from the follicle was due to lack of cell cycle proteins, such as p34^{cdc2} and cyclin B (Chesnel and Eppig, 1995) needed to allow meiotic progression.

The cultured oocytes were capable of being fertilized, resulting in normal offspring (Schroeder and Eppig, 1984). Studies were performed on the synthesis of abundant proteins such as the ones forming the zona pellucida–the extracellular matrix that surrounds growing oocytes, ovulated eggs and early embryos. This was composed of three sulfated glycoproteins: ZP1, ZP2, and ZP3, which were found to be critically involved in fertilization, the postfertilization block to polyspermy and protection of the preimplantation embryo and were coordinately synthesized (Epifano et al., 1995). There were detailed studies of the synthesis and control of multiple constituents of the developing egg which, of course, continued into the 21st Century. Thus, a review on the molecular control of oogenesis written in (Sanchez and Smitz, 2012) had over 250 references, almost all of which were in the new century.

Transmission ratio distortion (trd), a major sub-topic of this chapter, can also occur in mouse females and in this case it really is "meiotic drive", for example, an imbalance of the gametes produced at meiosis (meiotic drive is generally used for all cases of trd but many prefer the latter term since it more readily fits the postmeiotic events usually involved). Amplified and rearranged DNA segments on mouse chromosome 1 creating homogenously staining regions (HSRs) were isolated from a wild population of mice from Yakutsk Siberia (Agulnik et al., 1990).

When mated to CBA mice, 86.4% of the offspring carried the double HSRs, a result of preferential segregation into the oocyte instead of the polar bodies at meiosis. More broadly, it is now known that differences in centromere "strength" affect spindle interactions during the extended meiotic arrest, thus driving preferential segregation of specific centromeres to the oocyte or the polar body (Henikoff and Malik, 2002). The asymmetry of the female spindle that permits this process is mediated by differential tyrosination of the spindle microtubules, with tyrosinated microtubules located nearer the cortex (directing polar body segregation) and de-tyrosinated microtubules towards the centre of the oocyte (Dumont and Desai, 2012). This process of non-Mendelian segregation of "strong" vs "weak" centromeres in female meiosis is also believed to be responsible for non-Mendelian inheritance of chromosomal rearrangements such as Robertsonian fusions, in which the fused and unfused versions of the karyotype are distinguished by the molecular events associated with the centric fusion (Chmatal et al., 2014).

Also, in the 21st Century it became clear that a final event in the sequence of signaling for the resumption of meiosis was a rise in levels of retinoic acid which was inhibited by Fibroblast Growth Factor 9 (Bowles et al., 2010). In the testes, progressive pulses of retinoic acid occurred along the seminiferous tubules with the highest levels at stage VII/VIII and these initiated meiosis in the spermatogonia in the middle layer of the association (Hogarth et al., 2015).

Although it was dogma that no oocytes developed after the embryonic period, in the 21st Century, it was shown that they could arise postnatally in mice (Johnson et al., 2004). These germ cell derived oocyte precursors could not only replace lost oocytes following busulfan (a chemotherapeutic agent killing mitotic cells) treatment but, when marked with green fluorescent protein, invade transplanted wild-type ovaries to form follicles.

Spermatogenesis

Although of intense interest to large animal husbandry, this was a very popular subject to study in mice. It was especially relevant to reproductive physiology and mutational studies. The complex development of the sperm was highly amenable to study since the adult testes had abundant material and the relevant cell types could be separated. One approach was simply to compare the whole testis of mice as they aged and new stages of spermatogenesis appeared—new mRNAs or proteins were almost certainly due to the new class of cells which had appeared. The field rapidly grew and in 1993, it was stated that "a book will soon be required if one is to cover the subject fully" (Erickson, 1993). In fact, one book of collected papers had appeared : **Edinburgh symposium on the genetics of the spermatozoon, Beatty and Glueksohn-Waelsch (1972).**

a) *The stages of spermatogenesis*

Although there were many classical, histological studies of spermatogenesis in multiple species (Leblond and Clermont, 1952), the precise time spent at various stages needed to be known for mutation studies. As previously discussed, the interval from a mutagenic exposure to when the spermatozoa resulting from that treatment would be inseminated was essential (see Chapter 6). Valerio Monesi[1], at that time on leave from the Embryology group of the Biology Division of the National Committee for Nuclear Energy located in Frascati, Italy, used periodic acid/Schiff staining on testis sections to classify twelve different stages of development of the sperm head in the spermatid (Monesi, 1962). These indicated unique, vertical cell associations, with earlier cells lower and later cells higher, of a few, but separated stages in the whole seminiferous cycle. (One can think of this as rosettes [cross-sections of the seminiferous tubule] of columns of developing cells with unique vertical associations of several of the stages. Thus, an earlier stage would be internal with a stage several stages further along external for several layers of the column). This columns of associations could be followed as a "wave" along the seminiferous tubule. The wave indicated a cycle of progression from one set of cell associations through the 12 stages until the re-appearance of this association. This work, along with that of Oakberg and Huckins (Oakberg, 1971; Huckins and Oakberg, 1978), allowed the timing of the cycling from the A_s slow-dividing, spermatogonial stem cell whose division forms another A_s and the A_{1s}, which is committed to multiple divisions, the later of which leave cytoplasmic bridges between the resulting cells. These stages are indicated in Fig. 8.1. These early studies

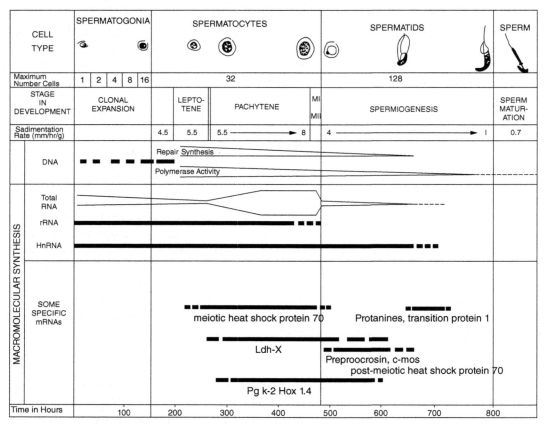

FIG. 8.1 Schematic representation of spermatogenesis. Qualitative data from macromolecular synthesis are indicated by solid bars; uncertainties about onset or duration are indicated by breaks. Quantitative variations for a single substance are indicated by the relative heights of the margined areas; no comparison from one substance to another is intended. From Erickson (1990), with permission.

suggested that the spermatogonial stem cell population divides 23 times per year following puberty in human males: a relatively high level of proliferation that was believed to underlie the observed ~3:1 ratio of *de novo* mutations arising in male vs female gametogenesis. However, more recent data show that stem cells alternate between dividing and non-dividing states, such that the net rate of division is less than one per year. Other studies show a very good correlations between the proliferation and mutational rates (reviewed in Goldmann et al., 2019). It now appears that DNA damage and its repair is a more significant source of such mutations than previously appreciated, and that the mutation frequency and site specificity spectrum does not depend purely on replicative errors (Agarwal and Przeworski, 2019; Gao et al., 2019).

b) *DNA synthesis during spermatogenesis*

The timing of DNA synthesis, and its modifications by DNA methylase were important for mutational studies and for a general understanding of the genetics of gametogenesis. In mice, as in other diploid eukaryotes, major replicative DNA synthesis occurred in pre-leptotene spermatocytes (Kofman-Alfaro and Chandley, 1970). This was associated with high levels of DNA polymerase activities, which declined to low levels during spermiogenesis but did not decline further during sperm maturation (Daentl et al., 1977). DNA methylation, however, was not completely replicative at this time (see Chapters 9 and 10 for the importance of cytosine methylation for gene control). While many single copy gene were methylated throughout spermatogenesis (Rahe et al., 1983), repetitive genes such a satellite sequences remained undermethylated throughout spermatogenesis (Ponzetto-Zimmerman and Wolgemuth, 1984). The pattern of methylation for genes expressed during spermatogenesis changed during the process: Transition Protein 1 became progressively less methylated, protamine 1 and 2 genes became progressively more methylated (Trasler et al., 1990) and *Pgk-2* was demethylated and then remethylated during spermatogenesis (Ariel et al., 1991).

A low level of DNA synthesis occurred during pachytene (Meistrich et al., 1975) which also could be a target of mutagenesis. This was thought to be related to alterations in gene expression or to repair and/or meiotic pairing (Smyth and Stern, 1973). Several DNA-repair synthesis-related enzymes reached their peak of specific activity at pachytene during mouse spermatogenesis (Orlando et al., 1984) while DNA helicase was also at its peak at this time (David et al., 1982). Hotta et al. (1985) found meiosis-specific transcripts of the fraction of DNA that has delayed replication and which could have a role in chromosome-pairing. The properties of these transcripts was surprising as they were apparently diverse, unique sequences, yet they maintained significant homology between plants and animals suggesting ancient homologies. These patterns of DNA synthesis are indicated in Fig. 8.1.

c) *RNA synthesis during spermatogenesis*

Studies on RNA synthesis started with histochemical descriptions and then progressed to detailed investigations of specific base sequences that are expressed. The application of autoradiography, following the administration of isotopic precursors, to studies of testicular RNA synthesis at first concluded that little RNA was transcribed after meiosis, since early mouse spermatids exposed to short pulses of H^3-uridine showed only a very small peak of incorporation when examined by autoradiography (Monesi, 1967, 1971[1]; Soderstrom and Parvinen, 1976).

RNA polymerase activity was also present in spermatids (Moore, 1971). Quantitative autoradiographic studies demonstrated that the rate of RNA synthesis per cell decreased fourfold during meiosis, so the RNA synthesis/DNA ratio was unchanged (Loir, 1972; Geremia et al., 1978). Studies on separated testicular cells confirmed that there are high rates of RNA synthesis in postmeiotic cell stages (Meistrich et al., 1981). Direct visualization of transcription by electron microscopy indicated that both ribosomal and heterogeneous nuclear RNA were synthesized in spermatocytes, but nucleolus-like ribosomal RNA transcription patterns were not found after meiosis (Kierzenbaum and Tres, 1975) although this was controversial (Schmid et al., 1977). Thus, the new RNA synthesis after meiosis was not merely ribosomal RNA. Sucrose gradient and electrophoretic characterization of newly synthesized RNA from spermatids demonstrated heterogeneous, presumptive mRNA (Geremia et al., 1978), summarized in Fig. 8.1.

d) *Postmeiotic gene expression during spermatogenesis*

It was generally thought that gene expression did not occur in animal gametes. Postmeiotic gene expression appeared to run counter to the evolutionary dogma that genetic selection should only be on the zygote: gene expression in gametes could result in phenotypic differences in spermatozoa affecting function and could potentially be subject to positive selection even though detrimental to the zygote. Such postmeiotic expression was found nearly a century ago in plants (Brink and Burnham, 1927), where one male pronucleus is activated during pollen formation. While the occurrence of distorted transmission ratios related to selection for different alleles was described then in plants, a direct demonstration of new RNA synthesis in the pollen tube, the description of pollen tube-specific isozymes, and the recording of the frequency of postmeiotic gene expression are more recent (reviewed in Frova et al., 1987).

In contrast to the long process of sperm development and function after meiosis, the time interval between the meiotic divisions of oogenesis and fertilization is so short (sometimes the meiotic divisions are triggered by fertilization) that there is little or no time for postmeiotic, gamete-limited gene expression during egg development. The overall features of spermatogenesis were discussed above.

At the same time that the first evidence for postmeiotic gene expression in plants was being obtained, quite an opposite conclusion about gene expression in animal gametes was reached by Muller (Muller et al., 1927). He found that sperm nullisomic for about 1/40 of the *Drosophila* genome, due to an unbalanced translocation, could fertilize eggs normally if the missing material was contributed to the zygote by eggs disomic for the missing material. The results were extended to most of the *Drosophila* genome in multiple experiments and similar conclusions were derived from similar use of translocations in mice (Lyon et al., 1972), but in both cases the authors were unaware of the syncytial nature of spermatogenesis. In mammals, large (~ 1μm) intercellular bridges can connect over a hundred spermatids, while in *Drosophila* 64 spermatids develop without interposed cell membranes (reviewed in Erickson, 1973). These bridges had been originally described from electron microscopic studies and interpreted as being important for synchronous development of spermatids (Fawcett et al., 1959; Gondos, et al., 1973). It was only much later that their unique biochemical features were described (Tes et al., 1996). Such cytoplasmic continuity could allow the products of any genes that might be expressed postmeiotically to be shared among haploid nuclei and, thus, not be needed from the egg. The only products that might not be shared could occur in the case of the mRNA for a membrane protein almost immediately attached to the ribosomes of the endoplasmic reticulum (ER) for intra-ER synthesis, with the protein then progressing to the Golgi apparatus for processing and from there directly into the spermatid plasma membrane (Erickson et al., 1981). The specializations of the cytoskeleton maintaining the ring-like openings between the spermatids might prevent diffusion in the membrane to adjacent spermatids. An example of spermatid limited expression is provided by the deficient transmission of Robertsonian chromosomal translocations Rb (6:15) and Rb (6:16) determined by the the Spam1 antigen. This sperm adhesion molecule, then called PH-20, was first found in guinea pigs (Cowan et al., 1986) and characterized as a testis-expressed molecule in humans (SPAM1; Jones et al., 1995). It was studied by Patricia Martin-DeLeon's lab (Chayko and Martin-DeLeon, 1992; Zheng and Martin-DeLeon, 1997a,b) in regards to trd of these translocations. Equal numbers of the two kinds of segregant spermatozoa occurred but the normal sperm fertilized 2-3 times as many eggs as did the translocation spermatozoa. *Spam1* was only expressed at about 30% of wild type levels from the translocation chromosome, suggesting that less binding led to lower rates of fertilization (Zheng and Marin-DeLeon, 1997). As was confirmed early in the 21st Century, the haploid-expressed *Spam1* encoded a membrane hyaluronidase whose mRNA and protein were compartmentalized with different allelic forms in the separated spermatozoa (Zheng et al., 2001).

Although there were suggestions of postmeiotic expression of sperm autoantigens (Radu and Voisin, 1975), an early demonstration of postmeiotic gene expression in mice involved a protamine-like histone (Erickson et al., 1980). Several studies of stage-specific expression of newly-synthesized, in sedimentation-separated and by radio-labelling, proteins showed unique ones in spermatids confirming this finding (Boitani et al., 1980; Kramer and Erickson, 1982; O'Brien and Bellve, 1980). In contrast to the infrequent detection of postmeiotic transcription in *Drosophila* (Gould-Somero and Holland, 1974), a plethora of such transcription has now been found in mammals - so much so that one wonders why it should have been thought not to occur. As more tools of molecular biology were developed and applied to the characterization of postmeiotically synthesized RNA, the results were surprising: Analysis by two-dimensional gel electrophoresis of the products of in vitro translation of RNA purified from separated spermatocytes and spermatids showed twice as many spermatid-specific as spermatocyte-specific gene products, with only a relatively small number of proteins synthesized in both cell types (Fujimoto and Erickson, 1982, see Fig. 8.2). Assays for the mRNA for specific proteins (by in vitro translation from purified RNA) demonstrated that mRNA for protamine and phosphoglycerate kinase-2 (PGK-2) increased after meiosis (Erickson et al., 1980).

However, since the mRNAs might have been transcribed earlier, and only processed postmeiotically, these results did not yet conclusively prove that there was postmeiotic transcription. The finding that heterogenous, poly (A)-containing RNA was synthesized postmeiotically strongly supported the conclusion of postmeiotic synthesis (Erickson et al., 1980). Thereafter, several groups made cDNA libraries and found clones for specific mRNAs

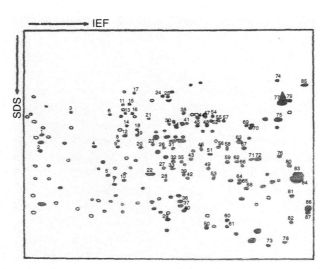

FIG. 8.2 Two-dimensional gel patterns of the translation products of RNA purified from spermatocytes compared to those from RNA purified from spermatids. Spermatid-specific proteins are indicated by filled, numbered traces, spermatocyte-specific proteins by unfilled, unnumbered traces and proteins common to both cell types by filled, unnumbered traces. IEF, isoelectric focusing; SDS, direction of electrophoresis in denaturing gel. From Erickson (1990), with permission.

that increased, or first appeared, after meiosis (Kleene et al., 1983; Dudley et al., 1984; Fujimoto et al., 1984). A survey of testicular cDNAs showed that about half increased in abundance after meiosis (Thomas et al., 1989). About half of these (a quarter of the total) first appeared after meiosis (ibid.). Thus, ample confirmation of postmeiotic gene expression had been obtained. Sharing of these products between the connected spermatids was specifically shown in the case of a postmeiotically transcribed transgene (Braun et al., 1989). When the flow cytometer was applied to separating X- and Y-bearing sperm[2], they were shown to have identical complements of proteins, again showing sharing of gene products (Hendriksen et al., 1996).

e) *Classes of postmeiotically transcribed genes*

Of the many genes now known to be transcribed postmeiotically, this pattern of transcription makes sense for some but not others. It is easily understood for sperm-specific proteins whose transcription occurs near the time of translation-the usual scenario in development. For example, the protamines, which are a class of histone-related, DNA-binding proteins involved in the special compacting of DNA in the dense male sperm nucleus, are synthesized postmeiotically (reviewed in Hecht, 1990). They provided another example of the sharing of transcripts between spermatids (Caldwell and Handel, 1991). In addition, transgenic mice with 4.8 kb of the 5' portion of the protamine I gene (Peschon et al., 1987) or 859 bp of protamine 2 gene (Stewart et al., 1988) were shown to elicit spermatid expression of these 2 genes. Another example was the mRNA for sperm-specific alpha-tubulin, perhaps needed for cytoskeletal reorganization during the dramatic structural changes that occur during spermiogenesis, or for the sperm flagellum itself, does not even appear until the late spermatid stage (Distel el al., 1984). This may also be the reason for postmeiotic transcription of gamma-actin, which was identified as the product of the gene that also codes for the gamma-actin of smooth muscle (Slaughter et al., 1989).

Protamine, transition protein 1, and testicular histone 2B, as chromatin proteins were clearly involved in the massive reorganization of DNA in the sperm nucleus. One would predict that they would not be needed until after meiosis and it was repeatedly demonstrated that they are transcribed after meiosis (Kleene et al., 1983; Heidaran et al., 1988; Moss et al., 1989). Postmeiotic transcription of histone 2B was confirmed by nuclear run-off experiments (Moss et al., 1989). Pre-pro-acrosin is a precursor of a sperm enzyme involved in fertilization and the transcript does not appear until after meiosis (Adhman et al., 1989). One testicular haploid expressed gene was involved in spermatid-sertoli cell interaction (Nayernia et al., 1999) while another coded a unique heat shock protein (Tunekawa et al., 1999).

For other postmeiotically expressed genes, the explanation was less obvious. Lactate dehydrogenase-X, the sperm-specific isozyme of the abundant lactate dehydrogenases which maps to an autosome, not the X chromosome, first appeared in the testes of 19-day old mice—the time when spermatocytes are first observed (Goldberg and Hawtrey, 1967). It was proposed that the abundant expression of LDH-X reflects a need for its altered substrate specificity in spermatozoa. However, its expression might be related to its ability to bind single-stranded DNA, which could play a role in chromatin reorganization. Radio-pulse, immunoprecipitation experiments performed on fractionated testis cells indicated that LDH-X synthesis continued postmeiotically with about half of the total synthesized in spermatids (Meistrich et al., 1977). Studies with the cloned LDH-X gene confirmed and extended these findings. Sequencing of an initially unidentified testicular cDNA that had been shown to have postmeiotic transcription by the very sensitive method of nuclear run-off (Fujimoto et al., 1984) found it to code for LDH-X (Tanaka and Fujimoto, 1986). Studies with this and other LDH-X clones demonstrated a high abundance of LDH-X message in round spermatids that decreased in later spermatids (Fujimoto et al., 1988; Jen et al., 1990).

The expression of Pgk-2 was thought to compensate for the extinction of expression of X-linked Pgk-1 due to postmeiotic X-inactivation. The developmental appearance of Pgk-2 initially indicated that the protein was not synthesized until after meiosis (Van de Berg et al., 1976; Kramer and Erickson, 1981). Immunofluorescent studies confirmed the localization of Pgk-2 antigen to postmeiotic cells (Kramer, 1981; Bluthmann et al., 1982; Rudolph et al., 1982). When the X-linked PGK-1 gene was cloned it was found to cross-hybridize to PGK-2 and thus was used to study transcription in the testis—demonstrating transcription at pachytene with further transcription occurring after meiosis (Erickson et al., 1985; Robinson and Simon, 1991). Little or none of the pre-meiotically transcribed message is found on polysomes while the specific mRNA is abundantly found on polysomes after meiosis (Gold et al., 1983). Genomic cloning of PGK-2 demonstrated that it was a reverse transcriptase (i.e., retroposon) mediated, processed gene insert (McCarry and Thomas, 1987). This must have been a sporadic event that was then maintained by selection, because, as mentioned,

PGK-2 can compensate for the extinction of expression of X-linked PGK-1 with X-inactivation during spermatogenesis. Only 323 bp 5′ to the coding region was sufficient to control proper expression in transgenic mice (Robinson et al., 1989).

The reasons for the abundant postmeiotic expression of genes such as the oncogenes c-*myc*, *c-abl* (Ponzetti and Wolgemuth, 1985), or truncated forms of *c-kit* (Albanesi et al., 1996), homeobox genes (Propst et al., 1988), and even a putative sex-determining gene, zinc finger Y (*Zfy*; Kalikin et al., 1989), are less clear. A catalog of many protein expressed at different stages of spermatogenesis, including postmeiotic, was presented in 1991 (Wolgemuth and Watrin, 1991). Testicular cell protein 1 (TCP 1) was known to be expressed at low levels in other tissues. It became apparent that its greatly increased synthesis during spermiogenesis, which is included in the phase of postmeiotic transcription, was probably related to its role in Golgi complex function (Willison et al., 1989), since the Golgi complex becomes greatly enlarged (and apparently more active) in preparation for acrosome formation. An understanding of its role was expanded when it was found to have a sequence related to that of a molecular chaperone from a thermophilic archaebacterium (Trent et al., 1991) and that its chaperone function was essential for tubulin synthesis (Yaffe et al., 1992). (It was found not to have a role in *t*-allele transmission ratio distortion [Sanchez et al., 1985], see below). The expression of another chaperone, a unique postmeiotic heat shock protein 24 was thought to be related to the unique temperature sensitivity of mammalian spermatogenesis (the reason why mammalian testes are held in the external scrotum). One feature in common among Zfy, which was transcribed postmeiotically (Kalikin et al., 1989), and some of the oncogenes and developmental genes, was the potential to encode DNA-binding proteins. Zfy, contains the zinc finger motif found in transcription factor IIIA and other DNA-binding proteins; it was conceivable that Zfy, the *ret* finger protein rfp (Takahashi et al., 1988) and Hox 1.4 (Wolgemuth et al., 1987) might be genes that played a role in nuclear DNA reorganization. However, the postmeiotic transcription of other oncogenes was surprising, given that the sperm is a terminally differentiated cell type, and bespoke the lack of knowledge about the function of cellular oncogenes in normal development.

f) *Regulation of postmeiotic gene expression*

Considerable evidence for altered transcriptional and posttranscriptional regulation in germ cells was soon found. The pre-proenkephalin mRNA appeared in spermatocytes but was found at higher levels in spermatids (Yoshikawa and Aizawa, 1988). The major transcript in mouse testes was larger than that found in the pituitary (Kilpatrick and Millette, 1986). The larger testicular transcript was the result of alternative RNA splicing in which an alternative acceptor site in intron A was used so that the germ cell-specific pre-proenkephalin transcript had a 491 bp untranslated leader (Garett et al., 1989). The germ cell promoter for pre-proenkephalin was found to have a GC-rich stretch lacking a TATA sequence but containing a consensus SP1-binding site downstream (Kilpatrick et al., 1990) and some of this sequence shared homology with the germ cell promoter of rat cytochrome C_T, the spermatogenic cytochrome c. Angiotensin-converting enzyme provided another example of a testicular protein with an altered size. Cloning studies demonstrated that the carboxyterminal sequence of the somatic and testicular isozymes were identical (Kumar et al., 1989). The difference was found to be due to the testicular isozyme's transcript being initiated in the 12th intron of the somatic angiotensin-converting enzyme gene (Howard et al., 1990).

Gold and Hecht (1981) studied compartmentalization of Poly (A)$^+$ RNA in total mouse testes and found that an unusually high proportion was non-polysomal, i.e., not being actively transcribed. This observation suggested that polyadenylation of mRNA could occur by two different mechanisms during spermatogenesis The first was the "classic" shortening of Poly (A)$^+$ tracts that was generally seen with mRNAs during translation. These changes are mediated by the poly(A)-binding protein, which determined mRNA stability in vitro. It was required for 60S ribosomal subunit-dependent translation initiation (Bernstein et al., 1989; Sachs and Davis, 1989). Examples of this process were protamine 1 and transition proteins 1 and 2 which became de-polyadenylated and located on ribosomes only in elongating spermatids (Kleene et al., 1984). The second mechanism involved proteins such as LDH-X which showed an increase in polyadenylation about the time of meiosis without major changes in translation rates at that time and cytochrome C_T which had a highly polyadenylated form while on polysomes and showed decreased adenylation later. Cytochrome C_T had a cytoplasmic polyadenylation element (Hake et al., 1990) and it was probable that LDH-X also had one. Many postmeiotic messages were stored in complexes with RNA/DNA-binding proteins (Kwon et al., 1993).

These hypotheses were explored in transgenic mice (see Chapter 9). A construct of the fusion of 156 nucleotides of 3'-untranslated sequence from the mouse protamine gene to the human growth hormone recorder gene was used in transgenic mice. It delayed the recorder gene's translation from the early in spermatogenesis to the elongating spermatid stage, the time when the protamine 1 gene is normally translated (Braun et al., 1989b). In addition, whereas the control gene product was located in the acrosome, the product of the fusion gene was intracellular. Such studies showed important species differences in regulation of spermatogenic transcription: 86 bp proximal to the angiotensin converting enzyme from rabbits did not lead to correct expression (298 bp did lead to correct expression; Erickson et al., 1996) while 91 bp of the mouse promoter was sufficient (Howard et al., 1993). Sequence comparisons of the 2 promoters showed quite an amount of divergence and emphasized the importance of a cAMP response element for the correct pattern of expression (Kessler et al., 1998). By the beginning of the 21st Century, so much was known about spermatogenic promoters that an online data base was set up for easy access to all the information about them (Lu et al., 2006).

Spermatozoal RNA

It was established in many species that the spermatozoal mitochondria did not contribute to the zygote and that mitochondrial inheritance was maternal (one species of mussels was the exception to prove the rule, as was 1 human with lifelong intolerance to exercise; Schwartz and Vissing, 2002). Although some paternal DNA could survive, and included 4-10 copies of mitochondrial DNA (compared to thousands of copies of maternal DNA; Gyllensten et al., 1991), that was a minute amount. As techniques improved towards the end of the century, the designation of sperm mitochondria for destruction by ubiquination (co-valent coupling of ubiquitin as a degradation signal) was established (Sudovskky et al., 1999).

The question remained as to whether the spermatozoa carried any RNA into the zygote which could influence its development. Early work using radioactive precursors first detected the incorporation of amino acids into proteins in purified bull (Bhargava, 1957), rabbit (Busby et al., 1974) and mouse spermatozoa (Bragg and Handel, 1979). As mRNA was discovered, the result was extended to the detection of the incorporation of radioactive phosphate and 2 of the 4 nucleic acid bases into RNA in such isolated spermatozoa and bacterial contamination was excluded (Abraham and Bhargava, 1961,1963) both in the head and tails of the spermatozoa (Fuster et al., 1977; Hecht and Williams, 1978). RNA sufficient to be studied by electrophoresis in gels was detected in highly purified spermatozoa (Betlach and Erickson, 1973). Such RNA could be localized to the sperm nucleus (Pessot et al., 1989) and included small nuclear RNAs (Concha et al., 1993). A number of the studies on RNA in spermatozoa were performed with human sperm. RNA finger-printing was used (Miller et al., 1994) and a complex population of RNAs found (Miller et al., 1999). In addition, haploid transcripts were found to persist in human spermatozoa (Wykes et al., 1997) and a transcription factor was found in mouse spermatozoa (Herrada and Wolgemuth, 1997).

In the 21st Century many advances were made in the study of sperm RNA, involving many more workers (40 co-authors on one paper in 2016; Sharma et al., 2016). Late synthesis of nuclear-encoded protein by mitochondrial type ribosomes were found while sperm were in the female reproductive tract (Gur and Breitbart, 2006). Strong evidence was found for non-Mendelian, paternal inheritance of an epigenetic change in the offspring's coat color due to postmeiotic accumulation of a genetically altered *Kit* gene (encoding a tyrosine kinase receptor for the mast cell/stem cell growth factor involved in pigment cell migration; Rassoulzadegen et al., 2006). It was found that vesicles from epididymal cells can fuse with the developing spermatozoa, carrying tRNA fragments. The spermatozoal constitution can be influenced by parental diet and, therefore, influence the resulting offspring (Sharma et al., 2016), another example of inherited epigenetic change (see Chapter 10). It was suggested that sperm RNA was essential for embryonic development (Conine et al., 2018) with sperm-lacking epididymally-acquired RNA failing to implant but these results were in contradiction to other results with maternal uni-parental liveborns (see Chapter 10)and not confirmed by others (Zhou et al., 2019).

The genetics of spermatozoa

The initial studies in mid-century used morphological characters which could be quantitated. There were clear differences between inbred strains in sperm head dimensions (Beatty, 1971) which were robust enough to distinguish the strain source of 2 kinds of spermatozoa in chimeras (Burgoyne, 1975) The differences were highly heritable ($h^2 \sim 0.9$) and similar to the heritability of the length of midpiece ($h^2 = 0.97+/-0.36$; Beatty and Wooley, 1967) while a minimum of 2 genes were needed to explain the variation in the F_2 (Illisson, 1969). The varying lengths of the midpiece could be selectively bred and increases in midpiece length were found to be accounted for by increases in the number of mitochondria in the midpiece (Woolley, 1970). Inbred strains of mice also showed great variation in in vivo fertilization frequencies (Krzanowska, 1960) which turned out to be due to genetic variation in both sperm and eggs. Sperm genotype affected rates of penetration (Fraser and Drury, 1976) while egg genotype did not (Maudlin and Fraser, 1978).

In addition to the metric differences between spermatozoa, there were differences in appearance (Braden, 1958a) and numbers of sperm (Krzanowska, 1962) which were strain specific. The latter indicated genetic differences in the kinetics of spermatogenesis which were documented (Bruce et al., 1973). There was a quite large variation in the number of abnormal sperm between different inbred stains of mice (Krzanowska, 1966). The C57BL/6J inbred strain had a surprisingly high percentage of abnormal spermatozoa (21%) which was largely accounted for by the genetic content of the Y chromosome (Brozek, 1970), i.e. the source of the Y was a major variable, as was the percentage of abnormal spermatozoa found in crosses between other inbred strains (Krzanowska, 1969). Particular mutations, such as the coat color mutation pink-eyed dilution, were found to influence sperm morphology (Wolfe et al., 1977). Although these were deletion alleles and other, nearby genes influencing sperm morphology might have been deleted, complementation studies between the various alleles did not suggest that this was the case (ibid.).

As biochemical genetics became more advanced, there started to be biochemical genetic studies of spermatozoa in the 1970s. Strain differences were found in the rate of removal of the cumulus oophorus and the zona pellucida (Krzanowska, 1972) and in the rate of sperm capacitation (Hoppe, 1980) which might be explained by differences in the enzymatic content of the spermatozoa, among other things. Differences in quantity of enzyme among inbred strains were found in unwashed spermatozoa for DOPA-oxidase, leucine aminopeptidase and lactic dehydrogenase (see above in regards to the sperm specific form, LDH-X) but the differences could have been due to the remaining seminal plasma (Mathur and Beatty, 1974). Many lysosomal enzymes (acid hydrolases) were found in spermatozoa and it became realized that the acrosome, the portion of the head of the sperm which dissolves with fertilization, was a lysosomal derivative (Alllison and Hartree, 1970). Hyaluronic acid is a major component of extracellular matrices such as those between cumulus cells and variations in sperm hyaluronidase levels between inbred strains suggested a possible relationship between these enzyme activities and the number of supplemental sperm found surrounding the egg (ones that made it through the cumulus, only one having penetrated the egg, Erickson and Krzanowska, 1974). Studies of electrophoretic variants of glutamate oxaloacetic transaminase disclosed hybrid bands in spermatozoa of the F_1 between inbred strains with such variants, arguing against postmeiotic expression of this enzyme (assuming intra-molecular chain association only at synthesis; Erickson, 1974). The lysosomal hydrolase, beta-glucuronidase was also present in spermatozoa (Erickson, 1974) and, again, there was no evidence for its postmeiotic expression (Erickson, 1977).

There were also major variations among inbred strains for levels of enzymes and metabolites related to energy metabolism. These included cyclicAMP (Erickson et al., 1979) and other adenine nucleotides and related enzymes (Erickson et al., 1987). It was in the latter part of the 80s that molecular techniques were developed and began to be applied to spermatozoal genetics. These allowed better attempts to study the role of postmeiotic gene expression (see above).

X-inactivation during spermatogenesis and sex chromosome transmission ratio distortion

While X-inactivation in somatic cells is considered in detail in Chapter 10, the mechanism and implication are different for male germ cells. An early review was that of Lifschytz (1972). This phenomenon explained three unusual aspects of male meiosis: 1) male sterility of X-autosome translocations (Russell and Montgomery, 1969) which were sometimes associated with XY chromosome dissociation which also is a cause of sterility (Beechey, 1973). X-inactivation was probably the cause of abnormal RNA synthesis in the sex vesicle (the aggregated DNA seen in postmeiotic male germ cells thought to be the condensed XY chromosome pair) which was seen in some autosomal-autosomal translocations (Speed et al., 1986); 2) the difference between XX, *Sxr* (sex-reversed; see Chapter 11) mice, which are aspermatogenic, and XO, *Sxr* mice which undergo spermatogenesis but immotile sperm result (Cattanach et al., 1971). The lack of inactivation of a second X was also the likely explanation of infertility in human Klinefelter syndrome, XXY; and 3) the recruitment of autosomal copies of X-linked genes by retroposon activities which were only expressed during spermatogenesis. These included *Pgk-2* (see above), *Zfa* (see Chapter 11) and a pyruvate dehydrogenase subunit (Dahl, et al., 1990). An early apparent exception seemed to be the X-linked glucose-6-phosphate dehydrogenase (Erickson, 1976b) and the gene remained unmethylated on the inactive X in spermatogenic cells (Grant et al., 1992). However, in 1997, a retroposon-captured *G6pd* gene expressed only in spermatozoa (Hendriksen et al., 1997) was identified, thus demonstrating that it was not an exception.

The subject is paired with sex ratio transmission distortion (srtd) since the mechanism of srtd is based on competing expression of X and Y linked spermatogenic genes, i.e., there is a special class of genes on the X which are expressed during spermatogenesis. The conflict between X and Y chromosomes is easily understood in multiparous species where insemination of 1 litter can occur by multiple males, as happens with mice in the wild (Dean et al., 2006). It is in the male's genetic interest that his genotype is the most abundant among the offspring while it is in the female's interest that hers are the most abundant. However, an allele driving srtd cannot go to fixation since a population of 100% males or 100% females is non-viable. Thus, balancing effects between the X and Y genes and their products are expected. "Skewed sex ratios also mean that suppressors of drive will automatically be selected for, not only on the opposite chromosome but also on the autosomes, with that selection growing stronger the more the population sex ratio deviates from 1:1 (the autosomal optimum)" **(Burt and Trivers, p. 60)**. Sex ratio deviations included ones involving mating between different subspecies of mice (Hurst and Pomiankowski, 1991) . In this case, hybrid sterility resulted as predicted by Haldane's Law which suggested that such hybrid sterility is due to an unbalancing of sex ratio distorter genes and their suppressors (see Chapter 2).

As the role of Xist in female X-inactivation (Chapter 10) became clear, it was found that it was expressed during X-inactivation during spermatogenesis as well (Ayoub et al., 1997) but it was not essential for male spermatogenic X-inactivation (Marahrens et al., 1997) and the question of its expression remains moot. Thus, the mechanism of inactivation was quite different. It was found that while un-synapsed chromosomes, in general and in sex chromosomes, express some of their genetic content (McKee and Handel, 1993), unsynapsed regions of the genome become transcriptionally silenced during pachytene (see update by Turner et al., 2006). This area of research was very active in the beginning of the 21st Century. This process was called meiotic silencing of unsynapsed chromatin (MSUC) or meiotic sex chromosome inactivation (MSCI) in the special case of the X and Y. In XYY males where the 2 Ys paired, meiotic arrest still occurred despite Y genetic expression (Royo et al., 2010). Transgenic mice expressing both *Zfy1* and *Zfy2* (see Chapter 11) were sterile, showing that expression of one or both of these genes was enough to block meiosis (Burgoyne et al., 2011). Thus, MSCI is essential to repress toxic genes on the Y chromosome that otherwise lead to meiotic arrest.

In addition to meiotic silencing, a second form of sex chromosome silencing operates during spermatogenesis, with silencing of unsynapsed regions being partially maintained into postmeiotic stages. This later stage of silencing is known as postmeiotic sex chromatin, PMSC and operates via a suite of associated histone modifications (Baarends et al., 2005; Moretti et al., 2016). However, PMSC is less complete than MSUC and some genes in asynaptic regions can reactivate in spermatids. The purpose of PMSC was initially unclear, however as with MSCI/MSUC it is becoming increasingly clear that it serves to defend the genome against deleterious elements present on the sex chromosomes, in particular sex ratio distorters.

Since Yq deletions caused sperm defects (Styrna et al., 1991), it was understood that the region contained genes the expression of which was essential for spermatogenesis (Burgoyne et al., 1992). In the 21st Century, it was shown that the Y chromosomal genes are predominantly expressed after spermatogenic meiosis while X-linked genes are dominant before this meiosis (Khil et al., 2004). At first, skewing of the sex ratio in offspring was not noticed, but also in the 21st Century, it was and mechanisms linking PMSC and srtd were elicited. Deletions on mouse Yq removing multiple copy genes (Conway et al., 1994) were shown to lead to offspring sex ratio distortion in favor of females, which in turn was found to be associated with upregulation of multiple X- and Y-linked transcripts in spermatids (Ellis et al., 2005). A multicopy gene, *Sly* (>40 copies), gene's products were found to repress gene expression of both male and female sex chromosomes in the male mouse germline after meiosis (Cocquet et al., 2009). Because of its multiple copies, it could only be downregulated by small interfering RNAs (produced by transgenesis) and it, along with its X-linked counterpart *Slx/Slxl1*, were found to be crucial opposing regulators of PMSC (Cocquet et al., 2009). Sequence studies of *Sly* among a number of species of mice (with the sequence of rats as an outlier) revealed that it evolved with great amplification of sex-linked, spermatid-expressed genes involved in srtd: X-linked ones suppressing male sperm success while Y-linked ones compensated for it (Ellis et al., 2011). The end result was spermatozoa of differing motility, as was found for *t*-allele trd (Rathje et al., 2019; see below).

Also, at the end of the 20th Century, and the beginning of the 21st, knockout technology was being widely applied with the finding that many knockouts resulted in male, but not female sterility. Some of these were easily understood such as with the loss of transition nuclear protein 1 (Yu et al., 2000) or testicular germ-cell proteases PC4 (Mbikay et al., 1997). Others were more obscure, for instance why bone morphogenetic protein 8B (Zhao et al., 1996), *Bclw*, a death-protecting member of the *Bcl2* family (Ross et al., 1998), apolipoprotein B (Huang et al., 1996) or the Na^+-K^+-$2Cl^-$cotransporter (Pace et al., 2000) caused male, but not female sterility were much more obscure.

Separation of X- and Y-bearing spermatozoa

Another major advance was the development of tools for separating cells of the testis into those of differing stages of spermatogenesis by sedimentation velocity. Such separations depended on the differing ratios of nuclear to cytoplasm. This could be accomplished in a large diameter, especially made clear plastic chamber with a shallow funnel for its bottom which provided the outlet. This was usually "home made" and separation was at one gravity for about 6 hours. The cells were usually separated by trypsin enzymatic digestion for a brief period and the suspension gently overlaid above a physiological salts and protein solution in a thin layer. Fractions of about 75 % pure pachytene spermatocytes and about 75% pure round spermatids could be obtained (Meistrich et al., 1973). Such separations could also be performed with a commercially available rotor for high speed centrifuges. This rotor had a complex valve system allowing fluid ingress and egress while the rotor was spinning. Separation took place in a funnel-shaped elutriation chamber. A suspension of cells was pumped at a preset flow rate from outside the centrifuge into the rotor to the narrow end of the elutriation chamber. As suspended cells were introduced into the chamber, they migrated according to their sedimentation rates to positions in the gradient where the effects of the two forces upon them are balanced. To harvest the cells, the fluid flow into the chamber was increased with the various cell fractions then leaving the chamber while fractions were collected. The small round spermatids with low sedimentation velocities were the first to be eluted into the collection chamber and, as the collection tubes were changed and the flow rates increased, larger pachytene cells would be collected.

Such methods were also applied to the problem of separating male and female determining sperm—a major goal for the dairy industry which mostly wanted females and of human in vitro fertilization programs which mostly wanted males (especially for China where selective abortion had a disastrous effect on the sex ratio). The "gold standard" for success was marked improvement in one direction of the ratio of the 2 sexes in fetuses or live borns after artificial insemination. Artificial insemination is somewhat difficult in mice[2].

A very slight excess of females was found when bovine semen was allowed to separate by 1 g sedimentation and then used in artificial insemination (Krzanowski, 1970). The difference was accredited to a higher propensity of the X-bearing sperm to aggregate rather than on the slighter greater weight (due to the X versus Y) difference which might be 2.4% in sperm which are primarily a nucleus of DNA). One of the first apparently successful attempts was in rabbits and the goal was to see if LDH-X was X-linked (it was not, see text; Stambaugh and Buckley,

1971). They used centrifugation in discontinuous density gradients and found that the 0,3% of sperm which had remained in the supernatant fraction had the highest LDH-X activity, in human and Rhesus monkey sperm as well, but when this fraction of rabbit sperm was used for artificial insemination, it resulted in a 2:1 ratio of males to females in 100 offspring (ibid.), opposite to the hypothesis. A controversy over the success or failure of separations of human sperm on the basis of motility of the sperm in bovine serum albumin gradients occurred in the pages of *Nature* with success being claimed (Ericcson et al., 1973) and denied (Ross et al., 1975). A number of investigators used quinacrine staining of the Y in sperm fractions separated after passage through various media (Goodall and Roberts, 1976; David et al., 1977). In general it was found that Y-bearing sperm were slightly more motile and could be somewhat enriched by this technique. Adding laminar flow in the solution to orient the sperm somewhat enhance the results (Bhattacharya et al., 1977; Sarkar et al., 1984). Density gradient centrifugation (Rohde et al., 1975) and Sephadex chromatography (Quinlivan et al., 1982) could also somewhat enhance the results but successful commercial exploitation was not in the offing.

The advent of the flow sorter led to a new era of separation attempts. The instrument was devised for separating various classes of immune cells attained with antibodies (see Chapt. 7) but was soon applied to sperm (Otto et al., 1979; Garner et al., 1983; Johnson et al., 1987; Penfold et al., 1998). A major problem was that these studies involved staining the sperm with DNA dyes which killed them! Another problem was the variable orientation of the flat sperm head to the laser beam which made a huge difference in the signal at the detector. This was largely overcome by the use of a special nozzle which oriented the stream of fluid (Rens et al., 1999). The method allowed the confirmation of different motility between X- and Y-bearing sperm (Keeler et al., 1983) and in rabbits, successful sex selection was achieved with 94% females and 81% males from the respective fractions after artificial insemination (Johnson et al., 1989).

A variety of other methods were tried with varying degrees of success. Ion exchange column chromatography, tested with artificial insemination in rabbits was unsuccessful (Downing et al., 1976), aqueous 2-phase partition by thin-layer counter current distribution achieved 80% Y sperm (Cartwright et al., 1993), while density gradient separation in sodium metrizoate resulted in about 80% Y-body (fluorescent body of the Y chromosome) positive human sperm (Shastry et al., 1977). Ninety-four % X-bearing spermatozoa were achieved with differential velocity sedimentation in Percoll density gradients (the # without Y-bodies) (Kaneko et al., 1984a) while free-flow electrophoresis achieved pure X-bearing sperm (also negative for the Y-body; Engelmann et al., 1988) but the resultant sperm had decreased motility. The different electrophoretic motility may have been due to different sialic acid content of the 2 kinds of sperm (Kaneko et al., 1984b). Well separated X- and Y- bearing sperm showed no difference in reactivity with an anti-H-Y antibody (Hendriksen et al., 1993) although elimination of the Y-bearing sperm with such antibodies had been tried with some apparent slight success (Bennet and Boyce, 1973). An apparent final success in separating spermatozoa of the two sexes depended on finding a haploid-expressed membrane protein as was described in the text for the transmission distortion-related Spam1 antigen. An X-encoded cell surface receptor had this property and activating its ligand suppressed the activity of the X-bearing spermatozoa, allowing greater than 80-20 preferential sex ratios in live born offspring (Umehara et al., 2019) but there has been no confirmations of this result as of 2020.

t-allele transmission ratio distortion

As seen in Chapter 4, the essential attributes of the T/t-complex with its multiple inversions are prezygotic transmission ratio distortion (trd) and crossover suppression, not the developmental recessive mutations leading to embryonic arrest. Males, but not females, heterozygous for a *t* allele usually transmit it to many more of their progeny than the 50% dictated by Mendel's laws. As well stated by Burt and Trivers **(2006, p. 19):**

> *"The t has existed for at least 3 million years. It has grown steadily in size since its origin to become more than 1% of the mouse genome. Its spread has generated adaptations on both sides, improvements in t action, as well as evolution of countermeasures by the rest of the genome. The t has been studied for the past 75 years and a great deal is known about it, although unfortunately on several key points the information is contradictory".*[3]

Many *t* alleles recently recovered from wild populations had segregation ratios greater than 95% (see Chapters 2 and 5). Laboratory populations are usually characterized by lower segregation distortion ratios. The ratios vary considerably between individual males carrying a particular *t* allele as well as between different *t* alleles (Braden and Weiler, 1964; Braden, 1972; Erickson, 1978); there are low distorters as well. However, males which are compound heterozygotes for 2 different *t* alleles that complement each (for viability) are sterile.

The studies of the mechanisms behind trd, and its related sterility, might be said to begin with Vernon Bryson's histological studies of spermatogenesis in various combinations of T/t complex genes which was performed in L.C. Dunn's laboratory. It was his doctoral thesis and was published in 1944 with him as sole author (Bryson, 1944).[4] Although providing a thorough description of spermatozoa and differential fertilization with various combinations of T and t^n, "None of these will explain adequately the effect of t^0 and t^1 upon ratios and fertility" (ibid.) and were eventually to be found in error as will be seen in the following.

Braden and Gluecksohn-Waelsch (1958) greatly advanced the study of the number and qualities of spermatozoa from both the sterile male combinations of two *t*-alleles or the trd positive spermatozoa of *t*-+ allele mice. They found that the proportion of abnormal spermatozoa, "the number of sperm ejaculated and their motility in the uterus or *in vitro* were not appreciably less than those of males of fertile genotypes" (ibid). They used artificial insemination (AI) to test whether the "abnormal" spermatozoa would have a toxic effect on the "normal" spermatozoa and found no such effect. They did not use AI, as was later done, to test the trd of precise mixtures of spermatozoa looking for trd competition with normal spermatozoa rather than with the co-developed spermatozoa. They found that the spermatozoa of the sterile males did not pass the utero-tubal junction implying decreased motility or, perhaps, decreased ability to penetrate the Fallopian fluid (ibid.).

The trd could be accounted for by one of two different mechanisms: either unequal numbers of two kinds of spermatozoa were present or equal numbers of functionally different spermatozoa were produced. Selective death during spermatogenesis could be hypothesized to be involved in the generation of unequal numbers of functionally different spermatozoa. In such an hypothesis, unequal numbers of spermatozoa could result because some chromosomes preferentially segregated to cytoplasm that was destined to die. However, there is a deficiency of only about 13% of the cells expected on the basis of mouse spermatogenic kinetics (Oakberg, 1956), far short of the number needed for, say 90% trd. Thus, nonmendelian ratios of the magnitude seen with *t* allele trd cannot be explained by this mechanism.

Alternatively, there was evidence that suggested that the trd controlled by *t* alleles involved functional difference between spermatozoa bearing different *t* alleles. Delayed mating, in which fertilization occurs as soon as the spermatozoa reach the fallopian tube, compared with normal mating, which requires that the spermatozoa remain in the female reproductive tract for several hours before fertilization, nullifies the trd in the case of multiple *t* alleles (Braden, 1958b; Yanagisawa et al., 1961; Erickson, 1973). The fact that the *t* allele effect can be altered by changing the time of insemination relative to that of ovulation suggested that there were two classes of spermatozoa with unequal physiological characteristics which would arise as the result of haploid gene expression. Hammerberg and Klein (1975) supported this notion by showing, using a Robertsonian translocation to mark the wild type allele, that there was equal segregation of *t* and + alleles at metaphase I.

As early as 1965, Yanagisawa suggested that there might be two functional classes of spermatozoa by showing a 2-part curve of loss of motility of T/t^1 spermatozoa *in vitro* over time, an apparent break appearing at about 3 hours (Yanigasawa, 1965). However, the ratio of the number of sperm which differ in their viability *in vitro* to the trd of the particular male was not great. Ginsberg and Hillman (1974) found increased aerobic respiration in pools of epididymal sperm which contained a trd *t* allele over pools which did not, but only with lactate, pyruvate or succinate as substrates but not with glucose where the respiration was decreased. The relevance seemed unclear since spermatozoa are mostly in a nearly anaerobic environment. This lab also found differences in spermatozoal cAMP levels but correctly pointed out that these might have been due to the closely linked *H-2* locus which had been associated with alterations in cAMP (Nadijcka and hillman, 1980). Dooher and Bennett (1974), using E.M., found abnormal microtubular systems in mouse spermatids in t^{w2} homozygotes which anticipated microtubular protein genes found later to map to the *T/t* complex. Olds-Clarke and Carey (1978) found more rapid egg penetration by t^{w32}-bearing sperm in vitro. Katz et al. (1979), using high speed cinematography, found that beat frequency was unimodal in inbred strains and their F1s while there were two, roughly equal classes of spermatozoa characterized by different beat frequencies of the sperm tail in *t*-allele heterozygotes. This finding was extended by Olds-Clarke and Johnson (1993) who found abnormalities of sperm flagellar function which were most severe in sterile t^n compound heterozygotes and less severe in the heterozygotes with wild type alleles. Molecular probes confirmed the presence of equal numbers of two kinds of spermatozoa, even when recovered from the female reproductive tract (Silver and Olds-Clark, 1984). The trds for a number of *t* alleles found with in vitro fertilization were similar to those found with delayed mating, i.e., Mendelian (McGrath and Hillman, 1980, 1981; Garside and Hillman, 1989). Artificial insemination and chimera experiments were interpreted as showing that the function of the non-*t*-bearing sperm was impaired (Olds-Clark and Peitz, 1985; Seitz and Bennett, 1985). Counts of spermatozoa flushed from the uteri and Fallopian tubes of naturally inseminated females by transmitting males were normal but those of sterile males homozygous for *t*-alleles (or heterozygous for 2 different *t*-alleles) were reduced 100-fold (Tucker, 1980). The sperm could still function if injected into the egg: intracytoplasmic sperm injection (ICSI), a procedure used in humans for some cases of male infertility (Kuretake et al., 1996).

A cell biological hypothesis to explain the multiple effects of *t* alleles predicted that proteins coded in the *T* region were related to membrane transport, hormone receptors, adenylate cyclase, or other membrane features affecting sperm function (Erickson, 1978). Such membrane proteins might not be shared between spermatids by the intracellular bridges. The importance of these for general lack of of postmeiotic expression was described above.

There was a search for such membrane proteins. Early studies used two-dimensional gel electrophoresis to focus on protein products expressed in the testes that were encoded by this region (Silver et al., 1983). Two of these were unique to testis and expressed strongly in postmeiotic cells (Silver et al., 1987). One of these attracted particular attention, Tcp-1. Sanchez et al. (1985a) demonstrated nonequivalent expression of wild-type and *t*-complex encoded forms of Tcp-1 in germ cell fractions isolated subsequent to ^{35}S-methionine labelling, and this was confirmed by Silver et al. (1987). Tcp-1 expression was not limited to testes and was found in a variety of cell culture lines, tissues, and developmental stages with the exception of two-cell embryos (Sanchez and Erickson, 1985b; Silver et al., 1983) The identification of Tcp-1 as a Golgi membrane-associated protein whose synthesis is enhanced with the enlargement of the Golgi complex during spermatogenesis, contributing to the synthesis of the acrosome, explained the wide tissue distribution of this gene product and the reason for enhanced synthesis postmeiotically (see above; Willison et al., 1989).

Microdissection of the proximal half of chromosome 17 isolated probes (Rohme et al., 1984) that were used to isolate cosmids (see Chapter 9) from the *T/t* complex region. Further search of the cosmids for CpG-rich islands led to the discovery of one postmeiotically expressed gene from the *Tcd-3* (distorter/sterility region of the *T/t* complex, see below (Rappold, et al., 1987). Other cosmids (large subclones of DNA) derived by this approach from the *t* complex region were studied for a possible role in sex determination with negative results (Durbin et al., 1989). Differential screening of DNA libraries to look for testes-specific clones mapping to chromosome 17 detected an unusual stretch of DNA with overlapping open reading frames and a stretch of alternating purine and pyrimidine residues (Sarvetnick et al., 1989) and a gene coding for a novel protein with a predicted coiled-coil structure mapping near the *t^{h20}* deletion (Mazarakis et al., 1991). Other developmentally interesting genes, such as *Oct-4*, coding for a transacting factor later found to be very important for reverting differentiation towards a pluripotent state (Niwa et al., 2009) were mapped to the *T/t* complex (Scholer et al., 1990).

Lyon (1984) put forth a model to explain transmission ratio distortion (trd) in which three transacting distorters would act in a harmful way on the wild-type form of *Tcr*, the *t*-complex response element, while the *Tcr* in the *t* complex (*Tcr*[1]) would be resistant to this effect. ("Responder insensitive" in the vocabulary used for the general class of trd due to "gamete killers" (**Burt and Trivers, 2006, Chapter 2**). She also suggested (Lyon, 1986) that the sterility of males homozygous for *t*-alleles (or heterozygous for 2 different *t*-alleles) was due to homozygosity (or heterozygosity) for *Transmission distorting genes, Tcds*. Thus, the *tcds* acted as "classic" hybrid sterility loci, located near *H-2*, and found in many populations of wild mice (e.g. France [Bonhomme et al., 1982] and Czechoslovakia) which did not display trd and which had been separately studied, especially by the Czech group, e.g. Forejt and Ivanyi (1975). In fact, Laurence Hurst had suggested that the *T/t complex* had evolved from hybrid sterility loci (Hurst, 1993). Mary Lyon hypothesized that the negative effect of the *Tcd* product on the wild type sperm function would reciprocally inactivate the sperm expressing another *Tcd*. The observation that deletion of the distorter *Tcd1* functioned similarly in trd with *Tcr* strongly suggested that the product was a null or hypomorph (Lyon, 1992). Lyon (1991) also pointed out the negative effect of *Tcr* alone, i.e., in the absence of *Tcds*, resulting in a low transmission ratio compared to wild type sperm, a condition called "drag" in the nomenclature used with "gamete killers" (**Burt and Trivers, 2006, Chapter 2**). Increased dosage of both *Tcrs* and *Tcds* was needed to increase trd (Silver, 1989).

Lyon had summarized a large number of experiments using various translocation chromosomes that provided "pieces" of the *t* complex to study their effects on trd. As Lyon herself admitted (Lyon and Zenthon, 1987) her model did not agree with the important data of Hammerberg (1982) mapping such genes and these difficulties in mapping loci in an area of cross-over suppression made the eventual cloning of candidate genes for trd difficult. In further work by Lyon and Zenthon (1987), a number of *cis* effects on the trd were found. Others found major effects of the genetic background (Gummere et al, 1986; Sanchez and Erickson, 1986) and more regions of the *t* complex with effects on trd were found (Silver and Remis, 1987). There was tremendous variation between males of one genotype and at various time intervals in those males (Erickson, 1978).

Molecular cloning of the region between chromosome breakpoints that define the hypothetical *Tcr* locus was performed (Rosen et al., 1990) and genes named T66 that encode male germ cell-specific transcripts were found to be localized in this segment (Bullard and Schimenti, 1990). One of the genes from the region, *Tcp10b'*, underwent alternative splicing during spermatogenesis such that an allele-specific and haploid-specific transcript occurred (Cebra-Thomas et al., 1991). Mice transgenic for the gene showed transmission ratio distortion, one construct with an alternatively spliced form of the gene showed positive trd while one with the alternatively spliced form showed low trd (Snyder and Silver, 1992). Further studies of the *Tcb10'* region suggested that a protein product of the *Tcr* might not be involved in trd (Schimenti, 1999). Willison et al. (1990) also identified a postmeiotic transcript from this region of chromosome 17. However, transgenic mice bearing the gene did not show trd and the targeted knock out did not eliminate trd (Ewulonu et al., 1996).

There was a flurry of cloning of the trd region of chromosome 17 at the end of the Century. Bernhard Herrmann, who had been involved in the initial cloning of the region from microdissected chromosome 17 (Rohme et al., 1984) in Hans Lehrach's laboratory, became the leader of this effort. Genes, such as *Tctc2*, were found which were differentially spliced in species that form sterile hybrids with laboratory mice, and these were deleted in *t*-alleles which showed trd (Braidotti and Barlow, 1997). A candidate gene for *Tcd1*, *Tctx1*, encoded a dynein chain involved in flagella and had *t*-allele specific variants (Lader et al., 1989; Harrison et al., 1998). Adjacent were two acetyl-CoA transferase genes transcribed in the opposite direction (Ashworth, 1993). A gene cloned as a candidate for *Hybrid sterility 6, Hst 6*, was another gene encoding a dynein chain and was a candidate gene for *Tcd2* (Fossela et al., 2000). *Tctex2* was a candidate gene for *Tcd3* and encoded another dynein chain and also had *t*-allele specific mutations (Huw et al., 1995; Patel-King et al., 1997). *Rsk3*, a gene mapping to the region of the *Tcr* (Kispert et al., 1999) was a candidate for, or marker of nearby genes for, involvement in trd. At the very end of the Century, the latter gene's involvement, and a potential mechanism for trd was obtained with the cloning of a Smok kinase (Sperm motility kinase, a homologue of MARK kinases which phosphorylate microtubule associated proteins such as dynein)/Rsk3 fusion gene (Hermann et al., 1999). This candidate *Tcr* gene had a reduced kinase activity, which could allow it to counterbalance a signaling impairment caused by the products of the *Tcd* loci whose mutations encoded hypomorphic alleles (above). Mice transgenic for the cloned gene had non-mendelian transmission of chromosomes on which they were carried (ibid.).

It took further work in the first decade of the new Century to determine that at least some the prior candidates for *Tcd*s were in error and to work out the details of a general mechanism for the trd genes. The importance of Rho proteins in the control of sperm motility was demonstrated in *Bovidae* (Hinsch et al., 1993) and they were found in mouse sperm flagella (Fujita et al., 2000) and midpiece (Nakamura et al., 1999). Their importance became apparent when, instead of phosphorylation substrate flagellar dyneins, GTPase activating proteins which could act upstream of Tcr were found at *Tcd1*a (Bauer et al., 2005) and *Tcd2* (Bauer et al., 2007). Thus, segregation distortion was seen as competition between strengths of signaling pathways: the *t*-allele distorters were generally diffusible hypomorphs affecting all the sperm and reducing motility, i.e. the decreased dosage of the products of their loci, half from the wild allele but little from the *Tcd* allele diffused through the intracellular bridges to all developing spermatozoa, while the motility was rescued in the *t*-allele-bearing spermatozoa by the *Tcr*-encoded dominant-negative SMOK variant which would be limited to them. Postmeiotic expression with limitation to haploid cells was soon established for Tcr (Veron et al., 2009).

An important open question in testis biology is whether there are any other genes that escape sharing between syncytial spermatids, in the manner of the *t* complex distorter and *Spam1*. Such genes are potent substrates for the evolution of "driving" alleles that cause transmission ratio distortion (sex ratio distortion if sex-linked). A preprint published during the drafting of this chapter, the authors of which performed single cell RNA sequencing on spermatid cells, suggests that escape from transcript sharing is a quantitative rather than a qualitative

FIG. 8.3 A) Plaque in "aula" dedicating it to Valerio Monesi **B)** Entry way to "aula" with Mones's name on the door. *Photograph courtesy of Dr. Carla Boitani*

phenomenon, and that up to 1/3 of the spermatid transcriptome may show some degree of divergence from fully equal sharing (Bhutani et al., 2019). Such data confirm earlier analyses of human pedigrees which showed excess genetic sharing among siblings requiring many loci with skewed transmission would be required of the observed genome-wide shift (Zollner et al., 2004). If further confirmed, this appears likely to have dramatic implications for our understanding of the degree to which specific genes are placed under haploid versus diploid selection, and the potential for conflict between germline and somatic expression of such genes.

Notes

1. We have met Valerio Monesi in Chapter 6 as the individual at the Oakridge National Laboratory who helped work out the stages of mouse spermatogenesis essential for relating the time of mutagen exposure to the time when spermatozoa potentially carrying induced mutations would appear. He returned to Italy where he became Professor of Histology at the University of Rome, La Sapienza. His success depended in large part on his social, as well as intellectual, prowess. As his colleague, and good friend of mine, Dr. Franco Mangia, succinctly puts it: "In Italy, first you do academic politics, then you do science". It was with Franco that my wife and I had dinner at Valerio's elegant apartment in Rome in the mid-70s. He was very charming and gracious with old world manners, as was his wife. He was highly respected in Italy and a new, free-standing lecture hall ("aula") at the Istituto di Istologia and Embriologia on the campus of the Università di Roma, La Sapienza is dedicated to him **Fig. 8.3**.

2. I spent a month at the Jackson Laboratory in 1972 in Eva Eicher's laboratory unsuccessfully trying to do it for a sperm mixing experiment related to *t*-allele transmission ratio distortion.

3. **Genes in Conflict** by Austin Burt and Robert Trivers is the definitive early 21st Century summary of the status of knowledge on the biology of selfish genetic elements. In a text of 475 pages devoted to these phenomena (and about 1600 references) in all Kingdoms of species, about 20 pages are devoted to *t*-allele segregation distortion, compared to only 6 pages for the well known *Segregation Distorter* system of *Drosophila*.

4. At this time, it was the tradition, especially in Germany, that the doctoral thesis was published only under the name of the student. Thus, it was upsetting to many that Hans Spemann added his name to the thesis publication of Hilde Mangold on the embryonic organizer which became the basis for his Nobel prize. At that time, especially because she was a woman, there probably was no question of her sharing it, as has also happened with two young female astronomers, one of whom discovered the universal cosmic microwave background radiation and the other pulsars.

References

Abraham, K.A., Bhargava, P.M., 1961. Nucleic acid metabolism of spermatozoa. J. Reprod. Fertil. 2, 195.

Abraham, K.A., Bhargava, P.M., 1963. Nucleic acid metabolism of spermatozoa. Biochem, J. 86, 298–303.

Adam I.R., McLaren A. 2002. Sexually dimorphic development of mouse primordial germ cells: switching from oogenesis to spermatogenesis. Devel. 129, 1155-1164.

Adham, I.M., Klemm, U., Maier, W.M., Hoyer-Fender, S., Tsaosidou, S., Engel, W., 1989. Molecular cloning of preproacrosin and analysis of its expression pattern during spermatogenesis. Eur. J. Biochem. 1822, 563–568.

Agarwal, I., Przeworski, M., 2019. Signatures of replication timing, recombination, and sex in the spectrum of rare variants on the human X chromosome and autosomes. Proc. Natl. Acad. Sci. U.S.A. 116, 17916–17924.

Agulnik, S.I., Agulnik, A.I., Ruvnsky, A., 1990. Meiotic drive in female mice heterozygous for the HSR inserts on chromosome 1. Genet. Res. 55, 97–100.

Allison, A.C., Hartree, E.F., 1970. Lysosomal enzymes in the acrosome and their possible role in fertilization. J. Reprod. Fertil. 21, 501–515.

Ariel, M., McCarry, J., Cedar, H., 1991. Methylation patterns of testis-specific genes. Proc. Nat'l. Acad. Sci. U.S.A. 88, 2317–2321.

Ashworth, A., 1993. Two acetyl-CoA acetyltransferase genes located in the *t*-complex region of the mouse chromosome 17 partially overlap the *Tcp-1* and *Tcp-1x* genes. Genomics 18, 195–198.

Ayoub, N., Richler, C., Wahrman, J., 1997. Xist RNA is associated with the transcriptionally inactive XY body in mammalian male meiosis. Chromosoma 106, 1–10.

Baarends, W.M., Wassenaar, E., van der Laan, R., Hoogerbrugge, J., Sleddens-Linkels, E., Hoeijmakers, J.H., de Boer, P., Grootegoed, J.A., 2005. Silencing of unpaired chromatin and histone H2A ubiquitination in mammalian meiosis. Mol. Cell Biol. 25, 1041–1053.

Baker, T.G., Neal, P., 1973. Initiation and control of meiosis and follicular growth in ovaries of the mouse. Annal. de Biol. Anim. Biochim. 13, 137–149.

Bauer, H., Willert, J., Koschorz, B., Herrmann, B.G., 2005. The t complex-encoded GTPase-activating protein Tagap1 acts as a transmission ratio distorter in mice. Nat. Genet. 37, 969–973.

Bauer, H., Veron, N., Willert, J., Herrmann, B.G., 2007. The t-complex-encoded guanine nucleotide exchange factor Fgd2 reveals that two opposing signaling pathways promote transmission ratio distortion in the mouse. Genes Dev. 21, 143–147.

Beatty, R.A., 1971. The genetics of size and shape of spermatozoan organelles. In: Beatty, R.A., Gluecksohn-Waelsch, S. (Eds.), Edinburgh Symposium on the Genetics of the Spermatozoon. Edinburgh, 97–115.

Beatty, R.A., Wooley, D.M., 1967. Inheritance of midpiece length in mouse spermatozoa. Nature 215, 94.

Beechy, C.V., 1973. X-Y chromosome dissociation and sterility in the mouse. Cytogenetic. Cell Genet. 12, 60–67.

Bennett, D., Boyce, E.A., 1973. Sex ratio in progeny of mice inseminated with sperm treated with H-Y antiserum. Nature 246, 308–309.

Bernstein, P., Peltz, L., Ross, J., 1989. The poly (A)-binding protein complex is a major determinant of mRNA stability in vitro. Mol. Cell Biol. 9, 659–670.

Betlach, C.J., Erickson, R.P., 1973. A unique RNA species from maturing mouse spermatozoa. Nature 242, 114–115.

Bhargava, P.M., 1957. Incorporation of Radioactive Amino-Acids in the Proteins of Bull Spermatozoa. Nature 179, 1120–1121.

Bhattacharya, B.C., Shome, P., Gunther, A.H., 1977. Successful Separation of X and Y Sprmatozoa in Human and Bull Semen. Int. J. Fertil 22, 30–35.

Bhutani, K., Stansifer, K., Ticau, S., Bojic, L., Villani, C, Joanna Slisz, J., Cremers, C., Roy, C., Donovan, J.S., Fiske, B., Friedman, R., 2019. Widespread haploid-biased gene expression in mammalian spermatogenesis associated with frequent selective sweeps and evolutionary conflict. https://www.biorxiv.org/content/10.1101/846 253v1.

Bluthmann, H., Cicurel, L., Luntz, G.W.K., Haedenkamp, G., Illmensee, K., 1982. Immunochemical localization of mouse testis-specific phosphoglycerate kinase 2 (PGK-2) by monoclonal antibodies. EMBO J. 1, 479–484.

Boitani, C., Geremia, R., Rossi, R., Monesi, V., 1980. Electrophoretic pattern of polypeptide synthesis in spermatocytes and spermatids of the mouse. Cell Differ. 9, 41–49.

Bonhomme, F., Guenet, J.-L, Catalan, J., 1982. *Presence d'un facteur de sterilite male,* Hst-2, *segregeaanat dans les crosisements interspecifiques,* M. musculus L. X *m. spretus Lastasste et lie a* Mod-1 *et* Mpi-1 *sur le chromosome 9.* C.R. Acad. Sc. Paris 295, 691–693.

Bowles, J., Feng, C.-W., Spiller, C., Davidson, T.-L., Jackson, A., Koopman, P., 2010. FGF9 suppressed meiosis and promotes male germ cell fate in mice. Development. Cell 19, 440–449.

Bowles J., Retinoid Signaling Determines Germ Cell Fate in Mice. Science 312, 596–600.

Braden, A.W.H., 1958a. Strain differences in the morphology of the gametes of the mouse. Aust. J. Biol. Sci. 12, 65–71.

Braden, A.W.H., 1958b. Influence of time of mating on the segregation ratio of alleles at the *T* locus in the house mouse. Nature 181, 786–789.

Braden, A.W.H., 1972. T-locus in mice; segregation distortion and sterilitiy in the male. In: Beatty, R.A., Gluecksohn-Waelsch, S. (Eds.), (1972) The Genetics of the Spermatozoon: Proceedings of an Internation Symposium held at the University of Edinburgh, Scotland on August 16-20, 1971. University of Edinburgh. Edinburgh, New York, 289–305.

Braden, A.W.H., Gluecksohn-Waelsch, S., 1958. Further studies of the effect of the T locus in the house mouse on male fertility. J. Exp. Zool. 138, 431–452.

Braden, A.W.H., Weiler, H., 1964. Trnsmission ratios a the *T*-locus in the mouse: inter- and intra-male heterogeneity. Aust. J. Biol Sci. 17, 921–934.

Bragg, P.W., Handel, M.A., 1979. Protein Synthesis in Mouse Spermatozoa. Biol. of Repro. 20, 333–337.

Braidotti, G., Barlow, D.P., 1997. Identification of a male meiosis-specific gene, *Tctc2*, which is differentially spliced in species that form sterile hybrids with laboratory mice and deleted in *t* chromosomes showing meiotic drive. Devel. Biol. 186, 85–99.

Braun, R.E., Behringer, R.R, Peschon, J.J., Brinster, R.L., 1989a. Genetically haploid spermatids are phenotypically diploid. Nature337: 373-376.chormosoes showing meiotic drive. Devel. Biol. 186, 85–99.

Braun, R.E., Peschon, J., Behringer, R.R, Brinster, R.L., Palmiter, RD, 1989b. Protamine 3'-untranslated sequences regulate temporal translational control and subcellular localization of growth hormone in spermatids of transgenic mice. Genes Dev. 3, 793–802.

Brink, R., Burnham, C.R., 1927. Differential action of the sugary gene in maize on two alternative classes of male gametophytes. Genetics 12, 348–378.

Brozek, C., 1970. Proportion of morphologically abnormal spermatozoa in two inbred strains of mice, their reciprocal F_1 and F_2 crosses. Acta Biol. Crac. XIII, 189–198.

Bruce, W.R., Furrer, R., Goldberg, R.B., Meistrich, M.L., Mintz, B., 1973. Genetic control of the kinetics of mouse spermatozoa. Genet. Res. 22, 155–169.

Bryson, V., 1944. Spermatogenesis and fertility in Mus Musculus as affected by factors at the T locus. J. Morphol. 74, 131–179.

Bullard, D.C., Schimenti, J.C., 1990. Molecular cloning and genetic mapping of the *t complex responder* candidate gene family. Genetics 124, 957–966.

Burgoyne, P.S., 1975. Sperm phenotype and its relationship to somatic and germ line genotype: a study using mouse chimeras. Devel. Biol. 44, 63–76.

Burgoyne, P.S., Mahadevaiah, S.K., Sutcliffe, M.J., Palmer, S.J., 1992. Fertility in mice requires X-Y pairing and a Y-chromosomal 'spermiogenesis' gene mapping to the long arm. Cell 71, 391–398.

Burgoyne, P.S., Vernet, N., Mahadevaiah, S., Szot, M., Turner, J.M.A., Royo, H., Taketo, T., Mitchell, M., Logepied, G., 2011. Involvement of Y-Encoded Zinc Finger Transcription Factors in Male and Female Infertility in Mice. Biol. Reprod. 85, 102.

Busby Jr., W.F., Hele, P., Chang, M.C., 1974. Apparent amino acid incorporation by ejaculated rabbit spermatozoa. Biochim. Biophys. Acta 335, 246–259.

Caldwell, K.A., Handel, M.A., 1991. Protamine transcript sharing among postmeiotic spermatids. Proc. Nat'l Acad., U.S.A. 88, 2407–2411.

Caldwell, K.A., Handel, M.A., 1991. Bearing spermatozoa in human semen samples after passage through bovine serum albumin. J. Reprod. Fertil. 50, 37–379.

Cartwright, E.J., Harrington, P.M., Cowin, A., Sharpe, P.T., 1993. Separation of Bovine X and Y sperm based on surface differences. Molec. Reprod. Devel. 34, 323–328.

Cattanach, B.M., Pollard, C.E., Hawkes, S.G., 1971. Sex reversed mice: XX and XO males. Cyogenet 10, 318–337.

Cebra-Thomas, J.A., Decker, C.L., Snyder, L.C., Pilder, S.H., Silver, L.M., 1991. Allele specific- and haploid-specific product generated by alternative splicing from mouse *t complex responder* locus candidate. Nature 349, 239–241.

Chayko, C.A., Martin-DeLeon, P.A., 1992. The murine Rb(6:16) translocation: alterations in the proportion of alternate sperm segregants effecting fertilization in vitro and in vivo. Human. Genet. 90, 79–85.

Chesnel, F., Eppig, J.J., 1995. Synthesis and accumulation of p34^{cdc2} and cyclin B in mouse oocytes during acquisition of competence to resume meiosis. Molec. Reprod. Devel. doi:10.1002/mrd.1080400414.

Chiquione, A.D., 1954. The identification, origin, and migration of the primordial germ cells in the mouse embryo. Anat. Rec. doi:10.1002/ar.1091180202.

Chmatal, L., Gabriel, S.I., Mitsainas, G.P., Martinez-Vargas, J., Ventura, J., Searle, J.B., Schultz, R.M., Lampson, M.A., 2014. Centromere strength provides the cell biological basis for meiotic drive and karyotype evolution in mice. Curr. Biol. 24, 2295–2300.

Cocquet, J., Ellis, P.J.I., Yamauchi, Y., Mahadevaiah, S.K., Affara, N.A., Ward, M.A., Burgoyne, P.S., 2009. The multicopy gene *Sly* represses the sex chromosomes in the male mouse germline after meiosis. PLoS Biol 7, e1000244.

Concha, I.I., Urzua, U., Yanez, A., Schroeder, R., Pessot, C., Burzio, L.O., 1993. U1 and U2 Snrna are localized in the sperm nucleus. Exp. Cell Res. 204, 378–381.

Conine, C.C., Sun, F., Song, L., Rivera-Perez, J.A., Rando, O.F., 2018. Small RNAs Gained during Epididymal Transit of Sperm Are Essential for Embryonic Development in Mice. Devel. Cell 46, 470–480.

Conway, S.J., Mahadevaiah, S.K., Darling, S.M., Capel, B., Rattigan, A.M., Burgoyne, P.S., 1994. Y353/B: a candidate multiple-copy spermiogenesis gene on the mouse Y chromosome. Mamm. Genome 5, 203–210.

Cowan, A.E., Primakoff, P., Myles, D.G., 1986. Sperm exocytosis increases the amound of PH-20 antigen on the surface of guinea pig sperm. J. Cell Biol. 103, 1289–1297.

Daentl, D., Erickson, R.P., Betlach, C.J., 1977. DNA synthetic capabilities of differentiating sperm cells. Differentiation 8, 159–166.

Dahl, H.-H.M., Brown, R.M., Hutchison, W.M., Maragos, C., Brown, G.K., 1990. A testis-specific form of the human pyruvate dehydrogenase EI subunit is coded for by an intronless gene on chromosome 4. Genomics 8, 22–232.

David, G., Jeulin, C., Boyse, A., Schwartz, D., 1977. Motility and percentage of Y- and YY -bearing spermatozoa in human semen samples after passage through bovine serum albumin. Reproduction 50, 3677–3679.

David, J.C., Vinson, D., Loir, M., 1982. Developmental changes of DNA ligase during ram spermatogenesis. Exp. Cell Res. 141, 357–364.

Dean, M.D., Ardlie, K.G., Nachman, M.W., 2006. The frequency of multiple paternity suggests that sperm competition is common in house mice (Mus domesticus). Mol. Ecol. 15, 4141–4151.

Distel, R.J., Kleene, K.C., Hecht, N.B., 1984. Haploid expression of a mouse testis alpha-tubulin gene. Science 124, 68–70.

Dooher, G.B., Bennett, D., 1974. Abnormal microtubular systems in mouse spermatids associated with a mutant gene at the T-locus. Devel 32, 749–761.

Downing, D.C., Black, D.L., Carey, W.H., Delahanty, D.L., 1976. The effect of ion-exchange column chromatography on separation of XX and Y chromosome-bearing human spermatozoa. Feril. Steril 27, 1187–1190.

Dudley, K., Potter, J, Lyon, M.E., Willison, K.R., 1984. Analysis of male sterile mutations in the mouse using haploid stage expressed cDNA probes. Nucleic Acids Res. 12, 4281–4293.

Dumont, J., Desai, A., 2012. Acentrosomal spindle assembly and chromosome segregation during oocyte meiosis. Trends Cell Biol. 22, 241–249.

Durbin, E.J., Erickson, R.P., Craig, A., 1989. Characterization of GATA/GACA-related sequences on proximal chromosome 17 of the mouse. Chromosoma 97, 301–306.

Ellis, P.J.I., Clemente, E.J., Bali, P., Toure, A., Ferguson, L., Turner, J.M.A., Loveland, K.L., Affara, N.A., Burgoyne, P.S., 2005. Deltions on mouse Yq lead to upregulation of multiple X- and Y-linked transcripts in spermatids. Hum. Mol. Genet. 14, 2705–2715.

Ellis, P.J.I., Bacon, J., Affara, N.A., 2011. Association of *Sly* with sex-linked gene amplification during mouse evolution: a side effect of genomic conflict in spermatids. Hum. Mol. Genet. 20, 3010–3021.

Engelmann, U., Krassnigg, F., Schatz, H., Bernhard-Wolf, S., 1988. Separation of human X and Y spermatozoa by free-flow electrophoresis. Gamete Res. 19, 151–159.

Epiphano, O., Liang, O.L.F., Familari, M., Moos, M.C.J, 1995. Coordinate expression of the three zona pellucida genes during mouse oogenesis. Devel. Biol. 121, 1946–1957.

Eppig, J., 1991. Maintenance of Meiotic Arrest and the Induction of Oocyte Maturation in Mouse Oocyte-Granulosa Cell Complexes Developed in Vitro from Preantral Follicles. Biol. of Reprod. 45, 824–830.

Ericsson, R.J., Langevin, C.N., Nisino, M., 1973. Isolation of fraction rich in human Y sperm. Nature 246, 421–424.

Erickson, R.P., 1973. Haploid gene expression versus meiotic drive: the relevance of intercellular bridges during spermatogenesis. Nature New Biol. 243, 210–212.

Erickson, R.P., 1974. Genetic control of glutamate oxaloacetic transaminase in murine spermatozoa. Exptl. Cell Res. 86, 429–430.

Erickson, R.P., 1976. Glucose-6-phosphate dehydrogenase activity changes during spermatogenesis: possible relevance to X-chromosome inactivation. Devel. Biol. 53, 134–137.

Erickson, R.P., 1977. Strain variation in spermatozoa beta-glucuronidase in mice. Genet. Res. 28, 139–145.

Erickson, R.P., 1978. t-alleles and the possibility of postmeiotic expression during mammalian spermatogenesis. Fed. Proc. 37, 2517–2521.

Erickson, R.P., 1990. Postmeiotic gene expression. Trends in Genet 6, 264–269.

Erickson, R.P., 1993. Molecular genetics of mammalian spermatogenesis. Genes in Mammalian Reproduction. Wiley-Liss, New York, pp. 1–26.

Erickson, R.P., Krzanowska, H., 1974. Differences in hyaluronidase activities among inbred strains of mice and their possible significance for variations in fertility. J. Reprod. Fert. 39, 101–104.

Erickson, R.P., Butley, M.S., Martin, S.R., Betlach, C.J., 1979. Variation among inbred strains of mice in adenosine 3′:5′ cyclic monophosphate levels of spermatozoa. Genet. Res. 33, 129–136, 1979.

Erickson, R.P., Erickson, J.M., Betlach, C.J., Mesitrich, M.L., 1980. Further evidence for haploid gene expression during spermatogenesis: heterogeneous, poly(A)-containing RNA is synthesized postmeiotically. J. Expt'l Zool. 214, 13–19.

Erickson, R.P., Kramer, J.M., Rittenhouse, J., Salkeld, A., 1980. Quantitation of mRNAs during mouse spermatogenesis: protamine-like histone and phosphoglycerate kinase-2 mRNAs increase after meiosis, Proc. Nat'l. Acad. Sci. U.S.A. 77, 6086–6090.

Erickson, R.P., Lewis, S.E., Butley, M., 1981. Is haploid gene expression possible for sperm antigens. J. Reprod. Immunol. 3, 195–217.

Erickson, R.P., Michelson, A.M., Rosenberg, M.P., Sanchez, E., Orkin, S.H., 1985. Postmeiotic transcription of phosphoglycerate kinase 2 in mouse testes. Biosci. Rep. 1091–1097.

Erickson, R.P., Harper, K.J., Hopkins, S.R., Brewer, G.J., 1987. Adenine nucleotides and other factors indicative of glycolytic metabolism in murine spermatozoa. J. Hered. 78, 407–409.

Erickson R.P. Kessler, S., Dremling, H., Sen, G.C., 1996. Species variation in the testicular angiotensin converting enzyme promoter studied in transgenic mice. Mol. Reprod. Devel. 44 324–331.

Ewolonu, U.K., Schimenti, K., Kuemerle, B., Magnuson, T., Schimenti, J., 1996. Targeted mutagenesis of a candidate *t* complex responder gene in mouse *t* haplotypes dos not eliminate transmission ratio distortion. Genetics 144, 785–792.

Fawcett, D.W., Ito, S., Slauttterack, D., 1959. J. Cell. Biol. 5, 453–460.

Forejt, J., Ivanyi, P., 1975. Genetic studies on male sterility between laboratory and wild mice. Genet. Res. 24, 189–206.

Fossella, J., Samant, S.A., Silver, L.M., King, S.M., Vaughan, K.T., Olds-Clarke, P., Johnson, K.A., Mikami, A., Vallee, R.B., Pilder, S.H., 2000. An axonemal dynein at the Hybrid Sterility 6 locus: Implications for t haplotype-specific male sterility and the evolution of species barriers. Mamm. Genome 11, 8–15.

Fraser LR Drury, L.M., 1976. Mouse sperm genotype and the rate of egg penetration in vitro. J. Exp. Zool. 197, 13–20.

Frova, C., Binelli, G., Ottaviano, E., 1987. Isozyme and HSP gene expression during male gametophyte development in maize. Curr. Top. Biol. Med. Res. 15, 92–120.

Fujimoto, H., Erickson, R.P., 1982. Functional assays for mRNA detect many new messages after male meiosis in mice. Biochem. Biophys. Res. Commun. 108, 1369–1375.

Fujimoto, H., Erickson, R.P., Quinto, M., Rosenberg, M.P., 1984. Postmeiotic transcription in mouse testes detected with spermatid cDNA clones. Biosci. Rep. 4, 1037–1044.

Fujimoto, H., Erickson, R.P., Tone, S., 1988. Changes in polyadenylation of lactate dehydrogenase-X MRNA during spermatogenesis in mice. Mol. Reprod. Dev. 1, 27–34.

Fujita, A., Nakamura, K., Kato, T., Watanabe, N., Ishizaki, I., Kimura, K., Mizoguchi, A., Narumiya, S., 2000. Ropporin, a sperm-specific binding protein of rhophilin, that is localized in the fibrous sheath of sperm flagella. J. Cell Sci. 113, 103–112.

Fuster, C.D., Farrell, D., Stern, Hecht, N.B., 1977. RNA polymerase activity in bovine spermatozoa. J. Cell Biol. 74, 698–706.

Gao, Z., Moorjani, P., Sasani, T.A., Pedersen, B.S., Quinlan, A.R., Jorde, L.B., Amster, G., Przeworski, M., 2019. Overlooked roles of DNA damage and maternal age in generating human germline mutations. Proc. Natl. Acad. Sci. U.S.A. 116, 9491–9500.

Garner, D.L., Glehill, B.L., Pinkel, D., Lake, S., Stphenson, D., Van Dilla, M.A., Johnson, L.A., 1983. Quantification of the X- and Y-chromosome bearing spermatozoa of domestic animals by flow cytometry. Biol. of Reprod. 28, 312–321.

Garrett, J.E., Collard, M.E., Douglass, J.O., 1989. Translational control of germ cell-expressed mRNA imposed by alternative splicing: opiod peptide expression in rat testis. Mol. Cell Biol. 9, 4381–4389.

Garside, W., Hillman, N., 1989. The transmission ratio distortion of the *th2*-haplotye in vivo and in vitro. Genet. Res. 53, 25–28.

Geremia, R., D'Agostino, A., Monesi, V., 1978. Biochemical evidence of haploid gene activity in spermatogenesis of the mouse. Exp. Cell. Res. 111, 23–30.

Ginsberg, L., Hillman, N., 1974. Meiotic drive in *tn*-bearing spermatozoa: a relationship between aerobic respiration and transmission frequency. J. Reprod. Fertil. 38, 157–163.

Gold, B., Hecht, N.B., 1981. Differential compartmentalization of messenger ribonucleic acid in murine testis. Biochem 20, 4871–4877.

Gold, B., Fujimoto, H., Kramer, J.M., Erickson, R.P., Hecht, N.B., 1983. Haploid accumulation and translational control of phosphoglycerate kinase 2 messenger RNA during mouse spermatogenesis. Devel. Biol. 98, 392–399.

Goldberg, E., Hawtrey, C., 1967. The ontogeny of sperm specific lactate dehydrogenase in mice. J. Exp. Zool. 164, 309–316.

Goldmann, J.M., Veltman, J.A., Glissen, C., 2019. *De Novo* Mutations Reflect Development and Aging of the Human Germline. Trends in Genet 35.

Gondos, B., Zemjanis, R., 1959. Fine structure of spermatogonia and intercellular bridges in Macaca nemestrina. J. Morphol. 131, 431–446.

Goodall, H., Roberts, A.M., 1976. Differences in motility of human X and Y-bearing spermatozoa. J. Reprod. Fertil. 48, 433–436.

Gould-Somero, M., Holland, L., 1974. The timing of RNA synthesis for spermiogenesis in organ cultures of *Drosophila melanogaster* testes. Wilhem Roux Arch. 174, 133–148.

Grant, M., Zuccotti, M., Monk, M., 1992. Methylation of CpG sites of two X–linked genes coincides with X–inactivation in the female mouse embryo but not in the germ line. Nature Genet 2, 161–166.

Gummere, G.R., McCormick, P.J., Bennett, D., 1986. The influence of genetic background and homologous chromosome 17 on t-haplotype transmission ratio distortion in mice. Genetics 114, 235–245.

Gur, Y., Breitbart, H., 2006. Mammalian sperm translate nuclear-encoded proteins by mitochondrial-type ribosomes. Genes and Devel. 20, 411–416.

Gyllensten, U., Wharton, D., Josefsson, A., Wilson, A.C., 1991. Paternal inheritance of mitochondrial DNA in mice. Nature 352, 255–257.

Hake, L.E., Alcivar, A.A., Hecht, N.B., 1990. Changes in mRNA length accompany translational regulation of the somatic and testis-specific cytochrome c genes during spermatogenesis in the mouse. Development 110, 249–257.

Hammerberg, C., Klein, J., 1975. Evidence for post-postmeiotic effect of *t* factors causing segregation distortion in mouse. Nature 253, 137–138.

Hammerberg, C., 1982. The effects of the t-complex upon male reproduction are due to complex interactions between its several regions. Genetical Research 39, 219–226.

Harrison, A., Olds-Clarke, P., King, S.J., 1998. Identification of the *t* complex-encoded cytoplasmic dynein light chain Tctex1 in inner are I1 supports the involvement of flagellar dyneins in meiotic drive. J. Cell Biol. 140, 1137–1147.

Hecht, N.B., Williams, J.L., 1978. Synthesis of RNA by Separated Heads and Tails from Bovine Spermatozoa. Biol. of Reprod. 19, 573–579.

Hecht, N.B., 1990. Regulation of "haploid expressed genes" in male germ cells. J. Reprod. Fert. 88, 679–693.

Heidaran, M.A., Showman, R.M., Kistler, W.S., 1988. A cytochemical study of the transcriptional and translational regulation of nuclear transition protein 1 (TP1), a major chromosomal protein of mammalian spermatids. J. Cell Biol. 106, 1427–1433.

Hendriksen, P.J.M., Tieman, M., van der Lende, T., Johnson, L.J., 1993. Binding of H-Y monoclonal antibodies to separated X and Y chromosome bearing porcine and bovine sperm. Molec. Repro. Devel. 35, 189–196.

Hendriksen, P.J.M., Hoogerbrugge, J.W., Baarends, W.B., de Boer, P., Vreeburg, J.T.M., Vos, E.A., der Lende, T., Grootegoed, J.A., 1997. Testis-Specific Expression of a Functional Retroposon Encoding Glucose-6-phosphate Dehydrogenase in the Mouse. Genomics 41, 350–359.

Hendriksen, P.J.M., 1999. Do X and Y spermatozoa differ in proteins. Theriogenol. 52, 1295–1307.

Henikoff, S., Malik, H.S., 2002. Centromeres: selfish drivers. Nature 417, 227.

Herrmann, B., Koschorz, B., Wertz, K., McLaughlin, K., Kispert, A., 1999. A protein kinase encoded by the t complex responder gene causes nonmendelian inheritance. Nature 402, 141–146.

Herrada, G., Wolgemuth, D.J., 1997. The mouse transcription factor Stat4 is expressed in male germ cells and is present in the perinuclear theca of spermatozoa. J. Cell Sci. 110, 1543–1553.

Hinsch, K.-D., Habermann, B., Just, I., Hinsch, E., Pfisterer, S., Schill, W.-B. Aktories, 1993. ADP-ribosylation of Rho proteins inhibits sperm motility. FEBS Lett 334, 32–36.

Hogarth, C.A., Arnold, S., Kent, T., Mitchell, D., Isoherranen, N., Griswold, M.D., 2015. Processive pulses of retinoic acid propel asynchronous and continuous sperm production. Biol of Repro 92, 1–11.

Hoppe, P.C., 1980. Genetic influences on mouse sperm capacitation in vivo and in vitro. Gamete Res. 3, 343–349.

Hotta, Y., Tabata, S., Stubbs, L., Stern, H., 1985. Meiosis-specific transcripts of a DNA component replicated during chromosome pairing: homology across the phylogenic spectrum. Cell 40, 785–793.

Howard, T.E., Balogh, R., Overbeek, P., Bernstein, K.E., 1993. Sperm-specific expression of angiotensin-converting enzyme is mediated by a 91-base pair promoter containing a CRE-like element. Mol. Cell Biol. 13, 18–27.

Howard, T.E., Shai, S.-Y., Langford, K.G., Martin, B.M., Bernstein, K.E., 1990. Transcription of testicular angiotensin-converting enzyme (ACE) is initiated within the 12th intron of the somatic ACE gene. Mol. Cell Biol. 10, 4294–4302.

Huang, L.S., Voyiaziakis, E., Chen, H.L., Rubin, E.M., Gordon, J.W., 1996. A novel functional role for apolipoprotein B in male infertility in heterozygous apolipoprotein B knockout mice. Proc. Nat'l Acad. Sci., U. S. A. 93, 10903–10907.

Huckins, C., Oakberg, E.F., 1978. Morphological and quantitative analysis of spermatogonia in mouse testes using whole mounted seminiferous tubules. I. The normal testes. Anat. Rec. doi:10.1002/ar.1091920406.

Hurst, L.D., Pomiankowski, A., 1991. Causes of Sex Ratio Bias May Account for Unisexual Sterility in Hybrids: A New Explanation of Haldane's Rule and Related Phenomena. Genetics 128, 841–858.

Hurst, L.D., 1993. A model for the mechanism of transmission ratio distortion and for *t*-associated hybrid sterility. Proc. R. Soc. Lond. (Biol) 253, 83–91.

Huw, L.Y., Goldsborough, A.S., Willison, K., Artzt, K., 1995. Tctex2: a sperm tail surface protein mapping to the *t*-complex. Dev. Biol. 170, 183–194.

Illisson, L., 1969. Spermatozoal head shape in two inbred strains of mice and their F_1 and F_2 progenies. Aust. J. biol. Sci. 22, 947–963.

Jen, J., Deschepper, C.F., Shackleford, G.M., Lee, C.Y.G., Lau, Y.-F.C., 1990. Stage-specific expression of the lactate dehydrogenases-X in the adult and developing mouse testis. Mol. Reprod. Devel. 25, 14–21.

Johnson, L.A., Flook, J.P., Look, M.V., Pinkel, D., 1987. Flow sorting of X and Y chromosome-bearing spermatozoa into two populations. Gamete Res. 16, 1–9.

Johnson, J., Canning, J., Kaneko, T., Pru, J.K., Tilly, J.L., 2004. Germline stem cells and follicular renewal in the postnatal mammalian ovary. Nature 428, 145–150.

Johnson, L.A., Flook, J.P., Hawk, H.W., 1989. Sex Selection in Rabbits: Live Births from X and Y Sperm Separated by DNA and Cell Sorting. Biol. of Reprod. 41, 199–203.

Jones, M.H., Davey, P.M., Aplin, H., Affara, N.A., 1995. Expression analysis, genomic structure, and mapping to 7q31 of the huan sperm adhesion molecule gene SPAM1. Genomics 29, 796–800.

Kallikin, L.M., Fujimoto, H., Witt, M.P., Verga, V., Erickson, R.P., 1989. A genomic clone of *Zfy-1* from a Y_{DOM} mouse strain detects postmeiotic gene expression of Zfy in testes. Biochem. Biophys. Res. Comm. 165, 1286–1291.

Kaneko, S., Oshio, S., Kobayashi, T., Mohri, H., Iizuka, R., 1984a. Selective isolation human X-bearing sperm by differential velocity sedimentation in Percoll density gradients. Biomed. Res. 5, 187–194.

Kaneko, S., Oshio, S., Kobayashi, T., Iizuka, R., Mohri, H., 1984b. Human X- and Y-bearing sperm differ in cell surface sialic acid content. Biochem. Biophys. Res. Commun. 124, 950–955.

Katz, D., Erickson, R.P., Nathanson, M., 1979. Beat frequency is bimodally distributed in spermatozoa from *T/t*12 mice. J. Exp. Zool. 210, 529–535.

Keeler, K.D., Mackenzie, N.M., Dresser, D.W., 1983. Flow microfluorometric analysis of living spermatozoa stained with Hoechst 33342. J. Reprod. Fertil. 68, 205–212.

Kessler, S.P., Rowe, T.M., Blendy, J.A., Erickson, R.P., Sen, G.C., 1998. A cyclic AMP response element in the angiotensin-converting enzyme gene and the transcription CREM are required for transcription of the mRNA for the testicular isozyme. J Biol. Chem. 273, 9971–9975.

Khil, P.P., Smirnova, N.A., Romanienko, P.J., Camerini_Otero, R.D., 2004. The mouse X chromosome is enriched for sex-biased genes not subject to selection by meiotic sex chromosome inactivation. Nature Genet. 36, 642–646.

Kierszenbaum, A.L., Tres, L.L., 1975. Structural and transcriptional features of the mouse spermatid genome. J. Cell. Biol. 65, 258–270.

Kilpatrick, D.L., Millette, C.F., 1986. Expression of proenkephalin messenger RNA by mouse spermatogenic cells. Proc. Nat'l Acad. Sci, U.S.A. 83, 5015–5018.

Kilpatrick, D., Zinn, S.A., Fitzgerald, M., Higuiei, H., Sabol, S.L., Meyerhardt, J., 1990. Transcription of the rat and mouse proenkephalin genes is initiated at distinct sites in spermatogenic and somatic cells. Mol. Cell Biol. 10, 3717–3726.

Kispert, A., Stoger, R.J., Caparros, M., Herrmann, B.G., 1999. The mouse *Rsk3* gene maps to the Leh66 elements carrying the *t*-complex responder *Tcr*. Mammalian Genome 10, 784–802.

Kleene, K.C., Distel, R.J., Hecht, N.B., 1983. cDNA clones encoding cytoplasmic poly (A)+ RNAs which first appear at detectable levels in haploid phases of spermatogenesis in the mouse. Dev. Biol. 98, 453–464.

Kleene, K.C., Distel, R.J., Hecht, N.B., 1984. Translational regulation and deadenylation of a protamine mRNA during spermiogenesis in the mouse. Devel. Biol. 105, 71–79.

Kofman-Alfaro, S., Chandley, A.C., 1970. Meiosis in the male mouse: An autoradiographic investigation. Chromosoma 31, 404–420.

Kramer, J.M., 1981. Immunoflourescent localization of PGK-1 and PGK-2 isozymes within specific cells of the testis. Dev. Biol. 87, 30–36.

Kramer, J.M., Erickson, R.P., 1981. Developmental programs of PGK-1 and PGK-2 isozymes in spermatogenic cells of the mouse: Specific activities and rates of synthesis. Dev. Biol. 87, 37–45.

Kramer, J.M., Erickson, R.P., 1982. Electrophoretic pattern of polypeptide synthesis in spermatocytes and spermatids of the mouse. J. Reprod. Fertil. 64, 139–144.

Krzanowska, H., 1960. Studies in heterosis. II. Fertilization rate in inbred lines and their crosses. Folia. Biol. Krako 8, 267–269.

Krzanowska, H., 1962. Sperm quantity and quality in inbred lines of mice and their crosses. Acta. Biol. Crac. V, 279–291.

Krzanowska, H., 1966. Inheritance of reduced male fertility, connected with abnormal spermatozoa, in mice. Acta Biol Crac IX, 61–70.

Krzanowska, H., 1969. Factor responsible for spermatozoan abnormality located on the Y chromosome in mice. Genet. Res. 13, 17–24.

Krzanowska, H., 1972. Rapidity of removal *in vitro* of the cumulus oophorus and the zona pellucida in different strains of mice. J. Reprod. Bertil. 31, 7–14.

Krzanowski, M., 1970. Dependence of primary and secondary sex ratio on the rapidity of sedimentation of bull semen. J. Reprod. Fert. 23, 11–20.

Kumar, R.S., Kusari, J., Roy, N., Soffer, R.L., Sen, G.C., 1989. Structure of testicular converting enzyme. J. Biol. Chem. 264, 16752–16758.

Kuretake, S., Maleszewski, M., Tokumasu, A., Fujimoto, H., Yanagimachi, R., 1996. Inadequate function of sterile *tw5/tw5* spermatozoa overcome by intracytoplasmic sperm injection. Molec. Reprod. Develop. 44, 230–233.

Kwon, K.Y., Hecht, N.B., 1991. Cytoplasmic protein binding to highlyconserved sequences in the 38 untranslated region of mouse prot-amine 2 mRNA, a translationally regulated gene of male germ cells. Proc. Nat'l Acad. Sci., USA 88, 3584–3588.

Lader, E., Ha, H.S., O'Neill, M., Artzt, K., Bennett, D., 1989. *tctex-1*: a candidate gene family for a mouse *t* complex sterility locus. Cell 58, 969–979.

Leblond, C.P., Clermont, Y., 1952. Spermiogenesis of rat, mouse, hamster and guinea pig as revealed by the "periodic acid-fuchsin sulfurous acid" technique. Amer. J. Anat. 90, 167–215.

Lifschytz, E., 1972. X-chromosome inactivation: An essential feature of normal spermatogeneis in male heterogametic organisms. In: Beatty, R.A., Glueck20hn-Waelsch, S. (Eds.), Proc. Internat'l Symp. On the genetics of the spermatozoon. Edinburgh, 223–232.

Loir, M., 1972. Metabolisme de L'acide nucleic et des proteines dans les spermatocytes et les spermatides du Belier (Ovis araies). Ann. Biol. Anim. Biochim. 12, 211–219.

Lu, Y., Plaatts, A.D., Ostermeier, G.C., Krawetz, S.A., 2006. K-SPMM; a database of murine spermatogenic promoters, modules, and motifs. BMC Bioimormatics 7, 238.

Lyon, M.F., 1984. Transmission ratio distortion in mouse *t*-haplotypes is due to multiple distorter genes acting on a responder locus. Cell 37, 621–628.

Lyon, M.F., 1986. Male sterility of the mouse *t*-complex is due to homozygosity of the distorter genes. Cell 44, 357–363.

Lyon, M.F., 1991. The genetic basis of transmission ratio distortion and male sterility due to the t complex. Am Nat 137, 349–358.

Lyon, M.F., 1992. Deletion of mouse *t*-complex distorter-1 produces an effect like that of the *t*-form of the distortor. Genet. Res. 59, 27–33.

Lyon, M.F., Glennister, P.H., Hawker, S.G., 1972. Do the H-2 and T-loci of the mouse have a function in the haploid phase of sperm? Nature 240, 152–153.

Lyon, M.F., Zenthon, J., 1987. Differences in or near the responder region of complete and partial *t*-haplotypes. Genet. Res. 50, 29–34.

McCarrey, J.R., Thomas, K., 1987. Human testis-specific PGK gene lacks introns and possesses characterisstics of a processed gene. Nature 326, 501–505.

McCoshen, J.A., McCallion, D.J., 1975. A study of the primordial germ cells during their migratory phase in steel mutant mice. Experientia 31, 589–590.

McGrath, J., Hillman, N., 1980. The invitro transmission frequency of the *t12* mutation in the mouse. J. Embyol. Exp. Morphol. 60, 141–151.

McGrath, J., Hillman, N., 1981. The invitro transmission frequency of the *t6* allele. Nature 243, 479–481.

McKee, B.D., Handel, M.A., 1993. Sex chromosomes, recombination, and chromatin conformation. Chromosoma 102, 71–80.

Mangia, F., Epstein, C.J., 1975. Biochemical studies of growing mouse oocytes: Preparation of oocytes and analysis of glucose-6-phosphate dehydrogenase and lactate dehydrogenase activities. Devel. Biol. 45, 211–220.

Mangia, F., Erickson, R.P., Epstein, C.J., 1976. Synthesis of LDH-1 during mammalian oogenesis and early development. Dev. Biol. 54, 146–150.

Marahrens, Y., Panning, B., Dausman, J., Strauss, W., Jaenisch, R., 1997. Xist-deficient mice are defective in dosage compensation but not in spermatogenesis. Genes Dev. 11, 156–166.

Mathur, R.S., Beatty, R.A., 1974. Enzyme activity of spermatozoa from the ductus deferens of inbred mice. J. Reprod. Fertil. 39, 23–27.

Maudlin, I., Fraser, L.R., 1978. The effect of sperm and egg genotypes on the incidence of chromosomal anomalies in mouse embryos fertilized *in vitro*. J. Reprod. Fert. 52, 107–112.

Mazarakis, N.D., Nelki, D., Lyon, M.F., Ruddy, S., Evans, E.P., Freemont, P., Dudley, K., 1991. Isolation and characterization of a testis-expressed developmentally regulated: gene from the distal inversion of the mouse *t*-complex. Development 111, 561–571.

Mbikay, M., Tadros, H., Ishida, N., Lerner, C.P., De Lamirande, E., et al., 1997. Impaired fertility in mice deficient for the testicular germ-cell protease PC4. Proc. Nat'l Acad. Sci., U. S. A. 94, 6842–6846.

Meistrich, M.L., Bruce, W.R., Clermont, Y., 1973. Cellular composition of fractions of mouse testis cells following velocity sedimentation separation. Exp. Cell Res. 79, 213–227.

Meistrich, M.L., Reid, B.O., Barcellona, W.J., 1975. Meiotic DNA synthesis during mouse spermatogenesis. J. Cell Biol. 64 211-222.

Meistrich, M.L., Trostle, P.K., Fraport, M., Erickson, R.P., 1977. Biosynthesis and localization of lactate dehydrogenase X in pachytene spermatocytes, and spermatids of mouse testes. Dev. Biol. 60, 428–441.

Miller, D., Tang, P.Z., Skinner, C., Lilford, R., 1994. Differential RNA fingerprinting as a tool in the analysis of spermatozoal gene expression. Hum. Reprod, 84–869.

Miller, D., Briggs, D., Snowden, H., Hamlington, J., Rollinson, S., Lilford, R., Krawitz, S.A., 1999. A complex population of RNAs exists in human ejaculate spermatozoa: implications for understanding molecular aspects of spermiogenesis. Gene 237, 385–392.

Monesi, V., 1962. Relation between X-ray sensitivity and stages of the cell cycle in spermatogonia of the mouse. Rad. Res 17, 809–838.

Monesi, V., 1967. Ribonucleic acid and protein synthesis during differentiation of male germ cells in the mouse. Arch. d'Anat. Microscop. et de Morphol. Exp. 56, 61–74.

Monesi, V., 1971. Chromosome activities during meiosis and spermiogenesis. J. Reprod. Fertil., Suppl. 13, 1–14.

Moore, G.P.M., 1971. DNA-dependent RNA synthesis in fixed cells during spermatogenesis in mouse. Expt'l Cell Res. 68, 462–465.

Moretti, C., Vaiman, D., Tores, F., Cocquet, J., 2016. Expression and epigenomic landscape of the sex chromosomes in mouse post-postmeiotic male germ cells. Epigenetics and Chromatin 9. doi:10.1186/s13072-016-0099-8.

Moss, S.B., Challoner, P.B., Groudine, B., 1989. Expression of a novel histone 2B during mouse spermatogenesis. Dev. Biol. 133, 83–92.

Muller, H.J., Settles, E., 1927. The non-functioning of the genes in spermatozoa. Z. Induk. Abstain. Vererb. 43, 285–312.

Nadijcka, M., Hillman, N., 1980. Differences in cAMP levels in tn and non-tn-bearing mouse spermatozoa. Biol. Reprod. 22, 1102–1105.

Nakamura, K., Fujita, A., Murata, T., Watanabe, G., Mori, C., Fujita, J., Watanabe, N., Ishizaki, T., Yoshida, O., Narumiya, S., 1999. Rhophilin, a small GTPase Rho-binding protein, is abundantly expressed in the mouse testis and localized in the principal piece of the sperm tail. FEBS Lett. 445, 9–13.

Nayernia, K., von Mering, M.H.P., Kraszucka, K., Burfeind, P., Wehrend, A., Köhler, M., Schmid, M., Engel, W., 1999. A Novel Testicular Haploid Expressed Gene (THEG) Involved in Mouse Spermatid-Sertoli Cell Interaction. Biol. Reprod. 60, 1488–1496.

Niwa, H., Miiyiazaki, J., Smith, A.G., 2009. Quantitative expression of Oct-3/4 defines differentiation, dedifferentiation or self-renewal of ES cells. Nature Genet. 24, 372–376.

Norris, R.P., Ratzan, W.J., Freudzon, M., Mehlmann, L.M., Krall, J., Movsesian, M.A., Wang, H., Ke, H., Nikolaev, V.O., Jaffe, L.A., 2009. Cyclic GMP from the surrounding somatic cells regulates cyclic AMP and meiosis in the mouse oocyte. Development 136, 1869–1878.

Oakberg, E.F., 1956. A description off spermiogenesis in the mouse and its use in analysis of the cycle of the seminiferous epithelium and germ cell renewal. Am. J. Anat. 99, 391–413.

Oakberg, E.F., 1971. A new concept of spermatogonial stem-cell renewal in the mouse and its relationship to genetic effects. Mutation Res. 11, 1–7.

O'Brien, D.A., Bellve, A.R., 1980. Protein constituents of the mouse spermatozoon: II. Temporal synthesis during spermatogenesis. Devel. Biol. 75, 4005–418.

Ohno, S., Smith, J.-B., 1964. Role of fetal follicular cells in meiosis of mammalian oocytes. Cytogenetics 3, 324–333.

Olds-Clarke, P., Carey, J., 1978. Rate of egg penetration in vitro accelerated by T/t locus in the mouse. J. Exp. Zool. 206, 323–332.

Olds-Clark, P., Peltz, B., 1985. Fertility of sperm from $tf+$ mice: evidence that +-bearing sprm are dysfunctional. Genet. Res. 47, 49–52.

Oldes-Clarke, P., Johnson, L.R., 1993. t haplotypes in the mouse compromise sperm flagellar function. Devel. Biol. 155, 14–25.

Orlando, P., Grippo, P., Geremia, R., 1984. DNA repair synthesis-related enzymes during spermatogenesis in the mouse exp. Cell Res. 154, 499–505.

Otto, F.J., Hacker, U., Zante, J., Schumann, J., Gohde, W., Meistrich, M.L., 1979. Flow Cytometry of Human Spermatozoa. Histochem. 61, 249–254.

Pace, A.J., Lee, E., Athirakul, K., Coffman, T.M., Deborah, A., O'Brien, D.A., Koller, B.H., 2000. Failure of spermatogenesis in mouse lines deficient in the Na⁺-K⁺-2Cl⁻ cotransporter. J. Clin. Invest. 105, 441–450.

Patel-King, R.S., Benashski, S.E., Harrison, A., King, S.M.A., 1997. Clamydomonas homologue of the putative murine *t* complex distorter Tctex-2 is an outer arm dynein light chain. J. Cell Biol. 137, 1081–1090.

Penfold, L.M., Holt, C., Holt, W.V., Welsch, G.R., Cran, D.G., Johnson, L.A., 1998. Comparitive motility of X and Y chromosome-bearing bovine sperm separated on the basis of DNA content by flow sorting. Mol. Reprod. Develop. 50, 323–327.

Peschon, J.J., Behringer, R.R., Brinster, R.L., Palmiter, R.D., 1987. Spermatid-specific expression of protamine 1 in transgenic mice. Proc. Nat' Acad. Sci. U.S.A. 84, 5316–5319.

Pessot, C.A., Brito, M., Figueroa, J., Concha, I.I., Yanez, A., Burzio, L.O., 1989. Presnece of RNA in the sperm nucleus. Biochem. Biophys. Res. Comm. 272–278. 158.

Ponzetto-Zimmerman, Wolgemuth, D.J., 1984. Methylation of satellite sequences in mouse spermatogonia and somatic DNAs. Nucleic Acids Res. 12, 2807–2822.

Ponnzetto, C., Wolgemuth, D.J., 1985. Haploid expression of a unique c-abl transcript in the mouse male germ line. Mol. Cell Biol. 5, 1791–1794.

Probst, F., Rosenberg, M.P., Oskasson, M.K., Russsell, L.B., Nguyen-Huu, M.C., Nadeau, J., Jenkins, N.A., Copeland, N.G., Vande Woude, G.F., 1988. Genetic analysis and developmental regulation of testis-specific RNA expression of *Mos, Abl,* actin and *Hox-1-4*. Oncogene 2, 227–233.

Quinlan, W.L.G., Preciado, K., Long, T.L., Sullivan, H., 1982. Separation of human X and Y spermatozoa by albumin gradients and sephadex chromatography. Fertil. & Steril. 37, 104–107.

Rahe, B., Erickson, R.P., Quinto, M., 1983. Methylation of unique sequence DNA during spermatogenesis in mice. Nucleic Acids Res 11, 7947–7959.

Rappold, G.A., Stubbs, L., Labeit, S., Crkvenjakov, R.B., Lehrach, H., 1987. Identification of a testis-specific gene from the mouse *t*-complex next to a CpG island. EMBO J. 6, 1975–1980.

Rassoulzadegan, M., Grandjean, V., Gounon, P., Vincent, S., Gillot, I., Cuzin, F., 2006. RNA-mediated non-mendelian inheritance of an epigenetic change in the mouse. Nature 441, 469–474.

Rathje, C.C., Johnson, E.E.P., Drage, D., Patinioti, C., Silvestri, G., Affara, N.A., Ialy-Radio, C., Cocquet, J., Skinner, B.M., Ellis, P.J.I., 2019. Differential sperm motility mediates the sex ratio drive shaping mouse sex chromosome evolution. Curr. Bio. doi:10.2139/ssrn.3396496.

Rens, W., Welsch, G.R., Johnson, L.A., 1999. Improved flow cytometric sorting of X- and Y- chromosome bearing sperm: substantial increase in yield of sexed semen. Mol. Reprod. Develop. 52, 50–56.

Roblinson, M.O., Simon, M.I., 1991. Determining transcript number using the polymerase chain reaction: Pgk-2 , mP2 , and PGK-2 transgene mRNA levels during spermatogenesis. Nucleic Acids Res. 19, 1557–1562.

Roblinson, M.O., McCarry, J.R., Simon, M.I., 1989. Transcriptional regulatory regions of testis-specific PGK2 defined in transgenic mice. Proc. Nat'l Acac. Sci. 86, 8437–8444.

Rohde, W., Porstmann, T., Prehn, S., Dorner, G., 1975. Gravitational pattern of the Y-bearing human spermatozoa in density gradient centrifugation. J. Reprod. Fertil. 42, 587–591.

Rohme, D., Fox, H., Hermann, B., Frischauf, A.-M., Edstrom, J.E., Mains, P., Silver, L.M., Lehrach, H., 1984. Molecular clones of the mouse *t* complex derived from microdissected metaphase chromosomes. Cell 36, 783–788.

Ross, A., Robinson, J.A., Evans, H.J., 1975. Failure to confirm separation of X- and Y-bearing human sperm using BSA gradients. Nature 253, 354–355.

Ross, A.J., Waymire, K.G., Moss, J.E., Parlow, A.F., Skinner, M.K., Russell, L.D., MacGregor, G.R., 1998. Testicular degeneration in Bclw-deficient mice. Nature Genet. 18, 251–256.

Rosen, L.L., Bullard, D.C., Silver, L.M., Schimenti, J.C., 1990. Molecular cloning of the *t complex responder* genetic locus. Genomics 8, 134–140.

Royo, H., Pollikiewica, G., Mahadevaiah, K., Prosser, H., Mitchell, M., Bradley, A., de Rooij, D.G., Burgoyne, P.S., Turner, J.M.A., 2010. Evidence that meiotic sex chromosome inactivation is essential for male fertility. Current Biol. 20, 2017–2023.

Rudolph, N.S., Shovers, J.B., VandeBerg, J.L., 1982. Spermatogenesis and sperm-specific hosphoglyceratekinaseB and lactatedehydrogenaseC4 isoenzymes in sex-reversedmice. J. Reprol. Fertil. 65, 39–44.

Russell, L.B., Montgomery, C.S., 1969. Comparative studies on X-autosome translocations in the mouse I. Origin, viability, fertility, and weight of five T(X;1) 'S'. Geneics 63, 103–120.

Sachs, A.B., Davis, R.W., 1989. The poly (A) binding protein is required for poly (A) shortening and 60S ribosomal subunit-dependent translation initiation. Cell 58, 857–867.

Sanchez, E.R., Erickson, R.P., 1985. Expression of the *Tcp-1* locus of the mouse during early embryogenesis. J. Embryol. Expt'l Morphol. 89, 113–122.

Sanchez, F., Smitz, J., 2012. Molecular control of oogenesis. Biochem. Biophys. Acta 1822, 1896–2012.

Sanchez, E.R., Erickson, R.P., 1986. Wild-derived Robertsonian translocation in mice: Chromosome 17, Rb (16:17)7, shows novel interactions with *t*-alleles. J. Hered. 77, 290–294.

Sanchez, E.R., Hammerberg, C., Erickson, R.P., 1985. Quantitation of two-dimensional gel proteins reveals unequal amount of *Tcp-1* gene products during mouse spermatogenesis but no correlation with transmission ratio distortion J. Embryol. Exp. Morph. 89, 121–131.

Sarkar, S., Jolly, D.J., Friedmann, T., Jones, O.W., 1984. Swimming behavior of X and Y human sperm. Differentiation 27, 120–125.

Saravetnick, N., Tsai, J.-Y., Fox, H., Pilder, S.H., Silver, L.M., 1989. A mouse chromosome 17 gene encodes a testes-specific transcript with unusual properties. Immunogenet 30, 34–41.

Schimenti, J.C., 1999. ORFless, intronless, and mutant transcription units in the moue *t* complex responder (*Tcr*) locus. Mammalian Genome 10, 969–976.

Scholer, H.R., Dressler, G.R., Balling, R., Rohdewohld, H., Gruss, P., 1990. *oct-4* : a germline-specific transcription factor mapping to the mouse *t*-complex. EMBO J. 9, 2185–2195.

Schroeder, A.C., Eppig, J.J., 1984. The developmental capacity of mouse oocytes that matured spontaneously *in vitro* is normal. Devel. Biol. 102, 493–497.

Schwartz, M., Vissing, J., 2002. Paternal inheritance of mitochondria. New Engl. J. of Med. 347, 576–580.

Seitz, A.W., Bennett, D., 1985. Transmission distortion of t-haplotypes is due to interactions between meiotic partners. Nature 313, 143–144.

Sharma, U., Conind, C.C., Shea, J.M., Boskovic, A., Derr, A.G. 35 more, 2016. Biogenesis and function of tRNA fragments during sperm maturation and fertilization in mammals. Science 351, 391–396.

Shastry, P.R., Hedge, U., Rao, S.S., 1977. Use of Ficoll-sodium metrizoate density gradient to separate human X- and Y-bearing spermatozoa. Nature 269, 58–60.

Silver, L.M., 1989. Gene dosage effects on transmission ratio distortion and fertility in mice that carry *t* haplotypes. Genetical Res 54, 221–225.

Silver, L.M., Uman, J., Danska, T., Garrels, J.I., 1983. A diversified set of testicular cell proteins specified by genes within the mouse *t* complex. Cell 35, 35–45.

Silver, L.M., Olds-Clarke, P., 1984. Transmission ratio distortion of mouse *t* haplotypes is not a consequence of wild-type sperm degeneration. Dev. Biol. 105, 205–212.

Silver, L.M., Kleene, K.C., Distel, R.J., Hecht, N.B., 1987. Synthesis of mouse *t* complex proteins durin haploid stages of spermatogenesis. Dev. Biol. 119, 605–608.

Silver, L.M., Remis, D., 1987. Five of the nine genetically defined regions of mouse*t*-haplotypes are involved in transmission ratio distortion. Genet. Res. 49, 51–56.

Slaughter, G.R., Meistrich, M.L., Means, A.R., 1989. Expression of RNAs for calmodulin, actins, and tubulins in mouse testis cells. Biol. Reprod. 49, 395–405.

Smyth, D.R., Stern, H., 1973. Repeated DNA synthesis during pachytene in *Lilium henyryi*. Nature New Biol 245, 94–95.

Snyder, L.C., Silver, L.M., 1992. Distortion of transmission ratio by a candidate *t* complex responder locus transgene. Mammal. Gen. 3, 588–596.

Soderstom K.O., Parvinen M., 1976. Incorporation of (3H)uridine by the chromatoid body during rat spermatogenesis. J. Cell Biol. 70, 239-246.

Sorenson, P.A., Wassarman, P.M., 1976. Relationship between growth and meiotic maturation of the mouse oocyte. Devel. Biol. 50, 531–536.

Speed, R.M., 1986. Abnormal RNA synthesis in sex vesicles of tertiary trisomic male mice. Chromosoma 93, 267–270.

Spiegelman, M., Bennett, D., 1973. A light- and electron-microscopic study of primordial germ cells in the early mouse embryo. Development 30, 97–118.

Stambaugh, R., Buckley, J., 1971. Association of the lactic dehydrogenase X, isozyme with male-producing rabbit spermatozoa. J. Reprod. Fert. 25, 275–278.

Stewart, T.A., Hecht, N.B., Hollingshead, P.G., Johnson, P.A., Leong, J.C., Pitts, S.L., 1988. Haploid-specific transcription of protamine-*myc* and protamine-T-antigen fusion genes in transgenic mice. Mol. Cell. Biol. 8, 1748–1755.

Styrna, J., Klag, J., Moriwaki, K., 1991. Influence of partial deletion of the Y chromosome on mouse sperm phenotype. J. Reprod. Fertil. 92, 187–195.

Sutovsky, P., Morena, R.D., Ramalho-Santos, J., Dominko, T., Simerly, C., Schatten, G., 1999. Ubiquitin tag for sperm mitochondria. Nature 402, 371–372.

Takahashi, M., Imaguma, Y., Hiai, H., Hirose, G., 1988. Developmentally regulated expression of a human "finger"-containing gene encoded by the 5' half of the *ret* transforming gene. Mol. Cell Biol. 8, 1853–1856.

Tanaka, S., Fujimoto, H., 1986. A postmeiotically expressed clone encodes lactate dehydrogenase isozyme X. Biochem. Biophys. Res. Commun. 136, 760–766.

Thomas, K.H., Wilke, T.M., Tomashevsky, P., Bellve, A.R., Simons, M.I., 1989. Diferentail gene expression during Mouse Spermatogenesis. Biol. Reprod. 41, 729–739.

Trasler, J.M., Hake, L.E., Johnson, P.A., Alcivar, A.A., Millette, C.F., Hecht, N.B., 1990. DNA methylation and de-methylation events during meiotic prophase in the mouse testis. Mol. Cell Biol. 10, 1828–1834.

Trent, J.D., Nimmesern, E., Wall, J.S., Hartl, F.-U., Horwich, A.L., 1991. A molecular chaperone from a thermophilic archaebacterium is related to the eukaryotic protein t-complex polypeptide-1. Nature 354, 490.

Tsunekawa, N., Matsumoto, M., Tone, S., Nishida, T., Fujimoto, 1999. The Hsp70 homolog gene, Hsc70t, is expressed under translational control during mouse spermiogenesis. Molec. Reprod. Develop. 52, 383–391.

Tucker, M.J., 1980. Explanation of sterility in *tx/ty* male mice. Nature 288, 367–368.

Turner, J.M., Mahadevaiah, S.K., Ellis, P.J., Mitchell, M.J., Burgoyne, P.S., 2006. Pachytene asynapsis drives meiotic sex chromosome inactivation and leads to substantial post-postmeiotic repression in spermatids. Dev Cell 10, 521–529.

Umehara, T., Tsujita, N., Shimada, M., 2019. Activation of Toll-like receptor 7/8 encoded by the X chromosome alters sperm motility and provides a novel simple technology for sexing sperm. PLoS Biol. doi:10.1371/journal.pbio.3000398.

Van de Berg, J.L., Cooper, D.W., Clossse, P.J., 1976. Testis specific phosphoglycerate kinase B in mouse. J. Exp. Zool. 198, 231–240.

Véron, N., Bauer, H., Weise, A.Y., Lu 'der, G., Werber, M., Herrmann, B.G., 2009. Retention of gene products in syncytial spermatids promotes non-Mendelian inheritance as revealed by the t complex responder. Genes and Develop. 23, 2705–2710.

Willison, K., Lewis, V., Zuckerman, K.S., Cordell, J., Dean, C., Miller, K., Lyon, M.F., Marsh, M., 1989. The *t* complex polypeptide (TCP-1) is associated with the cytoplasmic aspect of Golgi membranes. Cell 57, 621–632.

Willison, K.R., Hynes, G.G., Davies, P.A., Lewis, V.A., 1990. Expression of three *t*-complex genes, *Tcp-1*. D17Le-h117c3, and Leh66, in purified spermatogenic cell populations. Genet. Res. 56, 193–201.

Wolfe, H.G., Erickson, R.P., Schmidt, L.C., 1977. Effects on sperm morphology by alleles at the pink-eyed dilution locus in mice. Genetics 85, 303–308.

Wolgemuth, D.J., Viviano, C.M., Gizang-Ginsberg, E., Frohman, M.A., Joyner, A.L., Martin, F.R., 1987. Differential expression of the mouse homeobox-containing gene *Hox-1-4* during male germ cell differentiation and embryonic development. Proc Nat'l Acad. U.S.A. 84, 5813–5817.

Wolgemuth, D.J., Watrin, F., 1991. List of cloned mouse genes with unique expression patterns during spermatogenesis. Mammal. Genome 1, 283–288.

Wooley, D.M., 1970. Selection for the length of the spermatozoon midpiece in the mouse. Genet. Res. 16, 261–275.

Wykes, S.M., Visscher, D.W., Krawetz, S.A., 1997. Haploid transcripts persist in mature human spermatozo. Mol. Hum. Reprod. 3, 15–19.

Yaffe, M.B., Farr, G.W., Miklos, D., Horwich, A.L., Sternlicht, M.L., Sternlicht, H., 1992. TCP-1 complex is a molecular chaperone in tubulin biogenesis. Nature 358, 245–252.

Yanagisawa, K., Dunn, L.C., Bennett, D., 1961. On the mechanism of abnormal transmission ratios at *T* locus in the house mouse. Genetics 46, 1635–1644.

Yanagisawa, K., 1965. Studies on the mechanism of abnormal transmission ratios at the *T*-locus in the house mouse: test for physiological differences between *t*- and *T*-bearing sperm manifested *in vitro*. Japan. J. Genet. 40, 87–92.

Yoshikawa, K., Aizawa, T., 1988. Enkephalin precursor: gene expression in post-postmeiotic germ cells. Biochem. Biophys. Res. Commun. 151, 664–671.

Yu, Y.E., Zhang, Y., Unni, E., Shirley, C.R., Deng, J.M., Russell, L.D., Weil, M.M., Behringer, R.R., Meistrich, M.L., 2000. Abnormal spermatogenesis and reduced fertility in transition nuclear protein 1-deficient mice. Proc. Naat'l Acad. Sci., U. S. A. 97, 4683–4688.

Zhao, G.Q., Deng, K., Labosky, P.A., Liaw, L., Hogan, B.L., 1996. The gene encoding bone morphogenetic protein 8B is required for the initiation and maintenance of spermatogenesis in the mouse. Genes and Devel. 10, 1657–1669.

Zheng, Y., Martin-DeLeon, P.A., 1997a. The murine *Spam1* gene:RNA expression pattern and lower steady-state levels associated with the Rb(6:16) translocation. Mol. Reprod. Develop. 46, 252–257.

Zheng, Y., Martin-DeLeon, P.A., 1997b. Characterization of the genomic structure of the murine *Spam1* gene and its promoter: evidence for transcriptional regulation by a cAMP-responsive element. Mol. Reprod. Devel. 54, 8–16.

Zheng, Y., Deng, X., Martin-DeLeon, P.A., 2001. Lack of Sharing of Spam 1 (Ph-20) among mouse spermatids and transmission ratio distortion. Biol of Reprod. 64, 1730–1738.

Zhou, D., Suzuki, T., Asami, M., Perry, A.C.F., 2019. Caput epididymal mouse sperm support full development. Develop. Cell 50, 5–6.

Zollner, S., Wen, X., Hanchard, N.A., Herbert, M.A., Ober, C., Pritchard, J.K., 2004. Evidence for Extensive Transmission Distortion in the Human Genome. Am. J. Hum. Genet. 74, 62–72.

9

The cloning era and the cloning of *Brachury* and other *T/t* complex genes

Introduction

Much has been written about the history of molecular biology and recombinant DNA. Portugal and Cohen's **"A century of DNA"** is particularly complete about Miescher's discovery of DNA. They provide the details of Miescher's problems in purifying it from pus and finding a unique phosphorus to nitrogen ratio (compared to proteins which were the focus of most biochemical studies on macromolecules at the time [1869]). The need to work in the cold, and the inability to replicate his findings by many investigators, was almost certainly due to copurified RNA which would be rapidly degraded at room temperature by ubiquitous RNAses. They also provide an overview of Todd's work on the chemical structure of the components of DNA (deoxyribonucleotides) for which he won the Nobel prize and the early work on X-ray diffraction studies of DNA.

The prejudice that only proteins could have the specificity required to be the genetic material long hampered progress. Thus, the clear demonstration that the pneumococcal transforming factor principal, prepared from 1 serotype and used to treat and change the serotype of bacteria with a different serotype, consisted of DNA (Avery et al., 1944) was not generally accepted. The details of the continuing refinements of the proof that the hereditary material was indeed DNA by a number of chemical and biological experiments, involving a team which included the protein chemist Rollin Hotchkiss[1], at the Rockefeller Institute, are well detailed in Rene Dubos' book, **"The Professor, The Institute and DNA."** The conclusion that DNA was the hereditary material was only generally accepted after the Watson-Crick model of the structure of DNA provided a mechanism for its replication (Watson and Crick, 1953); their final sentence ("It has not escaped our notice that the specific pairing we have postulated immediately suggests a possible copying mechanism for the genetic material") being, perhaps, one of the greatest understatements in science. The story of this discovery, with its frank acknowledgement of stealing Rosalind Franklin's X-ray diffraction data, is detailed in James Watson's **"The Double Helix"** (already mentioned in The Haldane-Mitchison Clan for its dedication to Naomi Mitchison). Franklin's contribution is detailed in her biography by Anne Sayre, **"Rosalind Franklin & DNA."** She should have shared the Nobel prize with Watson and Crick (her boss, Wilkins, got a share of the prize for the essential contributions of X-ray diffraction data to determining the structure), but her untimely death prevented even the consideration (in those days of misogyny). The protein chemist Christian B Anfinsen's (he was to soon receive the Nobel prize for his work on

Twentieth Century Mouse Genetics. DOI: https://doi.org/10.1016/B978-0-12-824016-8.00003-9

sequencing proteins) book, **"The Molecular Basis of Evolution,"** was seminal in convincing his colleagues that proteins were not the hereditary material[2]. (As pointed out in The Haldane-Mitchison Clan, JBS Haldane thought it was the DNA-associated, arginine-containing proteins that were the hereditary material as late as 1954.)

A full history of the development of molecular biology is presented in Judson's magisterial book, **"The Eighth Day of Creation."** A darker picture, emphasizing the scientist's need to control the use, and potential misuse, of these discoveries, including suppressed experiments, is presented by John Lear in his **"Recombinant DNA: The Untold Story."** Another highly readable history is Sean Carroll's joint biography of Albert Camus and Jacques Monod[3] who were devoted friends; **"Brave Genius: A scientist, a philosopher, and their daring adventures from the French resistance to the Nobel prize."** As well as a thorough account of the Resistance movement in France during the German occupation and the Hungarian Uprising in 1956, it includes much about Francois Jacob's role in the Resistance as well. It presents the scientific developments by Monod and Jacob leading to the Operon as a model of regulation of cis-genes in bacteria. Francois Jacob was given a Funeral of State as a Resistance hero, one of the one thousand to receive the *Compagnon de la Liberation* medal, not as a Nobel laureate.

Many of these molecular biologists tried to apply their new methods and paradigms to higher organisms, including mice. Francois Jacob wasn't very successful (see Chapter 7, especially footnote 10) but people whom he recruited to mouse genetics, and continued to inspire, like Jean-Louis Guenet were. Others, like William Franklin Dove of the University of Wisconsin, who had spent a sabbatical year in Jacob's lab, left phage genetics and became highly successful in mouse cancer genetics (see Chapter 6 for his lab's contributions to mutational studies of the *T/t* complex). The French science historian, Michel Morange has analyzed some of the difficulties in this transition from microorganisms to more complex ones, focusing on the attempts of these molecular biologists to understand development. His article is titled "The transformation molecular biology on contact with higher organisms, 1960-1980: from a molecular description to a molecular explanation" (Morange, 1997). He sees the movement from a focus mostly on gene regulation to more complex understanding of cell-cell interactions, intracellular signaling, and compartments in developing organisms, as seen in *Drosophila,* as examples of the new concepts developed to lead to a much broader understanding of the molecular basis of cell and developmental biology. The saturation mutagenesis experiments in *Drosophila* of Nusslein-Vollhard and Weischaus was an important input with the many new pathways of development found (see Chapter 4). The history of the application of these advances to mice is covered in the second part of Kenneth Paigen's article, "One Hundred Years of Mouse Genetics: An Intellectual History. II. The Molecular Revolution (1981–2001)" (Paigen, 2003).

The development of recombinant DNA techniques in which DNA from higher organisms could be "pasted" into bacterial plasmids using restriction enzymes to make cohesive ends, multiplied in bacteria to billions of copies, purified, and then cutting out the cloned fragment, with the same restriction enzyme, was rapidly applied to genes from higher organisms (Cohen et al., 1973). Then the problem was to find the DNA one wanted to clone among the billions of bases in the higher organism. The most abundant RNAs could be reverse transcribed into DNA and cloned after the reverse transcriptase of RNA viruses was discovered (Baltimore, 1970;

Temin and Mizutani, 1970). Thus, the genes for hemoglobin, since over 90% of reticulocyte mRNA coded for hemoglobin (van den Berg et al., 1978) was one of the first genes cloned in mice while the very abundant ribosomal 18S gene was one of the first genes to be cloned in humans (Wilson et al., 1978). Since the transformed bacteria contained plasmids with other fragments of DNA as well as the desired ones, plating the bacteria on agar plates and choosing the colonies with the correct DNA fragments for subsequent growth and purification of the plasmid of interest was an important step. Duplication of the colonies to a second plate so that one of the plates could be treated and DNA transferred to a membrane overlaid on the colonies for hybridization to a probe (e.g., radioactive ribosomal RNA) to identify the clones was needed. As cloning in phage, e.g., lambda, allowing larger fragments to be cloned; designing degenerate probes from protein sequences; seeking tissues and organisms with high levels expression of the desired gene (for reverse transcription of the messages to cDNA for cloning); with homologous cloning in a different species using cross-hybridization; and with the development of expression vectors allowing antibodies to be used to detect the desired bacterial clones were invented; the cloning of many more genes was possible. Chromosomal location information could be used to clone a region around or near a gene of interest by either "scratching" DNA from a particular location on the stained, spread chromosomes of a karyotype slide, as was done for the *T/t* complex (Rohme et al., 1984; Fig. 9.1; see Chapter 8) or "walking" along the chromosome from a closely linked gene.

Another approach was to start with fine mapping of the gene and then identifying DNA with the flanking markers. For instance, in one approach to cloning the *Spm, Sphingomeylinosis* (Niemann-Pick C1) gene, mutations in which caused an animal model of this neurodegenerative disease, such mapping was done (Erickson et al., 1997). Examination of nearly 1,000 mice from the appropriate cross identified an 0.32 CM interval flanked by two markers (D18Mit110 and D18Mit 67). A yeast artificial chromosome was found spanning the two markers (Hsu and Erickson, 2000) but further identification was not needed as a competing group cloned the gene by homology to the human gene cloned at the same time (Loftus et al.,1997). Alternatively, if the gene was unstable (high reversion rate), it was likely that the mutation involved a transposon or tandem duplication and might be found by "genome scanning" (looking for stretches of DNA that differed in the mutant and in the inbred strain from which the mutant arose) as was done for two coat color mutations, *dilute* (Jenkins et al, 1989) and the *pink-eyed unstable* mutation (Brilliant et al., 1991). Thus, many mouse genes were soon cloned.

In addition to the re-purposing of older molecular biologists, and the recruitment of new mouse geneticists, many of the established mouse geneticists learned new molecular techniques. Kenneth Paigen (2003) dates the introduction of trangenesis in 1981 as the beginning of mouse molecular genetics. Gordon et al. (1980), a team working in Frank Ruddle's laboratory at Yale, were the first to succeed in getting gene expression from introduced DNA which had abnormal chromosomal locations. Their method involved injecting 10–20 thousand copies of a cloned fragment of DNA into the pronucleus of the 1 cell embryo by microinjection. The apparatus used had both a holding, blunt micropipette on which the embryo was held by mild suction with the pronucleus correctly positioned and a sharply pointed micropipette, especially pulled on a pipette forge, for injection of the DNA, working on a stage with an inverted

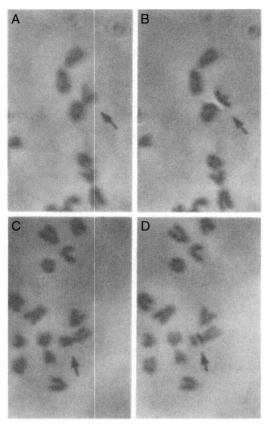

FIG. 9.1 The "scraped off" portion of mouse chromosome 17 used to clone the first *T/t* complex gene. A) and C), chromosome 17 on spread before "scrape," be and D), after the "scrape." From Rohme et al. (1984), with permission.

micoscope. Both were attached to micromanipulators to allow the very minute movements needed and a special pump impelled the small amount of solution (pl.s) into the pronucleus. Embryos with marked granularity of the cytoplasm were preferred for the greater visibility of the clear pronucleus. For this reason, many of the transgenic mice started as F1s of A/J by C57Bl6/J(BL6) since the standard background strain, BL6 did not have many granules in its cytoplasm. Thus, the mice were continuously backcrossed to BL6 to again provide that uniform background. Wagner et al. (1981) showed normal tissue-specific and temporal expression of the β-globin gene by a transgene. Control of the expression of one gene by the regulatory region of another was shown dramatically by the giant mice produced when the rat growth hormone was fused to the mouse metallothionine promoter and injected into the pronuclei of the 1 cell embryos (Palmiter et al., 1982). Some general enhancers such as that from the SV40 virus were widely used (Serfling et al., 1985; Fig. 9.2). Other examples of the use of transgenes for studies of development are given in Chapter 4. Of course, these advances depended on the previous cloning of the involved DNA segments.

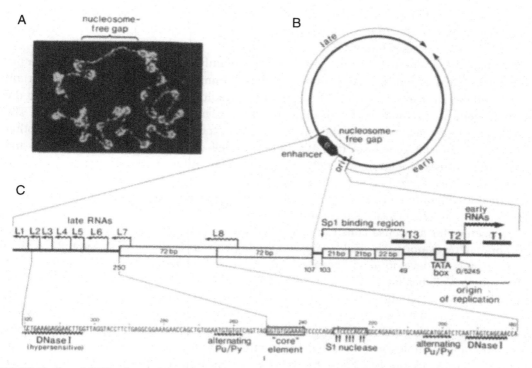

FIG. 9.2 Cloning of SV40 enhancer for use in transgenesis. A) Electron micrograph of the SV40 virus with a nucleosome free gap containing the enhancer. B) A cartoon of the SV40 genome with the enhancer widened. C) Breakdown of sequence motifs and D) sequence of the 72 bp repeat and the unique 5′ regions of the enhancer. From Serfling et al. (1985), with permission.

In addition to pronuclear injection of DNA, transgenic mice could be made by transforming pluripotent stem cells and injecting selected cells with integrated DNA into the blastocyst (where they can integrate into the inner cell mass becoming part of the embryo and its germ line; Gossler et al., 1986). Also, viruses could be used to carry the mouse DNA, initially with leukemia virus (Jaenisch, 1976) and later with lentiviruses which could "carry" longer segments of DNA (Pfeiffer et al., 2002). Successful transgenesis with cosmids (Bruggemann et al., 1991), and eventually YACs (yeast artificial chromosomes; Peterson et al.,1995), allowed larger and larger amounts of contiguous DNA to be used although fragmentation to smaller pieces frequently occurred. The multiple copies frequently inserted in tandem, head-to-tail arrays and complex genomic re-arrangements at the site of integration, which was where a break in the chromosomal DNA was being repaired, were common. The number of copies inserted could be estimated by quantitative PCR but this wasn't always accurate and very large tandem arrays could expand or contract by "slippage" during DNA replication. Because the location of the insertion could not be controlled, expression was also affected by "position effects"—a heterochromatic localization could lead to silencing while placement near enhancing regions could lead to expression not merely controlled by the transgene and any controlling elements it contained. It was found that constructs containing an intron were better expressed (Brinster et al., 1988). About 5–10% of the time, the transgene insertion caused a mutation and Miriam Meisler (1992) has reviewed many of these.

Although the major use of transgenes was to identify and characterize the regulatory sequences of various genes, making animal models of human disease was another important use. Sometimes over-expression of the normal gene would cause a phenotype similar to the loss of the gene as was found with *Foxc2* in mice, causing lymphedema (Noon et al., 2006). The importance of gene expression in specific tissues was explored (Zhang, et al, 2008). Another important use of transgenes was to make mouse models of cancer (reviewed in Adams and Cory, 1991). Transgenics for many single oncogenes could induce particular cancers, e.g. *c-H-ras* inducing pancreatic cancers (Quaife et al., 1987; this model showed cancerous cells circulating in the blood before there was visible cancer in the pancreas—an important observation in regards to the early metastasis of this tumor in humans).

Transgenics for antisense constructs provided an approach for the inactivation of genes. This approach was not successful in *Xenopus* embryos from fertilization to the mid-blastula transition stage of development, because an RNA-unwinding activity was present (Brass and Weintraub, 1987; Rebagliati and Melton, 1987). However, Bevilacqua et al. (1989) demonstrated that such an RNA-unwinding activity was not present in mouse embryos, and they successfully inactivated an enzyme (beta-glucuronidase), which is normally expressed in them. Its regulatory region could also be elucidated by antisense constructs (Bevilacqua and Erickson, 1989). Developmental examples include the demonstration that E-cadherin is important for compaction (Ao and Erickson, 1992) and that *Wnt1* is not essential for spermatogenesis (Erickson et al., 1993). Milder phenotypes included abnormal skin which occurred in mice expressing an antisense CD44 transcript (Kaya et al., 1997) and the *shiverer* phenotype with antisense to myelin basic protein (Katuski et al., 1988). The use of transgenic "short hairpin" interfering RNAs (iRNA) was less successful (Cao et al., 2005).

During the 21st Century, great advances were made with antisense therapies for human genetic disease. In 2016, a drug called Spinraza (nusinersen) was approved for the autosomal recessive disorder, spinal muscular atrophy, a devastating disease affecting communication between muscles and nerves. Of course, transgenesis was not used but inhibitory oligoribonucleotides targeted to an intronic inhibitory sequence element, named ISS-N1 (for "intronic splicing silencer"), located in the SMN2 gene (Singh et al., 2010) were. The inhibition of this site results in the splicing of SMN2 pre-mRNA to include exon 7 which results in elevated expression of the SMN protein which greatly improved patient performance. The drug has to be given by spinal tap several times a year and is very expensive.

The next major advance was the use of homologous recombination in embryonic stem cells to make knockouts. As discussed in Chapter 6 where their use as targets for mutagenesis was discussed, embryonic stem (ES) cells are pluripotent (meaning they can develop into almost all cell types but not those of the placenta) cells that can be cultured from early embryos (Evans and Kaufman, 1981).[4] They can be maintained and manipulated in culture and subsequently introduced into blastocysts, sometimes contributing to the germ line. Initially, they needed to be maintained on a "feeder" layer of cells secreting growth factors but eventually it was found that the leukemia inhibitory factor (LIF) would substitute for this requirement (Williams et al., 1988). Thus, in vitro-selected mutants or genes introduced into ES cells can be "cloned" into mice. These embryonic stem cells, when injected into the blastocoele cavity, frequently fused with the inner cell mass and contributed to the embryo. Also, they frequently contributed to the male germ line and thus, offspring which were the product of the embryonic cell line and carrying the altered sequences would result. Smithies et al. (1985) first demonstrated that trans-fected cloned genomic sequences would align sequences with, and replace by recombination, the endogenous gene. This was soon performed using the selectable marker, hypoxanthine guanine phosphoribosyl transferase (HGPRTase) and introduction into embryos resulted in HGPRTase deficient mice (Hooper et al., 1987) which resulted in an imperfect model of Lesch-Nyhan syndrome. Schwartzberg et al. (1989) were the first to" knockout" a gene with a non-selectable marker while Doetschman et al. (1987) showed that a mutation could be reversed, i.e. they made a "knockin" which requires a knockout first. The examples of the use of these techniques are too numerous to list. These exciting developments have attracted wide inter-est and have been used in an economic analysis of the development of ideas in academia and their later commercialization (Aghion et al., 2010). See Figure 6.2 as an example of their use for mutagenic studies.

There were many surprises in store from the knockouts. Genes that were thought to be essential sometimes did not produce a phenotype and similar genes substituting for their func-tion were found, e.g. *MyoD* and *Myf-5* (Rudnicki et al., 1992). Major effects of modifiers were also found, e.g. the Tgf-beta1 knockout caused a severe neuroinflammatory disease and death at 3–4 weeks on the 129/J background (Shull et al., 1992) while it caused pre-natal lethality with disrupted vasculogenesis and hematopoiesis on a partial C57BL/6J inbred strain background (2 backcross generations; Dickson et al., 1995). Even more surprising was the finding that it became a preimplantation lethal when fully on the C57BL/6J background (Kallapur et al., 1999).

The development of temporal and tissue specific expression of the inserted genes was then soon developed. The first approach used a tissue specific promoter to drive a toxic protein and "kill" the targeted tissue/cells. Thus, the potent diphtheria toxin was expressed under the control of a lens-crystallin promoter to cause microphthalmia (Breitman et al., 1987). Another approach was to target the Herpes Simplex virus thymidine kinase to particular tissues with the appropriate promoters. Then when the cancer drug ganciclovir is given which, when phos-phorylated by the enzyme, kills the targeted cells (Heyman et al., 1989). Great specificity of the promoter was required or wider fields of cell death would occur. Another early controllable promoter was that for heat-shock protein 68. Rossant's group demonstrated that this promoter (−664 to +113) was specifically inducible by short-term treatments with heat or sodium arsenate in preimplantation and gestational stage of development (Kathory et al., 1989).

A more general system was that of the Cre/LoxP recombinase system (Sauer et al., 1992). In this system, the target gene which had been added to the genome by transgenesis or embry-onic stem cell methodology was flanked by loxP sites from the bacteriophage P1. These short sequences would undergo recombination when the Cre recombinase was expressed from another exogenous gene, which would use specific promoters to then delete the interven-ing DNA in specific tissues creating a knockout in those tissues. With this methodology and homologous recombination to insert the construct in place of the normal gene, genes which, when absent in the embryo, would be lethal, could be studied at later stages in development. A variant was to use the tetracycline-responsive *tet* repressor which allowed downstream genes to be specifically turned off or on by the presence of tetracycline in the drinking water (Furth et al., 1994).

In the 21st Century, gene editing, either to correct or make mutations made fantastic advances with the CRISPR (clustered regularly interspaced short palindromic repeats)-Cas9 system. This system uses synthetic RNA strands to guide the Cas9 nuclease to induce targeted double strand breaks (DSBs), or possibly single strand nicks (SSNs) after injection into single-cell embryos. At the same time, a large excess of a synthetic repair template, designed to replace the original sequence with the desired sequence is supplied. The DSBs or SSNs will have activated the homology-directed repair, which uses homologous recombination between chromosomal DNA and the supplied foreign DNA for the repair, i.e., replacement. There are many reviews of the subject, e.g. Wade, 2015. There has been great concern about off-target modifications (Wu et al., 2014) and ethical dilemmas, which has prevented to date (except one ill-considered trial in China) its use in humans.

The cloning of Brachury *and other T/t complex genes*

The cloning of *T* was discussed briefly in Chapter 4 and is here expanded. As mentioned above, Rohme et al. (1984) used the laborious techniques of scraping DNA from spread chromosomes on glass slides in order to obtain clones from the *T* region of chromosome 17. These provided the starting material to find many more DNA probes from the region which were mapped using a number of recombinant *T/t* complex chromosomes, many provided from Lee Silver's lab. These probes were first used to identify inversions in the *T* region (see Chapter 3) as well as new probes mapping close to *T* (Hermann et al., 1986; 1987). Screening of a library made from reverse-transcribed RNA of 8.5 day embryos, a time when T is expressed, with these nearby-mapping probes led to further probes. A genomic library from DNA of a *T*^{Wis}/+ mutant mouse, a spontaneously arising *T* mutation, was also made in the lambda vector EMBL3. Clones were isolated which covered the *T*^{Wis} region and a related embryonic clone which then identified a gene encoding 456 amino acids which had been modified in this mutant (Herrmann et al., 1990) and was properly expressed in the embyo (Wilkinson, et al, 1990). The modification involved a retroposon insertion as were many spontaneous mutants (see Chapter 6).

When Brachury protein was made by in vitro translation and allowed to bind to 26-mer DNA oligonucleotides, a 10 bp element was identified as the target for its binding to genomic DNA (Kispert and Herrmann, 1993). X-ray crystallographic examination of its binding to DNA identified a new mode of binding as a dimer with the carboxy terminal in the minor grove without bending the DNA ((Muller and Hermann, 1997). Using degenerate primers matching a 7 amino acid stretch of the central region of the gene (one with homology also to a *Drosophila* gene which showed a conserved relationship to *Brachyury*), Lee Silver's group identified a family of related genes with a "T-box" motif (Bollag et al., 1994). They identified at least 3 *T-box* genes in the vertebrate lineage with unique temporal and spatial expression patterns. Soon the number of T-box genes was extended to 6 and the evolutionary history extended back to an early metazoan ancestor (Agulnik et al., 1995). Physical mapping genetic linkage, and phylogenetic studies demonstrated that unequal crossing-over had provided duplication of the T-box gene ancestor which was then dispersed to multiple chromosomal locations (Agulnik et al., 1996) as were a number of other genes suggesting an ancient duplication event (Ruvinsky and Silver, 1997). Homologues were found in *C. elegans* (Wattler et al., 1998) and shown to have a role in head formation in *Hydra* (Technau and Bode, 1999). Further T-box genes were found to bring the number to 18 eventually. The evolutionary lineage of a many of them from *C. elegans* to mammals was reviewed by Papioanou and Silver (1998). Included was the gene, *Eomesodermin,* which, like *T,* played an essential role in the trophoblast and in mesoderm formation (Russ et al., 2000). Considered overall, the family of *T-box* genes appeared to be first involved in defining the primitive blastopore and this involvement led to their role in mesoderm later in evolution (Technau, 2001). In addition to the cloning of the "classic" *T,* a gene with no homology to *T* but mapping very near to, and interacting with it to cause only a stump of a tail , *Brachyury the second,* was cloned (Rennebeck et al., 1998).

At the end of the 20th, and as the 21st Century started, the human homologues of these T-box genes were soon mapped, cloned and found to be involved in genetic diseases. Studies of expression in mice and mouse knockouts were much used in elucidating the roles of *TBX* genes in disease. A collection of chapters in **Erickson and Wynshaw-Boris, ed.s, 2016, pp. 821–842,** "D. The *T-BOX* Gene Family," provides further details.

Chieffo et al. (1997) cloned the human *TBX1* gene, which mapped to the center of 22q11.2 which had 3 well known syndromes associated with relatively frequent deletions of the region. These included the DiGeorge sequence, which was also called the III-IV pharyngeal pouch complex defect, with absence or hypoplasia of the thymus and/or parathyroid glands and cardiovascular anomalies, ear abnormalities and micrognathia (DiGeorge, 1965); the Velo-Cardio-Facial syndrome with a combination of an unusual "pear" shaped nose and other facial anomalies including a short mandible, frequent heart defects, cleft palate, and delayed development (Shprintzen et al.,2021); and a syndrome with distinctive facies associated with pulmonary atresia with ventriculoseptal defect (Seaver et al., 1994). The cardiac defects found in these syndromes can be explained by the TBX1 deficiency found in almost all cases of the 3 syndromes–its role in cardiac development was well studied in mice (Merscher et al., 2001). The mouse TBX1 proteins share 98% amino acid identity with the human, only differing for 2 residues within the T-box domain. The heterozygous *Tbx1* KOs have a high incidence of cardiac outflow tract anomalies while the homozygous KO show hypoplasia of the thymus and parathyroid glands, cleft palate, abnormal vertebrae and facies as well as the cardiac defects (Guris et al., 2001; Jerome and Papaioannou, 2001). The explanation of the distinctiveness of the 3 syndromes related to *TBX1* deletions may be explained by the extent of deletions, i.e., there may be genes related to different sorts of congenital malformations coded in 22q and, depending on the extent of the deletion, different cardiac malformations will be found as well as other associated findings.

TBX2 was soon found but establishing its role in development and human disease took longer than it had taken for *TBX1*. It was mapped to mouse chromosome 11 in a region of synteny to human 17q. This allowed its cloning in a YAC (Law et al., 1995). In a study of the early expression of a number of *Tbx* genes, it was found to be expressed in otic and optic vessels, the developing pharynx, the cartilage primordia of the ribs, and parts of the brain and hind limbs (Chapman et al., 1996a). It was one of several *Tbx* genes cloned in a retinal library, so its expression was studied in the retina (Sowden et al., 2001) but also in the heart (Harrelson et al., 2004). However, the phenotype of a syndrome with compound heterozygosity for mutations in *TBX2* would not have been well predicted by these developmental studies. Such *TBX2* mutations were found in a recently described syndrome: vertebral anomalies and variable endocrine and T-cell dysfunction (VETD; Liu et al., 2018). It was also found that *TBX2* was amplified in a subset of primary human breast cancers suggesting a possible role in their development (Jacobs et al., 2000).

Although *Tbx3* showed a large overlap in expression with *Tbx2*, mutations in *TBX3* were found to be associated with a very different syndrome, the Ulnar-mammary Syndrome (UMS) characterized by apocrine abnormalities including absent sweating and breast defects and malformations of the posterior elements of the upper limb including duplications and/or splitting. In the first description, delayed growth and onset of puberty, obesity, hypogenitalism and diminished sexual activity were also included (Schnizel, 1987). Bamshad et al. (1995) described a large Utah kindred and mapped the gene to a location near to that responsible for the Holt-Oram syndrome (see below). They soon found the gene responsible to be the one encoding *TBX3* (Bamshad et al., 1997). The close linkage was explained by the above mentioned duplication and dispersal process leaving the cognate pairs *Tbx2/Tbx3* and *Tbx4/Tbx5* and the linked pairs *Tbx3/Tbx5* and *Tbx2/Tbx4*. *Tbx3* is expressed in a wide variety of areas in mouse (Chapman et al., 1996a) which greatly overlap those of *Tbx2*. This overlap may explain the finding that most of the tissues and organs in which it is expressed are normal in individuals with UMS.

Tbx4 and *Tbx5* have related patterns of expression in mice. In the developing heart, *Tbx5* has a longer and wider expression than *Tbx4* and is expressed in the developing optic vesicle while *Tbx4* is expressed early in the genital papilla and tail mesenchyme (Chapman et al., 1996a). A major difference in their expression concerns the developing limbs where *Tbx5* is only expressed in the forelimbs while *Tbx4* has some expression in the forelimbs but is much more highly expressed in the hindlimbs (Gibson-Brown et al., 1996). Interestingly, *Tbx4* is lacking in sea horses which do not have pelvic fins and the KO in zebra fish leads to their loss of pelvic fins (Lin et al., 2016). These differences in patterns of expression are reflected in the human disorders associated with mutation in the human homologues. Mutations in *T-BOX 4* are associated with the small patella syndrome (Bongers et al., 2004), also known as ischiocoxopodopatellar syndrome, which can be associated with pulmonary hypertension (Kerstjens-Frederikse et al.,2013). Variants in it can also be associated with isolated pulmonary hypertension, especially in children (Levy et al., 2016), and null mutations may be associated with the fetal syndrome of amelia (absent limbs; Kariminejad et al., 2019).

Mutations in *TBX5* result in the dominantly inherited Holt-Oram syndrome of upper limb defects and cardiac abnormalities. Polydactyly or a triphalangeal thumb with atrial septal defect are "classic" findings, but there is great variability, even in the same family. The association with *TBX5* mutations was simultaneously reported by 2 laboratories (Basson et al., 1997;

Li et al., 1997). Missense mutations could cause predominantly heart or limb defects while null mutations created more severe heart and limb defects (Basson et al., 1999). Developmental studies in mice of Tbx5 function were important for understanding limb specification and heart development. It had been thought that presumptive limb fields were due to localized increased cell division but Gros and Tabin (2014) showed that it was an earlier epithelial-to-mesenchyme transition of the coelomic epithelium, in part mediated by Tbx5, which specified the limb fields. When *Tbx5* was knocked out specifically in the developing ventricle, only a single chamber developed (Koshiba-Takeuchi et al., 2009), while if it was knocked out in the developing cardio-myocytes, conduction defects occurred (Zhu et al., 2008). Interestingly, although mouse studies showed a role in the developing eye (Koshiba-Takeuchi et al., 2000), visual problems have not been reported—a possible man/mouse difference.

Mutations in *TBX6* are associated with spondylocostal dysostosis (Sparrow et al., 2013), a syndrome which can be caused by mutations in a number of other skeletally expressed genes (Gucev et al., 2010). It is also frequently mutated in cases of congenital scoliosis (Wu et al., 2015). Chapman et al. (1996b) had shown a role for *Tbx6* in paraxial mesoderm formation and its knockout, surprisingly, had 3 neural tubes (Chapman and Papaioannou, 1998)!

The very rare Cousin syndrome (Cousin et al., 1982) was described in 2 children of a first-cousin marriage. The two showed mild mental retardation and dwarfism, facial dysmorphism with ear abnormalities and bilateral agenesis of the alae of the scapula and hip dislocation. This syndrome has been associated with mutations in *TBX15* in one, also, consanguineous family (Lausch et al., 2008). The knockout of *Tbx15* mimicked the *droopy ear* mutant of mice and included small scapulae, moderate shortening of long bones, and a dysmorphic cranial bones and cervical vertebrae (Kuipjer et al., 2005).

Mutations in *TBX18* have been found in one family with congenital anomalies of the kidneys and urinary tract (Vivante et al., 2015). It had been found to be expressed in the anterior somite halves of but also in the developing heart, the mandibular/maxillary region, the urogenital ridge, and the limb buds (Kraus et al., 2001). The nulls died shortly after birth with severe skeletal abnormalities (Bussen et al., 2004), but their renal development was abnormal (Airik et al., 2006).

TBX18 is better known as *TPIT* since it is essential for normal pituitary development and is mutated in a number of endocrine disorders involving the pituitary (Couture et al., 2012). Studies in mice demonstrated that it activated the transcription of proopiomelanocortin (POMC) in cooperation with homeobox proteins (Lamolet et al., 2001).

Another T-box gene found to be mutated in humans was the X-linked *TBX-22* gene. Mutations in it were found to be associated with the X-linked, male-limited cleft palate frequently associated with ankyloglossia (tongue-tie) described by Lowry (1970). Stanier's lab at University College, London, first reported 6 inactivating mutations (Braybrook et al., 2001). Although its expression pattern included the developing eye and otic vessel (Paxton et al., 2010) the knockout mice lacking *Tbx22* only had cleft palate and suggested that its role was in the development of the bony palate rather than in palatal closure, half the affected mice died due to orofacial abnormalities (Pauws et al., 2009). Mutations in it were also found in the rare Abruzzo-Erickson syndrome which had variably penetrant cleft palate associated with coloboma, deafness, short stature and urogenital anomalies i.e., abnormalities affecting other areas of Tbx22 expression (Abruzzo and Erickson, 1977). The mutation was an unusual splice site junction abnormality and dominant negative effect may have been involved (Pauws et al., 2013).

Notes

1. Rollin Hotchkiss went on to be one of the major contributors to the biochemistry of nucleic acids. My wife and I got to know this lovely gentleman in 1965-1967 when she worked for Priscilla Ortiz, a member of Rollin's laboratory at the Rockefeller Institute. Rollin characterized his contribution to the pneumococcal transforming studies as absolutely proving that the DNA preparations contained no protein. He was also a great friend of Salome Gluecksohn-Waelsch (see minibiography, Chapter 4).

2. I had the pleasure of spending my obligatory military time (essentially all doctors were drafted during the Vietnam War) working as a researcher in Anfinsen's lab at NIH. Daily luncheon conversations in the conference room were a stimulating mix of science and politics. Many of the great names in protein chemistry were visitors and gave seminars.

3. I had a brief encounter with Jacques Monod at a molecular biology meeting at Cold Spring Harbor meeting in Sept., 1969, which he attended with Susan Bourgoise as close company. I presented a paper on the molecular weight of highly purified E. coli β-galactosidase in solutions of highly concentrated guanidine hydrochloride by ultracentrifugation with meniscus-depletion sedimentation equilibrium which came up with a much lower estimate than the accepted, by gene size, weight of ~160,00 Daltons (Erickson, 1970). He came up to me and said that my data was a "red herring." Appropriately, he was correct as amino acid sequencing of β-galactosidase later showed.

4. Matthew Kaufman and Martin Evans at the University of Cambridge and the American Gail Martin independently cultured embryonic stem cells from mouse embryos in 1981. In 2007 Martin Evans, still at the University of Cambridge, shared the Nobel Prize with the Americans, Mario Capecchi and Oliver Smithies (see Chapter 3) for developing the methods for modifying genes in embryonic stem cells. In 2012, John B Gurdon, then also at the University of Cambridge, and Shinya Yamanaka of Japan shared the Nobel Prize for showing that mature cells could be reverted to pluripotent cells.

References

Abruzzo, M., Erickson, R.P., 1977. A new syndrome of cleft palate associated with coloboma, hypospadias, deafness, short stature, and radial synostosis. J. Med. Genet. 14, 76–80.

Adams, J.S., Cory, S., 1991. Transgenic models of tumor development. Science 254, 1161–1167.

Aghion, P., Dewatripont, M., Koley, J., Murray, F., Stern, S., 2010. The Public and Private Sectors in the Process of Innovation: Theory and Evidence from the Mouse Genetics Revolution. Amer. Econ. Rev.: Papers Proc. 100, 153–158.

Agulnik, S.I., Bollag, R.J., Silver, L.M., 1995. Conservation of the T-box gene family from Mus musculus to Caenorhabditis elegans. Genomics 25, 214–219.

Agulnik, S.I., Garvey, N., Hancock, S., Ruvinsky, I., Chapman, D.L., Agulnik, I., Bollag, R., Papaioannou, V., Silver, L.M., 1996. Evolution of mouse T-box genes by tandem duplication and cluster dispersion. Genetics 144, 249–254.

Airik, R., Bussen, M., Singh, M.K., Petry, M., Kispert, A., 2006. Tbx18 regulates the development of the ureteral mesenchyme. J. Clin. Invest. 116, 663–674.

Ao, A., Erickson, R.P., 1992. Injection of antisense RNA specific for E-cadherin demonstrates that E-cadherin facilitates compaction, the first differentiative step of the mammalian embryo. Antisense Res. Develop. 2, 153–159.

Avery, O.T., MacLeod, C., McCarty, M., 1944. Studies on the Chemical Nature of the Substance Inducing Transformation of Pneumococcal Types : Induction of Transformation by a Desoxyribonucleotide acid Fraction Isolated from Pneumococcus Type III. J. Exp. Med. 79, 137-15.

Baltimore, D., 1970. RNA-dependent DNA polymerase in virions of RNA tumour viruses. Nature 226, 1209–1211.

Bamshad, M., Krakowiak P., A., Watkins W., S., Root, S., Carey J., C., Jorde L., B., 1995. A gene for ulnar-mammary syndrome maps to 12q23-q24.1. Hum. Mol. Genet. 4, 1973–1977.

Bamshad, M., Lin, R.C., Law D., J., Watkins W., S., Krakowiak P., A., et al., 1997. Mutations in human TBX3 alter limb, apocrine and genital development in ulnar-mammary syndrome. Nature Genet 16, 311–315.

Basson, C.T., Bachinsky, D.R., Lin, R.C., Levi, T., Elkins, J., et al., 1997. Mutations in human TBX5 cause limb and cardiac malformation in Holt-Oram syndrome. Nature Genet 15, 30–35.

Basson, C.T., Huang, T, Lin, R.C., Bachinsky, D.R., Weremowicz, S., et al., 1999. Different TBX5 interactions in heart and limb defined by Holt-Oram syndrome mutations. Proc. Nat. Acad. Sci. 96, 2919–2924.

Bevilacqua, A., Loch-Caruso, R., Erickson, R.P., 1989. Abnormal development and dye coupling produced by antisense RNA to gap junction protein in mouse preimplantation embryos. Proc. Natl Acad. Sci. USA 86, 5444–5448.

Bevilacqua, A., Erickson, R.P., 1989. Use of antisense RNA to help identify a genomic clone for the 5′ region of mouse β-glucuronidase. Biochem. Biophys. Res. Commun 160, 937–941.

Bollag, R.J., Siegfried, Z., Cebra-Thomas, J.A., Garvey, N., Davison, E.M., Silver, L.M., 1994. An ancient family of embryonically expressed mouse genes sharing a conserved protein motif with the T locus. Nature Genet 7, 383–389.

Bongers, E.M.H.F., Duijf, P.H.G., van Beersum, S.E.M, Schoots, J., van Kampen, A., et al., 2004. Mutations in the human TBX4 gene cause small patella syndrome. Am. J. Hum. Genet. 74, 1239–1248.

Brass, B.L., Weintraub, H., 1987. A developmentally regulated activity that unwinds RNA duplexes. Cell 48, 607–613.

Braybrook, C., Doudney, K., Marcano, A.C.B., Arnason, A., Bjornsson, A., Patton M., A., Goodfellow P., J., Moore G., E.Stanier, 2001. The T-box transcription factor gene TBX22 is mutated in X-linked cleft palate and ankyloglossia. Nature Genet 29, 179–183.

Breitman, M.L., Clapoff, S., Rossant, J., Tsui, L.C., Glode, L.M., Maxwell, I.H., Bernstein, A., 1987. Genetic ablation: targeted expression of a toxin gene causes microphthalmia in transgenic mice. Science 238, 1563–1565.

Brilliant, M.H., Gondo, Y., Eicher, E.M., 1991. Direct Molecular Identification of the Mouse Pink-Eyed Unstable Mutation by Genome Scanning. Science 252, 566–569.

Brinster, R.L., Allen, J.M., Behringer, R.R., Gelinas, R.E., Palmiter, R.D., 1988. Introns increase transcriptional efficiency in transgenic mice. Proc. Nat'l Acad. Sci., U.S.A. 85, 836–840.

Bruggemann, M., Spicer, C., Buluwela, L., Rosewell, I., Barton, S., Surani, M.A., Rabbitts, T.H., 1991. Human antibody production in transgenic mice: expression from 100 kb of the human IgH locus. Eur. J. Immunol. 23, 1321–2326.

Bussen, M., Petry, M., Schuster-Gossler, K., Leitges, M., Gossler, A., Kispert, A., 2004. The T-box transcription factor Tbx18 maintains the separation of anterior and posterior somite compartments. Genes Dev 18, 1209–1221.

Cao, W., Hunter, R., Strnatka, D., McQueen, C.A., Erickson, R.P., 2005. DNA constructs designed to produce short hairpin, interfering RNAs in transgenic mice sometimes show early lethality and an interferon response. J. Appl. Genet. 46, 217–225.

Chapman, D.J., Garvery, N., Hancock, S., Alexious, M., Agulnik, S.I., Gibson-Brown, J.J., Cebra-Thomas, J., Bollag, R.J., Silver, L.M., Papaioannou, V.E., 1996a. Expression of the T-box genes, Tbx1-Tbx5, during early mouse development. Develop. Dyn. 206, 379–390.

Chapman, D.L., Agulnik, I., Hancock, S., Silver, L.M., Papaioannou, V.E., 1996b. Tbx6, a mouse T-box gene implicated in paraxial mesoderm formation at gastrulation. Dev. Biol. 180, 534–542.

Chapman, D.L., Papaioannou, V.E., 1998. Three neural tubes in mouse embryos with mutations in the T-box gene Tbx6. Nature 391, 695–697.

Chieffo, C., Garvey, N., Gong, W., Roe, B., Zhang, G., Silver, L., Emanuel, B.S., Budarf, M.L., 1997. Isolation and characterization of a gene from the DiGeorge chromosomal region homologous to the mouse Tbx1 gene. Genomics 43, 267–277.

Cohen, S., Chang, A., Boyer, H., Helling, R., 1973. Construction of biologically functional bacterial plasmids in vitro. Proc. Nat'l Acad. Sci. U.S.A. 70, 3240–3244.

Cousin, J., Walbaum, R., Cegarra, P., Huguet, J., Louis, J., Pauli, A., Fournier, A., Fontaine, G., 1982. Dysplasie pelvi-scapulaire familiale avec anomalies epiphysaires, nanisme et dysmorphies: un nouveau syndrome? Arch. Franc. Pediat. 39, 173–175.

Couture, C., Saveanu, A., Barlier, A., Carel, C., Fassnacht, M., et al., 2012. Phenotypic homogeneity and genotypic variability in a large series of congenital isolated ACTH-deficiency patients with TPIT gene mutations. J. Clin. Endocr. Metab. 97, E486–E495.

Dickson, M.J., Martin, J.S., Cousins, F.M., Kulkarni, A.B., S., Karlson, Akhurst, R.J., 1995. Defective hematopoiesis and vasculogenesis in transforming growth factor beta 1 knock out mice. Develop 121, 1845–1854.

DiGeorge A., M., 1965. Discussion of a new concept of the cellular basis of immunology. J. Peds 67, 907–912.

Doetschman, T., Gregg, R.G., Maeda, N., Hooper, M.L., Thompson, S., Smithies, O., 1987. Targeted correction of a mutant HPRT gene in mouse embryonic stem cells. Nature 330, 576–578.

Erickson, R.P., 1970. Molecular weight of Escherichia coli β-galactosidase in concentrated solutions of guanidine hydrochloride. Biochem. J. 120, 255–261.

Erickson, R.P., Lai, L.-W., Grimes, J., 1993. Creating a conditional mutation of Wnt-1 by antisense transgenesis provides evidence that Wnt-1 is not essential for spermatogenesis. Dev. Genet. 14, 274–281.

Erickson, R.P., Aviles, R.A., Zhang, J., Kozloski, M.A., Garver, W.S., Heidenreich, R.A., 1997. High-resolution mapping of the spm (Niemann-Pick Type C) locus on mouse Chromosome 18. Mamm. Genome 8, 355–356.

Furth, P.A., L., St.Onge, Boger, H., Gruss, P., Gossen, M., Kistner, A., Bujard, H., L., Hennighausen, 1994. Temporal control of gene expression in transgenic mice by a tetracycline-responsive promoter. Proc. Nat'l Acad. Sci.,U.S.A 91, 9302–9306.

Evans, M.J., Kaugman, M.H., 1981. Establishment in culture of pluripotential cells from mouse embryos. Nature 292, 154–156.

Gibson-Brown, J.J., Agulnik, S.I., Chapman, D.L., Alexiou, M., Garvey, N., Silver, L.M., Papaioannou, V.E., 1996. Evidence for a role for T-box genes in the evolution of limb morphogenesis and specification of forelimb/hindlimb identity. Mech. Devel. 56, 93–101.

Gordon, J.W., Scangos, G.A., Plotkin, D.J., Barbosa, J.A., Ruddle, F.H., 1980. Genetic transformation of mouse embryos by microinjection of purified DNA. Proc. Natl. Acad. Sci. USA 77, 7380–7384.

Gossler, A., Doetschman, T., Korn, R., Serfling, E., Kemler, R., 1986. Transgenesis by means of blastocyst-derived embryonic stem cell lines, Proc. Nat'l Acad. Sci., U.S.A., 83, 9065–9069.

Gros, J., Tabin C., J., 2014. Vertebrate limb bud formation is initiated by localized epithelial-to-mesenchymal transition. Science 343, 1253–1256.

Gucev, Z.S., Tasic, V., Pop-Jordanova, N., Sparrow, D.B., Dunwoodie, S.L., Ellard, S., Young, E., Turnpenny, P.D., 2010. Autosomal dominant spondylocostal dysostosis in three generations of a Macedonian family: negative mutation analysis of DLL3, MESP2, HES7, and LFNG. Am. J. Med. Genet. 152A, 1378–1382.

Guris, D.L., Fantes, J., Tara, D., Druker, B.J., Imamoto, A., 2001. Mice lacking the homologue of the human 22q11.2 gene CRKL phenocopy neurocristopathies of DiGeorge syndrome. Nature Genet. 27, 293–298.

Harrelson, Z., Kelly R., G., Goldin S., N., Gibson-Brown J., J., Bollag R., J., Silver L., M., Papaioannon V., E., 2004. Tbx2 is essential for patterning the atrioventricular canal and for morphogenesis of the outflow tract during heart development. Development 131, 5041–5052.

Herrmann, B., Bucan, M., Mains, I.E., Frischauf, A.-M., Silver, L.M., Lehrach, H., 1986. Genetic analysis of the proximal portion of the mouse t complex: evidence for a second inversion within t haplotypes. Cell 44, 469–478.

Herrmann, B.G., Barlow, D.F., Lehrach, H., 1987. A Large Inverted Duplication Allows Homologous Recombination between Chromosomes Heterozygous for the Proximal t Complex Inversion. Cell 48, 813–825.

Herrmann, B.G., Labeit, S., Poustka, A., King, T., Lehrach, H, 1990. Cloning of the T gene required in mesoderm formation in the mouse. Nature 343, 617–622.

Heyman, R.A., Borrelli, E., Lesley, J., Anderson, D., Richman, D.D., Baird, S.M., Hyman, R., Evans, R.M., 1989. Thymidine kinase obliteration: creation of transgenic mice with controlled immune deficiency. Proc. Nat'l Acad. Sci., U.S.A. 86, 2698–2702.

Hooper, M., Hardy, K., Handyside, A., Hunter, S., Monk, M., 1987. HPRT-deficient (Lesch-Nyhan) mouse embryos derived from germline colonization by cultured cells. Nature 326, 292–295.

Hsu, S.J., Erickson, R.P., 2000. Construction of the long-range YAC physical map of the 10 cM region between the markers D18Mit109 and D18Mit68 on mouse proximal chromosome 18. Genome 43, 427–433.

Jacobs J. J., L., Keblusek, P., Robanus-Maandag, E., Kristel, P., Lingbeek, M., Nederlof P., M., van Welsem, T., van de Vijver M., J., Koh E., Y., Daley G., Q., van Lohuizen, M., 2000. Senescence bypass screen identifies TBX2, which represses Cdkn2a(p19ARF) and is amplified in a subset of human breast cancers. Nature Genet 26, 291–299.

Jaenisch, R., 1976. Germ line integration and Mendelian transmission of the exogenous Moloney leukemia virus. Proc. Nat'l Acad. Sci., U.S.A. 73, 1260–1264.

Jenkins, N.A., Strobel, M.C., Seperack, P.K., Kingsley, D.M., Moore, K.J., Mercer, J.A., Russell, L.B., Copeland, N.G., 1989. A retroviral insertion in the dilute (d) locus provides molecular access to this region of mouse chromosome 9. Prog Nucleic Acid Res Mol Biol 36, 207–220.

Jerome L., A., Papaioannou V., E., 2001. DiGeorge syndrome phenotype in mice mutant for the T-box gene, Tbx1. Nature Genet 27, 286–291.

Kallapur, S., Ormsby, I., Doetschman, T., 1999. Strain dependency of TGFbeta1 function during embryogenesis. Mol. Reprod. Develop. 52, 341–349.

Kariminejad, A., Szenker-Ravi, E., Lekszas, C., Tajsharghi, H., Moslemi, A.-R., et al., 2019. Homozygous null TBX4 mutations lead to posterior amelia with pelvic and pulmonary hypoplasia. Am. J. Hum. Genet. 105, 1294–1301.

Kathory, R., Clapof, S., Darling, S., Perry, M.D., Moran, L.A., Rossant, J., 1989. Inducible expression of an hsp68-lacZ hybrid gene in transgenic mice. Develop 105, 707–714.

Katsuki, M., Sato, M., Kimura, M., Yokoyama, M., Kobayashi, K., Nomura, T., 1988. Conversion of normal behavior to shiverer by myelin basic protein antisense cDNA in transgenic mice. Science 241, 593–595.

Kaya, G., Rodriguez, I., Jorcano, J.L., Vassalli, P., Stamenkovic, I., 1997. Selective suppression of CD44 in keratinocytes of mice bearing an antisense CD44 transgene driven by a tissue-specific promoter disrupts hyaluronate metabolism in the skin and impairs keratinocyte proliferation. Genes Develop 11, 996–1007.

Kerstjens-Frederikse, W.S., Bongers, E.M.H.F., Roofthooft, M.T.R., Leter, E.M., Menno Douwes, J., et al., 2013. TBX4 mutations (small patella syndrome) are associated with childhood-onset pulmonary arterial hypertension. J. Med. Genet. 50, 500–506.

Kispert, A., Herrmann, B.G., 1993. The *Brachyury* gene encodes a novel DNA binding protein. EMBO J 12, 3211–3220.

Koshiba-Takeuchi, K., Mori, A.D., Kaynak B., L., Cebra-Thomas, J., Sukonnik, T., et al., 2009. Reptilian heart development and the molecular basis of cardiac chamber evolution. Nature 461, 95–98.

Koshiba-Takeuchi, K., Takeuchi, J.K., Matsumoto, K., Momose, T., Uno, K., Hoepker, V., Ogura, K., Takahashi, N., Nadamura, H., Yasuda, K., Ogura, T., 2000. Tbx5 and the retinotectum projection. Science 287, 134–137.

Kraus, F., Haenig, B., Kispert, A., 2001. Cloning and expression analysis of the mouse T-box gene Tbx18. Mech. Dev. 100, 83–86.

Kuijper, S., Beverdam, A., Kroon, C., Brouwer, A., Candille, S., Barsh, G., Meijlink, F, 2005. Genetics of shoulder girdle formation: roles of Tbx15 and aristaless-like genes. Development 132, 1601–1610.

Lamolet, B., Pulichino, A.-M., Lamonerie, T., Gauthier, Y., Brue, T., Enjalber, A., Drouin, J., 2001. A pituitary cell-restricted T box factor, Tpit, activates POMC transcription in cooperation with Pitx homeoproteins. Cell 104, 849–859.

Lausch, E., Hermanns, P., Farin, H.F., Alanay, Y., Unger, S., Nikkel, S., Steinwender, C., Scherer, G., Spranger, J., Zabel, B., Kispert, A., Superti-Furga, A., 2008. TBX15 mutations cause craniofacial dysmorphism, hypoplasia of scapula and pelvis, and short stature in Cousin syndrome. Am. J. Hum. Genet. 83, 649–655.

Law, D.J., Gebhur, T., Garvey, N., Agulnik, S.I., Silver, L.M., 1995. Identification, characterization, and localization to chromosome 17q21-22 of the human TBX2 homolog, member of a conserved developmental gene family. Mammalian Genome 6, 793–797.

Levy, M., Eyries, M., Szezepanski, I., Ladouceur, M., Nadaud, S., Bonnet, D., Soubrier, F., 2016. Genetic analyses in a cohort of children with pulmonary hypertension. Europ. Resp. J. 48, 1118–1126.

Li, Q.Y., Newbury-Ecob, R.A., Terrett, J.A., Wilson, D.I., Curtis AR, J., et al., 1997. Holt-Oram syndrome is caused by mutations in TBX5, a member of the Brachyury (T) gene family. Nature Genet 15, 21–29.

Lin, Q., Fan, S., Zhang, Y., Xu, M., Zhang, H., et al., 2016. The seahorse genome and the evolution of its specialized morphology. Nature 540, 395–399.

Liu, N., Schoch, K., Luo, X., Pena L. D., M., Bhavana V., H., Kukolich M., K., Stringer, S., Powis, Z., Radtke, K., Mroske, C., Deak, K., McDonald M., T., et al., 2018. Functional variants in TBX2 are associated with a syndromic cardiovascular and skeletal developmental disorder. Hum. Molec. Genet. 27, 2454–2465.

Loftus, S.K., Morris, J.A., Carstea, E.D., Gu, J.Z., Cummings, C., et al., 1997. Murine Model of Niemann-Pick C Disease: Mutation in a Cholesterol Homeostasis Gene. Science 277, 232–235.

Lowry R., B., 1970. Sex-linked cleft palate in a British Columbia Indian family. Pediatrics 46, 123–128.

Meisler, M.H., 1992. Insertional mutation of 'classical' and novel genes in transgenic mice. Trends Genet 8, 341–344.

Merscher, S., Funke, B., Epstein J., A., Heyer, J., Puech, A., et al., 2001. TBX1 is Responsible for Cardiovascular Defects in Velo-Cardio-Facial/DiGeorge Syndrome. Cell 104, 619–629.

Morange, M., 1997. The transformation molecular biology on contact with higher organisms, 1960-1989: from a molecular description to a molecular explanation. Hist. Phil. Life Sci. 19, 369–393.

Muller, C.W., Hermann, H.G., 1997. Crystallographi structure of the T domain-DNA complex of the Brachury transcription factor. Nature 389, 884–888.

Noon, A., Hunter, R.J., Witte, M.H., Kriederman, B., Bernas, M., Rennels, M., Percy, D., Enerbäc, S., Erickson, R.P., 2006. Comparative lymphatic, ocular, and metabolic phenotypes of FOXC2 haploinsufficient and AP2-FOXC2 transgenic mice. Lymphology 39, 84–94.

Paigen, K, 2003. One hundred years of mouse genetics: an intellectual history. II. The molecular revolution (1981-2002). Genetics 165, 1227–1235.

Palmiter, R.D., Brinster, R.L., Hammer, R.E., Trumbauer, M.E., Rosenfeld, M.G., et al., 1982. Dramatic growth of mice that develop from eggs microinjected with metallothionein-growth hormone fusion genes. Nature 300, 611–615.

Papaioannou, V.E., Silver, L.M., 1998. The T-box family. BioEssays 20, 9–19.

Pauws, E., Hoshino, A., Bentley, L., Prajapati, S., Keller, C., Hammond, P., Martinez-Barbera, J.-P., Moore, G.E., Stanier, P., 2009. Tbx22(null) mice have a submucous cleft palate due to reduced palatal bone formation and also display ankyloglossia and choanal atresia phenotypes. Hum. Molec. Genet. 18, 4171–4179.

Pauws, E., Peskett, E., Boissin, C., Hoshino, A., Mengrelis, K., Carta, E., Abruzzo, M.A., Lees, M., Moore, G.E., Erickson, R.P., Stanier, P, 2013. X-linked CHARGE-like Abruzzo-Erickson syndrome and classic cleft palate with ankyloglossia result from TBX22 splicing mutations. Clin. Genet. 83, 352–358.

Paxton, C.N., Bleyl, S.B., Chapman, S.C., Schoenwolf, G.C., 2010. Identification of differentially expressed genes in early inner ear development. Gene Expr Patterns 10, 31–43.

Peterson, K.R., Li, Q.L., Clegg, C.H., Furukawa, T., Navas, P.A., Norton, E.J., Kimbrough, T.G., Stamatoyannopoulos, G., 1995. Use of yeast artificial chromosomes (YACs) in studies of mammalian development: production of beta-globin locus YAC mice carrying human globin developmental mutants. Proc. Nat'l Acad. Sci. U.S.A. 92, 5655–5659.

Pfeiffer, A., Ikawa, M., Dayn, Y., Verma, I.M., 2002. Transgenesis by lentiviral vectors: Lack of gene silencing in mammalian embryonic stem cells and preimplantation embryos. Proc. Nat'l Acad. Sci., U.S.A. 99, 2140–2145.

Quaife, C.J., Pinkert, C.A., Ornitz, D.M., Palmiter, R.D., Brinster, RL, 1987. Pancreatic Neoplasia Induced by ras Expression in Acinar Cells of Transgenic Mice. Cell 48, 1023–1034.

Rebagliati, M.R., Melton, D.A., 1987. Antisense RNA injections in fertilized frog eggs reveal an RNA duplex unwinding activity. Cell 48, 599–605.

Rennebeck, G., Lader, E., Fujimoto, A., Lei, E.P., Artzt, K, 1998. Mouse Brachury the Second (T2) is a gene next to classical T and a candidate gene for tct. Genetics 150, 1125–1131.

Rohme, D., Fox, H., Herrmann, B., Frischauf, A.-M., Edstrom, J-E, Mains, P, Silver, LM, Lehrach, H, 1984. Molecular clones of the mouse t complex derived from microdissected metaphase chromosomes. Cell 36, 783–786.

Rudnicki, M.A., Braun, T., Hinuma, S., Jaenisch, R., 1992. Inactivation of MyoD in mice leads to up-regulation of the myogenic HLH gene Myf-5 and results in apparently normal muscle development. Cell 71, 383–390.

Ruvinsky, I., Silver, L.M., 1997. Newly identified paralogous groups on mouse chromosome 5 and 11 reveal the age of a T-box cluster duplication. Genomics 6, 262–266.

Russ, A.P., Wattler, S., College, W.H., Aparacio, S.A.J.R., Carlton, M.B., Pearce, J.J., Barton, S.C., Surani, M.A., Ryan, K., Nehls, M.C., Wilson, V., Evans, M.J., 2000. Eomesodermin is required for mouse trophoblast development and mesoderm formation. Nature 404, 95–98.

Sauer, B., Mosinger Jr., B., Lee, E.J., Manning, R.W., Yu, S.H., Mulder, K.L., Westphal, H., 1992. Targeted oncogene activation by site-specific recombination in transgenic mice. Proc. Nat'l Acad. Sci., U.S.A. 89, 6232–6236.

Schinzel, A., 1987. Ulnar-mammary syndrome. J. Med. Genet. 24, 778–781 1987.

Schwartzberg, P.L., Goff, S.P., Robertson, E.J., 1989. Germ line transmission of a c-abl mutation produced by targeted disruption in ES cells. Science 246, 799–803.

Seaver L., H., Pierpont J., W., Erickson R., P., Donnerstein R., L., Cassidy S., B., 1994. Pulmonary atresia associated with maternal 22q11.2 deletion: possible parent of origin effect in the conotruncal anomaly face syndrome. J Med Genet 31, 830–834.

Serfling, E., Jasin, M., Schaffn, W., 1985. Enhancers and eukaryotic gene transcription. Trend in Genet. 1, 224–230.

Shprintzen, R.J., Goldberg, R.B., Lewin, M.L., Sidoti, E.J., Berkman, M.D., Argamaso, R.V., Young, D., 2021. learning disabilities: velo-cardio-facial syndrome. Cleft Palate J 15, 56–62.

Shull, M.M., Ormsby, I., Kier, A.B., Pawlowski, S., Ribald, R.J., Yin, M., Allen, R., Sidman, C., Proetzei, O., Calvin, D., Annunziata, N., Deutschman, T., 1992. Targeted disruption of the mouse transforming growth factor-beta1gene results in multifocal inflammatory disease. Nature 359, 693–699.

Singh R.N., Singh N.N., Singh N.K., Androphy E.J. (2010) Spinal muscular atrophy (SMA) treatment via targeting of SMN2 splice site inhibitory sequences. US Patent 7838657B2.

Smithies, O., Gregg, R.G., Boggs, S.S., Koralewski, M.A., Kucherlapti, R.S., 1985. Insertion of DNA sequences into the human chromosome beta-globin locus by homologous recombination. Nature 317, 230–234.

Sowden J., C., Holt J. K., L., Meins, M., Smith H., K., Bhattacharya S., S., 2001. Expression of Drosophila omb-related T-box genes in the developing human and mouse neural retina. Invest. Ophthal. Vis. Sci. 42, 3095–3102.

Sparrow, D.B., McInerney-Leo, A., Gucev, Z.S., Gardiner, B., Marshall, M., et al., 2013. Autosomal dominant spondylocostal dysostosis is caused by mutation in TBX6. Hum. Molec. Genet. 22, 1625–1631 2013.

Technau, U., Bode, H.R., 1999. HyBra, a Brachyury homologue, acts during head formation in Hydra. Development 126, 999–1010.

Technau, U., 2001. Brachyury, the blastopore and the evolution of the mesoderm. BioEssays 23, 788–794.

Temin, H.M., Mizutani, S., 1970. RNA-dependent DNA polymerase in virions of Rous sarcoma virus. Nature 226, 1211–1213.

van den Berg, J., van Ooyen, A., Mantei, N., Schambock, A., Grosveld, G., Flavell, R.A., Weissmann, C., 1978. Comparison of cloned rabbit and mouse β-globin genes showing strong evolutionary divergence of two homologous pairs of introns. Nature 276, 37–44.

Vivante, A., Kleppa, M.-J., Schulz, J., Kohl, S., Sharma, A., et al., 2015. Mutations in TBX18 cause dominant urinary tract malformations via transcriptional dysregulation of ureter development. Am. J. Hum. Genet. 97, 291–301.

Wade, M., 2015. High-Throughput Silencing Using the CRISPR-Cas9 System: A Review of the Benefits and Challenges. J. Biomole. Screening 20, 1027–1039.

Wagner, T.E., Hoppe, P.C., Jollick, J.D., Scholl, D.R., Hodinka, R.L., et al., 1981. Microinjection of a rabbit beta-globin gene into zygotes and its subsequent expression in adult mice and their offspring. Proc. Natl. Acad. Sci. USA 78, 6376–6380.

Watson, J.D., Crick, F.H.C., 1953. A Structure for Deoxyribose Nucleic Acid. Nature 171, 737–738.

Wattler, S., Russ, A., Evans, M., Nehls, M., 1998. A combined analysis of genomic and primary protein structure defines the phylogenetic relationship of new members of the T-box family. Genomics 48, 24–33.

Wilkinson, D.G., Bhatt, S., Herrmann, B.G., 1990. Expression pattern of the mouse T gene and its role in mesoderm formation. Nature 343, 657–659.

Williams, R.L., Hilton, D.J., Pease, S., Wilson, T.A., Steward, C.L., et al., 1988. Myeloid leukemia inhibitory factor maintains the developmental potential of embryonic stem cells. Nature 336, 684–687.

Wilson, G.N., Hollar, B.A., Waterson, J.R., Schmickel, R.D., 1978. Molecular analysis of cloned human 18S ribosomal DNA segments. Proc Nat'l Acad. Sci., U.S.A. 75, 5367–5371.

Wu, N., Ming, X., Xiao, J., Wu, Z., Chen, X., et al., 2015. TBX6 null variants and a common hypomorphic allele in congenital scoliosis. New Eng. J. Med. 372, 341–350 2015.

Wu, X., Kriz, A.J., Sharp, P.A., 2014. Target specificity of the CRISPR-Cas9 system. Quant. Biol. 2, 59–70.

Zhang, M., Strnatka, D., Donohue, C., Hallows, J.L., Vincent, I., Erickson, R.P., 2008. Astrocyte-Only Npc1 Reduces Neuronal Cholesterol and Triples Life Span of Npc1-/- Mice. J. Neurosci. Res. 86, 2848–2856.

Zhu, Y., Gramolini, A.O., Walsh, M.A., Zhou, Y.-Q., Slorach, C., et al., 2008. Tbx5-dependent pathway regulating diastolic function in congenital heart disease. Proc. Nat. Acad. Sci. 105, 5519–5524.

X-inactivation, epigenetics, and imprinting, including of the *T/t* complex

X-inactivation

The need for differential gene control of the 1 dose (male) versus 2 doses (females) of the X chromosome creates a major problem for gene expression. The X chromosome is large and mostly euchromatic (likely to contain active genes) while the Y chromosome is small and mostly heterochromatic (condensed and without much potential for gene expression). Mammals are highly intolerant of altered gene dosage. Humans who are trisomic for the smallest chromosome, 21 (22 actually has slightly more DNA), can survive to adulthood, most with severe limitations, but the only other 2 trisomies that can survive to birth (13 and 18) don't live very long. Most of these 3, and all other, trisomies are lost in utero. Thus, to maintain generally equivalent gene expression from the X in males and females, either the single male X needs to be up-regulated or one or both of the female Xs needs to be down-regulated. It is the latter approach, which mammals use.

Mary Lyon has described the discovery and development of the "Lyon hypothesis" of X-inactivation (Lyon, 1992). In the early 50s she had noted the variegated coat colors of the X-linked genes, *mottled* (Fraser et al., 1953) and *tabby*. In the ensuing years Barr (and Bertram) described a deeply staining condensation seen in neurons of female, but not male, cats that they were studying (Barr and Bertram, 1949) which became known as the "Barr body". This finding was soon extended to many species and many tissues. The thought-provoking cytogeneticist, Susumu Ohno, demonstrated that this "object" was one of the X-chromosomes in a condensed state (Ohno et al., 1959). The discovery of another X-linked color gene with mottled appearance in females (Cattanach et al., 1961) led Lyon to formulate her hypothesis (Lyon, 1961). Possibly independently, Russell (1961) "suggested a similar, but rather less complete, explanation for variegation in female mice carrying X-autosome translocations" (Lyon, 1992). [1] Beutler, et al (1962) extended the observation to biochemical studies by invoking X-inactivation to explain mosaicism for glucose-6-phosphate dehydrogenase in the red blood cells of female carriers for its deficiency. Lyon's paper had appeared in April while Russell's appeared in June and, thus, there was a question of priority, which might be the reason that the discovery did not earn a Nobel prize. (Beutler's article appeared in January of the following year and cited Lyon's but not Russell's papers; he was very grateful to Susumu Ohno for helpful discussions.) Mary Lyon received many other honors (see mini-biography).

Mary F. Lyon, 1925–2014

Mary Lyon (see Fig. 10.1) presents much of her scientific autobiography in her "A personal history of the mouse genome" (2002). A brief account with her full bibliography to 2006 is that of Jane Gitschier (2006) which was based on interviews with her. She was born the eldest of 3 children in Norwich, the cathedral city for the county of Norfolk and not too far from Cambridge. Her father was in the Army Medical Corp during WWI and served in the Near East and then in France. Afterwards, he was a civil servant as an inspector of taxes. Her mother was a school teacher until her marriage. The family moved around somewhat and it was in Birmingham at King Edward VI High School for Girls that she developed her interest in biology. She won an essay contest with a prize of four books on wild flowers, birds and trees. Her interest was strengthened by a biology teacher and a maternal relative who was a very well-known animal scientist (Gitschier, 2006). She was an undergraduate at Cambridge University from 1943-1946 and majored in zoology. She lived at, and was a member of, Girton College, the first women's college which was located quite far from the center of Cambridge where all the men's colleges were located.[1] It was not formally admitted to the University as one of the colleges until 1948. Although Waddington (see Chapter 3 and Salome Glueksohn-Walsh mini-biography) was absent on wartime service, she read, and was impressed by, his writings and came to believe "that genes clearly must underlie embryological development. This seems blindingly obvious now but at that time it was a relatively new idea" (Lyon, 2002). R A Fisher was the professor of genetics at Cambridge at the time—"he gave a course of lectures, which were largely incomprehensible" (ibid.). When she graduated she was only awarded a "titular" degree since Cambridge degrees were not yet awarded to women.

As mentioned in Chapter 3, she started a Ph.D. in Fisher's department and was co-author of a very theoretical paper on sex linkage with him (and Owen; Fisher et al., 1947). She remained in Cambridge long enough to discover that mice with the *pallid* mutation, studied by her because of abnormal postural reflexes, had no otoliths in the inner ear (Lyon, 1951). She also studied the causes of variable penetrance of *pallid*, showing that maternal age and litter size were important (Lyon, 1953). However, as stated in chapter 3, she thought that mapping genes was not enough to do for a thesis and moved to Edinburgh where Waddington became Professor of Genetics after the war. She finished her Ph.D. with D. S. Falconer, but this work does not seem to have been published. She was offered a position with Toby Carter who was funded to study radiation-induced mutations in mice (see Chapter 6) and was soon publishing with him (Carter et al., 1957). When Carter transferred his group to the newly established MRC Radiobiology lab at Harwell, close to Oxford, she went with him and was to remain there for her entire career (excluding a short sabbatical in Cambridge). As Lyon reports in her "personal history" (Lyon, 2002), the discovery of a series of sex-linked coat color mutants was particularly fortuitous. *Tabby* was found in 1953 (Falconer, 1953), as was *Mottled* (Fraser et al., 1953). In 1960, she found a sex-linked mutation which was provisionally named *Dappled-2* but was found to be indistinguishable from *Mottled* (Lyon, 1960). It was the variegated coat patterns of all 3 mutants that led her to wonder "why most X-linked mutants showed a variegated effect in heterozygotes and why the pattern of mottling resembled that seen in somatic mosaics" (Lyon, 2002). These speculations led to her formulation of the Lyon hypothesis as seen in Chapter 10.

In 1970–1971 she spent a sabbatical year in Cambridge with Richard Gardner to study mechanisms of X-inactivation in the embryo. In the early sixties, her colleagues, Tony Searle and Bruce Cattanach had begun to characterize X-autosome translocations generated in chemical or radiation mutagenesis experiments (Cattanach, 1961; Searle, 1962; see Chapter 6). She realized that these could be used to study the time of X-inactivation. Since Gardner was expert at embryo surgery, she wanted to inject single cells from embryos heterozygous for 2 coat color markers located on translocations of the relevant chromosomes to their X chromosomes. The recipient embryos had a third, and different coat color. Thus, if X-inactivation had already occurred, the embryo would be mosaic for 2 coat colors, if not, for three. They found that the resultant mice had 3 coat colors indicating a later X-inactivation (Gardner and Lyon, 1971). They used cells from all the early stages before implantation, 4.5 days, and still did not find X-inactivation. Although the failure of X-inactivation to fully spread into autosomal translocation was not yet appreciated and might have obscured the result, it was confirmed later by biochemical assays on single embryos (Epstein et al., 1978; see Chapter 10).

Mary Lyon was drawn to her extensive studies of the *T/t* complex because of the apparent high rate of mutation t^{ns} to other *tns*, which, of course was really due to rare crossing overs in the inverted regions (see Chapter 3). Her group's

---

[1] The "Girton Grind", as it was called, is the about a mile's bike ride into the center of Cambridge and the other colleges. There was a small hill where once there had been a castle by the edge of the river Cam and this needed some energy on the return to Girton.

extensive work on the genetics of the *T/t* complex (Chapter 3), the development of the mutant embryos (Chapter 4) and the mechanisms of *t*-allele transmission rate distortion (Chapter 8) have been extensively presented. "Lyon's elucidation of *t*-complex genetics and biology was a true testament of her love of mammalian genetics and development….With her incisiveness and knowledge of genetics, Lyon simplified the structure and function of the *t*-complex , both of which had stymied scientists for years" (Kalantry and Mueller, 2015).

At Harwell, she had become head of the genetics division in 1962 and this division became the independently funded Mammalian Genetics Unit of the Medical Research Council in1995. As discussed in Chapter 1, the Mouse News Letter was an important source of shared information and she edited it from 1956-1970. She then chaired the Committee on Standardized Genetic Nomenclature for Mice from 1975-1990. As the mass of mouse genetic information increased, organizing it and making it available became an important function. Margaret Green was the first editor of "Genetic Variants and Strains of the Laboratory Mouse", 1981. This 1st edition has 476 pages. When Lyon edited it, with Rastan and Brown (Lyon et al., 1996) it had expanded to 1807 pages.

As the number of mutants increased, the cost of maintaining them was becoming prohibitive. Thus, Lyon collaborated with David Whittingham (who had contributed greatly to the development of media for the culture of embryos, Chapter 4) on finding optimal conditions for freezing them (Whittingham et al., 1977a, 1977b). They were concerned about the potential long-term effects of background radiation and, so, exposed the embryos, frozen and maintained in liquid nitrogen, to about 100 times the annual dose of background radiation for 2 years. The frozen embryos developed beautifully when thawed and placed in pseudopregnant females and, thus, Harwell started a national, major storage facility for mutant embryos.

Mary Lyon was much appreciated by many colleagues and friends. "Although quiet and pensive, on a personal level May Lyon was considered warm and was interested in the well-being of her friends and colleagues and their families. She was an inspiration to and a supporter of young scientists" (Kalantry and Mueller, 2015)[2]. She was so careful about scientific communication that conversation could sometimes be difficult. "Because everything Mary said was so carefully thought through, she could be difficult to talk to on the phone, it was easy to think you had been cut off. She did not suffer fools gladly, but was a great supporter of the bright young scientist, often eschewing authorship of publications to enhance the profile of junior collaborators" (Rastan, 2015). As a young molecular biologist, Murray Brilliant worked with Mary (and Eva Eicher) to identify and clone the gene encoded by classic coat color mutation, pink-eyed dilution, *p*, (Gardner et al., 1992) that together with albino, *c*, formed the first linkage group in the mouse (Haldane et al., 1915, see Chapter 3) Mary's contributions to this effort were essential, by providing a series of radiation-induced alleles and insights into complementation of these alleles (Lyon et al., 1992). Murray Brilliant recalls Mary's response on learning that he and collaborators had finally identified the gene: "Mary spontaneously uttered 'Oh' in a voice few decibels louder than her usual very soft and quiet voice. Knowing Mary's understated emotions, I immediately realized that this was equivalent to any other person shouting from the rooftops!"[3].

Mary Lyon received many honors. She was honorary treasurer of the Genetical Society (1968-1976) and served as its vice president from 1976 to 1979. She was a convener and Mammalian Genetics Group Member of the European Molecular Biology Organization. She chaired Committee 4 of the International Commission for Protection against Environmental Mutagenesis and Carcinogens (1977–1984). She was elected to the Royal Society in 1973 (the 28th women to be selected as a fellow) and was an ad hoc member of its Group of Human Fertilization and Embryology from 1982-1984. She received the Royal Society's highest honor of the Royal Medal in 1984. She was a foreign associate of the US National Academy of Science. Other awards included the Wolf Prize in Medicine (1997), the William Allan Award of the American Society of Human Genetics (1988), the March of Dimes Prize in Developmental Biology (2004), and the annual Pearl Meister Greengard Prize given to an outstanding women scientist by the Rockefeller University (2006). She received Honorary Doctorates from Princeton University and the Massachusetts Institute of Technology.

---

[2] My postdoctoral fellow, Dr. Susan E. Lewis, shared a room with Mary Lyon at the 1976 Mouse *t* complex meeting. She commented "I was awed by sharing a room with her. She didn't seem shy at all. She was very friendly and invited me to visit Harwell which I did some years later and she remembered me warmly. She, herself, didn't spend much time on her morning toilet but made sure I was up in time to get to the meeting at it's start—this was very important to her".

[3] Quote from Dr. Murray Brilliant, 2020.

FIG. 10.1 Mary Lyon on a visit to Tucson, Arizona in 1998, visiting Dr. Murray Brilliant. Photos courtesy of Dr. Murray Brilliant.

An early concern was the randomness of X-inactivation. X-inactivation was always paternal in marsupials (Sharman, 1971) and was found to be paternal in the extraembryonic tissues, such as trophectoderm and extra-embryonic endoderm (West et al., 1977; Takagi, 1978). The paternal X was also found to be preferentially inactivated in interspecies hybrids such as mules (Hamerton et al., 1969). Both Xs were found to be active in female germ cells, where meiotic pairing might require their "openness", and in the early embryo (Epstein et al., 1978), while inactivation occurred randomly in the early embryonic ectoderm. On the other hand, the X chromosome was always inactive during male spermatogenesis (Lifschytz and Lindsley, 1972; Erickson, 1976). The spermatogenic X-inactivation was clearly due to different mechanisms than the inactivation occurring in female somatic cells. This was demonstrated by the fact that DNA prepared from extraembryonic membranes where the paternal X is inactive (Kratzer et al., 1983) would readily transvect HGPRT⁻ cells to grow in HAT media, while DNA prepared from female somatic cells when only the inactive X is HGPRT⁺ will not. Of course, this is pre-gonadal and the X inactivation during spermatogenesis might be different. The history of X-chromosomal activity during spermatogenesis is further discussed in Chapter 8.

The answer to questions of an inactivation center and the stimulus for its activation were then sought. While the two active X-chromosomes in oocytes and early female embryos were inactivated, this did not occur in tetraploids, suggesting that the X:autosome ratio was being counted (Webb et al., 1992), as also seemed to be the case of very early sex determination (see Chapter 11). The location of a single X inactivation center was found by the study of many X:autosomal translocations which located a single inactivation center to band D of the X in mouse and Xq13 in humans (Russell and Montgomery, 1965). When the cloning era came, a gene from this chromosomal site, the *Xist* gene, was cloned in man (Brown et al., 1991) and mice (Brocksdorff et al., 1992). It was found to only be expressed from the inactive X; the timing of its expression strongly suggested a role in X-inactivation (Kay et al., 1993). Impressively, the gene was capable of inducing inactivation wherever it resides (as long as it can remain transcriptionally active)—even when inserted into an autosome (Lee et al., 1996; Lee and Jaenisch, 1997). Surprisingly, it was also expressed from the single X-chromosome of the male in embryonic stem cells but not when they were differentiated (Panning and Jaenisch,1996). Hypomethylation caused by knocking out DNA methyltransferase reestablished expression (ibid., see below, it was lethal at mid-gestation [Li et al., 1992]). The knockout of the gene prevented X-inactivation but not chromosome counting or random X-inactivation suggesting that multiple mechanisms were involved in the process of X-inactivation (Penny, el al, 1996) and did not involve sex determination (Mroz, et al, 1999). Human mutations in *XIST* were able to skew X-inactivation (Plenge et al., 1997).

It soon became apparent that there were regions of the "inactive X" that were not inactivated and these frequently had active homologs on the Y chromosome. These were from regions of homology between the X and Y chromosomes (Burgoyne, 1982; Polani, 1982). It was predicted that there would be a region of recombination between the X and Y chromosomes and that this region would not be inactivated on the X chromosome. These were named the "pseudoautosomal" regions (PAR). Genes later found to be in these regions were initially considered to be autosomal, e.g. steroid sulfatase (Erickson et al., 1983). Soon Goodfellow (1986) and associates cloned a gene from the human PAR region and many more were to follow. Some non-PAR genes, such as *Smcx,* (see Chapter 11) were found that showed tissue specific variation in the degree of escape from X-inactivation (Disteche, 1995; Carre et al., 1996; Sheardown et al., 1996). It was also found that a number of genes which are inactivated on the human X are not inactivated on the mouse X. Only genes with X and Y homologs were pseudoautosomal and human and mouse differed for this as well; Fig. 10.2.

This is one a possible explanation for the great difference in the phenotype of XO individuals in the 2 species: XO mice are initially fertile while XO humans are sterile (Ashworth et al., 1991). Alternatively, since XO mice end oogenesis at about 6 months of age, about the same chronological time that XO humans do so in utero, there is not a difference in absolute time which seems important for some pathological processes (Erickson, 1989) Fig. 10.3.

Mechanisms for the maintenance of X-chromosome inactivation were found slowly. Riggs (1975) suggested early that the inactive X was heavily CpG methylated. Although he suggested this in 1975, when writing a review on DNA methylation 5 years later, he had no certainty that such methylation was involved in gene silencing: "the biological role of m^5Cyt in eukaryotic DNA is not known, but its ubiquity suggests some important function" (Razin and Riggs, 1980). The notion only became more credible as it was established that inactive genes had great increases in CpG methylation (Bird et al., 1985; Bird, 1986). Monk, et al (1987) showed early demethylation in the preimplantation mouse embryo with subsequent and differential

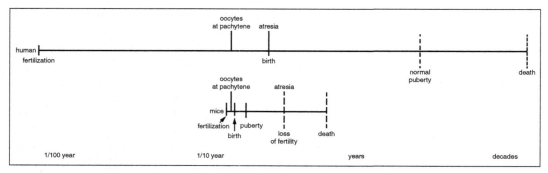

FIG. 10.2 A comparison of oogenesis in XO women and XO mice on a logarithmic timescale so the discrepant life spans can be easily depicted. The 2 life spans are aligned at the point where oocytes enter the pachytene stage of meiosis. From Erickson, 1989.

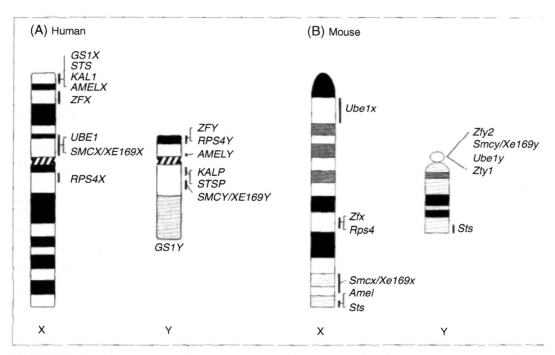

FIG. 10.3 Nonpseudoautosomal genes with X-Y homology compared for mouse and human. Gene positions indicated on right. If unknown, they are listed below the drawing. The mouse steroid sulfatase gene is pseudoautosomal while the human steroid sulfatase gene is not. Ellis, et al., 1989. With permission.

methylation in embryonic, extraembryonic and germ cell lineages (Groudine and Conklin, 1985). The repetitive DNA sequences were undermethylated in extraembryonic cell lineages but highly methylated in the embryonic lineage (Chapman et al., 1984). Pseudoautosomal genes were under methylated (Mondello, et al., 1988). When the *Xist* promoter was cloned in front of the luciferase gene, for easy detection, and used for transgenesis, DNA demethylation was found to be correlated with its expression in embryos and the testis (Goto et al., 1998) where

DNA methylation is extensive (Chapter 8). However, DNA methylation came to be seen as secondary for X-inactivation. Methylation of the inactive X did not occur in the extra-embryonic ectoderm and only was found sometime after X-inactivation had occurred (Lock et al., 1987). This fit a model developed by Stanley Gartler's group which found that large chromosomal domains were late replicated and transcriptionally inactive while transcription was blocked at a local level by methylation (Hansen et al., 1996).

An alternative modification of the inactive X chromosome was identified when changes in the DNA-binding histones, which regulate gene expression, were found. These included a lack of histone H4 acetylation (Jeppeson and Turner, 1993) which occurs in somatic female tissues but not during X-inactivation during spermatogenesis (Armstrong et al., 1997). Underacetylation of other histones was soon found (Belyaev et al., 1996), but only in promoter regions (Gilbert and Sharp, 1999). Antibodies to a unique histone found only on the inactive X chromosome allowed an easy cytogenetic detection of it (Costanzi and Pehrson, 1998). However, a conditional deletion of *Xist* disrupted histone macroH2A localization, but not the maintenance of X inactivation (Csankovszki et al., 1999). A complication originally thought limited to mice was an antisense gene to *Xist, Tsix*, which was transcribed from both Xs, (Lee et al., 1999). The human homolog, *TSIX*, had not been found because it was a truncated version (Migeon et al., 2001). The deletion of *Tsix* led to parent-of-origin effects with normal male transmission but mostly lethal female transmission (Lee, 2000). This was due to X-inactivation of both Xs in female embryos and suggested that *Tsix* contained an imprinting center.

The 21st Century brought levels of complexity not previously imagined. Both Xs were found to differ from autosomes by the presence of a large number of unique LINE-1 repeats (Bailey et al., 2000) which were hypothesized to be involved in the mechanism of X-inactivation by Mary Lyon (1998). A polycomb protein, a class of proteins much studied in *Drosophila* for their role in position effects, *Eed*, was found to be needed for X-inactivation (Mager et al., 2003). Analyses of chromatin conformational changes mediated by polycomb proteins (de Napoles et al., 2004) furthered the understanding of the inactivation mechanisms (reviewed in Pinheiro and Heard, 2017). A non-polycomb protein, APEN, works instead by its involvement with the transcription machinery (Dossin,et al, 2020). It appears as *Xist* transcription starts and is recruited to enhancers and promoters and has a domain which interacts with factors involved in the regulation of transcription initiation and elongation.

The end of the 20th Century and the beginning of the 21st Century also brought greatly improved understanding of genetic regulation by DNA methylation and by the "histone code". Methyl-binding proteins (MBDs) were found to interact with histone deacetylase (HDAC) linking methylation to histone-mediated condensation of chromosomal regions (Ng et al., 1999; Wade et al., 1999). The neologism, "methylome", was introduce by Andrew Feinberg in 2001 (Feinberg, 2001), since methylated cytosine could be identified during massively parallel DNA sequencing by prior treatment of the DNA with sodium bisulfite which converts cytosines to uracil while methylcytosine is resistant. It was soon highly studied, especially in cancer (Jackson-Grusby et al., 2001). While originally only found to be associated with inactive genes, CpG methylation was also found on the active allele during imprinting (see below). Specifically, the mice resulting from embryos with the KO of DNA methyltransferase I (DMTI) had H19 expressed (an imprinted gene, more below) from both parental alleles while the usually uniparentally expressed *Igf2* and *Igf2r* were completely silenced (Li et al., 1993; Shemer et al., 1997). The target region for these methylation events was a stretch of about 60 CpG dinucleotides located 2-4 kb upstream of the *H19* start site (Bartolomei, et a, 1993). A specific isozyme of DMT1, DMTo1, was found to only be active at the 8-cell stage and provided maintenance methylation at imprinted genes (Martinet et al., 1998). The understanding of the mechanism of action of Dmt1 was enhanced with the finding that UHRF1 (Ubiquitin-like protein containing Phd and ring finger domains) showed strong binding to hemi-methylated DNA as well as to Dmt1 (Bostick et al., 2007; Kim et al., 2007). Surprisingly, the enzyme isocitrate dehydrogenase, which is important in the citric acid cycle, was found to sometimes have mutations which impaired histone methylation by resulting in a modification of enzyme function so that it produced 2-hydroxyglutarate (Lu et al., 2012; Turcan et al., 2012).

Imprinting

Imprinting refers to heritable changes in gene expression limited to particular regions of the genome, leading to only one parental chromosome being expressed from that area in the offspring. It is an example of epigenetics in the broad sense (see below). It is better termed genomic imprinting to distinguish from the psychologist's use of the term for a phase-specific learning. Ernst Hadorn, in his masterful summary of developmental genetics to 1955, **Developmental Genetics and Lethal Factors,** devoted a chapter to "Maternal influence and cytoplasmic effects". Some of his examples, especially *frizzled* in fowls, were probably due to imprinting but his explanation was to emphasize maternal cytoplasmic factors in oogenesis. D L Nanney, seeking explanations for persisting state of altered phenotypes in *Paramecia*, used an example of Brink's work on the R locus in *Maize,* to show that chromosomal change could be involved: "Certain chromosomally localized alterations are specifically induced when particular homologous genetic elements are brought together in heterozygotes; these modifications may persist indefinitely after the inducing conditions are removed" (Nanney, 1958). In the early sixties it was established that, in certain insects, chromosomes transmitted through the female contain different information by being so transmitted, that is, compared to when they are transmitted through the male (Crouse, 1960) and she coined the term "imprinting". In this case, the imprinting results in differential elimination of the chromosomes in the male.

Maternal effects found at the *T/t* complex

Some of the earliest maternal effects found in mammals were those of mutations in the *T/t* complex. Two mutation in this complex showed maternal effects expressed by differences in the phenotypes of offspring from reciprocal crosses. *Fused (Fu)* heterozygotes had a tendency not to manifest and to be phenotypically normal, even though it was considered a dominant gene. The overlap with wild type also occurred sometimes with the homozygotes. Breeding experiments excluded modifying genes as the cause (Reed, 1937) but he noticed large differences in reciprocal crosses. If *Fu* came from the father, the percentage of normal phenotypes was 17% while if it came from the mother, it was 57%. This suppressive effect of maternal transmission was confirmed by Gluecksohn-Waelsch and Dunn (1954). Reed (1937) repeatedly back-crossed *Fu* to a normal strain and these differences persisted, ruling out cytoplasmic effects in the original strain. While **Gruneberg (1952)** suggested that the intrauterine environment could be responsible, the egg transfer experiments to test the hypothesis were not performed.

There were also maternal effects during the transmission of *T* and $t^{1^{\text{"Br"}}}$, a *t* allele which did not enhance the effects of *T*, i.e., *T/* $t^{1^{\text{"Br"}}}$ was not tailless. In regards to the former, the penetrance of *T* was less when the DBA variant was provided by the mother than with the father, when the cross was between the DBA and "Black Spot" strain (Wittman and Hamburgh, 1968). There was a much higher penetrance when the female was DBA. This influence of t^{12} on the penetrance of *T*-linked gene was confirmed for *Fused* and *Kinky* (Ruvinsky et al., 1988; Ruvinsky and Agulnik, 1990). Thus, the maternal genotype influenced the development in a different way than did the paternal. Similar effects were found for *hairpin-tail* (T^{hp}) by David Johnson (1975). Eventually this region of chromosome 17 was found to be an imprinted region (see below). It was also found that the closely linked *H-2* complex affected expressivity of *T* but only in the male (Mickova and Ivanyi, 1974).

Mouse genetics and rare human diseases lead to the notion of gametic imprinting

In the early 1980s, as embryo culture techniques and micromanipulation of embryonic nuclei become possible, it soon became apparent that embryos would not survive more than a few days without pronuclei of both paternal and maternal origins—parthenogenesis wasn't possible (McGrath and Solter, 1984; Surani et al., 1984). Azim Surani re-introduced the word "imprinting" for this phenomenon. Meanwhile, the mouse geneticists, especially Bruce Cattanach, were studying the biological properties of the many chromosomal translocations produced by their mutagenesis experiments (see Chapter 6). They found that some chromosomally balanced offspring showed a variety of abnormalities including slowed or advanced growth rates or lethality (in contrast to the aneuploid offspring which never survived, see Chapter 3). (Cattanach and Kirk, 1985; Cattanach, 1986). Cattanach concluded that some chromosomes had to be inherited from the mother, others from the father and some from both parents. Again, parental-specific imprinting was implicated. Parent-of-origin effects on specific transgenes was soon found, their insertion sites identifying single loci, which could were imprinted (Swain et al., 1987; Sapienza, 1989). The effects were frequently modified by strain background (reviewed in Chaillet, 1994).

Such imprinting was inportant in humans, too. Although the first case identified of uniparental inheritance in man in 1988 was bimaternal inheritance of a mutant cystic fibrosis allele (Spence et al., 1988), as cytogenetics gradually found smaller deletions after banding was introduced, the importance of uniparental disomy and imprinting became much more apparent. The not too rare Prader-Willi syndrome of mild dysmorphism, growth retardation, and delayed development was associated with chromosome 15 deletions (Ledbetter et al., 1981). It soon became apparent that these deletions were of paternal origin (Butler et al., 1983) but the notion of imprinting was not yet familiar to the human genetic' community. This was suggested to be due to imprinting effects by Nicholls, et al in 1989 (reviewed in Lalande, 1996). In contrast, deletions of the same region of paternal origin were found to be causal in the very different Angelman syndrome, a disorder of much greater delayed development (Knoll et al., 1989). Genetic imprinting provided the explanation for rare cases such as that of a trisomy 15 conceptus diagnosed by amniocentesis which was allowed to go to term (since trisomy 15 was always lethal and the fetus was growing well) and at birth the child was found to have Prader-Willi syndrome (Cassidy et al., 1992). During the resolution of the trisomy in the fetus (extraembryonic tissues having been sampled in the amniocentesis), it was the paternal chromosome 15 that was lost, leaving 2 copies of the maternal chromosome.

Imprinting was soon hypothesized to be the cause of genetic anticipation in which the descendants of one sex parent would transmit a more severe form of a dominant disease than the other, e.g. Huntingtons chorea where the offspring of fathers frequently had much earlier onset (Erickson, 1985). This was erroneous as the anticipation was due to greater expansions of the CAG repeats (leading to polyglutamine) in the *Huntingtin* gene when transmitted by the male than when transmitted by the female.

Using the *H19-Igf2* region of mouse chromosome 7 which is syntenic to human chromosome 15 and is homologous to the Prader-Willi imprinted region as an example, studies in mice found a 2.3 kb non-transcribed RNA encoded by *H19* (Leighton et al., 1995). Antisense transcripts were similarly found in the human *SNRPN* gene central to Angelman syndrome (Dittrich et al., 1996). Thus, as with X-inactivation and *Xist*, non-coding RNAs were involved in this genetic control. Small deletions identified controlling elements for imprinting in the human Prader-Willi/Angelman region (Dittrich et al., 1996). Yeast artificial chromosomes (YACs; see Chapter 9) were used to identify them in mice. A 300 kb YAC containing *Igfr2* showed appropriate imprinting when inherited from the mother, but not the father (Wutz et al., 1997). Deletion of a differentially methylated intronic CpG island within it abolished the imprinting (ibid.). Ainscough, et al (1997) used a YAC containing both *Igfr2* and *H19* and found appropriate mutually exclusive expression when transmitted by the different sexes. A 14 kb transgene containing only controlling elements, a 5' region with a differentially methlylated CpG island and a 3' un-transcribed segment showed sex-correct imprinting (Elson and Bartolomei, 1997). If either of these sequences was deleted from the transgene, imprinting no longer occurred (ibid.). Towards the end of the century, it was realized that imprinted regions were always in regions of late DNA replication which provided another means of detecting them (Simon et al., 1999). However, late replication was not the determining factor as a mouse mutation, *Mnf,* (*Minute*), at the *Igf2-H19* locus, altered imprinting without altering replication timing (Cerrato et al., 2003). Excellent reviews of the status of knowledge of imprinting at the end of the 20th Century are those of Constancia et al., (1998) and Dean and Ferguson-Smith (2001).

In the 21st Century, more than 20 regions and more than 100 genes were found to be imprinted in the mouse. There were tissue specific differences in rates of detection which led to the discovery of even more imprinted genes, e.g. Wang, et al (2008) for neonatal brain genes. Bacterial Artificial Chromosome (BAC) microarrays (Ching et al., 2005) and massively parallel sequencing of the "methylome" (see above) identified Differentially Methylated Regions

(DMRs) which corresponded to imprinted regions (Yu et al., 2012). Further details of the mechanisms involved were unraveled in the 21st Century. Regulation of the chromosome 7 (*Igfr2* and *H19*) imprinted region (syntenic to the human chromosome 15p, Prader-Willi/Angelman region) involved DMRs, methyl-CpG-binding proteins, histone deacetylases and the formation of chromatin insulator complexes (Sasaki et al., 2000). One particular maternal protein, GPC7/Stella, was found which prevented demethylation of a number of imprinted genes and interacted with a nuclear transport protein, RanBP5 (Nakamura et al., 2006). It was an essential protein and its loss led to early embryonic lethality. Irreversible inactivation by lysine9 methylation of histone H3 was found for *Oct-3/4,* one of the transforming factors needed for embryonic stem cell pluripotency (Feldman et al., 2006). Of course, it was found that this could be undone with exogenous supply of the gene, and a few others, to induce pluripotent stem cells (see Chapter 4).

Expression of *H19* and *Igf2* was studied in uniparental embryos made by artificial activation of oocytes showed that in androgenetic embryos, these genes are not inactivated (Sotomaru et al., 2003). Thus, some genes are not normally silenced in uniparental embryos suggesting that both parental genomes are required to establish maternal specific expression by trans-acting mechanisms (ibid). While the knockout of *H19* led to overgrowth of fetuses with upregulation of *Igf2*, over-expression of *H19* as a transgene re-established normal levels suggesting that the *H19* transcript is a trans-regulator of gene expression (Gabory et al., 2009).

The role of non-coding RNAs, specifically long non-coding RNAs (lncRNAs) was explored by progressive shortening of one (*Air* involved in *Igfr2* silencing; Sleutels et al., 2002) to show that only the portion over-lapping the promotor, thus preventing transcription was required (Latos et al., 2012).

These imprinted regions are re-set during gametogenesis. This occurs as the primordial germ cells enter the developing gonads by day E13 with re-methylation of some genes at E15 (Sanford et al., 1987; Kafri et al, 1992). This involved removing of the associated histones and other proteins maintaining the silenced state and demethylation of the cytosines (reviewed in Ferguson-Smith, 2011). However, most genes in the female are not re-methylated until the growing oocyte stage after birth when Dnmt1 is at very high levels in the nucleus (Martineit et al., 1998). When sperm were eventually induced from embryonic stem cells, the lack of the appropriate epigenetic reprogramming led to inviable offspring (Nayernai et al., 2006).

Epigenetics

As originally defined by Waddington (1942), the term referred to all secondary changes in gene expression. He developed these ideas while working on *Drosophila*. "Waddington became the first person to study Drosophila organogenesis through the systematic analysis of mutations and also perhaps the first to realize that embryogenesis is the external manifestation of genetic regulation" (Bard, 2008). More recently, a second meaning has limited the term to mean only changes in gene expression that can be inherited which has gradually replaced the original definition. Haig (2004) has argued that the disparate definitions had two, semi-independent origins, one from developmental geneticists and one from molecular biologists. The history of these two somewhat disparate definitions is further amplified by Deans and Maggert (2015). It wasn't until the 20th Century that the latter definition became dominant (Berger et al., 2009) but both definitions are still valid.

At the beginning of the 21st Century studies of transgenerational epigenetic inheritance were found at the A^{vy} *(viable yellow) allele* of the *agouti* locus. This mutation was due to the insert of a transposon, an Intra-cisternal A Particle (IAP) into a non-coding region of the *agouti* locus resulting in the expression of agouti being controlled by the degree of methylation of the cytosines in the Long Terminal Repeat (LTR) of the transposon. The degree of expression of the gene was very variable with coat color ranging from yellow to black when the allele was on a *black* background (*b/b*-see Chapter one; Argeson et al., 1996). There was a large difference in the degree of expression of the gene which remained variable when coming from the father but was related to the degree of expression in the mother, i.e., when the mother was yellow, more of the offspring were like her but her sons would again express the range of colors. This was an example of a transgenerational epigenetic change and the degree of it could be selected by choosing succeeding generations of either the mostly black or mostly yellow females (Cropley et al., 2012). While it seemed that putting the mothers on a "high-methyl donor" diet increased the number of yellow offspring, it was found that the locus was completely unmethylated in the blastocyst and that epigenetically inherited changes in chromatin configuration must be involved since hemizygosity for a Mel18, a polycomb group protein involved in chromatin structure, led to transgenerational inheritance of the degree of coat color by males (Blewitt, el al, 2006). Histone H3.3 insertion into chromatin was particularly important (Ng and Gurdon, 2008) and the interaction of the polycomb repressive complex 1 (PCR1) with the *Suv39h* H3K9 trimethyltransferases was found to specify parental asymmetry for constitutive heterochromatin (Puschendorf et al., 2008). A further modification of 5-methylcytosine to 5-hydroxymethyl cytosine by the Tet3 dioxygenase, another essential maternal protein, was involved in epigenetic reprogramming (Gu et al., 2011). However, the "guru" of genetic regulation, Mark Ptashne, was particularly critical of the notion that methylation was a primary event in gene regulation (Ptashne , 2014). It was also found that retarding the growth of embryos by in vitro culture from 1 cell to the blastocyst led to mice with hypomethylation of the locus (mechanism not established for the hypomethylation, however) and more yellow coat in the A^{vy} offspring (Morgan et al., 2008). This demonstration of an environmental stress leading to transgenerational inheritance was confirmed in humans. A search for transgenerational inheritance at other IAPs in the mouse genome found little evidence for other genes with this property (Kazachenka et al., 2018).

It was in the 21st Century that epidemiological studies implicated transgenerational epigenetic inheritance in a large number of human diseases. The first, and perhaps clearest study was that of the offspring of women pregnant during in The Netherlands during the winter of 1943–1944 when the Germans imposed a near starvation limitation of food for the rebellious population. The offspring were frequently obese and had metabolic syndrome (Ravelli et al., 1976). It was as if epigenetic changes activated genes in the fetus maximizing caloric retention later in life, but such was not the interpretation at the time. The effects were confirmed by relating infant birth weights to adult cholesterol levels by Barker, et al (1993). D J Barker become a major proponent of these effects and the phenomenon was known as "The Barker Hypothesis". Recent studies implicate dietary effects on the maternal gut microbiome as being a mediator of such epigenetic changes (Ma et al., 2014). Further examples include maternal influences on mental illnesses such as schizophrenia (reviewed in Bohacek et al., 2013), asthma and allergies (reviewed in Miller and Ho, 2008), and aging where even identical twins differ in their changing DNA methylation patterns (Fraga et al., 2005). Although most of these effects were thought to occur during in utero development or early childhood, there was also evidence for modifications of gene expression during adolescence (Gouin et al., 2017).

There is also evidence for paternal effects. A genomic screen using ENU (see Chapter 6) to screen for genes involved in reprogramming of the epigenome led to the finding that heterozygosity for the mutations resulted in paternal, but not maternal, effects in the offspring (Chong et al., 2007). A stress-induced, depression-like state in male mice influenced their daughter's investments in maternal care (Dietz et al., 2011), while paternal obesity influenced the metabolism of first and some second generation offspring (Fullston et al., 2013). Such paternal imprinting is the likely reason for the old observation of an effect of the father's drinking, independent of the mother's, on the weight of the newborn in 1987, before imprinting was well known (Little and Sing, 1987). Paternal epigenetic inheritance of many traits, through sperm RNA rather than by DNA modifications, is reviewed by Chen, et al (2016). However, sperm RNA's effects require such modifications since they are abrogated in the absence of cytosine methyltransferase (Kiani et al., 2013). Such results provided an explanation for the much earlier observation of RNA in spermatozoa (Betlach and Erickson, 1973) as discussed in Chapter 8.

Chromosomal 17 imprinting and the T/t complex

One of the early discoveries of maternal effects found before the understanding of imprinting was (see other examples above) the T^{Hp} mutation that can be transmitted through the male, but not through the female (Johnson,1975). When the mutation is transmitted through the female, the embryos die perinatally, whereas when transmitted by the male, the mutation survives with little fetal loss. Nuclear transplantation experiments demonstrate that the difference is not in the cytoplasm of the egg but is an inherent property of the mutation having been transmitted by an oocyte derived nucleus rather than by a spermatogenesis-derived nucleus (McGrath and Solter, 1984). Intriguingly, there are parallels in the imprinted genes of the *T* region of chromosome 17 in mice (*Igfr2*) with that of the Prader-Willi and Angelman syndromes region of chromosome 15p in humans (*IGF2*) This example, as well as all or nearly all imprinted regions, involve growth factors and are involved in the maternal vs. paternal genome competition where, with multiple male inseminations of 1 litter, it is in the mother's interest to invest resources in all the offspring and the father's interest to have the investment only in his own; Moore and Haig, 1991). Thus, imprinting might not be selected for in monogamous species. Indeed, hybrids between a monogamous and a polyandrous species of *Peromysci* showed parent-of-origin growth defects (Vrana et al., 1998). This notion of parental conflict is also supported by the results of experiments in which imprinted regions of the genome were deleted. It only took the deletion of 3 imprinted regions in parthenogenetic haploid ESCs to restore normal growth and fertility of bimaternal mice (Kawahara et al., 2007). On the other hand, it required the deletion of 7 imprinted regions to get bipaternal mice and they died at birth (Li, et al, 2018). An alternative view, based on an analysis of the evolution of imprinting in mammals, suggests that imprinting of many genes which are expressed in the brain could indicate that effects on behavior are important (John and Surani, 2000). Certainly. many of the imprinted genes are implicated in human disorders of neural development (Nicholls, 2000). Nonetheless, the effects could still be on growth through the brain/pituitary axis. For instance, the maternal-only, chromosome 15 disorder, Prader-Willi syndrome, has short stature corrected by treatment with human growth hormone, a pituitary hormone.

The imprinting of *Igfr2* was studied in detail by Barlow et al. (1991). It maps into the deletion of t^{Lub2} along with *Plg, Tcp1*, and *Sod2* and was only transcribed from the maternal allele while the others were expressed biallelically. This is in contrast to the non-transcribed *H19* and *Igf2* on chromosome 7 which are only transcribed from the paternal allele, as are these in the syntenic stretch of human chromosome 15 affected in Prader-Willi syndrome (see above). Thus, t^{Lub2} could only be viably transmitted by the father. At the time, general implications of growth factors in imprinting was not fully appreciated and *Igfr2* imprinting was probably responsible for the effect of the *T-associated maternal effect (Tme)* locus mapped to this region (Tsai and Silver, 1991) which, however, could be abrogated by a paternally derived duplication of the locus. At the turn of the Century, Wutz, et al (2001) created a non-imprinted version of *Igfr2* by using knockout/knockin (see Chapter 9) to create an allele lacking the Imprinting Control Region (ICR) and demonstrated that it could rescue maternally inherited t^{Lub2}.

A number of new modalities were applied to identifying imprinted genes in the *T/t* complex. As discussed in Chapter 7, the F9 antigen was a putative antigen encoded by t^{12}, a notion which was found to be false. In the work of Erickson and Lewis (1980) which found F9 antigen to be expressed in t^{12}/t^{12} embryos, an interesting effect of the parental origin of *T* was found in the timing of it's expression: when *T* came from the father there was delayed expression of the antigen so that the percentage of the embryos expressing F9 was low at day 2 ½ while 100% expressed the antigen by day 3 ½. The delay of antigen expression was seen in the controls as well, although there the effect was not as large although statistically just as significant. This was not interpreted as due to imprinting at the time. Two-dimensional gel electrophoresis of proteins radioactively labelled in the developing blastocyst identified Tcp-1a and Tcp-1b allelic products (see Chapter 8) among them and showed delayed expression of the paternal allele (Sanchez and Erickson, 1985).

Many other genes from the *T/t* complex region may yet be found to be involved in the imprinting of the region. Surprisingly, since it is only 1.5% of the genome, 6.5% of extraembryonic transcripts identified in a search for genes possibly involved in implantation mapped to the region (Ko, et al 1998). Another fascinating finding was that of random monoallelic expression of three genes clustered very closely to the *T/t* complex (Sano et al., 2001). These were not imprinted genes since which allele was inactive could alternate with each cell division in clonal bone marrow stem cells which the authors established.

Notes

1. Susumu Ohno, in his seminal book, **Sex chromosomes and Sex-linked genes (1967)** certainly gives Mary Lyon full credit for the idea: "The elucidation of the dosage compensation mechanism of mammals, however, we owe to Mary F. Lyon (1961). Ingenuity of Lyon's hypothesis lies in her assumption that random inactivation of one or the other X occurs early in the embryonic development of the female....once an X-chromosome has been inactivated, it remains inactivated in all descendants of that cell" **(Ohno, 1967, p.p. 97–98)**.

2. I met Hans-Beat Hadorn, a pediatric gastroenterologist and then Chairman of Pediatrics at the University of Graz, at a pediatrics meeting. He was very surprised that I knew of his father's work-I guess that developmental genetics was not yet part of pediatric genetics in German-speaking countries.

References

Ainscough, J.F.-X., Koide, T., Tada, M., Barton, S.C., Surani, M.A., 1997. Imprinting of Igf2 and H19 from a 130 kb YAC transgene. Develop 124, 3621–3632.

Argeson, A.C., Nelson, K.K., Siracusa, L.D., 1996. Molecular basis of the pleiotropic phenotype of mice carrying the hypervariable yellow (Ahvy) mutation at the agouti locus. Genetics 142, 557–567.

Armstrong, S.J., Hulten, M.A., Keohane, A.M., Turner, B.M., 1997. Different strategies of X inactivation in germinal and somatic cells: histone H4 underacetylation does not mark the inactive X chromosome in the mouse male germline. Exp Cell Res 230, 399–402.

Ashworth, A., Rastan, S., Lovell-Badge, R., Kay, G., 1991. X-chromosome inactivation may explain the difference in viability of XO humans and mice. Nature 351, 406–408.

Bailey, J.A., Carrel, L., Chakravarti, A., Eichler, E.A., 2000. Molecular evidence for a relationship between LINE-a elements and X chromosome inactivation: The Lyon repeat hypothesis. Proc. Nat'l Acad. U.S.A. 97, 6634–6639.

Bard, J.B.L., 2008. Waddington's Legacy to Developmental and Theoretical Biology. Biol. Theory. 3, 188–197.

Barker, D.J., Martyn, C.N., Osmond, C., Hales, C.N., Fall, C.H., 1993. Growth in utero and serum cholesterol concentrations in adult life. Brit. Med. J. 307, 1524. doi:10.1136/bmj.307.6918.1524.

Barlow D.P., Stoger R., Hermann M.L., Saito K., Schweifer N., (1991). The mouse insulin-like growth factor type-2-receptor is imprinted and closely linked to the TME locus. Nature 349, 84–87.

Barr, M.L., Bertram, E.G., 1949. A morphological distinction between neurones of the male and female, and the behaviour of the nucleolar satellite during accelerated nucleoprotein synthesis. Nature 163, 677–679.

Bartolomei, M.S., Webber, A.L., Brunkow, ME, Tilghman, SM, 1993. Epigenetic mechanisms underlying the imprinting of the mouse *H19* gene. Genes & Devel 7, 1663–1673.

Belyaev, N.D., Keohane, A.M., Turner, B.M., 1996. Differential underacetylation of histones H2A, H3 and H4 on the inactive X chromosome. Human Genetics 97, 573–578.

Berger, S.L., Kouzarides, T., Shiekhattar, R., Shilatifard, A., 2009. An operational definition of epigenetics. Genes & Devel 23, 781–783.

Beutler, E., Yeh, M., Fairbanks, V.F., 1962. The normal human female as a mosaic of X-chromosome activity: studies using the gene for G-6-PD-deficiency as a marker. Proc. Nat'l Acad. Sci., U.S.A. 48, 9–16.

Betlach, C.J., Erickson, R.P., 1973. A unique RNA species from maturing mouse spermatozoa. Nature 242, 114–115.

Bird, A., Taggart, M., Frommer, M., Miller, O.J., MacLeod, D., 1985. A fraction of the mouse genome that is derived from islands of nonmethylated, CpG-rich DNA. Cell 40, 91–99.

Bird, A.P., 1986. CpG-rich islands and the function of DNA methylation. Nature 321, 209–213.

Blewitt, M.E., Vickaryous, N.K., Paldi, A., Koseki, H., Whitelaw, E., 2006. Dynamic Reprogramming of DNA Methylation at an Epigenetically Sensitive Allele in Mice. PLOS Genet. doi:10.1371/journal.pgen.0020049.

Bohacek, J., Gapp, K., Bechara, J., Saab, I., Mansuy, M., 2013. Transgenerational epigenetic effects on brain function. Biol. Psychiat. 73, 313–330.

Bostick, M., Kim, J.K., Esteve, P.-O., Clark, A., Pradhan, S., Jacobsen, S.E., 2007. UHRF1 plays a role in maintaining DNA methylation in mammalian cells. Science 317, 1760–1764.

Brockdorff, N., Ashworth, A., Kay, G., McCabe, V., Norris, D.P., Cooper, P., Swift, S., Rastan, S., 1992. The product of the mouse *Xist* gene is a 15 kb inactive X-specific transcript containing no conserved ORF and located in the nucleus. Cell 71, 515–526.

Brown, C.J., Balabio, A., Rupert, J.L., Lafreniere, R.G., Gromple, M., Tonlorenzi, R., Willard, H.F., 1991. A gene from the region of the human X inactivation centre is expressed exclusively from the inactive X chromosome. Nature 349, 38–44.

Burgoyne, P.S., 1982. Genetic homology and crossing over in the X and Y chromosomes of mammals. Hum. Genet. 61, 85–90.

Butler, M.G., Palmer, C.G., 1983. Parental origin of chromosome 15 deletion in Prader-Willi syndrome. Lancet 1, 1285–1286.

Carter, T.C., Lyon, M.F., Phillips, R.J.S., 1957. Induction of Sterility in Male Mice by Chronic Gamma Irradiation. Brit. J. Radiol. doi:10.1259/0007-1285-27-320-418.

Cassidy, S.B., Lai, L.-W., Erickson, R.P., Magnuson, L., Thomas, E., Gendron, R., Hermann, J., 1992. Trisomy 15 with loss of paternal 15 as a cause of Prader-Willi Syndrome due to maternal disomy. Am. J. Hum. Genet. 51, 701–708.

Carrel, L., Hnt, P.A., Willard, H.F., 1996. Tissue and lineage-specific variation in inactive X chromosome expression of the murine *Smcx* gene. Human Mol. Genet., 1361–1366.

Cattanach, B.M., 1961. A chemically-induced variegated-type position effect in the mouse. Zeit. Vererbungsl. 92, 165–182.

Cattanach, B.M., 1986. Parental origin effects in mice. J. Embyol. Exp. Morphol. 97, S137–S150.

Cattanach, B.m., Kirk, M., 1985. Differential activity of maternally and paternally derived chromosome regions in mice. Nature 315, 496–498.

Cerrato, F., Dean, W., Davies, K., Kagotani, K., Mitsuya, K., Okumura, K., Riccio, A., Reik, W., 2003. Paternal imprints can be established on the maternal *Igf2-H19* locus without altering replication timing of DNA. Hum. Mol. Genet. 12, 3123–3132.

Chaillet, J.R., 1994. Genomic imprinting: lessons from mouse transgenes. Mut. Res. 307, 441–449.

Chapman, V., Forrester, L., Sanford, J., Hastie, N., Rossant, J., 1984. Cell lineage specific undermethylation of mouse repetitive DNA. Nature 307, 284–286.

Chen Q Yan, W., Duan, E., 2016. Epigenetic inheritance of acquired traits through sperm RNAs and sperm RNA modifications. Nat. Rev. Genet. 17, 733–743.

Ching, T.-T., Maunakea, A.K., Jun, P., Hong, C., Zardo, G., and 7 more, 2005. Epigenome analyses using BAC microarrays identify eolutionary conservation of tissue-specific methylation of *SHANK3*. Nature Genet 37, 645–651.

Chong, S., Vickaryous, N., Ahe, A., Zamudio, N., Youngson, N., and 9 more, 2007. Modifiers of epigenetic reprogramming show paternal effects in the mouse. Nature Genet 39, 614–622.

Constancia, M., Pickard, B., Kelsey, G., Reik, W., 1998. Imprinting Mechanisms. Genome Res 8, 881–900.

Costanzi, C., Pehrson, J.R., 1998. Histone macroH2A1 is concentrated in the inactive X chromosome of female mammals. Nature 393, 599–601.

Cropley, J.E., Dang, T.H., Martin, D.I., Suter, S.M., 2012. The penetrance of an epigenetic trait in mice is progressively yet reversibly increased by selection and environment. Proc. Bio. Sci. B 279, 2347–2353.

Crouse, H.V., 1960. The controlling element in sex chromosome behaviour in Sciara. Genetics 45, 1429–1443.

Csankovszki, G., Panning, B., Bates, Pehrson, J.R., Jaenisch, J., 1999. Conditional deletion of *Xist* disrupts histone macroH2A localization but not maintenance of X inactivation. Nature Genet 22, 323–324.

Dean, W., Ferguson-Smith, A., 2001. Genomic imprinting: mother maintains methylation marks. Current Biol 11, R527–R530.

Deans, C., Maggert, K.A., 2015. What do you mean, "epigenetic. Genetics 199, 887–896.

de Napoles, M., Mermoud, J.E., Wakao, R., Tany, Y.A., Endoh, M., Appanah, R., Nesterova, T.B., Arie, J.S., Otte, A.P., Vidal, M., Koseki, H., Brockdorff, N., 2004. Polycomb Group Proteins Ring1A/B Link Ubiquitylation of Histone H2A to Heritable Gene Silencing and X Inactivation. Devel. Cell 7, 663–673.

Dietz, D.M., La Pland, Q., Watts, E.L., Hodes, G.H., Russo, S.J., Feng, J., Oosting, R.S., Vialou, V., Neslier, J., 2011. Paternal transmission of stress-induced pathologies. Biol. Psychiat. 70, 408–414.

Disteche, C.D., 1995. Escape from X-inactivation in human and mouse. Trends in Genet 11, 18–22.

Dittrich, B., Buiting, K., Korn, B., Rickard, S., Buxton, J., and 6 more, 1996. Imprint switching on human chromosome 15 may involve alternative transcripts of the SNRPN gene. Nature Genet 14, 163–170.

Dossin, F., Pinheiro, I., Zyclicz, J.J., Roensch, J., Collombet, S., and 10 more, 2020. SPEN integrates transcriptional and epigenetic control of X-inactivation. Nature 578, 455–460.

Dunn, L.C., Gluecksohn-Waelsch, S., 1954. A genetical study of the mutation 'fused' in the house mouse, with evidence concerning its allelism with a similar mutation 'kink. J. Genet. 52, 383–391.

Ellis, N., Goodfellow, P.N., 1989. The mammalian pseudoautosomal region. Trends in Genet. 5, 406–410.

Elson, D.A., Bartolomei, M.S., 1997. A 5′ differentially methylated sequence and the 3′-flanking region are necessary for H19 transgene imprinting. Mol. Cell Biol. 17, 309–317.

Epstein, C.J., Smith, S., Travis, B., Tucker, G., 1978. Both X chromosomes function before visible inactivation in female embryos. Nature 274, 500–503.

Erickson, R.P., 1976. Glucose-6-phosphate dehydrogenase activity changes during spermatogenesis: Possible relevance to X-chromosome inactivation. Dev. Biol. 53, 124–137.

Erickson, R.P., Lewis, S.E., 1980. Cell surfaces and embryos: Expression of the F9 teratocarcinoma antigen in *T*-region lethal, other lethal, and normal pre-implantation embryos. J. Reprod. Immunol. 2, 293–304 1980.

Erickson, R.P., Harper, K., Kramer, J.M., 1983. Identification of an autosomal locus affecting steroid sulfatase activity among inbred strains of mice. Genetics 105, 181–189.

Erickson, R.P., 1985. Chromosomal imprinting and the parent transmission specific variation in expressivity of Huntington disease. Am. J. Hum. Genet. 37, 827–829.

Erickson, R.P., 1989. Why isn't a mouse more like a man? Trends in Genetics 5, 1–3.

Falconer, D.S., 1953. Total sex linkage in the house mouse. Zeit. Indukt. Abstammungs-Vererbungsl. 85, 210–219.

Feinberg, A.P., 2001. Methylation meets genomics. Nature Genetics 2, 9–10.

Feldman, N., Gerson, A., Li, E., Zhang, Y., Shinkai, Y., Cedar, H., Bergmann, Y., 2006. G9a-Mediated irreversible epigenetic inactivation of *Oct-3/4* during early embryogenesis. Nature Cell Biol 8, 188–194.

Ferguson-Smith, A.C., 2011. Genomic imprinting: the emergence of an epigenetic paradigm. Nature Rev. Genet. 12, 565–575.

Fisher, R.A., Lyon, M.F., Owen, A.R.G., 1947. The sex chromosome in the house mouse. Heredity 1, 355–365.

Fraga, M.F., Ballesta, E., Paz, M.F., Ropero, S., Setien, F., and 15 more, 2005. Epigenetic differences arise during the lifetime of monozygotic twins. Proc. Nat'l Acac. Sci., U.S.A. 102, 10604–10609.

Fraser, A.S., Sobey, S., Spicer, C.C., 1953. Mottled, a sex-modified lethal in the house mouse. J. Genet. 51, 217–221.

Fullston, T., Teague, E.M.C.O., Palmer, N.I., DeBlasio, M.J., Mitchell, M., Corbett, M., Print, C.G., Owens, J.A., Lane, M., 2013. Paternal obesity initiates metabolic disturbances in two generations of mice with incomplete penetrance to the F2 generation and alters the transcriptional profile of testis and sperm microRNA content. FASEB J 10, 4226–4243.

Gabory, A., Ripoche, M.-A., Le Digarcher, A., Watrin, F., Ziyyat, A., Forne, T., Jammes, H., Ainscough, J.F.X., Surani, M.A., Journot, L., Dandolo, L., 2009. *H19* acts as a trans regulator of the imprinted gene network controlling growth in mice. Develop 346, 3413–3421.

Gardner, J.M., Nakatsu, Y., Gondo, Y., Lee, S., Lyon, M.F., King, R.A., Brilliant, M.H., 1992. The mouse pink-eyed dilution gene: association with human Prader-Willi and Angelman syndromes. Science 257, 1121–1124.

Gardner, R.I., Lyon, M.F.F., 1971. X-chromosome inactivation studied by injection of a single cell into the mouse blastocyst. Nature 231, 385–386.

Gilbert, S.L., Sharp, P.A., 1999. Promoter-specific hypoacetylation of X-inactivated genes. Proc. Nat'l Acad., U.S. A. 94, 13825–13830.

Gitschier, J., 2006. The gift of observation: an interview with Mary Lyon. PLOS Genet 6 (1), e1000813. doi:10.1371/journal.pgen.1000813.

Goodfellow, P.J., Darling, S.M., Thomas, N.S., Goodfellow, P.N., 1986. A pseudoautosomal gene in man. Science 234, 740–743.

Gouin, J.P., Zhou, Q.Q., Booij, L., Boivin, M., Cote, S.M., Hebert, M., Ouellet-Morin, I., Szyf, M., Tremblay, R.E., Turecki, G., Vitaro, F., 2017. Associations among oxytocin receptor gene (*OXTR*) DNA methylation in adulthood, exposure to early life adversity, and childhood trajectories of anxiousness. Sci. Reports 7, 7446. doi:10.1038/s41598-017-07950-x.

Goto, T., Christians, E., Monk, M., 1998. Expression of an *Xist* promoter-luciferase construct during spermatogenesis and in preimplantation embryos: regulation by DNA methylation. Mol. Reprod. Develop. 49, 356–367.

Groudine, M., Conkin, K.F., 1985. Chromatin structure and de novo methylation of sperm DNA: implications for activation of the paternal genome. Science 228, 1061–1068.

Gruneberg, H., 1952. The Genetics of the Mouse, 2nd Ed. Martinus Nijhoff, The Hague. Netherlands.

Gu, T.-P., Guo, F., Yang, H., Wu, H.-P., Xu, G.-F., and 12 more, 2011. The role of Tet3 DNA dioxygenase in epigenetic reprogramming by oocytes. Nature 477, 606–610.

Haig, D., 2004. The (Dual) Origin of Epigenetics. Cold Spring Harbor Symp. Quant. Biol. LXIX, 67–70.

Haldane, J.B.S., Sprunt, A.D., Haldane, N.M., 1915. Reduplication in mice. J. Genet. 5, 133–135.

Hamerton, J.L., Gianelli, F., Colllins, F., Hallett, J., Fryer, A., McGuire, V.M., Short, R.V., 1969. Non-random X-inactivation in the female mule. Nature 222, 1277–1278.

Hansen, R.S., Canfield, T.K., Fjeld, A.D., Gartler, S.M., 1996. Role of late-replication timing in the silencing of X-linked genes. Human Mol. Genet. 9, 1345–1353.

Jackson-Grusby, L., Beard, C., Possemato, R., Tudor, M., Fambrough, D., Csankoszki, G., Dausman, J., Lee, P., Wilson, C., Lander, E., Jaenisch, R., 2001. Loss of genomic methylation causes p53-dependent apoptosis and epigenetic deregulation. Nature Genet 27, 31–39.

Jeppeson, P., Turner, B.M., 1993. The inactive X chromosome in female mammals is distinguished by a lack of histone H4 acetylation, a cytogenetic marker for gene expression. Cell 74, 281–289.

John, R.M., Surani, M.A., 2000. Genomic imprinting, mammalian evolution, and the mystery of egg-laying mammals. Cell: 101, 585–588.

Johnson, D.R., 1975. Further observations on the hairpin-tail (*THp*) mutation in the mouse. Genet. Res 24, 207–213.

Kafri, T., Ariel, M., Brandeis, M., Shemer, R., Urven, L., Mc Carry, J., Cedar, H., Razin, A., 1992. Developmental pattern of gene-specific DNA methylation in the mouse embryo and germ line. Genes & Devel 6, 707–714.

Kalantry, S., Mueller, J.L., 2015. Mary Lyon : A Tribute. Amer. J. Hum. Genet. 97, 507–511.

Kawahara, M., Wu, Q., Takahashi, N., Morita, s., Yamada, K., Ito, M., Ferguson-Smith, A.C., Kono, T., 2007. High-frequency generation of viable mice from engineered bi-maternal embryos. Nature Biotech 25, 1025–1050.

Kay, G.F., Penny, G.D., Patel, D., Ashworth, A., Brockdorff, N., Rastan, S., 1993. Expression of Xist during mouse development suggests a role in X chromosome inactivation. Cell 72, 171–182.

Kazachenka, A., Bertozzi, T.M., Sjoberg-Herrera, M.D., Walker, N., Gardner, J., more, 5, 2018. Identification, Characterization, and Heritability of Murine Metastable Epialleles: Implications for Non-genetic Inheritance. Cell 175, 1259–1271.

Kiani, J., Grandjean, V., Liebers, R., Tuorto, F., Ghanbarian, H., Lyko, F., Cuzin, F., Rassoulzadegan, M., 2013. RNA-mediated epigenetic heredity requires the cytosine methyltransferase Dnmt2. PLoS Genet 9, e10003498. doi:10.1371/journal.pgen.1003498.

Kim, J.K., Estee, P.-O., Clark, A., Prakhan, S., Jacobssen, S.E., 2007. UHRF1 plays a role in maintaining DNA methylation in mammalian cells. Sci 317, 760–764.

Knoll, J.H.M., Nicholls, R.D., Magenis, R.E., Graham Jr., J.M., Lalands, M., Latt, S.A., Opitz, J.M., Reynolds, J.F., 1989. Angelman and Prader-Willi syndromes share a common chromosome 15 deletion but differ in parental origin of the deletion. Am. J. Med. Genet. 32, 285–290.

Ko, S.H., Threat, T.A., Wang, X., Horton, J.H., Cui, Y.12 more, 1998. Genome-wide mapping of unselected transcripts from extraembryonic tissue of 7.5-day mouse embryos reveals enrichment in the *t*-complex and under-representation on the X chromosome. Hum. Mol. Genet. 7, 1967–1978.

Kratzer, P.G., Chapman, V.M., Lambert, H., Evans, R.E., Liskay, R.M., 1983. Differences in the DNA of the inactive X chromosomes of fetal and extraembryonic tissues of mice. Cell 33, 37–42.

Lalande, M., 1996. Parental imprinting and human disease. Ann. Rev. Genet. 30, 173–195.

Latos, P.A., Pauler, F.M., Koerner, M.V., Senergin, H.B., Hudson, Q.J., and 8 more, 2012. *Airn* Transcriptional Overlap, But Not Its lncRNA Products, Induces Imprinted *Igf2r* Silencing. Science 338, 1469–1472.

Ledbetter, D.H., Riccardi, V.M., Airhart, S.D., Strobel, R.J., Keenan, B.S., Crawford, J.D., 1981. Deletions of chromosome 15 as a cause of the Prader—Willi syndrome. New Engl. J. Med. 304, 325–329.

Lee, J.T., 2000. Disruption of Imprinted X Inactivation by Parent-of-Origin Effects at *Tsix*. Cell 103, 17–27.

Lee, J.T., Strauss, W.M., Dausman, J.A., Jaenisch, R., 1996. A 450 kb Transgene Displays Properties of the Mammalian X-Inactivation Center. Cell 86, 83–94.

Lee, J.T., Davidow, L.S., Warshawsky, D., 1999. *Tsix*, a gene anti-sense to *Xist* at the X-inactivation centre. Nature Genet 21, 400–404.

Lee, J.T., Jaenisch, R., 1997. Long-range *cis* effects of ectopic X-inactivation centres on a mouse autosome. Nature 386, 275–279.

Leighton, P.A., Saam, J.R., Ingram, R.S., Stewart, C.L., Tilgman, S.M., 1995. Disruption of imprinting caused by deletion of the *H19* region of the mouse. Nature 375, 34–39.

Li, E., Bestor, T.H., Jaenisch, R., 1992. Targeted mutation of the DNA methyltransferase gene results in embryonic lethality. Cell 69, 915–926.

Li, E., Beard, C., Jaenisch, R., 1993. Role for DNA methylation in genomic imprinting. Nature 366, 362–365.

Li, Z.-K., Wang, L.-Y., Wang, L.-B., Feng, G.-H., Yuan, Y.-W., and 10 more, 2018. Generation of Bimaternal and Bipaternal Mice from Hypomethylated Haploid ESCs with Imprinting Region Deletions. Cell Stem Cell 23, 665–676.

Lifschytz, E., Lindsley, D.L., 1972. The role of X-chromosome inactivation during spermatogenesis. Proc. Nat'l Acad. Sci., U.S.A. 69, 182–186.

Little, R.E., Sing, C.F., 1987. Father's drinking and infant birth weight: Report of an association. Teratol 36, 59–65.

Lock, L.F., Takagi, N., Martin, G.R., 1987. Methylation of the *Hprt* gene on the inactive X occurs after chromosome inactivation. Cell 48, 39–46.

Lu, C., Ward, P.S., Kapoor, G.S., Rohle, D., Turcan, S., and 12 more, 2012. IDH mutation impairs histone demethylation and results in a block to cell differentiation. Nature 483, 474–478.

Lyon, M.F., 1951. Hereditary absence of otoliths in the house mouse. J. Physiol. 114, 410–418.

Lyon, M.F., 1953. Absence of otoliths in the mouse: an effect of the pallid mutant. J. Genet. 51, 638–650.

Lyon, M.F., 1960. A further mutation of the mottled types in the house mouse. J. Hered. 51, 116–121.

Lyon, M.F., 1961. Gene action in the X-chromosome of the mouse (*Mus musculus*). Nature 190, 372–373.

Lyon, M.F., 1992. Some milestones in the history of X-chromosome inactivation. Ann Rev. Genet. 26, 17–28.

Lyon, M.F., King, T.R., Gondo, Y., Gardner, J.M., Nakatsu, Y., Eicher, E.M., Brilliant, M.H., 1992. Genetic and molecular analysis of recessive alleles at the pink-eyed dilution (p) locus of the mouse. Proc. Natl Acad. Sci. U. S. A. 89, 6968–6972.

Lyon, M.F., 1998. X-chromosome inactivation: A repeat hypothesis. Cytogenet. Cell Genet. 80, 133–137.

Lyon, M.F., 2002. A personal history of the mouse genome. Ann. Rev. Genomics Hum. Genet. 3, 1–16.

Lyon M.F., Rastan S., Brown S.D.M. (1996) "Genetic Variants and Strains of the Laboratory Mouse" Oxford University Press. 3rd Ed., 1807 pp, Oxford.

Ma, J., Prince, A.L., Bader, D., Hu, M., Ganu, R.more, 2014. High-fat maternal diet during pregnancy persistently alters the offspring microbiome in a primate model. Nature Commun 5. doi:10.1038/ncomms4889.

Mager, J., Montgomery, N.D., Pardo-Manuel de Villena, F., Magnuson, T., 2003. Genome imprinting regulated by the mouse Polycomb group protein Eed. Nature Genet 33, 502–507.

McGrath, J., Solter, D., 1984. Completion of mouse embryogenesis requires both the maternal and paternal genomes. Cell 37, 179–183.

Mary F. Lyon (1925–1997) in Women in the Biological Sciences: A bibliographic source book, edited by Grinstein LS, Bierman CA, Rose RK, Greenwood Press.

Mertineit, C., Yoder, J.A., Taketo, T., Laird, D.W., Trasler, J.M., Bestor, T.H., 1998. Sex-specific exons control DNA methyltransferase in mammalian germ cells. Develop 125, 889–897.

Mickova M., Ivanyi P., (1974). Sex-dependent and *H-2*-linked influence of expressivity of the Brachyury gene in mice. J. Hered. 65, 369–372.

Migeon, B.R., Chowdhury, A.K., Dunston, J.A., McIntosh, I., 2001. Identification of *TSIX*, encoding an RNA antisense to human *XIST*, reveals differences from its murine counterpart: implications for X inactivation. Am. J. Human Genet. 69, 951–960.

Miller, R.L., Ho, S., 2008. Environmental epigenetics and asthma: current concepts and call for studies. Amer. J. Crit. Care and Respir. Dis. Med. 177. doi:10.1164/rccm.200710-1511PP.

Mondello, C., Goodfellow, P.J., Goodfellow, P.N., 1988. Analysis of methylation of a human X located gene which escapes X inactivation. Nucleic Acids Res 16, 6813–6824.

Monk, M., Boubelik, M., Lehnert, S., 1987. Temporal and regional changes in DNA methylation in the embryonic, extraembryonic and germ cell lineages during mouse embryonic development. Devel 99, 371–382.

Moore, T., Haig, D., 1991. Genomic imprinting in mammalian development: A parental tug of war. Trends Genet 7, 45–49.

Morgan, H.D., Jin, X.l., Li, A., Whitelaw, E., O'Neill, C.O., 2008. The Culture of Zygotes to the Blastocyst Stage Changes the Postnatal Expression of an Epigenetically Labile Allele, Agouti Viable Yellow, in Mice. Biol. of Reprod. 79, 618–623.

Mroz, K., Carrel, L., Hunt, P.A., 1999. Germ cell development in the XXY mouse: evidence that X chromosome reactivation is independent of sexual differentiation. Develop. Biol. 207, 229–238.

Nakamura T., Arai Y., Umehara H., Masuhara M., Kimura T and 8 more (2006). PGC7/Stella protects against DNA methylation in early embryogenesis. Nature Cell Biol. 9, 66–71.

Nanney, D.L., 1958. Epigenetic control systems. Proc. Nat'l Acad. Sci., U.S.A. 44, 712–717.

Nayernia, K., Nolte, J., Michelmann, H.W., Lee, J.H., Rathsack, K.10 more, 2006. In Vitro-Differentiated Embryonic Stem Cells Give Rise to Male Gametes that Can Generate Offspring Mice. Dev. Cell 11, 125–132.

Ng, H.-H., Zhang, Y., Hendrich, B., Johnson, C.A., Turner, B.M., Erdjument-Bromage, H., Tempst, P., Reinberg, D., Bird, A., 1999. MBD2 is a transcriptional repressor belonging to the MeCP1 histone deacetylase complex. Nature Genet 23, 58–61.

Ng, R.K., Gurdon, J.B., 2008. Epigenetic memory of an active gene state depends on histone H33 incorporation into chromatin in the absence of transcription. Nature Cell Biol 10, 102–109.

Nicholls, R.D., 2000. The impact of genomic imprinting for neurobehavioral and developmental disorders. J. Clin. Invest. 105, 413–418.

Ohno, S., Kaplan, D.W., Kinosita, R., 1959. Formation of the sex chromatin by a single X-chromosome in live cells of *Rattus norvegicus*. Exp. Cell Res. 18, 415–418.

Panning, B., Jaenisch, R., 1996. DNA hypomethylation can activate *Xist* expression and silence X-linked genes. Genes Develop 10, 1991–2002.

Penny, G.D., Kay, G.F., Sheardown, S.A., Rastan, S., Brockdorff, N., 1996. Requirement for *Xist* in X chromosome inactivation. Nature 339, 131–137.

Pinheiro I., Heard E. (2017) X chromosome inactivation: new players in the initiation of gene silencing. F1000Research,6(F1000 Faculty Rev.):344. doi:10.12688/f1000research.10707.1.

Plenge, R.M., Hendrich, B.D., Schwartz, C., Fernando Arena, J., Naumova, A., Sapieza, C., Winter, R.M., Willard, H.F., 1997. A promoter mutation in the *XIST* gene in two unrelated families with skewed X-chromosome inactivation. Nature Genet 17, 353–356.

Polani, P.E., 1982. Pairing of X and Y chromosomes, non-inactivation of X-linked genes and the maleness factor. Hum Genet 60, 207–211.

Puschendorf, M., Terranova, R., Boutsma, E., Mao, X., Isono, K-I., and 7 more, 2008. PRC1 and Suv39h specify parental asymmetry at constitutive heterochromatin in the early mouse embryos. Nature Genet 40, 411–420.

Ptashne, M., 2014. The chemistry of regulation of genes and other things. J. Biol. Chem. 289, 5417–5435.

Rastan, S., 2015. Mary F. Lyon (1925-2014) Grande dame of mouse genetics. Nature 518, 36.

Ravelli, G.-P., Stein, Z.A., Susser, M.W., 1976. Obesity in Young Men after Famine Exposure In Utero and Early Infancy. New Eng. J. Med. 295, 349–353.

Razin, A., Riggs, A.D., 1980. DNA Methylation and Gene Function. Science 210, 604–610.

Reed, S.C., 1937. The Inheritance and Expression of Fused, a New Mutation in the House Mouse. Genet 22, 1–13.

Riggs, A.D., 1975. X inactivation, differentiation and DNA methylation. Cytogenet., Cell Genet 14, 5–25.

Russell, L.B., 1961. Genetics of mammalian sex chromosomes. Science 133, 1795–1803.

Russell, L.B., Montgomery, C.S., 1965. The use of X-autosome translocations in locating the X-chromosome inactivation center. Genetics 52, 470–471.

Ruvinsky, A.O., Agulnik, A.I., Protopopov, A.I., Agulnik, S.I., Belyaev, D.K., 1988. Effecct of the *t¹²* haplotype on penetrance and inheritance of the Fused and Kinky genes in mice. J. Hered. 79, 141–146.

Ruvinskky, A.O., Aglulnik, A.I., 1990. Gametic imprinting and the manifestation of the Fused gene in the house mouse. Develop. Genet. 11, 263–269.

Sanchez, E.R., Erickson, R.P., 1985. Expression of the *Tcp-1* locus of the mouse during early embryogenesis. J. Embyol. Exp. Morph. 89, 113–122.

Sanford, J.P., Clark, H.J., Chapman, V.M., Rossant, J., 1987. Differences in DNA methylation during oogenesis and spermatogenesis and their persistence during early embryogenesis in the mouse. Genes & Devel 1, 1039–1046.

Sano, Y., Shimada, T., Nakashima, H., Nicholson, R.H., Eliason, J.F., Kocarek, T.A., Ko, M.S.H., 2001. Random Monoallelic Expression of Three Genes Clustered within 60 kb of Mouse *t* Complex Genomic DNA. Genome Res 11, 1833–1841.

Sapienza, C., Paquette, J., Tran, T.H., Peterson, A., 1989. Epigenetic and genetic factors affect transgene methylation imprinting. Develop 107, 165–168.

Sasaki, H., Ishihara, K., Kato, R., 2000. Mechanisms of *Igf2/H19* Imprinting: DNA Methylation, Chromatin and Long-Distance Gene Regulation. J. Biol. Chem. 127, 711–715.

Searle, A.G., 1962. Is sex-linked Tabby really recessive in the mouse? Heredity (Edin.) 17, 297.

Sharman, G.B., 1971. Late DNA replication in the paternally derived X chromosome of female kangaroos. Nature 230, 231–232.

Sheardown, S., Norris, D., Fisher, A., Brockdorff, N., 1996. The mouse *Smcx* gene exhibits developmental and tissue specific variation in degree of escape from X inactivation. Hman Mol. Genet. 5, 1355–1360.

Shemer, R., Birger, Y., Riggs, A.D., Razin, A., 1997. Structure of the imprinted mouse *Snrpn* gene and establishment of its parental-specific methylation pattern. Proc. Nat'l Acad. Sci., U. S. A. 94, 10267–10272.

Simon, I., Tenzen, T., Reubinoff, B.E., Hillmen, D., McCarrey, J.R., Cedar, H., 1999. Asynchronous replication of imprinted genes is established in the gametes and maintained during development. Nature 401, 929–932.

Sleutels, F., Zwart, R., Barlow, D.P., 2002. The non-coding *Air* RNA is required for silencing autosomal imprinted genes. Nature 415, 810–813.

Sotommaru, Y., Katsuzawa, Y., Hatad, I., Obata, Y., Sasaki, H., Kono, T., 2003. Unregulated Expression of the Imprinted Genes *H19* and *Igf2r* in Mouse Uniparental Fetuses. J. Biol. Chem. 277, 12474–12478.

Spence, J.E., Perciaccante, R.G., Greig, G.M., Willad, H.F., Ledbetter, D.H., Hejmancik, J.F., Pollack, M.S., O'Brien, W.e., Beaudet, A.L., 1988. Uniparental disomy as a mechanism for human genetic disease. Am. J. Hum. Genet. 42, 217–226.

Surani, M.A.H., Barton, S.C., Norris, M.L., 1984. Development of reconstituted mouse eggs suggests imprinting of the genome during gametogenesis. Nature 308, 548–550.

Swain, J.L., Steward, T.A., Leder, P., 1987. Parental legacy determines methylation and expression of an autosomal transgene: a molecular mechanism for parental imprinting. Cell 50. 719–727.

Takagi, N., 1978. Preferential inactivation of the paternally derived X chromosome in mice. In: Russell, LB (Ed.), Genetic Mosaics and Chimeras in Mammals. Plenum, New York, pp. 341–360.

Tsai, J.-Y., Silver, L.M., 1991. Escape from genomic imprinting at the mouse *T-associated maternal effect (TME)* locus. Genet 129, 1159–1166.

Turcan S., Rohle D., Goenka A., Walsh L.A., Fang F and 15 more (2012) IDH1 mutation is sufficient to establish the glioma hypermethylator phenotype. Nature, 483: 479–483.

Vrana, P.B., Guan, X.-J., Ingram, R.S., Tilgman, S.M., 1998. Genomic imprinting is disrupted in interspecific *Peromyscus* hybrids. Nature Genet 20, 362–365.

Waddington, C.H., 1942. The "epigenotype. Endeavour 1, 18–20.

Wade, P.A., Gegonne, A., Jones, P.L., Ballestar, E., Aubry, F., Wolffe, A.P., 1999. Mi-2 complex couples DNA methylation to chromatin remodelling and histone deacetylation. Nature Genet 23, 62–66.

Wang, X., Sun, Q., McGrath, S.D., Mardis, E.R., Soloway, P.D., Clark, A.G., 2008. Transcriptome-Wide Identification of Novel Imprinted Genes in Neonatal Mouse Brain. PLOS One. doi:10.1371/journal.pone.0003839.

Webb, S., DeVries, T.J., Kaufman, M.H., 1992. The differential staining pattern of the X chromosome in the embryonic and extra-embryonic tissues of post-implantation homozygous tetraploid mouse embryos. Genet. Res. 59, 204–214.

West, J.D., Frels, W.I., Chapman, V.M., Papainnou, V.E., 1977. Preferential expression of the maternally derived X chromosome in the mouse yolk sac. Cell 12, 873–882.

Whittingham, D.G., Lyon, M.F., Glenister, P.H., 1977a. Long term storage of mouse embryos stored at -196^{0}C: the effect of background radiation. Genet. Res. Camb. 29, 171–181.

Whittingham, D.G., Lyon, M.F., Glenister, P.H., 1977b. Re-establishment of breeding stocks of mutant and inbred strains of mice from embryos stored at -196°C for prolonged periods. Genet. Res. Camb. 30, 287–299.

Wittman, K.S., Hamburgh, M., 1968. The development and effect of genetic background on expressivity and penetrance of the Brachury mutation in the mouse: a study in developmental genetics. J. Exp. Zool. 168, 137–146.

Wutz, A., Smrzka, O., Schweifer, N., Schellander, K., Wagner, E.F., Barlow, D.P., 1997. Imprinted expression of the Igf2r gene depends on an intronic CpG island. Nature 389, 7455–7749.

Wutz, A., Theussi, H.C., Dausman, J., Jaenisch, R., Barlow, D.P., Wagner, E.F., 2001. Non-imprinted *Igfr2* expression decreases growth and rescues the *Tme* mutation I mice. Develop 128, 1881–1887.

Yu, M., Hon, G.C., Szulwach, K.E., Song, C.X., Zhang, I., Kim, A., Li, X., Dai, Q., Shen, Y., Park, B., Min, J.H., Jin, P., Ren, B., He, C., 2012. Base-resolution analysis of 5-hydroxymethylcytosine in the mammalian genome. Cell 149, 1368–1380.

11

Sex determination, sex differentiation, and the Y chromosome—a mostly last quarter of the century effort

Sex determination

Mankind has speculated on the cause(s) for sex determination for, probably, all of our evolution. Most theories said that the determination was by the dominant sex and related masculine power to having male offspring. Even Aristotle, who rejected many previous Greek theories, went along with this. In "Aristotle's...theory, movements of the semen encode maleness, movements in the menses, femaleness. Since female movements occur only when the males are weak, every little girl represents a failure in the father's semen" (**Lerois, 2014, p. 216**). This belief in male determinism with stronger males fathering sons persists into the 20th Century.[1] Very little progress was made until the beginning of the 20th Century and the discoveries of Mendelian genetics. As cited and translated by (**Mittwoch, 1973, p.1**), Cuenot could say in 1899 (Cuenot, 1899): "It is rather humiliating to state that as regards man and other mammals no advance has made since the time of the predecessors of Aristotle, even though a considerable amount of work has been expended in trying to solve this problem [of sex determination]; evidently, the wrong approaches were chosen." Sturtevant, in his "A History of Genetics," published in 1965, concluded his section on the history of studies of sex determination with the comment "It has sometimes been felt that the determination of sex offered the best opportunity for the study of the action of genes, and the results described here have contributed largely to our understanding; it now, however, appears that it will be more profitable to study simpler situations, and it is to these that attention is now more often turned," **Sturtevant, p 86.** Certainly, he didn't mean mice!

As discussed in Chapter 3 (subheading : General cytogenetics in mice) chromosomal studies in mice allowed the determination that the Y chromosome determined male sex since the XO sex chromosome karyotype (i.e., absence of the Y chromosome) produced basically a female phenotype (Welshons & Russell, 1959) and XXY produced a predominantly male phenotype (Cattanach, 1961). These observations demonstrated that the presence of a Y chromosome resulted in the development of a predominantly male phenotype independently of the number of X chromosomes and that the absence of the Y chromosome resulted in the development of a predominantly female phenotype. Thus, the Y chromosome appeared to possess a gene or genes the presence or absence of which determined the destiny of the bipotential gonad as a

Twentieth Century Mouse Genetics. DOI: https://doi.org/10.1016/B978-0-12-824016-8.00013-1

testis or ovary, respectively, and, through this, the development of the dimorphic secondary sex characteristics. In mice this hypothetical Y-chromosomal gene was named *testis-determining gene on the Y (Tdy)* while in humans it was named *testis-determining factor (TDF)*.

Prior to these discoveries, the hormonal basis of sexual development was the primary focus. Alfred Jost (Jost, 1947; summarized in Jost et al., 1973) performed surgical castration on the fetal rabbit, and found that, irrespective of the genetic sex of the fetus, in the absence of the gonad, secondary sexual development (i.e., development of the internal duct systems and the external genitalia) proceeded in a female pattern. In other experiments, testosterone was delivered into female fetuses and male-type development of both external and internal genitalia occurred. From these findings, Jost concluded that the testis promoted masculine secondary development by producing hormones which supported growth of internal and external male structures, and other hormone molecules which suppressed female organs. However, the triggers for the production of these "hormones" needed to be discovered and the search for the Y gene or genes started.

An alternative route of research explored sexual differentiation before the gonads developed and sexual hormones were synthesized. These investigations occurred predominantly at the end of the 20th Century and continued into the 21st—much later than the midCentury hormonal studies. There are sexually dimorphic transcripts and traits which exist long before the 12th day of mouse development when the gonads become distinguishable. Transcripts of two Y genes, *Sry* and *Zfy* can be detected at the blastocyst stage (Zwingman et al., 1993; and even earlier in humans at the 1-cell stage; Ao et al.,1994). The Y-coded H-Y antigen (see Chapter 5) is found as early as the 8-cell stage (Epstein et al., 1980). *Sxr* (a sex-determining fragment of the Y chromosome, see below) was associated with a fragile site (Hunt and Burgoyne, 1987) in pre-implantation embryos. Fragile sites are indicated by a portion of the chromosome which remains "thin" compared to the normal "thick" chromosomal appearance on a standard karyotype)It is now known that such fragile sites are often associated with transcription sites for huge genes (Glover et al., 2017). Burgoyne (1993) reported that the Y chromosome of some inbred strains had an accelerating effect on mouse preimplantation embryos growing in culture, suggesting the possibility of a Y-linked growth factor. This result was confirmed by PCR typing of the fast and slow growing mouse embryos (Valdivia et al., 1993). Thornhill and Burgoyne (1993) found that the paternal X had a retarding effect on post-implantation female mouse embryos which is almost certainly due to inactivation of X^p by imprinting (see Chapter 10). Peippo and Bredbacka (1995) reported that at day 3, female mouse embryos were more advanced than males, and in subsequent in vitro culture, cell increase was greater in males, suggesting that the reported increased rate in males might be artifacts of in vitro culture. However, in vivo studies confirmed the accelerated growth of the early male embryo. When embryos were classified as to rate of development by the time of blastocoel formation and then transferred into pseudo-pregnant females, the faster embryos were 71% males, while the slower developing ones were 80% females (Tsunoda et al., 1985). A similar result was found in humans (Pergament et al.,1994). Clearly, there are sexual difference in preimplantation growth. Sry's first manifestation in the developing gonad is also to accelerate cell proliferation (Schmahl et al., (2000). After the turn of the Century, it was postulated that a possible

"Y growth factor" effect could be mediated by insulin and its receptors, since male to female sex reversal occurs in triple knockouts of the three insulin receptor genes in mice (Nef et al., 2003) and these are expressed in the pre-implantation embryo (Erickson and Strnatka, 2011). Mittwoch showed as early as 1969 (Mittwoch, 1969) that the male gonadal volume was significantly greater than the female gonadal volume prior to the appearance of histological differences between the two types of gonads. These differences were not only the result of Y chromosomal factors but were also influenced by the X, autosomal factors, and maternal factors (Hunt and Mittwoch, 1987). She (Mittwoch, 1993) proposed that rapid growth rate itself might be the cause of testis formation. A clear example of the importance of a prehormonal stage of sexual development comes from marsupials. In the Tammar wallaby, the scrotum, gubernaculum, the processus vaginalis, and mammary glands appear before development of the gonads and independently of hormones (O et al., 1988). As the scholar of comparative sex determination, R. H. F. Hunter (**Hunter, 1995**) well summarized:

"embarkation upon the testicular pathway in the presence of a Y chromosome or portions thereof appears in some critical and essential manner to be linked to precocious development of the male embryos, a situation frequently demonstrable even as early as during the first few cleavage divisions and predictably so by the morula-blastocyst stage. This finding may imply quite specific interactions between sex determining genes and genes that influence the rate of early embryonic development. In any event, if male embryos consistently show an accelerated rate of cleavage, this would suggest that they already know what sex they are, perhaps as a consequence of *Sry* transcription at the 2-cell stage" (**p. 293**).

Efforts to find Tdy were concentrated in the last quarter of the Century and involved many people working both in mice and man and many parallel discoveries were made. Bruce Cattanach, visiting at the Jackson laboratory, found a sex reversal factor (*Sxr)* in mice which caused XX mice to develop as males (Cattanach et al., 1971). This visit represents the Anglo-American major contribution to studies of sex determination in mice with several British contributors: Bruce Cattanach, who then moved to Harwell (see minibiography of Mary Lyon); Paul Burgoyne and Elizabeth Simpson in London and Anne McLaren, initially in Edinburgh and then in London[2] and the American effort mostly at the Jackson Laboratory. The ratio of women to men was fairly typical for mouse genetics and much higher than in most branches of science. Women play a very prominent role in mouse genetics, perhaps because they have greater patience to work with material, which has several months per generation while *Drosophila, C. elegans,* bacteria, and phage have generation times in the hours to days.

It was not known if *Sxr* was a chromosomal or smaller mutation. The abundance of humans, of which not an insignificant number have variations in sexual determination and differentiation, provided material for cytogenetic studies. The discovery that 46,XX males usually have translocations of Yp to one of their Xps was a major breakthrough in the ability to map a single *TDF* on the Y chromosome (Evans et al., 1979; Magenis et al., 1982). At the same time it was discovered that *Sxr* is a small fragment of the Y carried as extra material on *Sxr*-carrier males which is translocated to the X during male meiosis (Evans et al., 1982; Hansmann, 1982; Shapiro et al., 1982; Singh and Jones, 1982) Fig. 11.1.

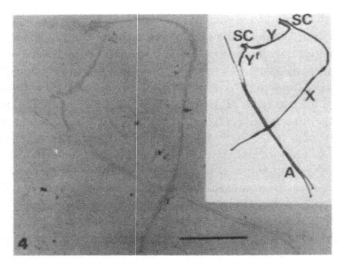

FIG. 11.1 Midpachytene spermatocyte of an Sxr, XY carrier, male mouse. The long arm of the X chromosome is paired with the long arm of the Y chromosome by means of a short synaptonemal complex (SC). The opposite centromeric end of the complete Y chromosome is associated through another SC and with an extra Y fragment (Y f), The X and Y chromosomes and the Yf fragment have thicker axial chromosomal cores than the autosomal bivalent. Whole-mount electron microscopy. Scale mark: 2/um. From Shapiro, et al, 1982.

Singh and Jones (1982) had detected the mouse Y chromosomal material translocated to the X chromosome using a probe that detected a highly repetitive sequence originally found in female elapid snakes (snakes, like birds, have a WZ sex determination with the female being the heterozygous sex), particularly in the banded krait (a large snake found in India). There was speculation that this sequence could be sex determining but it was not found on the human Y (Kiel-Metzger et al., 1985). The original *Sxr* supported testis development in many XX mice but X-inactivation spreading into the fragment caused variable penetrance (McLaren, 1986). It did not support spermatogenesis which was blocked at meiosis I. This was not due to defects in supporting cells or paracrine factors—the *Sxr* spermatocytes remained incapable of completing spermatogenesis in chimeric mice (see Chapt. 4) where they were surrounded by normal cells (Gordon,1976). However, it was determined that it was the 2 Xs which prevented spermatogenesis, both due to problems of chromosome pairing during meiosis and lack of X inactivation of both Xs (see X-inactivation during spermatogenesis, Chapter 8).

A derivative (*Sxr^b*), which still determined a male phenotype but was bereft of spermatogenesis, was soon found (McLaren et al., 1984; Burgoyne et al., 1986). This translocation variant also no longer had Histocompatibility Y (HY) antigen (McLaren et al., 1988a) which was part of the destruction of the hypothesis that HY was sex determining (see Chapter 5). McLaren et al subsequently identified a rare XX^Sxrb male with restored H-Y antigen expression, which proved to be due to a crossover event between the normal Y and *Sxr^b* in the X^SxrbY father. This allowed the determination that both *Sry* and *Hya* (coding for H-Y), were on the minute short arm of the Y chromosome (McLaren, et al 1988b). *Sxr^b* was soon found to have resulted from the deletion of DNA sequences from Sxr^a (Bishop et al., 1988; Mardon et al., 1989) which occurred between copies of Zfy. There was a duplication event in rodents which resulted in two copies of Zfy (see below; Simpson and Page, 1991). "interestingly the breakpoints of this deletion map within the Zfy genes themselves such that Sxr^b contains a fusion gene consisting of the 5' promoter region of Zfy-2 fused to the coding region of Zfy-1" (Gubbay and Lovell-Badge, 1994). The Sxr^b mice were sterile with a lack of spermatogonial proliferation. Since the Sxr^b fragment did not have proliferating spermatogonia, further loss of spermatogenic genes was suggested and the locus concerned was named Spy. The DNA deleted from Sxr^b was found to contain a gene encoding a ubiquitin activating-enzyme E1 which was only expressed in testis and, therefore, an excellent candidate for Spy (Mitchell et al., 1991). Although there is a syntenic, homologous locus in human, in them it is not X-inactivated while in mice it is (Disteche et al., 1992).

Anne McLaren (Anne Laura Dorinthea McLaren), 1927–2007

Anne McLaren was one of those people whose life story is almost too good to be true. Blessed with a superb intellect, a provident upbringing and magnificent education, she was both a highly honored, world renowned scientist and devoted mother. Many people marveled at how she could accomplish so much. I am certain that there were "hidden currents" as well. For instance, why was there a divorce from Donald Michie, although they continued to share a house and harmoniously raise their four children.

Anne came from a family of the new industrial and business magnates who were ennobled. Her father was Sir Henry Duncan McLaren CBE (1879-1953), the 2nd Baron Aberconway, a title that came to his Liberal MP father who had inherited many business interests. His great grandfather on his father's side had been a Scots crofter whose son became a very successful business man and a Liberal MP. The grandfather's second wife had been a well-known suffragette and published *The Women's Charter of Rights and Liberties*. Henry's father had not only inherited his father's business but his mother's father's mines and factories which included the Tredegar Iron and Coal Company, the Metropolitan Railway Company, and the shipbuilding firm John Brown & Company.[1] Henry was active in trade associations including the British Iron Trade Association and the Institute of Naval Architects. These many activities were summarized in his book, *Basic Industries of Britain*. His many business interest took him away frequently, according to Anne, on 2 or 3 overnight trips a week.[2]

Anne had two older brothers, an older sister and a younger brother. The children were spread out over 23 years. Anne, had a close relationship with her father who had been busier with his enterprises when the older children were growing up. She would spend early morning with him. "…I used to have breakfast with him and I opened his mail. I think he very much enjoyed my company. When I was quite small, I used to watch him shaving in the mornings, and he used a cutthroat razor and you cleaned your cut-throat razor in those days on a, sort of, piece of tissue. Not the soft stuff we have now, and he showed me how one could fold the toilet paper, make little cuts in it with the razor, and then you would open it out and you had a beautiful mat with all designs, patterns in. He was that sort of person."[2]

The family divided their time between London and the Estate in Wales—the trip there was with 37 cases and/or trunks; the nanny's and children traveled third class, while the parents went first class (there was no second class).[2] The summers at Bodenant were among famous horticultural gardens in which both her grandfather and father had a great interest. Anne would spend more time with her father visiting neighboring farms…"we would drive round and look at different farms and things and I was fascinated because I noticed that before we were going to meet somebody and he would say something to them, but he always rehearsed what he was going to say, under his breath, while he was driving before we got there, which suggested that he was rather a shy man and didn't find it all that easy spontaneously."[2]

The parents had a busy social life in London and Anne would be in attendance at many luncheon parties with luminaries. Her mother was a great friend of H.G. Wells which led Anne to audition for, and play a bit part in his film "Things

---

[1]Fuller details of Anne's fascinating family history can be obtained from the Clark and Johnson biography (3).

[2]Dame Anne McLaren DBE FRCOG FRS in interview with Dr Max Blythe Oxford, 3 July 1998, Interview One, Part One, The Royal College of Physicians and Oxford Brookes University Medical Sciences Video Archive MSVA 187.

A. Anne at the International Society of Developmental Biology in Basel in 1981 with Andrzej Tarkowski (courtesy of Marilyn Monk). **B.** Anne McLaren (right) with Marilyn Monk (left) and Asangla Ao (middle) at the latter's farewell, post-Ph.D. defense party (just before leaving for Bob Erickson's lab, courtesy of Asangla Ao).

to Come". In it, she is being taught by her "great grandfather". The subject was "mice around the moon and there was a video screen….It was a history lesson on this video screen, and we saw this rocket going off with the mice in and I said, 'Mice that have gone around the moon!' Then he said, 'Yes, and it was dreadful back in those early days because people had awful disease. They had things called colds and they sneezed.' And I had to say, 'Atishoo, atishoo,' deeply shocked."[2]

Her older brothers took her to other films, soccer games, and the ballet. Her enjoyment of the ballet led to her taking ballet lessons which were eventually interrupted by the war. She attended several different private schools and had tutors at home for French, painting, and calligraphy. This was all to end with the start of WW II. The family moved to Bodenant and the London house was closed up. Her schooling radically changed in Wales. Initially the local vicar tutored her and her brother, but that didn't work out and Anne's next 4 years of education were by correspondence course. She wanted to do practical things for the war effort, but didn't have any practical instruction. She did manage to make a hen house on her own using a blue print from the newspaper. She was also involved in child care of her nieces and nephew while her older brother was in the army. She did a lot of walking in the mountains on her own. Near the end of the war, as her ten-year old brother was sent off to prep school, she was sent off to a girl's school in Cambridgeshire. The school was stimulating and she made friends, but the curriculum was somewhat limited. When she wanted to attend University, there was some resistance because her father had never thought of women in the family going to university, but her mother was supportive and the times were changing. The choice was Oxford because that was where her father and two older brothers had gone. It was assumed that she would study English. "I was good at writing essays and so that was what people thought. When I revealed that I wanted to go to University, which was a bit of a shock, it was assumed that it would be to do English Literature, because that was what I had got good reports for."[2]

However, when she saw the curriculum she was supposed to have mastered for English studies at Oxford, she found herself woefully lacking. Thus, she chose biology, for which she felt better prepared. This was a surprise for the family. To prepare for entrance exams, she was sent to a prep school in Oxford and boarded with distant cousins on the suffrage grandparent's side with whom her family had had little contact. However, the relationship went well and she became close friends with these surrogate parents. This new world of education was a revelation for her. "These tutors, who opened all sorts of doors to me, that had never been opened before because conversation at home was gossip about people, to some extent talk about maybe paintings, maybe a little bit books, garden, land, business, that sort of thing. But never ideas and never any politics. Never any of the wider sociological things that you get in Oxford scholarship and entrance general papers. So I was coached in all of that and, of course, I found it tremendously exciting."[2] She did exceedingly well on the entrance exam and was awarded the Senior Scholarship of Lady Margaret Hall. Her first two years were spent in "Honor Mods" of Zoology, Physics and Maths and the last 2 years in Zoology. Among her tutors was the geneticist E.B. Ford and she found genetics fascinating. However, the embryology lecturer, Harold Pusey, was a dull lecturer and she swore "that whatever else I did in my scientific life, I would never do any embryology."[2] Of course, that was to become her major research interest!

While at Oxford, Anne cultivated a very different approach to life than that of her family's. She dressed down and became very active in politics. "She set up a peace committee called the Oxford and District Peace Committee, of which Dorothy Hodgkin FRS, was a member. In her second term at Oxford, Anne became a member of the Communist Party of Great Britain, a membership that continued and which initially proved problematic for Anne when visiting the USA. She continued to be a committed and active socialist for the rest of her life.[3]

She had decided to pursue a doctorate in genetics and applied for a Christopher Welch scholarship which required a mini research project. Her classmate friend, Av Mitchison (see The Haldane-Mitchison Scientific Clan) suggested his uncle J.B.S. Haldane. She performed a study of genetic predilection to mite infection during a month with Haldane at University College. With this accomplishment, she was the first women to receive a Christopher Welch scholarship which supported her for her first year of graduate work which was with Peter Medawar (see Chapter 5). He was her advisor in absentia as he had moved on to a Chair in Birmingham. He put her to a project on inducing eye abnormalities in fetus's by immunizing to the mother to lens crystalline. The project failed as the vivarium could not provide more than 2 pregnant rabbits! Luckily a project on neurotropic microRNA viruses with Kingsley Sanders and Medical Research Council support became available and she studied these in mice as a model of polio infection (McLaren, 1953; McLaren & Sanders, 1959). Her work was very biological and studied host:viral interactions. "This work was attractive to Anne because of its potential practical value, coming as it did at a time when there was an epidemic of polio-induced infantile paralysis, but before polio vaccines had been developed".[3]

It was during these two years of doctoral research that she came to know, and love, Donald Michie, who was married at the time. He had been part of the Turing code-breaking group at Bletchley Park during the war and came back as a veteran, entering Oxford as a medical student. He had become fascinated with genetics and moved to working for a PhD. with E.B. Ford. Anne shared his interest and they bred coat color variant mice in the housing they then shared. E.B. Ford was a butterfly geneticist, so Donald took the variously colored mice of his experiments to Cambridge to show to R.A. Fisher who was doing mouse genetics at this time (see Chapter 1). Michie and Fisher hit it off as Michie was sophisticated in math and statistics from his time at Bletchley Park. Anne and Donald received a small grant from the Royal Society to continue mouse genetic studies as they wrote up their Ph.D. theses.

The next phase of the couple's lives was as postdocs with Medawar who had moved to University College London from Birmingham. When Anne and Donald moved to London it was as independent researchers in the Dept. of Zoology which Medawar now headed. The Dept. was spaced next to Haldane's and it was a politically charged atmosphere as Haldane was a committed socialist and Medawar was conservative. The grant from the Agricultural Research Council was entitled "extra chromosomal inheritance in mice". They were interested in maternal, uterine effects on development which they applied to studies of the number of vertebrae using embryo transfer as an approach which Anne was to continue to use (see Chapter 4 for the introduction of this approach). They were thus able to separate the genetic from the

[3]Ann G Clark and Martin H Johnson, "Anne Laura Dorinthea McLaren DBE 26 April 1927 – 7 July, Elected as FRS in 1975", Memoirs of the Royal Society, in preparation

environmental effects. It took them a while to develop this technique. "...first of all, we had to get the embryo transfer technique going. In order to do that, we obviously didn't use inbred strains because they don't breed well, you get small litters, they are not very fertile. We used good, random bred strains and we induced ovulation with hormones, so as to get more eggs. And in the course of doing that, we did work on the actual technique, how the number of eggs shed related to the amount of hormone given, that was interesting to the Agricultural Research Council, who were paying us".[4]

This was very slow work and soon (after 3 years) the number of mice was so great that they had to move the mouse colony to the Royal Veterinary College. Luckily, the Agricultural Research Council supported them for 7 years. It was a highly productive period and resulted in 28 papers. Anne has also collaborated with Biggers on the culture of mouse embryos. "...while we were at the Royal Veterinary College, John Biggers was working there too and he had a nice culture system going for chick bone. He was working on chick embryo bones growth in culture and so, as we had these early mouse embryos, he cultured some of these embryos from eight-cell stage up to blastocyst and then I put them back into the uterus of a foster mother and they were born as normal, healthy mice. We mated them, and they were fertile and that was actually the first time it was shown that mammalian embryos could be culture[d] in vitro, outside the body, and they would develop normally thereafter"[3] . This work resulted in great publicity (McLaren and Biggers, 1958). She had known Robert Edwards as a student and knew that he was doing similar work in human reproduction. " I had a great admiration for him, so my guess was that he would pull it off and that one would, in the end, see Louise Brown being born. But, of course, I didn't know how long it would take"[4].

In the early 50s, they would frequently travel from London to Edinburgh.

"We were in London then, this was 1950s. We used to go up and visit because all the exciting genetics, from our point of view, was being done in Edinburgh and there was Moreton Hall, this, I don't quite know how to describe it but commune that Waddington organised because there was no available accommodation in Edinburgh for this team that he brought. Accommodation was very very difficult and so he put them all in this commune and it would have worked extremely well if it hadn't been that there was food rationing. And you know, there were families and wives who felt that Waddington was eating more of his fair share of butter that should have gone to their children. And there was no petrol for cars and the bus service was very bad. So it was really the problem of spouses, mainly wives, yes all wives because Ruth Clayton's husband was working in the Institute as well. But wives and children left behind in this commune. Eventually the thing folded. But while it was going, it was a great place for post-docs to go and visit. And there was a basement which was uninhabited where they stored all the spare mattresses and things. So one could simply dose down and stay as long as one wanted and there was sort of communal eating. So we saw quite a bit of Bob Edwards and Alan Gates who was a great embryo transfer, transferer[sic]. We learned a lot from Alan Gates[5]

This was also the period when the first of their children were born (Susan in 1955 and Jonathan in 1957). They were politically very active, travelling to Moscow in 1957 for a world youth conference for peace and international solidarity. They met Lysenko and were favorably impressed, probably because of their interest in environmental effects and lack of knowledge of the persecution of geneticists (see Chapter 1). They were active in the anti-nuclear movement and took part in demonstrations and meetings. However, they drifted apart and divorced, remaining good friends and raising the children together.

Donald was the first to move permanently to Edinburgh, to the Department of Surgical Sciences and Anne was to follow, finding a position supported by the Agricultural Research Council in Waddington's Institute of Animal Science in 1959 where she was to stay for 15 years.

This was an exciting scientific environment for which Anne credited Waddington (see Chapter 4). "It was the most wonderful scientific atmosphere I have come across anywhere, partly because of the intellectual excitement. What Waddington did was to get grants from all over the place, probably more than one hundred grants came in to that institute. He put them all into a bit [sic] pot, stirred them up and used them for whatever he thought was most in need. One of the things he did was to subsidize the canteen there and at that time, I think it was the only canteen on the campus and people used to come from all over, zoology, molecular biology, and there were these wonderful lunchtime conversations"[4] .

[4]Dame Anne McLaren DBE FRCOG FRS in interview with Dr Max Blythe Oxford, 3 July 1998, Interview One, Part Two, The Royal College of Physicians and Oxford Brookes University Medical Sciences Video Archive MSVA 187.

[5]Interview with Dr Anne McLaren, Gurdon Institute, Cambridge, 2 February 2007-07-16. Interviewers: Prof Martin Johnson, Prof Sarah Franklin

Here her work on ovulation continued and she was pre-occupied with determining the signals for implantation of each embryo, knowing that this is a step at which many human conceptions fail. She and Alan Beatty obtained a large grant from the Ford Foundation for their studies of reproduction. A "portable" building intended for several years use (but used for 30 years) was put up to house them. "Work in the Ford Hut was not a 5-day week. Anne brought in her children to the lab at weekends and in the holidays and while they amused themselves, work carried on as usual"[3]. During this time she had 9 graduate students and refused to put her name on many of their papers (numbering 29). Her ability to culture embryos also led to interesting experiments with chimeras (see Chapter 4). Among other analyses, she collaborated with Gruenberg (see Chapters 1 and 4) on different genetic contributions to the skeleton and teeth (Gruneberg & McLaren, 1972). Since embryos of different sexes were aggregated randomly, intersexes were expected ½ the time. There variable phenotypes led her to her interest in sex determination which is documented in Chapter 11.[6] Her interest in the new DNA approaches was stimulated by hearing talks at a congress of zoology in Washington ,D.C. in 1961. On her return to Edinburgh, she started visiting Peter Walker's lab and his graduate student, Ed Southern (of Southern blot fame) at the University. This work led her into DNA hybridization studies (Sherman, et al, 1972).

"The social life was also exceptional, with frequent parties held at Anne's house for Edinburgh's biological community and the numerous scientific visitors who passed through. On one occasion, when preparations started from scratch two hours before forty or so people were due to arrive, her daughter came into the kitchen and asked a question about one of Shakespeare's plays for her homework. Anne dropped everything and gave her undivided attention for twenty minutes or so, until the problem was solved. It was a lesson on the priorities of life. The party was ready on time"[3].

Anne published 99 papers in the fifteen year stay in Edinburgh, as well as attend many national and international meetings and stay heavily involved with raising her children. She took on many organizational tasks for a variety of societies. These included being a

"committee member of the British Society of Developmental Biology (1973-79) and later its Chair (1975-79), a member of the Council of Management of the Journal of Reproduction and Fertility (1971-74 and of the Cell Board of the Medical Research Council (1974-78). She was also on the Advisory Board for the World Health Organization's special programme on human reproduction (1971-74), and Chair of the Society for the Study of Fertility (1974-78). She was editor of Advances in Reproductive Physiology (1965-69), the Journal of Embryology and Experimental Morphology (now Development 1974-76) and an associate editor of the Journal of Reproduction and Fertility"[3].

As her children reached university age, and with having spent 15 years in Edinburgh, she was ready to move back to the London area. The Medical Research Council wanted to have a new unit of Mammalian Development and Anne was named head of it. It was initially located in Grueneberg's space (which she had visited when collaborating with him and knew to have adequate animal facilities) at University College. This space was in the Wolfson House located on a back alley near Euston Station which made it easy to access (once you knew the way).[7]

However, some of the space was taken away when needed for people from the Lister Institute which was closing and then things got crowded. According to Anne, "the Unit was never actually on the scale that I would have liked and that made problems with accommodation all the way through".

This unit became a mecca for mouse developmental biologists. Marilyn Monk, who was to stay 18 years, describes her "recruitment" vividly:

[6]It was in 1971 that I first met Anne at the "Edinburgh symposia on the genetics of spermatozoa" arranged by Allan Beatty and Salome Gluecksohn-Waelsch. She gave a paper titled "Germ cell differentiation in artificial chimaeras of mice" (McLaren, 1972). We remained acquaintances, but never good enough friends to discover our shared friendships with the Mitchisons. I was warmed by a communication from my former postdoc, Asangla Ao (Feb. 12/ 2019) who had done her Ph.D. with Marilyn Monk in Anne's unit in the Wolfson House (see below). She wrote "I have a very vivid memory of one conversation I had with Anne (I may have told you). When you had offered me the postdoc position, I went to her for advice. She told me that I should go and find out how science is done across the pond but she also told me to come back to the UK after a few years! Bob, she had high regards for you." Asangla, as with almost everybody who new her, had very high regards for Anne. "she was always so approachable even though she does not look like one when you see her. I do not remember anybody in that MRC unit who spoke ill of her- not one. Almost everybody admired her. When I first joined the unit, it was my first trip to any country outside India. I was quite apprehensive about everything but Anne made fee [sic] so accepted and psychologically it helped me a lot as she was our director. Of course, Marilyn and the rest of the people were also very welcoming and [I] never felt like an outsider. Those days there were not many international students around."

[7]I visited here several times and gave at least one seminar. The labs were very crowded but people were considerate of each other and work went on relatively smoothly.

"Right from the beginning Anne accepted me unconditionally. My first encounter with Anne was my phone call to her in Edinburgh in 1974. [My unit closed and] The MRC told me that I could relocate to another MRC unit that interested me and that would have me…. Harry Harris at the Galton Laboratory suggested I contact Anne McLaren as she was just about to move from Edinburgh to London to start up a new MRC Mammalian Development Unit at the Galton. I knew nothing at all about development - let alone mammalian development. A move to mice and their embryos would be a huge leap both intellectually and technically. In any case, I plucked up courage to phone Anne in Edinburgh in 1974. I remember everything about that moment when I phoned Anne because I was holding onto my last hopes of continuing as a scientist. I introduced myself, told her my problems, and asked her if she would consider taking me on in her new MRC Unit in London. I told her I knew nothing about mice – I had only worked with bacteria, viruses and amoebae. She said, 'Yes of course you can join me. You must!'. I was flabbergasted. So overjoyed I could not speak."[8]

Marilyn's previous experience brought new tools to the research program. Her ability to do very sensitive biochemical assays for X-linked and autosomal enzymes allowed direct studies of X-inactivation in the embryo (Monk & Kathuria, 1977). These tools were used with Anne in multiple studies (e.g., Monk & McLaren, 1981). These studies led to the eventual ability to perform preimplantation diagnosis for genetic disorders, an effort greatly advanced by the discovery of the polymerase chain reaction (PCR).

When Anne retired in 1992, the unit was closed. Handily, the Wellcome Trust/Cancer Research UK Gurdon Institute in Cambridge recruited her to head a research unit and she continued to be very productive, especially with her research on sex determination (Chapter 11). She moved to a small house in Cambridge and continued to be a hostess to many scientists. She became a Fellow of Christ's College and of King's College, where she later became an Honorary Fellow, as well as the women's college, Lucy Cavendish. As a retiree, and with the children long gone, her activities for various organizations greatly increased despite her scientific productivity. Her efforts on behalf of women included not only involved the Lucy Cavendish college. She was a founding member of the Association for Women in Science and Engineering and was its president in 1995. She was made

"the Fullerian Professor of Physiology [of the Royal Society], a member of the Royal Institution, President of the European Developmental Biology Organization, Member of the WHO Scientific and Technical Advisory Group Special Program on Human Reproduction and of the Governing Body of the Lister Institute of Preventative Medicine. Abroad she became the Raman Professor at the Indian Academy of Sciences, a Professorial Fellow at the Department of Zoology at the University of Melbourne and was a Member of the Royal Society for Asian Affairs and of the International Advisory Board of the Institute of Molecular Bioscience in Brisbane[3].

Her activities included the Presidency of the British Association for the Advancement of Science in 1993-1994. She was concerned with animal welfare and was on the Council and then a Trustee of the Institute of Zoology and the London Zoo. In her later years she was highly involved in biobanking of the DNA of threatened species—she founded the Frozen Ark project for this purpose.

"Her research interests in stem cells led her to make a notable contribution to the ethical debate about their generation and use, in particular those made from early human embryos. She became a member of the Nuffield Foundation's Bioethics Council and the European Group on Ethics that advises the European Commission on the social and ethical implications of new technologies. She was a long-term member and council member of the Pugwash Conferences. This group, which was awarded the Nobel Prize for Peace in 1995, is devoted to reducing the danger from all forms of armed conflict and seeks to raise awareness of the ethical issues that arise from scientific advance."[9]

In addition, she was appointed to the Warnock Commission, set up by the British parliament to give advice on reproductive biology, in 1982. She was the only basic scientist on the committee. There first report came out in 1984 and included the recommendation that research could be done on extra human embryos from assisted reproduction (AR, Anne insisted that it not be called artificial reproduction) up to day 14. The recommendations were not accepted by parliament but, with the establishment of a Voluntary Licensing Authority (VLA) (which Anne served on from 1984–1990), AR moved ahead. The VLA was replaced by Human Embryology and Fertility Authority in 1990 and she served on it from 1990 to 2001.

---

[8]Marilyn Monk, Tribute to Anne McLaren. British Library Archives.

[9]Obituary, Dame Anne McLaren, July 7, 2007.

[10]Paul Burgoyne, Obituary, Anne McLaren, 1927-2007.

Perhaps her greatest honors were being elected as the first female Foreign Secretary of the Society (the first woman officer in its 331 years existence), the Japan [Kyoto] Prize (along with Andrej K. Tarkowski)[10], only second to the Nobel Prize, and being made a Dame Commander of the British Empire. Her activities as the Foreign Sec'y of the Royal Society involved a large amount of international travel, especially with the breakup of the USSR which involved efforts to help the now unsupported scientists in the former satellite states.

"Anne was an inveterate traveler, heading off to meetings in all parts of the globe with only a small rucksack and a plastic bag of papers to read on the plane. During the cold war period, she made a number of visits to Eastern Europe and Russia; Mike Snow of the MDU recalls that on these occasions she also carried a small suitcase packed with items requested by her hosts behind the iron curtain. I remember one occasion when she went off in search of specific fly- fishing lures requested by the Russian cytogeneticist A.P. Dyban."[11,12]

Over all these years, she and Donald Michie had remained very good friends. They purchased a 3 floor house together in London with 1 floor for Donald, one for Anne and the floor in between to meet together. They made many trips to Scotland together and were killed together in a traffic accident on one of these trips. She was survived by her son Jonathan Michie, Professor of Management and Director of the Business School at the University of Birmingham, and her two daughters.

[11]Anne and Tarkowski were good friends: " I knew Tarkowski quite well. In fact, when he was working in Bangor and making the first chimeras, he came over to Edinburgh and stayed with me in Edinburgh and we got on very well." From (7). See Figure 11-1a.

[12]Robin Lovell-Badge, Obituary, Dame Anne McLaren, The Independent and The Independent on Sunday, July 12, 2007.

The *Sxr* chromosomal fragment allowed further proof for pre-hormonal influences on sexual differentiation. The length of the anogenital space (the distance from penis or clitoris to anus, greater in males than in females) is significantly smaller in *Sxr* than in normal males (Atkinson and Blecher, 1994). When adult, its epididymis lacks the important initial segment even though the *Sxr* male has testosterone levels as high or higher than the normal male (Lebarr and Blecher, 1986) and the abnormality is again present before the pubertal androgen surge (Lebarr et al., 1991). Since X*Sxr*0 (with *Sry* and 1 X) males have normal epididymides, the abnormalities in the *Sxr* male are caused by the presence of 3 functional copies of some gene(s) with homology on the X in a non-inactivated region or 2 copies of an X gene which escapes inactivation. "Thus it appears that epididymal development, and that of the anogenital space including the scrotum, are not solely dependent on testosterone and that sex-chromosomal sex-determining loci influence these traits of secondary sexual differentiation" (Blecher and Erickson, 2007).

Because of the large number of XX male clinical cases with fragments of the Y translocated to the X, research in humans led to cloning of *TDF*. Molecular analyses in such patients permitted the construction of a deletion map of the Y chromosome (Vergnaud et al., 1986; Nakahori et al., 1991) This small region of Yp adjacent to the Yp pseudoautosomal region is essential for testicular differentiation.. Regions at the proximal and distal end of X and Y chromosomes pair and allow crossing-over, thus, a gene located there appears to be neither X or Y linked and segregates as if it were autosomal. Finer mapping continued, and one candidate gene was mistakenly identified. David Page (1987) and colleagues, working at the Whitehead Institute in Cambridge, Massachusetts, found a highly conserved sequence that seemed to map with a male-determining Y fragment and named it *ZFY*[3]. However, one of their XX males did not have *ZFY,* and the finding of more such patients (Palmer et al., 1989; Verga et al., 1989) suggested that *ZFY* could not be *TDF.* Also, it was found that *Zfy* was not sex-linked in marsupials (Sinclair et al., 1988). The resumed search for *TDF* was targeted on 35 kilobases of DNA between the proximal (i.e. towards the centromere of the chromosome) limit of the pseudo-autosomal region of the Y chromosome and the breakpoint in XX males lacking *ZFY*. Peter Goodfellow, working at the Imperial Cancer Research Laboratories located in Lincoln Inn's Fields, London, and his colleagues, then took the lead. A subclone selected for its male specificity in mammals was found to encode a protein containing an 80-amino acid segment that was homologous to the mating-type protein of *Schizosaccharomyces pombe* (one common type of yeast) and to a conserved 80-amino acid, DNA-binding motif found in the high mobility group (HMG) transcription regulating proteins (Sinclair et al., 1990. Further analyses of a variety of tissues disclosed a transcript of this gene only in the testes. It was proposed that this gene, designated *sex-determining region, Y chromosome (SRY)* was *TDF.* Soon thereafter, mutations in the *SRY* gene were found in humans in some cases of XY, pure gonadal dysgenesis (Affara, et al 1993; Berta et al., 1990; Jager et al., 1990; Hawkins et al., 1992). The mouse *Sry* gene was cloned at the same time by homologous cloning since the sequence of the DNA-binding box was highly conserved (Gubbay et al., 1990; Hacker, et al, 1995). These results allowed the detailed experimental studies only possible in mice. Some mice transgenic (i.e. in which a gene has been artificially transferred) for a fragment containing *Sry* showed 40,XX sex reversal—that is, mice otherwise destined to be female had a male appearance (although they were sterile; Koopman et al., 1991). Testes of these mice were histologically similar to those of human XX males and *Sry* expression correlated with the initiation of testicular determination in the gonadal ridge (Koopman et al., 1990). However, sex reversal did not always occur. This incomplete penetrance may have been due to the influence of genetic modifiers present in different inbred strains of mice. Interestingly, human *SRY* does not cause sex reversal in transgenic mice; it is not nearly as conserved in overall sequence as some human genes, a few of which can work even in *Drosophila,* the fruit fly.

Indeed the "strength" of Sry seems quite variable. Two wild-derived Y chromosomes, which were perfectly good male determinants in their original background, worked weakly or not at all in the standard C57BL/6J inbred strain. One was derived from the *Mus poschiavinus* species and designated Y[POS] by Eva Eicher working at the Jackson Laboratory (Eicher et al., 1982). In a series of well-designed genetic crosses between inbred strains, Eicher and colleagues identified the variant allele in the C57BL/6J inbred strain and named it *testis-determining autosomal-1* (*Tda*-1). The C57BL/6J, non-functioning allele was designated *Tda-1[B6]*(Eicher and Washburn, 1983). Burgoyne and colleagues demonstrated that this mutation caused delayed expression of *Sry* such that the developing gonad was already committed to ovarian development (Palmer & Burgoyne, 1991). The other "weak" Y was derived from wild *Mus musculus domesticus* carrying a Robertsonian translocation which had been trapped in Alpie Orobie in Switzerland and it was designated Y[ORB] (Eicher et al., 1988). This difference in Y "strength" is not likely to be associated with coding sequence differences in the two species *Sry* genes. The sequence of the high mobility group box of the two inbred strains shows one conservative substitution at position 61 (threonine to isoleucine) in *Mus musculous domesticus* and *Mus Poschiavinus,* but this was found in both "strong" (*M. poschiavinus)* and "weak" (SJL/J) Ys (Graves and Erickson, 1992).

Since the Y[POS] works later (Palmer and Burgoyne, 1991), has reduced expression (Washburn et al., 2001) and has prolonged expression (Lee and Taketo, 1994), there may be differences in the two *Sry*'s promoters which are responsible for the differences. This is made likely since the

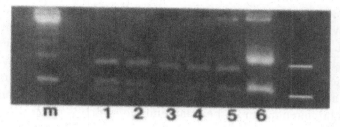

FIG. 11.2 Restriction fragment polymorphism showing the sequence difference between the SRY box of *musculus* **strain (lanes 1-5) and a** *domesticus* **strain (lane 6).** The difference between the presence of a threonine (*M.m.d.*) and isoleucine at position 61 in the amino acid sequence of the SRY box in the DNA sequence was sensitive to cutting by the *MboI* restriction enzyme and led to the larger fragment (lane 6) in the *domesticus* variant strain *poschiavanus*. From Graves and Erickson 1992.

Mus musculus Sry delivered as a transgene corrects the defect (Eicher et al., 1995). In fact, in one species of voles (Just et al., 1995) and in spinous country-rats (Sutou et al., 2001), *Sry* cannot be detected and, therefore, is unnecessary. Thus, the mechanism of action needed to be determined. Again, humans with the many physicians finding multiple cases of sexual ambiguity or reversal, led the way but only the experimental material mice provided allowed the elucidation of mechanisms and experimental verification.

In man, a rare, autosomal dominant disorder, campomelic dysplasia (a frequently lethal disorder with prominent bent bones as the most obvious manifestation) was found to be due to mutations in *SOX9* and is very frequently associated with ambiguous genitalia or sex reversal. Activation of *SOX9* is clearly one method by which an *Sry*-negative individual can develop a male phenotype. *Sox9* transgenics develop a normal male appearance in the absence of *Sry* (Vidal et al, 2001). A dominant insertional mutation, *Odsex,* which represents a 150 kb deletion that maps about 1 million base pairs upstream of *Sox 9* causes sex reversal (Bishop et al., 2000) due to activation of *Sox9*. Collectively these observations demonstrated that *Sox9* expression in developing gonads is both necessary and sufficient for testis development, independently of *Sry,* and it is therefore believed to be the most important (perhaps the only important) target of Sry. In the 21st Century, it was demonstrated that Sry regulates Sox9 expression by binding to a 3.2 kb. testis-specific enhancer sequence 13 kb upstream of the *Sox9* transcriptional start site (Sekido and Lovell-Badge, 2008). This binding is dependent on Sf1, the product of *Steroidogenic factor 1* gene, a protein expressed in the developing urogenital ridges of both sexes prior to the expression of Sry (de Santa Barbaa et al., 1998) and in which mutations can cause sex reversal. A core region of this testis-specific enhancer, TESCO, is conserved in humans, mice, rats, and dogs, and a 180 base pair element within TESCO is conserved in birds, reptiles, and amphibians (Sekido and Lovell-Badge, 2008; Bagheri-Fam et al., 2010).

While *Sox9* is downstream of *Sry,* another gene was found that might have a highly conserved role in sex determination. *Dmrt-1,* is a gene expressed in the gonadal ridge at day 9.5 before sexual determination becomes evident (Raymond, et al 1999) and it and may have a role in preparing the gonad for sexual determination. It is a mammalian homolog of a gene shared by *C. elegans* and *Drosophila,* which is involved in their sexual determination/differentiation.

This was particularly intriguing since conservation of sex determination pathways is the exception, not the rule. While mutations in this highly conserved gene are infrequent in human 46,XY females, it maps to a critical region in human 9p where deletions frequently cause sex reversal and a homologue, *DMRT-2* is part of the frequently deleted region. Initially expression of *Dmrt* is equal in the genital ridges of both sexes but gradually expression is limited to the male gonad (Moniot et al., 2000). *Dmrt-1* is also expressed in the genital ridge and in the Wolffian ducts before sexual differentiation in the avian embryo, which is not true of *SRY* (Raymond, et al 1999). *DMRT-1* expression is appropriately temperature-controlled in the turtle gonad before sexual determination (where sex reversal can be controlled by the temperature at which eggs are incubated[6]; Kettlewell et al., 2000) and is expressed during gonadal development and spermatogenesis even in fish (Marchand et al., 2000). Thus, as might be expected for a gene with homology to a *C. elegans* and *Drosophila* gene, *Dmrt-1* may have a more conserved role in sexual determination in vertebrates than does *Sry*.

Sexual differentiation

Jost's experiments (Jost et al., 1973), already described, established that, although sex-determining genes are responsible for directing the development of the gonad, secondary sexual differentiation of external genitalia and internal duct systems is under hormonal control in eutherian mammals. Indifferent (bipotential) external structures are initially present in males and females. The testis produces androgen, which masculinize the labioscrotal folds (to form a scrotum) and the genital tubercle (to form the penis). In the absence of the testis, and thus of high levels of androgen molecules, the labioscrotal folds fail to fuse and form labia, and the genital tubercle becomes the clitoris. Internally, the mesonephric (Wolffian) and paramesonephric (Mullerian) ducts develop into male or female structures. Testicular androgen promotes the development of the Wolffian ducts into the epididymides, vasa deferentia, and seminal vesicles, and absence of androgens lead to their degeneration. Also produced in the testis is Anti-Mullerian Hormone (AMH), which causes the degeneration of the Mullerian duct. In the female, in the absence of AMH, this duct becomes the uterine tubes, uterus, and upper part of the vagina.

The absence of certain genes causes a lack of any visible gonadal development and some of these genes are also implicated in renal development. For example, knockouts for *Emx2* do not have kidneys, ureters, or genital tracts (Miyamoto et al, 1997). Although Wilms Tumor 1 (*Wt1*) and other genes that are expressed normally in the metanephric mesenchyme were initially expressed normally in this KO mouse, subsequent expression was greatly reduced, and degeneration and apoptosis occurred (Miyamoto et al., 1997) Kidneys and gonads are also missing in *Lhx1* KOs (Shawlot and Behringer, 1995). This is not surprising, given that a very early role in gastrulation for had been found for *Lhx1*.

Another gene whose absence was found to prevent gonad formation was *Sf-1* (see above) or steroidogenic factor-1. *Sf-1* is homologous to the *Drosophila* gene *Fushi tarazu 1* factor and sometimes is referred to as *Ftz* F1; it is an orphan nuclear receptor gene that has been shown to be a key regulator of steroidogenic enzymes in adrenocortical cells (Lala et al., 1992; Sadovsky, et al, 1995). Developmental studies demonstrated that Sf-1 is expressed the urogenital ridge at embryonic day 9 to 9.5 in the mouse. This is at an early stage of organogenesis of the developing gonads and before *Sry* is expressed (Ikeda, et al 1994). The KO of *Sf-1* resulted in mice that died at 8 days and had a resulting deficiency in corticosterone that was the likely cause of death. Expression studies of *Sf-1* (Hatano et al., 1994; Shen et al., 1994) have demonstrated a sexually dimorphic expression pattern of *Sf-1* with persistence in males and a discontinuation of expression in females. This was thought to link its expression to the regulation of AMH since SF1 was shown to regulate the expression of AMH in vitro (Shen et al., 1994) and in vivo (Giuli et al., 1997).

The role of growth factors in the differentiation of the genital tubercle (GT) was well explored in mice (in addition to their role in sexual determination examined above). While *Fgf9* was implicated in gonadal differentation, *Fgf8* and *Fgf10* were implicated in external sexual differentiation. The knockout of *Fgf10* showed quite marked sexual ambiguity (Haraguchi et al., 2000). The interaction of a number of these factors, in a pathway initiated by Sonic hedgehog (Shh), was studied in organ cultures of murine GTs (Haraguchi et al., 2001). GTs explants were treated with beads soaked in Shh and gene expression was studied by in situ hybridization. Bone morphogenetic protein 4, Fgf10, Gli1, Hoxd13, and Ptch1 all showed increased expression in the Shh-treated outgrowths (ibid.). Antibodies to Shh and the *Shh* knockout were also studied. *Shh$^{-/-}$* embryos showed no external genitalia at 12.5 days gestation and showed downregulation of *Fgf8*. Cultured GT explants treated with antibody to Shh also showed marked down-regulation of *Fgf10*. Since Shh was known to influence three Gli transcription factors, Haraguchi et al (ibid) studied *Gli2* knockout mice which exhibited ventral malformation of the GT. The 21st Century was to see great progress in these studies, finding the genes responsible for ovarian development and, surprising to many of us, genes whose expression was needed to maintain testicular and ovarian differentiation (Greenfield, 2015)!

The mouse Y chromosome

The story of the mouse (and other mammalian Y chromosomes) evolved from the simple thought, based on evolutionary considerations, that it was mostly a "cipher" of maleness and only carried the Testis-determining factor compared to other chromosomes with many functional genes. The view changed as it became apparent that there were regions of homology between the X and Y chromosomes (Burgoyne, 1982; Polani, 1982). These, as already mentioned (see Chapter 10), were the pseudoautosomal regions (PAR). It was predicted that these would be a region of recombination between the X and Y chromosomes and that this region would not be inactivated on the X chromosome. As discussed in Chapter 10, soon Goodfellow and associated (1986) cloned a gene from this region and many more were to follow. The boundary of the PAR is spanned by a gene encoding a RING finger protein which suggests that this gene may be transitioning from the PAR to being X unique (Palmer et al., 1997). This is in contrast to the more general movement from the Y to autosomal locations for genes across mammalian evolution. Although the gene for ubiquitin activating enzyme E1 is expressed in spermatids, i.e. post-meiotically and, thus, presumably important for spermatozoal maturation (Odorisio et al., 1996), it does not need to be located on the Y as it is not on the Y elsewhere in the primary lineage (Mitchell et al., 1998). Another example is *tspy* which has multiple copies in humans and Bovidae but only 1, frequently nonfunctional copy in mice (Schubert et al., 2000). The reason for other autosomal genes with copies on the Y is sometimes less apparent. Originally autosomal *RhoA* doesn't have a significant open-reading frame in the Y copy (due to deletions) yet is expressed during spermatogenesis (Boettger-Tong et al., 1998).

Early studies of the mouse Y depended on clones for repeated elements. Lamar and Palmer (1984) used 3 randomly chosen Y-specific sequences to show that the Y chromosome exists in two polymorphic forms among inbred strains. It was found that these represented Ys of *Mus musculus domesticus* and *Mus musculus musculus* and that most laboratory strains carried the *musculus* Y (Bishop et al., 1985). Interestingly, there was very limited flow of these Ys across the hybrid zone between the two subspecies (see Chapt. 2; Vanlerberghe et al., 1986). Variation in another repeat, *Sxl* affected *Sry* expression, leading to XY female reversal in some cases and showing long range position effects on *Sry* expression (Capel et al., 1993). Similarly, work by Eicher and colleagues (1983) with such clones of repeat DNA identified 4 different regions of the mouse Y chromosome, one of which was assigned for "sperm motility." Such Y chromosomal repeats included both Murine Retrovirus (MRV) and intracisternal A Particles (IAP) in huge numbers, explaining the degree of heterochromatization of the Y (Fennely et al., 1996). It soon became apparent that multiple genes involved in spermatogenesis were encoded on the Y.

Partial deletions of the Y chromosome resulted in abnormal sperm morphology (Styrna et al., 1991), which decreased fertilizing capability (Xian et al., 1992). Fertility was found to require X-Y pairing and a "spermiogenesis" gene mapping to the Y short arm (Burgoyne, et al 1992) of which one was *Smcy* (Agulnik, et al 1994). These genes may be selected for a Y location, as there is a dense region of genes needed for spermatogonial proliferation which is syntenic to the human Y (Mazeyrat et al., 1998). In the 21st Century it was established that the mouse Y chromosomal long arm shares no genes with the human Y and is has hundreds of copies of a single repeat unit as well as the repeats mentioned above (Mazeyrat, et al, 2001).

The burgeoning amount of data required sharing and communication between the multiple laboratories. As with other chromosomes (see Chapter 3), there was a series of Y chromosome workshops concentrating on the human Y chromosome but collating information on other mammalian Ys (the third was in Heidelberg in 1997; Vogt et al., 1997). The data were on several web sites curated by members of the committee but this was superseded when a full genome sequence of C57BL/6J was completed in 2002.

T/t complex-associated sex reversal and sex effects on elements of the complex

The *t*-region was initially of interest for sex determination because Bkm-related sequences were localized near it on chromosome 17 (Kiel-Metzger and Erickson, 1984), a result confirmed by DNA sequence analyses (Durbin EJ et al., 1989; Nishioka et al.,1992). Although RNA hybridizing to sense and anti-sense Bkm clones could be detected in day-14 gonads, they were equally present in the gonads of both sexes (Erickson and Durbin, 1987). As it became apparent that Bkm was a "red herring" as far as sex determination is concerned (above), interest waned. However, given the many properties attributed to alleles of the *T/t* complex, the interest was renewed when Eicher's group at the Jackson laboratory found a T-associated sex-reversal. This occurred in mice heterozygous for T^{hp}, *T Hairpin tail* (Washburn and Eicher, 1983). A similar sex reversal also occurred with T^{Orl}, *T orlean* (Washburn and Eicher,1989). These sex-reversals had occurred when the *T*-alleles had been back crossed onto the C57BL/6J genetic background, as had been the case with Y^{POS}. However, the sex-reversal was associated with the AKR Y chromosome which had been introduced onto this particular C57BL/6J strain. It turned out that the AKR Y chromosome had also been derived from *Mus musculous domesticus* as had Y^{POS} and similarly delayed testicular development but not sufficiently to cause sex reversal (Washburn and Eicher, 1989). Further crosses indicated that a 3rd (Washburn et al, 1990) or more (Eisner et al., 1996) genes were also involved. It seems probable that the two *T*-deletions slow gonadal development as a lower amount of Sry was expressed (Washburn and Eicher, 2001), delaying mesonephric cell migration into the germ cords (Albrecht et al., 2000), just sufficiently to let ovarian development start as do several other mutants (*steel* and *white spotting* [both of which prevent germ cell migration]; Burgoyne and Palmer, 1991). It is also possible that *t*-linked *H-2* genes could explain this difference as a gene, *Ped,* which has alleles associated with fast or slow embryonic growth maps to the Q9 region of H-2 (Xu et al., 1994).

There has long been interest in genetic modifiers of the expression of *T* resulting in variation in length of the tail. Sex-dependent and *H-2*-influenced expression was described by Mickova and Ivani (1974)—the *H-2* linked effect was only found in males. When a quantitative trait locus (QTL) search was performed on the 24 shortest tailed and the 24 longest tailed of 328 mice from two reciprocal, two-generation backcrosses (Agulnik et al., 1997) a different sex effect was found. Using 98 microsatellite markers, chromosomal regions of influence were identified and the use of region-specific marker further localized the modifiers on Chromosomes 9 and 15 which had additive effects on tail length but only in females (Agulnik et al., 1998).

Notes

1. A retired physician patient of mine insisted this was true to me in the 60s.

2. Anne McLaren became Director of the Medical Research Council Mammalian Development Unit located in the Wolfson House near Euclid Station in central London in 1974. I was able to visit this unit and give seminars several times over the years where Anne was a gracious hostess. I was also able to recruit an excellent post-doctoral fellow, Asangla Ao, here. She was doing research with Marilyn Monk. Ph.D students here earned their degrees from University College, University of London. Usually there was no course work but 3 years of research. The written doctoral thesis was then evaluated by 2 external reviewers and no oral defense was required.

3. David Page became the leader in cloning and sequencing the human Y chromosome. He kindly shared with me his yeast artificial chromosome (YAC) library of the Y chromosome which was important for my laboratory's effort to clone a gene responsible for lymphedema located at a Y-16 translocation breakpoint. We didn't find a candidate gene on the Y side of the translocation and the search moved to the chromosome 16 side where colleagues at the University of Michigan played the major role in finding the gene (*FOXC2*) there.

4. Stan R Blecher; at that time at the University of Guelph, Ontario, Canada; spent a 6 month sabbatical in my laboratory at the University of Michigan in 1989-1990. He increased my understanding of these pre-gonadal differentiation differences in development.

5.   I enjoyed a sabbatical with Peter in 1983-84. This was prior to his exciting work leading to the cloning of *TDF* and I worked on other aspects of the human Y chromosome.

6.   Environmental control of sex determination frequently occurs in fish, reptiles, and worms—both marine and nematodes. For excellent descriptions and discussions of mechanisms, **James Bull's "Evolution of Sex Determining Mechanisms (1983)"** is recommended.

References

Affara, N.A., Chalmers, I.J., M.A., Ferguson-Smith, 1993. Analysis of the SRY gene in 22 sex-reversed XY females identifies four new point mutations in the conserved DNA binding domain. Hum. Mol. Genet. 2, 785–789.

Agulnik, A.I., Mitchell, M.J., Lerner, J., Woods, D.R., Bishop, CE, 1994. A mouse Y chromosome gene encoded by a region essential for spermatogenesis and expression of male-specific minor histocompatibility antigens. Hum. Molec. Genet. 6, 873–878.

Agulnik, A.I., Bishop, C.E., Lerner, J.L., Agulnik, S.I., Solovyev, V.V., 1997. Analysis of mutation rates in the *SMCY/SMCX* genes shows that mammalian evolution is male driven. Mamm. Genome 8, 134–138.

Agulnik, A.I., Agulnik, S.I., Saatkamp, B.D., Silver, L.M., 1998. Sex-specific modifiers of tail development in mice heterozygous for brachyury (*T*) mutation. Mammal. Genome 9, 107–110.

Albrecht, K.H., Aapel, B., Washburn, L.L., Eicher, E.M., 2000. Defective Mesonephric Cell Migration Is Associated with Abnormal Testis Cord Development in C57BL/6J*Mus domesticus*mice. Devel. Biol. 225, 26–36.

Ao, A., Erickson, R.P., Winston, R.M.L., Handyside, A.H., 1994. Transcription of paternal Y-linked genes in the human zygote as early as the pronucleate stage. Zygote 2, 281–287.

Atkinson, T.G., Blecher, S.R., 1994. Aberrant anogenital distance in XX*Sxr* ("sex-reversed") pseudoautosomal mice. J. Zool. Lond 233, 581–589.

Bagheri-Fam, S., Sinclair, A.H., Koopman, P., Harley, V.R., 2010. Conserved regulatory modules in the Sox9 testis-specific enhancer predict roles for SOX, TCF/LEF, Forkhead, DMRT, and GATA proteins in vertebrate sex determination. Int. J. Biochem. Cell Biol. 42, 472–477.

Berta, P., Hawkins, J.R., Sinclair, A.H., et al., 1990. Genetic evidence equating SRY and the testis-determining factor Nature 348, 448–450.

Bishop, C.E., Boursot, P., Baron, B., Bonhomme, F., Hatat, D., 1985. Most classical *Mus musculus domesticus* laboratory mouse strains carry a *Mus musculus musculus* Y chromosome. Nature 315, 70–72.

Bishop, C.E., Weith, A., Mattei, M.G., Roberts, C., 1988. Molecular aspects of sex determination in mice: an alternative model for the origin of the Sxr region. Phil. Trans. R. Soc., Lond B322, 119–124.

Bishop, C.E., Whitworth, D.J., Qin, Y., Agoulnik, A.I., Agoulnik, I.U., Harrison, W.R., Behringer, R.R., Overbeek, P.A., 2000. A transgenic insertion upstream of *Sox9* is associated with dominant XX sex reversal in the mouse. Nat. Genet. 26, 490–494.

Blecher, S.R., Erickson, R.P., 2007. Genetics of Sexual Development: A New paradigm. Am. J. Med. Genet. Part A 143A, 3054–3068.

Boettger-Tong, H.L., Agulnik, A.I., Ty, T.I., Bishop, CE, 1998. Transposition of *RhoA* to the murine Y chromosome. Genomics 49, 180–187.

Burgoyne, P.S., 1982. Genetic homology and crossing over in the X and Y chromosomes of mammals. Hum. Genet. 61, 85–90.

Burgoyne, P.S., Levy, E.R., A, McLaren, 1986. Spermatogenic failure in male mice lacking H-Y antigen. Nature 320, 170–172.

Burgoyne, P.S., Palmer, S.J., 1991. The genetics of XY sex reversal in the mouse and other mammals. Seminars Develop. Bio. 2, 277–284.

Burgoyne, P.S., Mahadevaiah, S.K., Sutclifffe, M.J., Palmer, SJ, 1992. Fertility in mice requires X-Y pairing and a Y-chromosomal "spermiogenesis" gene mapping to the long are. Cell 71, 391–398.

Burgoyne, P.S., 1993. A Y-chromosomal effect on blastocyst cell number in mice. Development 117, 34'–345.

Capel, B., Rasberry, C., Dyson, J., Bishop, C.E., Simpson, E., Vivian, N., Lovell-Badge, R., Rastan, S., Cattanach, B.M., 1993. Deletion of Y chromosome sequences located outside the testis determining region can cause XY female sex reversal. Nature Genet 5, 301–307.

Cattanach, B.M., 1961. XXY mice. Genet. Res. 2, 156–158.

Cattanach, B.M., Pollard, C.E., Hawkes, S.G., 1971. Sex-reversed mice: XX and XO males. Cytogen 10, 318–337.

Cuenot, L., 1899. Sur la determination due sexe chez les animaux. Bull. Sci. Fr. Belg. 32, 461–534.

de Santa Barbara, P., Bonneaud, N., Boizet, B., et al., 1998. Direct interaction of SRY-related protein SOX9 and steroidogenic factor 1 regulates transcription of the human anti-mullerian hormone gene. Mol. Cell Biol. 18, 6653–6665.

Disteche, C.M., Zacksenhaus, E., Adler, D.A., Bressler, S.L., Keitz, B.T., Chapman, V.M., 1992. Mapping and expression of the ubiquitin-activating enzyme E1 (*Ube1*) gene in the mouse. Mammal. Genome 3, 156–161.

Durbin, E.J., Erickson, R.P., Craig, A., 1989. Characterization of GATA/GACA-related sequences on proximal chromosome 17 of the mouse. Chromosoma (Berl) 97, 301–306.

Eicher, E.M., Washburn, L.L., Whitney III, J.B., Morrow, KE, 1982. *Mus poschiavinius* Y chromosome in C57BL/6J murine genome causes sex reversal. Science 217, 535–537.

Eicher, E.M., Phillips, S.J., Washburn, L.L., 1983. The use of molecular probes and chromosomal rearrangements to partition the mouse Y chromosome into functional regions ed.s. In: Messer, A, Porter, IH (Eds.), Recombinant DNA and Medical Genetics. Academic press, New York, pp. 57–71.

Eicher, E.M., Washburn, L.L., 1983. Inherited sex reversal in mice: identification of a new primary sex determining gene. J. Exp. Zool 228, 297–304.

Eicher, E.M., 1988. Autosomal genes involved in mammalian primary sex determination. Philos. Tran. Royal Soc., London, B 322, 109–118.

Eicher, E.M., Shown, E.P., Washburn, L.L., et al., 1995. Sex reversal in C57BL/6J-Y[POS] mice corrected by a *Sry* transgene. Proc. Trans. R.Soc. Lond. B. doi:10.1098/rstb.1995.0160 doi. org/.

Eisner, J.R., Eales, B.A., Biddle, F.G., 1996. Segregation analysis of the testis-determining autosomal trait, *Tda*, that differs between the C57BL/6J and DBA/2j mouse strains suggests a multigenic threshold model. Genome 39, 322–335.

Epstein, C.J., Smith, S., Travis, B., 1980. Expression of H-Y antigen on mouse preimplantation embryos. Tissue Antigens 15, 63–67.

Erickson, R.P., Durbin, E.J., Tres, L.L., 1987. Sex determination in mice: Y and chromosome 17 interactions. Develop., 101, 25–32 Suppl.

Erickson, R.P., Strnatka, D, 2011. Insulin Receptor-Related (Irr) Is Expressed in Pre-Implantation Embryos: A Possible Relationship to "Growth Factor Y" and Sex Determination. Molec Reprod Develop 78, 552.

Evans, H.J., Buckton, K.E., Spoward, G., Carothers, A.D., 1979. Heteromorphic X chromosomes in 46,XX males: evidence for the involvement of X-Y interchange. Hum. Genet. 49, 11–31.

Evans, E.P., Burtenshaw, M.D., Cattanach, B.M., 1982. Meiotic crossing-over between the X and Y chromosomes of male **mice** carrying the sex-reversing (Sxr) factor. Nature 300, 443–445.

Fennelly, J., Harper, K., S., Laval, Wright, E., Plumb, M., 1996. Co-amplification of tail-to-tail copies of MuRVY and IAPE retroviral genomes on the *Mus musculus* Y chromosome. Mammal. Genome 7, 31–36.

Giuili, G., Shen, W.-H., Ingraham, H.A., 1997. The nuclear receptor SF-1 mediates sexually dimorphic expression of Mullerian inhibiting substance. vivo. Devel 124, 1799–1807.

Glover, T.W., Wilson, T.E., Arlt, M.F., 2017. Fragile sites in Cancer: More Than Meets the Eye. Nature Rev.s Canc 17, 489–501.

Goodfellow, P.J., Darling, S.M., Thomas, N.S., Goodfellow, PN, 1986. A pseudoautosomal gene in man. Science 234, 740–743.

Gordon, J., 1976. Failure of XX cells containing the Sex Reversed Gene to Produce Gametes in Allophenic Mice. J. Exp. Zool 198, 367–374.

Graves, P.E., Erickson, R.P., 1992. An amino acid change in the DNA-binding region of Sry found in *Mus musculus domesticus* and other species, does not explain C57BL/6J-Y^DOM sex-reversal. Biochem Biophs Res. Comm 29, 310–316.

Gruneberg, H., McLaren, A., 1972. The skeletal phenotype of some mouse chimaeras. Proc. Roy soc. London, B 182, 9–23.

Greenfield, A., 2015. Understanding sex determination in the mouse: genetics, epigenetics and the story of mutual antagonism. J. Genet. 94, 585–590.

Gubbay, J., Collignon, J., Koopman, P., et al., 1990. A gene mapping to the sex-determining region of the mouse Y chromosome is a member of a novel family of embryonically expressed genes. Nature 346, 245–250.

Gubbay J., Lovell-Badge R. (1994) The Mouse Y chromosome. In Wachtel S.S, ed. "Molecular Genetics of Sex Determination," Academic Press, San Diego, etc., p.47.

Hacker, A., Capel, B., Goodfellow, P., Lovell-Badge, R., 1995. Expression of Sry, the mouse sex-determining region gene. Development 121, 1603–1614.

Hansmann, I., 1982. Sex Reversal in the Mouse. Cell 30, 331–332.

Haraguchi, R., Suzuki, K., Murakami, R., Sakai, M., Kamikawa, M., Kengaku, M., Sekine, K., Kawano, H., Kato, S., Ueno, N., Yamada, G., 2000. Molecular analysis of external genitalia formation: the role of fibroblast growth factor (Fgf) genes during genital tubercle formation. Develop 127, 2471–2479.

Haraguchi, R., Mo, R., Hue, C.C., Motoyama, J., Makino, S., Shirioishi, T., Gaffield, W., Yamada, G., 2001. Unique functions of Sonic hedgehog signaling during external genitalia development. Devel 128, 4241–4250.

Hatano, O., Takayama, K., Imai, T., Waterman, M.R., Takakusu, A., Omura, T., Morohasi, K.-I., 1994. Sex-dependent expression of a transcription factor, Ad4BP, regulating steroidogenic P-450 genes in the gonads during prenatal and postnatal rat development. Devel 120, 2787–2797.

Hawkins, J.R, Taylor, A., Goodfellow, P.N., et al., 1992. Evidence for increased prevalence of SRY in XY females with complete rather than partial gonadal dysgenesis. Am. J. Hum. Genet. 51, 979–984.

Hunt, P.A., Burgoyne, P.S., 1987. A tissue-specific fragile site associated with the sex reversed (*Sxr*) mutation in the mouse. Chromosoma 97, 67–71.

Hunt, S.E., Mittwoch, U., 1987. Y-hromosomal and other factors in the development of testis size in mice. Genet. Res. (Camb.) 50, 205–211.

Ikeda, Y., Shen, W.-H., Ingraham, H.A., Parker, K.L., 1994. Developmental expression of mouse steroidogenic factor 1, an essential regulator of the steroid hydroxylases. Mol. Endocrinol. 8, 654–662.

Jager, R.J., Anvret, M., Hall, K., Scherer, G., 1990. A human XY female with a frame shift mutation in the candidate testis-determining gene SRY. Nature 348, 452–454.

Jost, A., 1947. Recherches sur la differeciation sexuelle de l'embryon de lapin. III. Role des gonads foetales dans la differentiation sexuelle somatique. Arch. Anat. Morphol. Exp. 36, 271–315.

Jost, A., Vigier, B., Prepin, J., Perchellet, J.P., 1973. Studies on sex differentiation in mammals. Recent Prog. Horm. Res. 29, 1–41.

Just, W., Rau, W., Vogel, W., Akherverkian, M., Fredga, K., Graves, J.A., Evapunova, E., 1995. Absence of *Sry* in species of the vole *Ellobius*. Nat. Genet. 11, 117–118.

Kettlewell, J.R., Raymond, C.S., Zarkower, D., 2000. Temperature-dependent expression of turtle Dmrt-1 prior to sexual differentiation. Genesis 26, 174–178.

Kiel-Metzger, K., Erickson, R.P., 1984. Regional localization of sex-specific Bkm-related sequences on proximal chromosome 17 of mice. Nature 310, 579–581.

Kiel-Metzger, K., Warren, G., Wilson, G.N., Erickson, RP, 1985. Evidence that the human Y chromosome does not contain clustered DNA sequences (Bkm) associated with heterogametic sex determination in other vertebrates. New Engl. J. Med. 313, 242–245.

Koopman, P., Munsterberg, J., Capel, A., et al., 1990. Expression of a candidate sex-determining gene during mouse testis differentiation. Nature 348, 450–452.

Koopman, P., Gubbay, J., Vivian, N., et al., 1991. Male development of chromosomally female mice transgenic for Sry. Nature 351, 117–121.

Lala, D.S., Rice, D.A., Parker, K.L., 1992. Steroidogenic factor 1, a key regulator of steroidogenic enzyme expression, is the mouse homolog of fushi tarazu-factor 1. Mol. Endocrinol. 6, 1249—1258.

Lamar, E.E., Palmer, E., 1984. Y-encoded, species-specific DNA in mice: evidence that the Y chromosome exists in two polymorphic forms in inbred strains Cell 37, 171–177.

LeBarr, D.K., Blecher, S.R., 1986. Epdidymides of sex-reversed XX mice lack the Initial Segment. Dev. Genet. 7, 109–116.

LeBarr, D.K., Moger, W., Blecher, S.R., 1991. Development of the normal XY male and sex-reversed XXSxr pseudomale mouse epididymis. Mol. Reprod. Dev. 28, 9–17.

Lee, C.-H., Taketo, T., 1994. Normal onset but Prolonged Expression, of *Sry* Gene in the B6.Y^{DOM} Sex-Reversed Mouse Gonad. Develop. Biol. 165, 442–452.

Magenis, R.E., Webb, M.J., McKean, R.S., Tomar, D., Allen, K.J., Kammer, H., VanDyke, D.L., Louvien, E., 1982. Translocation (X:Y)(p22.33:p11.2) in XX males: etiology of male phenotype. Hum. Genet. 62, 271–276.

Marchand, O., Govoroun, M., D'Cotta, H., McMeel, O., Lareyre, J., Bernot, A., Laudet, V., Guiguen, Y., 2000. DMRT-1 expression during gonadal differentiation and spermatogenesis in the rainbow trout, *Oncorhynchus mykiss*. Biochem. Biophys Acta 1493, 180–187.

Mardon, G., Mosher, R., Disteche, C.M., Nishioka, Y., McLaren, A., Page, D.C., 1989. Duplication, deletion and polymorphism in the sex-determining region of the mouse Y chromosome. Science 243, 78–80.

Mazeyrat, S., Saut, N., Grigoriev, V., Mahadevaiah, S.K., Ojarikre, O.A., Rattigan, A., Bishop, C., Eicher, E.M., Mitchell, M.J., Burgoyne, P.S., 2001. A Y-encoded subunit of the translation initiation factor Eif2 is essential for mouse spermatogenesis. Nat. Genet. 29, 49–53.

Mazeyrat, S., Saut, N., Sargent, C.A., Grimmond, S., Longepied, G., Ehrmann, I.E., Ellis, P.S., Greenfield, A., Affara, N.A., Mitchell, M.J., 1998. The mouse Y chromosome interval necessary for spermatogonial proliferation is gene dense with syntenic homology to the human *AZFa* region. Hum. Molec. Genet. 7, 1713–1724.

McLaren, A., 1953. Protective effect of aged vaccines on a neurotropic virus infection in mice. Nature 172, 38.

McLaren A. (1972) Germ cell differentiation in artificial chimaeras of mice. The Genetics of the Spematozoon, ed. By R.A. Beatty and S. Gluecksohn-Waelsch, pp. 313-324.

McLaren, A., Biggers, J.D., 1958. Successful development and birth of mice cultivated in vitro as early as early embryos. Nature 182, 877–878.

McLaren, A., Sanders, K., 1959. The influence of the age of the host on local virus multiplication and on the resistance to virus infections. Epidemiol. and Infect. 57, 106–122.

McLaren, A., 1986. Sex Ratio and Testis Size in Mice Carrying *Sxr* and T(X;16)16H. Develop. Genet. 7, 177–185.

McLaren, A., Simpson, E., Tomonari, K., Chandler, P., Hogg, H., 1984. Male sexual differentiation in mice lacking H-Y antigen. Nature 312, 552–555.

McLaren, A., Hunt, R., Simpson, E., 1988a. Absence of any male-specific antigen recognized by T lymphocytes in X/XSxr' male mice. Immunology 63, 447–449.

McLaren, A., Simpson, E., Epplen, J.T., Studer, R., Koopman, P., Evans, E.P., Burgoyne, P.S., 1988b. Location of the genes controlling H-Y antigen expression and testis determination on the mouse Y chromosome. Proc. Nat'l Acad. Sci., U.S.A. 85, 6442–6445.

Mickova, M., Ivanyi, P., 1974. Sex-dependent and *H-2*-linked influence on expressivity of the Brachury gene in mice. J. Hered. 65, 369–372.

Mitchell, M.J., Woods, D.R., Tucker, P.K., Opp, J.S., Bishop, C.E., 1991. Homology of a candidate spermatogenic gene from the mouse Y chromosome to the ubiquitin-activating enzyme E1. Nature 354, 483–486.

Mitchell, M.J., Wilcox, S.A., Watson, J.M., Lerner, J.L., Woods, D.R., Scheffler, J., Hearn, J.P., Bishop, C.E., Marshall Graves, J.A, 1998. The origin and loss of the ubiquitin activating enzyme gene on the mammalian Y chromosome. Hum. Molec. Genet. 7, 429–434.

Mittwoch, U., 1969. Do genes determine sex? Nature 221, 446–448.

Mittwoch, U., 1993. Blastocysts prepare for the race to be male. Hum Reprod 8, 1550–1555.

Miyamoto, N., Yoshida, M., Kuratani, S., Matsuo, I., Aizawa, S., 1997. Defects of urogenital development in mice lacking. Emx2. Develop 124, 1653–1664.

Moniot, B., Berta, P., Shere, G., Sudbeck, P., Poulat, F., 2000. Male specific expression suggests role of DMRT1 in human sex determination. Mech. Dev. 91, 323–325.

Monk, M., Kathuria, H., 1977. Dosage compensation for an X-linked gene in pre-implantation embryos. Nature 270, 599–601.

Monk, M., McLaren, A., 1981. X-chromosome activity in foetal germ cells of the mouse. J. Embryol. Exp. Morphol. 63, 75–84.

Nakahori, Y., Tamura, T., Nagafuchi, S., et al., 1991. Molecular cloning and mapping of 10 new probes on the human Y chromosome. Genomics 9, 765–769.

Nef, S., Verma-Kurvari, S., Merenmies, J., Vassali, J.D., Efstratiadis, A, Accili, D, Parada, LF, 2003. Testis determination requires insulin receptor family function in mice. Nature 426, 291–295.

Nishioka, Y., Dolan, B.M., Fiorellino, A., Prado, V.F., 1992. Nucleotide sequence analysis of a mouse Y chromosomal DNA fragment containing Bkm and LINE elements. Genetica 87, 7–15.

O, W.-S., Short R., V., Renfree, M.B., Shaw, G., 1998. Primary genetic control of somatic sexual differentiation in a mammal. Nature 331, 716–717.

Odorisio, T., Mahadevaiah, S.K., mcCarrey, J.R., Burgoyne, P.S., 1996. Transcriptional Analysis of the Candidate Spermatogenesis Gene *Ube1y* and of the Closely Related *Ube1x* Shows That They Are Coexpressed in Spermatogonia and Spermatids but Are Repressed in Pachytene Spermatocytes. Develop. Biol. 180, 336–343.

Page, D.C., Mosheer, R., Simpson, E.M., 1987. The sex-determining region of the human Y chromosome encodes a finger protein. Cell 51, 1091–1104.

Palmer, M.S., Sinclair, A.H., Berta, P., et al., 1989. Genetic evidence that ZFY is not the testis-determining factor. Nature 342, 937–939.

Palmer, S.J., Burgoyne, P.S., 1991. The Mus musculus domesticus Tdy allele acts later than the Mus musculus musculus Tdy allele: a basis for XY sex-reversal in C57BL/6-YPOS mice. Develop 113, 709–714.

Palmer, S., Perry, J., Kipling, D., Ashworth, A., 1997. A gene spans the pseudoautosomal boundary in mice. Proc. Nat'l Acad. Sci., U.S.A. 94, 12030–12035.

Peippo, J., Bredbacka, P., 1995. Sex-related growth rate differences in mouse preimplantation embryos in vivo and in vitro. Mol. Reprod. Dev. 40, 56–61.

Pergament, E., Fiddler, M., Cho, N., Johnson, D., Holmgren, W.J., 1994. Sexual differentiation and preimplantation cell growth. Hum. Reprod. 9, 1730–1732.

Polani, P.E., 1982. Pairing of X and Y chromosomes, non-inactivation of X-linked genes and the maleness factor. Hum Genet 60, 207–211.

Raymond, C.S., Ketlewell, J.R., Hirsch, B., Bardwell, V.J., Zarkower, D., 1999. Expression of *Dmrt-1* in the genital ridge of the mouse and chicken embryo suggests a role in vertebrate sexual development. Dev. Biol. 215, 208–220.

Sadovsky, Y., Crawford, P.A., Woodson, K.G., Polish, J.A., Clements, M.A., Tourtellotte, L.M., Simburger, K., Milbrandt, J., 1995. Mice deficient in the the orphan receptor steroidogenic factor 1 lack adrenal glands and gonads but express P450 side-chain cleaving-enzyme in the placenta and have normal embryonic serum levels of corticosteroids. Proc. Nat'l Acad. Sci., U.S.A 92, 10939–10943.

Schmahi, J., Eicher, E.M., Washburn, L.L., Capel, B., 2000. *Sry* induces cell proliferation in the mouse gonad. Develop 27, 65–73.

Schubert, S., Dechend, F., Skawran, B., Kunze, B., Winking, H., Weile, C., Romer, I., Hemberger, M., Fundele, R., Sharma, T., Schmidtke, J., 2000. Silencing of the Y-chromosomal gene *tspy* during murine evolution. Mammal. Genome 11, 288–291.

Sekido, R., Lovell-Badge, R., 2008. Sex determination involves synergistic action of SRY and SF1 on a specific *Sox9* enhancer. Nature 453, 930–934.

Sharpiro, M., Erickson, R.P., Lewis, S., Tres, L.L., 1982. Serological and cytological evidence for increased Y-chromosome related material in *Sxr,*XY (sex-reversed carrier, male) mice. J. Reprod. Immunol. 4, 191–206.

Shawlot, W., Behringer, B.R., 1995. Requirement for *Lim1* in head-organizer function. Nature 374, 425–430.

Shen, W.-H., Moore, C.C.D., Ikeda, Y., Parker, K.L., Ingraham, H.A., 1994. Nuclear receptor steroidogenic facto 1 regulates the mullerian inhibiting substance gene: a link to the sex determination pathway. Cell 77, 651–661.

Sherman, M.I., McLaren, A., Walker, P.M.B., 1972. Mechanism of accumulation of DNA in giant cells of the mouse trophoblast. Nature New Biol 238, 175–176.

Simpson, E.M., Page, D.C., 1991. An interstitial deletion in mouse Y chromosomal DNA created a transcribed *Zfy* fusion gene. Genomics 11, 601–608.

Sinclair, A.H., Foster, J.W., Spencer, J.A., Page, D.C., Palmer, M., Goodfellow, P.N., Graves, J.A., 1988. Sequences homologous to ZFY, a candidate human sex-determining gene, are autosomal in marsupials. Nature 336, 780–783.

Sinclair, A.H., Berta, P., Palmer, M.S., et al., 1990. A gene from the human sex-determining region encodes a protein with homology to a conserved DNA-binding motif. Nature 346, 240–244.

Singh, L., Jones, K.W., 1982. Sex reversal in the mouse (Mus musculus) is caused by a recurrent nonreciprocal crossover involving the X and an aberrant Y chromosome. Cell: 28, 205–216.

Styrna, J., Klag, J., Moriwaki, K., 1991. Influence of partial deletion of the Y chromosome on mouse sperm phenotype. J. Reprod. Fert. 92, 187–195.

Sutou, S., Mitsui, Y., Tsuhiya, K., 2001. Sex determination without the Y chromosome in two Japanese rodents: *Todudaia osimensis osimensis* and *Tokudaia osimensis* spp. Mamm. Genome 12, 17–21.

Thornhill, A.R., Burgoyne, P.S., 1993. A paternally imprinted X chromosome retards the development of the early mouse embryo. Develop 118, 171–174.

Tsunoda, Y., Tokunaga, T., Sugie, T., 1985. Altered Sex Ratio of Live Young After Transfer of Fast- and Slow-Developing Mouse Embryos. Gamete Res 12, 301–314.

Valdivia, R.P.A., Kunieda, T., Azuma, S., Toyoda, Y., 1993. PCR Sexing and Developmental Rate Differences in Preimplantation Mouse Embryos Fertilized and Cultured In Vitro. Mol. Reprod. Devel. 35, 121–126.

Vanlerberghe, F., Dod, B., Boursot, P., Bellis, M., Bonhomme, F., 1986. Absence of Y-chromosome introgression across the hybrid zone between *Mus musculus domesticus* and *Mus musculus musculus*. Genet. Res., Camb. 48, 191–197.

Verga, V., Erickson, R.P., 1989. An extended long-range restriction map of the human sex-determining region on Yp, including ZFY, finds marked homology on Xp and no detectable Y sequences in an XX male. Am. J. Hum. Genet. 44, 756–765.

Vergnaud, G., Page, D.C., Simmler, M.C., et al., 1986. A deletion map of the human Y chromosome based on DNA hybridization. Am. J. Hum. Genet. 38, 109–124.

Vidal, V.P.I., Chaboissier, M.-C., de Rooij, D.G., Schedl, A., 2001. *Sox9* induces testis development in XX transgenic mice. Nature Genet 28, 216–217.

Vogt, P.H., Affara, N., Davey, P., Hammer, M., Jobling, M.A., Lau, Y.-F.C., Mitchell, M., Schempp, W., Tyler-Smith, C., Williams, G., Yen, P., Rappold, G.A., 1997. Report of the third international workshop on Y chromosome mapping 1997. Cytogenet. Cell Genet.

Washburn, L.L., Eicher, E.M., 1983. Sex reversal in XY mice caused by dominant mutation on chromosome 17. Nature 303, 338–340.

Washburn, L.L., Eicher, E.M., 1989. Normal testis determination in the mouse depends on genetic interaction of a locus on chromosome 17 and the Y chromosome. Genetics 123, 173–179.

Washburn, L.L., Lee, B.K., Eicher, E.M., 1990. Inheritance of T-associated sex reversal in mice. Genet. Res. 56, 185–191.

Washburn, L.L., Albrecht, K.H., Eicher, E.M., 2001. C57BL/6J-*T*-Associated Sex Reversal in mice Is Caused by Reduced Expression of a *Mus domesticus Sry* allele. Genetics 158, 1675–1681.

Welshons, W.J., Russell, L.B., 1959. The Y-chromosome as the bearer of male determining factors in the mouse. Proc. Nat'l Acad. Sci., USA 45, 560–566.

Xian, M., Azuma, S., Naito, K., Kunieda, T., Moriwaki, K., Toyoda, Y., 1992. Effect of a partial deletion of Y chromosome on in vitro fertilizing ability of mouse spermatozoa. Biol. Reprod. 47, 549–553.

Xu, Y., Ping, J., Mellor, A.N., Warner, C.M., 1994. Identification of the *Ped* gene at the molecular level: the Q9 MHC class I Transgene Converts the *Ped Slow* to the *Ped Fast* Phenotype. Biol Reprod 51, 695–699.

Zwingman, T., Erickson, R.P., Boyer, T., Ao, A., 1993. Transcription of the sex-determining region genes *Sry* and *Zfy* in the mouse preimplantation embryo. Proc. Natl. Acad. Sci. USA 90, 814–817.

12

Pharmacogenetics—a mostly last half of the century effort

Introduction

The history of pharmacogenetics is more one of human medical research than of mouse genetics but the high genetic variation of mice and the ability to modify the mouse genome have allowed studies in mouse pharmacogenetics to make major contributions to the field. Human variation in responses to ingested substances was noted as early as 510 B.C. when Pythagoras noted that the ingestion of Fava beans was lethal in some individuals but not most (Nebert, 1999), a result now known to be due to glucose-6-phosphate dehydrogenase deficiency. As discussed in Chapter 1, there is considerable disagreement about who first clearly enunciated the "one gene, one enzyme" concept. That Cuenot's "ferments", controlled by pigment genes, proceeded Archibald Garrod's enunciation of the concept for alkaptonuria seems clear (but also see Ernest Caspari's mini-biography for another pre-Beadle and Tatum postulation of the concept). However, it has been said that Garrod deserves the credit for enunciating the concept that enzymes were responsible for detoxification of foreign substances. According to **Weber, (1997, p. 5)**, this idea was suggested by a case of porphyria brought on by the ingestion of sulfonal, a hypnotic drug in use at the time. He does not provide a reference but careful reading of Garrod's pre-1900 papers, especially Garrod, 1896, presents multiple, frequently fatal cases but does not find any mention of enzymes. Garrod's concerns at this time were in showing that the amount of porphyrin in the urine is too great to be explained merely by destruction of blood. Only in the second edition of his book, **Inborn Errors of Metabolism, p. 8**, did he use the concept of drug conjugation when, following Emil Fischer, he pointed out that "the primary combination of the foreign substance is with glucose itself, and that, the aldehyde group being thus protected from change, oxidation to glycuronic acid occurs". The conjugation of many drugs with, as we would now say, glucuronic acid, is a major excretory pathway. The first identification of wider human genetic variation in response to xenobiotics were those concerned with olfaction (Blakeslee, 1918) and taste (Snyder,1932). Soon a range of variation in responses to a drug was reported for the response to succinylcholine due to mutant forms of its degrading enzyme, serum cholinesterase (Kalow and Gunn, 1957). Also in 1957, one of the fathers of medical genetics, Arno Motulsky reviewed the importance of genetic variation for the metabolism of drugs (Motulsky, 1957). Yet, even in 1959, Harris makes no mention of drug metabolism in his **Human Biochemical Genetics**.

While the midcentury studies emphasized the biochemical detoxification of drugs, there were many other aspects of drug metabolism which altered their length of stay in the body. Variations in uptake, exemplified by the role of Niemann-Pick C like 1 (NPCL1) in intestinal

Twentieth Century Mouse Genetics. DOI: https://doi.org/10.1016/B978-0-12-824016-8.00016-7

cholesterol uptake (Better and Yu, 2010), transport across endothelial and epithelial membranes (see below) and receptor affinities (see below), all had effects which were gradually elucidated. The application of these variations– such as the important common variants in cytochrome P450 CYP1A2 and the rate of warfarin metabolism (Kaminsky and Zhang, 1997) or in cytochrome P450 CYP3A4 and the risk of statin-induced myopathy (Ghatak et al., 2010)– to clinical use is only now (2020) being established, despite being well publicized, such as Erickson (1998). It has required inexpensive DNA sequencing and electronic medical records, where warnings on drug use can be "pop-ups" when a particular drug is ordered, to establish the use of such information more generally (Herr et al., 2015; Rasmussen-Torvik et al., 2014; reviewed by Weinshilboum and Wang (2017).

Some of the early studies in mice noted strain differences in teratogenic susceptibility (Frazer and Fainstat, 1951), seizure susceptibility to anesthetics (Davis and King, 1967; Vessell, 1968), depressive response to chlorpromazine (Huff, 1962), and response to poisons (Meier et al., 1962). Genetic variations in mice included variations in phase 1 (mostly P450s) and phase 2 enzymes involved in the disposal of xenobiotics and endogenous compounds as well as receptors and membrane transporters.

Xenobiotic detoxification

CytochromeP450s

Mouse studies were important for discovering the major phase 1 metabolizer of xenobiotics, the P450 system: Elliot Vessel's (1967) observation that aromatic bedding from soft wood trees placed in the mouse cages induced a number of microsomal drug metabolizing enzymes was the beginning of the understanding of the role of aryl hydrocarbon hydroxylase in xenobiotic metabolism. Aryl hydrocarbon (benzo[a]pyrene) hydroxylase (Ahh) induction was found to occur as a binary trait with responsive (e.g.C57BL/6J; BL6) and nonresponsive strains (DBA/2J; DBA). The *Ah* locus was discovered to code for a receptor inducing a number of cytochrome P450, xenobiotic oxidizing enzymes. It had similarities to steroid receptors and its induction of cytochrome P450s was blocked by inhibitors of transcription. Poland, et al. (1987) used a radioligand to characterize it in mice and found a third allele, mapping the locus to chromosome 12. It was soon found, however, that while induction by benzo[a]pyrene or zoxazolamine segregated with the *Ah* locus, the induction of metabolizing enzymes for other drugs, e.g. diphenylhydantoin or hexobarbital, while high in BL6 and low in DBA, did not segregate with the *Ah* locus (Robinson and Vessell, 1974). This suggested that there were inducers and/or other cytochrome P450s, a result seen in other crosses (Thomas and Hutton, 1973). This was soon confirmed biochemically and other cytochrome P450s, the microsomal monooxygenases, were discovered and considered the first phase of metabolism for many drugs (Ullrich and Kremers, 1977). Conjugation of these hydrophobic compounds with acetyl groups (N-acetylation by transferases) or glucuronic acid (by glucuronidase) to make them more aqueous-soluble was then the second phase.

The cytochrome P450s catalyzed the insertion of one atom of molecular oxygen into very hydrophobic substrates. Thus, monooxygenases were the initial rate-limiting step in the formation of water-soluble metabolites. The insertion of oxygen at either an aliphatic or aromatic unsaturated carbon-carbon bond resulted in the formation of an epoxide or arene oxide which are potent electrophiles. Thus, the products of these reactions were unstable reactive intermediates which could damage DNA and proteins and were potential carcinogens (Thorgeirsson and Nebert, 1977; Slater et al., 1980).

With time it was found that the non-responsive strains would respond with other aromatic compounds (Robinson et al., 1974) and the number of distinct P450s identified rapidly continued. It became apparent that different species had different arrays of cytochrome P450s, presumably selected to handle xenobiotics encountered in their respective environments. The competition between plants synthesizing new compounds to sicken or poison their exploiters and the exploiters making new cytochromes to detoxify these compounds has been seen by evolutionists as an example of a "Red Queen Race" where each competitor runs faster and faster, but stays in one place in regards to each other. All animals, including insects, have many cytochrome p450s and they vary between species reflecting natural compounds found in their environment. For instance, variation in them is found with diet (and other factors) among lepidoptera species, including moths (Calla et al., 2017)'

With the cloning era, the multiplicity of cytochrome P450s and their chromosomal dispersals began to be understood. A DNA clone from one member of the family identified 7 or 8 clustered, related genes with unique Southern blotting restriction patterns for different inbred strains (Meehan et al., 1988). The authors used a set of recombinant inbred lines to show that the members of this set were within 1 cM of each other and did not co-localize with other known cytochrome loci but did map near a gene controlling constitutive expression of them (ibid.). The genes were mapped to chromosome 19 by in situ hybridization. Eventually 40 loci were identified with many expressed early in embryo (Choudhary et al., 2003). This was not surprising as some cytochromes were involved in the metabolism of sex steroids, important for sexual development, see Chapter 11 or retinoic acid, important in many aspects of development, see Chapter 4.

N-acetyltransferase

The important phase 2, drug-inactivating enzyme, N-acetyl transferase (Nat, arylamine N-acetyltransferase, which catalyzes the acetylation of the extracyclic amino groups of aromatic amine and hydrazines) was extensively studied in mice. Its polymorphism had been studied in humans because of differential effects in the metabolism of isoniazid and sulfamethazine (Evans et al., 1960). Unlike the cytochromes, the "natural" substrates for them were not apparent. A well known example of their importance is the great sensitivity of dogs to chocolate or Tylenol—dogs have very low levels of N-acetylation, important for Tylenol and for theobromine, in the case of chocolate, detoxification.

A survey of inbred strains of mice found 17 to be rapid acetylators while only 3 were slow acetylators (Levy et al., 1990). The C57BL/6J strain was a rapid acetylator while the A/J strain was a slow acetylator. Congenic lines were made between the two to isolate the effects of the acetylator variation from all the other pharmacogenetically-relevant polymorphisms between the two strains (Mattano et al., 1988).[1] These congenic lines were C57BL/6.AJ.NatS (B6.A), where the C57BL6/J strain received the low-activity *Nat2*9* allele from the A/J strain and the converse line, AJ.B6.NatR(A.B6) strain where the A/J strain received the wildtype, high activity *Nat2*8* allele from the C57BL6/J strain. Although *Nat1*, because it is closely linked to *Nat2*, was also "carried in" with these congenics, there are no DNA sequence differences between the two *Nat1* genes. The resulting B6.A strain differed from the parental C57L/6J strain only in a 12 centiMorgan segment of chromosome 8 containing the *Nat2* and *Nat1* loci (ibid.). These congenics were used to study the Nat's role in the formation of DNA-adducts of 2-amino-3-methylimidazol[4-5f]quinoline (IQ) in a variety of tissues (Nerurkar et al., 1995). Transgenic over-expression of human *NAT1* (which is the human homologue of the mouse *Nat2* gene; Estrada-Rogers et al., 1998) was also performed on the A/J strain to create A/J mice with high acetylator levels and used, along with the congenics, to study 4-aminobiphenlyl toxicity in mouse liver (McQueen et al., 2003). However, high levels of exogenous Nat could not be achieved for reasons to be discussed below (Sim et al., 2003; Cao et al., 2005). A hypomorphic, ENU-induced mutation was also found (Erickson et al., 2008a).

Pharmacogenetics was also important for understanding the action of teratogens and Nat was studied in this regard. For instance, cortisone and related steroids can induce cleft lip and/or cleft palate (CL/P) in some inbred strains of mice (Fraser and Fainstat, 1951). The genetic control of this drug-induced CL/P in mice was much studied. Early on, the major histocompatibility locus (*H-2*, see Chapter 7) was implicated in the causation of steroid-induced cleft palate (CP). There was a report that inbred strains of mice differing only in this locus (and an adjoining chromosomal patch) in congenic lines (see Chapter 7) differed in steroid-induced CP frequency (Bonner and Slavkin, 1975). This report was confirmed and it was found that cyclic adenosine monophosphate measured in palatal shelf and tongue varied in a way that could link the genetic (*H-2*) and environmental (glucocorticoid treatment) factors in their effect (Erickson et al., 1979). Abundant evidence was provided that this H-2 influence on glucocorticoid-induced CP was not mediated via H-2 effects on the glucocorticoid receptor (Butley et al., 1978; Leach et al., 1982; Harper and Erickson,1983) which had been claimed (Goldman et al., 1983). However, there was no apparent association of spontaneous human CP with the human major histocompatibility locus, *HLA* (Pairitz et al., 1985).

Hydrocortisone (HC) was also studied with two other drugs/teratogens: 6-aminonicotinamide (6-AN) and phenytoin (Dilantin) in the causation of orofacial clefting among recombinant inbred (RI) lines (see Chapter 3). The results showed that similar genetic regions influenced HC and 6-AN caused CP (Erickson, et al, 2005), whereas Dilantin-induced clefting mapped to different regions (Erickson et al., 2005). The relevance of Nat was found when its nonpharmacological role was better understood. Endogenous substrates for Nat had been difficult to identify. Evidence was found that suggested folic acid was an endogenous substrate. A potential link of Nat2 with folic acid metabolism was found in 1995. It was discovered that a folate breakdown product, p-aminobenzoylglutamate, is an endogenous substrate of murine Nat2 and its functional equivalent human NAT1 (Ward et al., 1995). This suggested a mechanism for the role of murine Nat2 or human NAT1 in orofacial clefting because preconceptual folate supplementation for pregnant women to prevent neural tube defects had also decreased the incidence of orofacial clefting (Shaw et al., 1995). The previously described inability to achieve high levels of *NAT1* or *Nat2*, thus, could be due an effect in lowering endogenous levels of essential folic acid, which would select against any transgenic line showing high levels of expression. Even when a deoxycycline-inducible system was used to express the higher levels when development was completed, by giving the drug at adulthood to induce more activity, only a slight increase was achieved in kidney and none in liver (Cao et al., 2010).

The congenic lines described above were used to study this relationship of Nat to folic acid metabolism and teratogenicity, hypothesizing that lower Nat would correspond to higher endogenous folic acid levels which would be protective for drug-induced facial clefting. However, the congenic lines behaved more like the Nat2 donor lines, than the recipient lines, in their susceptibility to orofacial clefting, especially the AJ.B6.NatR for Dilantin-induced clefting which was lower (Karolyi et al., 1990). Since Nat had not been implicated in the metabolism of Dilantin (Atlas et al., 1980) and because there were many linked genes in the chromosomal segment containing Nat 1 and 2 in the congenic strains, a role for Nat in orofacial clefting had always been speculative.

Genetically modified mice were used to further study the possible relationship between Nat2, drug-induced orofacial clefting, and folic acid metabolism in the 21st century. The use of the transgenic mice described above which overexpressed human NAT1 (equivalent to mouse Nat2) at somewhat increased levels demonstrated that expression in target tissues markedly lowered, rather than increasing, the Dilantin-induced orofacial clefting and had a suggestive effect in lowering the HC-induced orofacial clefting in the A/J strain (Erickson et al., 2008a,b). This was surprising as was the finding that further lowering (ablating) the Nat2 activity with the Nat2 knockout in A/J mice had little effect on the teratogen-induced facial clefting (although it did lower sporadic clefting at a time when the control sporadic clefting was unexpectedly high; ibid.). Contrary to predictions, lowering Nat2 activity in all tissues of the C57BL/6J strain by replacing the wild-type allele with a knockout allele did not affect sporadic or Dilantin- or HC-induced orofacial clefting (ibid.). Thus, although the congenics had effects in both strains, the increased clefting found on C57BL/6J background when the A/J Nat2 allele was introduced (Karolyi et al., 1990) was not confirmed as due to altered Nat activity. Although the effect has been largest with HC induced clefting, these results over all argued against a major role for an effect of Nat on folic acid as playing a major role. In addition, the increase, especially found with glucocorticoids when the A/J low allele was introduced onto C57BL/6J, could be due to the introduction of other genes in the 12 to 20 centimorgans of chromosome transferred (Mattano et al., 1988) as well as the folic acid hypothesis being in error . However, the effect of low Nat2 activity in the A/J background had been confirmed by the results found when higher Nat2 (NAT1) activity was provided. It is formally possible that feedback mechanisms integrate the expression of the Nat2 gene with other genes engaged in similar metabolic pathways. In this case, introducing a null allele that generates a truncated form of Nat2 or no Nat2 protein (like the knockout used) might elicit compensatory changes in gene expression that would not be elicited by a full-length but slow-acetylating Nat2 protein. It is also possible that the manipulations were affecting folic acid levels in some tissues but not the embryo.

The relationship of N-acetyltransferase 2 and folate levels in red blood cells (RBCs) was therefore tested. The inbred mouse strain with higher N-acetyltransferase 2 levels, the C57BL/6J strain, had RBC folic acid levels about 17% lower than those of A/J and C57BL6.AJNats, the congenic strain with the C57BL/6J Nat2 replaced by the A/J Nat2 allele (Cao et al., 2010). In the liver, Nat2 enzymatic activity had been 50% lower in these 2 strains compared with C57BL/6J. Since the amount of Nat2 enzymatic activity in RBCs or their precursors was not known, it was possible that there was a more linear relationship between enzymatic activity and folate levels than the data suggested, for example, folic acid levels in liver might have been much higher but were not measured for technical difficulties. Studies of folic acid levels in developing palatal and lip tissues would have been the most relevant but were also not performed.

Higher folate levels rather than lower folate levels, with increased Nat enzymatic activity from the transgene on the A/J background, would fit the findings of Juriloff, 1995; Juriloff and Mah, 1995; Juriloff,et al, 2001) on the genetic causation of sporadic clefting in the A/J strain and its relationship to folate supplementation. Dr. Juriloff and coworkers (ibid.) have spent decades determining the genetic causation of the high incidence of CL(P) in the A/J and related inbred strains. They found that a recessive gene on distal chromosome 11 (*clf1*) and a second gene on chromosome 13 interacted, with a strong maternal effect, to increase the frequency of CL(P); ibid. Further mapping narrowed the candidate regions for the 2 clefting genes (Juriloff et al., 2004). An intracisternal A particle, retrotransposon (see Whitelaw, et al, 2001), was found inserted 6.6 kb 3' to *Wnt9b* (Juriloff et al., 2005; Juriloff et al., 2006), a gene previously implicated in midfacial morphogenesis (Lan et al., 2006). Since folate is already in the diet, such increased gene expression from the locus, presumably by enhancing methylation of CpG islands, is correlated with altered methylation of the intracisternal A particle retrotransposon, as was found in the case of the A^{vy} allele (Wolff et al., 1998; see Chapter 10 and this gene's transgenerational imprinting). Folate is essential for methyl-donor metabolism. Thus, decreased Nat activity, rather than the predicted increase, with concomitant increased folic acid in the developing palatal and lip tissues might also increase Wnt9b expression and decrease facial clefting.

A different class of genes affecting teratogenicity is exemplified by the anti-tumor gene, p53. This gene is involved in DNA repair and its lack affects both tumorigenicity of a variety of compounds and the teratogenicity of, for instance, benzo[α]pyrene (Nicol et al., 1995). The haplodeficiency of heterozygotes for the p53 knockout had several fold increased fetal resorptions and in utero death with exposure to benzo[α]pyrene.

Glucuronosyl transferase

Glucuronidation is a second major pathway in phase 2 detoxification pathway. It is catalyzed by UDP-glucuronosyl transferase (UDPGT) and is important for the conjugation of a number of endogenous substrates, such as bilirubin and steroids, as well as exogenous compounds. For instance, its delayed appearance in newborns is sometimes the cause of newborn jaundice. It also has an important role in the brain where serotonin is one of the major substrates (Ghersi-Egea et al., 1987). It was found that there were at least two functionally defined UDPGT activities toward phenolic substrates in the late fetal mouse liver with one of them being transplacentally inducible by beta-napthoflavone or methylcholanthrene (Chauhan et al., 1991). This Udpgt was only inducible transplacentally in the C57BL/6J strain but not in the DBA/2J strain (ibid.), probably reflecting the *Ah* locus difference between these 2 strains which had previously been shown to affect Udpgt inducibility (Owen, 1977). There was also a second Udpgt gene induced by phenobarbital which was expressed at low levels in the fetus, but it could not be induced through the placenta (ibid.) The structural gene for Udpgt was soon cloned (Lamb, et al 1994) and the 2 isoforms found to be 98% identical.

In the 21st Century, glucuronosyl transferases were studied much more extensively. Differing expression between the sexes and wide tissue variation were found (Buckley et al., 2007). The lack of UDP-glucuronyltransferase in humans cause unconjugated hyperbilirubinemia of generally benign consequences but of varying degrees of severity (Arias, 1969) which continued to be studied (Hsieh et al., 2007). Serotonin was found to play an important role in odorant detection (Petzold et al., 2009) and Udpgt further implicated in the levels of this neurotransmitter (reviewed in Heydel et al., 2010). A single amino acid residue difference was implicated in the preference for different substrates (Uchihashi, et al 2012). (See Fig. 12.1).

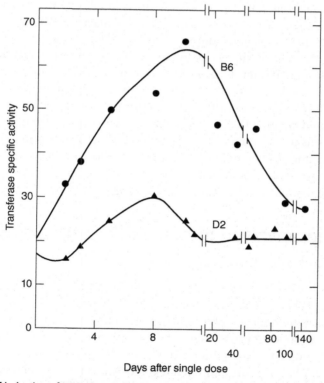

FIG. 12.1 Time course of induction of UDP-glucuronyl transferase activity by TCDD in C57BL/6N (B6) and DBA/2N (D2) mice. Each symbol represents mean transferase specific activity of from two to four individual livers from C57BL/6N or DBA/BN mice after a single dose of 50 pg of TCDD/kg of body weight to six- to eight-week-old mixed sex mice. From Owen IS (1977) with permission.

Glutathione S-transferases

The glutathione S-transferases (GSTs) are phase II metabolic enzymes and are key regulators of drug and toxin clearance. They catalyze the conjugation of electrophilic compounds, among which are many carcinogens, with reduced glutathione. Their importance is indicated by their number and degree of control. They fall into 4 classes consisting of α, μ, π and τ. These are coded by 7 loci and all seven family members are tightly clustered on mouse chromosome 3. They show great variation among tissues, including a spermatogenic specific isoenzyme (Fulcher et al., 1995), and show sex differences in expression levels (McLellan and Hayes, 1987). They also show circadian rhythms in expression (Xu et al., 2012), a variable requiring careful control of the time of day at which mice were sacrificed. Perhaps this large number of isoenzymes and the variables of expression is why little genetic variation was found in mice. In the cross of BALB/cA^{vy}/A X VY, agouti mice were less susceptible to 2-aminoacetylfluorene than their yellow siblings and had higher GST activities (Wolff et al., 1983). A 3-fold difference in the degree of induction by ethoxyquin hydrochloride was found among 5 inbred strains and, unlike a number of the drug metabolizing enzymes mentioned above, C57BL/6J and DBA2J did not differ significantly for this induction (Makary et al., 1989). In the 21st Century this complexity was assayed in Recombinant Inbred lines. Quantitative trait mapping from the hypothalami of 33 strains of inbred mice revealed 10,655 (!) trans associations and 31 cis eQTLs (Hayes,et al, 2009). A locus studied in detail found cis associations for the expression of Glutathione S-transferase, mu 5 (Gstm5).

The great variation in GST levels also resulted from the great number of inducing factors found. Levels increase with oxidative stress and one of the controlling elements is Antioxidant Response Factor (Rushmore et al., 1991). Another inducing protein is Nrf2 (Nuclear factor erythroid 2 p45-related factor 2; Hayes et al., 2000). This was studied with the knockout of Nrf2 which was highly sensitive to pulmonary injury by butylated hydoxytoluene Chan and Kan, 1999). Another regulator, studied in hepatic tumor and normal cells in culture, is the WNT/β–catenin pathway (Giera et al., 2010). Thus, there were many reasons for the complexities of the study of genetic regulation of the family of GSTs in mice.

Other enzymes involved in metabolic degradations
Alcohol and aldehyde dehydrogenases

The first step in alcohol degradation occurs when alcohol dehydrogenase (Adh) converts ethanol to acetaldehyde. While variation in the next step in its catabolism (by aldehyde dehydrogenase to acetate) is more important for human variation in alcohol tolerance (8-45% of Asian populations lack it), significant variation in Adh was found in mice and related to alcohol consumption (Eriksson and Pikkarainen, 1968). The differences were found to be in the levels of the enzyme (Balak et al., 1982) and the androgen regulation was much studied (Ohno et al., 1970). The genetic variation was not only important for alcohol metabolism but also for retinoic acid metabolism in the developing embryo (Ang et al, 1996). On the other hand, not much variation was found among the few inbred strains investigated, which varied in ethanol-induced sleep time, for aldehyde dehydrogenase (Belnap et al., 1972). (See Fig. 12.2).

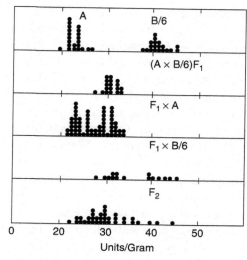

FIG. 12.2 Segregation of alcohol dehydrogenase activity in liver extracts of single mice from backcross and F2 progeny. A, B/6, and F1 refer to the parental strains A/J and C57BL/6/J and the (A/J X C57BL/6/J)F1 hybrid, respectively. Backcross and F2 progeny are also shown. The F2, with it's small numbers, does not clearly show the expected 3 peaks with ¼, ½ and ¼ proportions. From Balak KJ, Keith RH, Felder MR (1982), with permission

Thiopurine methyltransferase

Genetic variation in thiopurine methyltransferase (TPMT) is of great importance in humans as the rare individuals with a deficiency may have lethal bone marrow suppression when given the drug, 6-mercaptopurine, used for lymphoblastic leukemia and other malignancies. Genetic variation in mice was found with low levels in the C57BL/6J inbred strain and higher levels in the DBA/2J strain, a difference which segregated as a single gene trait (Otterness and Weinshilboum, 1987a). Careful characterization of the purified enzyme showed no difference in properties between the enzyme of the 2 strains indicating that the genetic variation was most likely to be regulatory (Otterness and Weinshilboum, 1987b).

Membrane transporters

Towards the end of the century, many drug transporters were found and characterized. These played roles in both the entry of compounds into cells and their export out of the cell. They were particularly important in the liver, playing a major role in the export of compounds into bile, a major excretory route for many xenobiotics and endogenous products. During periods of hepatic stress, such as sepsis or cholestasis, the downregulation of several sinusoidal uptake transporters, such as Na⁺-taurocholate co-transporting polypeptide (Ntcp), may be one of the compensatory mechanisms to protect the liver against the accumulation of toxic substances. Conversely, the upregulation of sinusoidal excretion transporters such multiple resistant protein 3 (Mrps, see below) may also compensate by exporting these compounds from the liver back into blood.

Other transporters included organic anion transporters (Oats) and solute carrier protein family members (Slcs). Most of these had multiple forms and were involved in a variety of physiological processes. Defects could lead to disease. For instance, the lack of SLC26A2, a calcium oxalate co-transporter located on the epithelial membrane of kidney and other tissues, in a mouse knockout led to kidney stones (Jiang et al., 2006). The knockout of *Ntcp* showed reduced body weights, a common observation for knockout mouse models with defects in bile acid transporters or bio-synthetic enzymes important for the enterohepatic circulation of bile acids (Slijepcevic et al., 2015). Alterations in the expression of several of these transporters were found in some mouse models of human diseases with liver involvement, e.g. Niemann-Pick C (Erickson et al., 2005).

Multiple drug-resistant proteins, which are involved in the transmembrane transport of cancer drugs for instance, are also important for determining the effects of many drugs. In the 21st Century, these could be studied in mice where knockouts (KOs) of some of them had been made. These proteins are members of the ATP-binding cassette (ABC) transporters which can transport drugs out of cells. They mostly have wide substrate specificity, allowing them to pump out several kinds of xenobiotics from the cell (Borst and Elferink, 2002). They are particularly important for cancer therapeutics as some cancers amplify the ABC-transporter gene involved for the chemotherapeutic agent being used and make the tumor resistant to it. They can also be important for protecting the fetus from teratogens. Using the knockout, Mdr1a was shown to be expressed in the fetus and placenta and its role in the fetus was shown to be protective for Dilantin-induced CP (Rawles et al., 2007). On the other hand, the KO for *Mdr2* did not change the frequency of Dilantin or 6-AN included CL(P) (ibid.).

Receptor variation

Drug and many other cell surface receptors generally fall into 3 classes: ion channels and ion transporters which generally depend on ATP as their source of energy, G-protein coupled receptors which use cAMP or phosphoinositol as an intracellular signal, and receptors which involve autophosphorylation. The G-protein receptors have an abundancy of about 1000 members and have been said to take up to 1% of the genome (Gilman, 1987). Variation in the G-protein receptors explained some mutations in mice. For instance, a missense mutation in the growth hormone-releasing hormone receptor explained the dwarfism of the *little* mouse (Godfrey et al., 1993) while variation in an endothelin B receptor explained the *piebald* mouse (Hosada et al., 1994) and variation in GPR143 was causative in OA1 (ocular albinism; Newton et al., 1996). At the end of the century and as the 21st started, the newer tools of transgenics and knockouts were used to explore the multiplicity of functions of this large class of receptors. Heart rate variability was decreased in mice overexpressing the β_1 adrenoreceptor (Mansier, et al., 1996) while the knockout of the G protein γ^3 are epileptic and don't gain weight (Schwindinger et al., 2004). The examples could be in the hundreds with so many G proteins of interest. Many of the knockout and transgenic models are reviewed by Wettschureck, et al (2004).

Variation in the affinity of receptors was often important in determining the therapeutic effect of a drug. Adrenergic receptors are one very important class of widely distributed receptors involved in a large variety to physiological processes. The knockout of 1 of the 3 $\alpha2$

adrenergic receptors did not show an apparent phenotype, presumably the other two examples of this receptor type could compensate for its loss (Link et al., 1995). In humans, variation in the β–adrenergic receptor, an equally important physiological receptor targeted by many agonists and antagonists, influences the severity of asthma and response to treatment with agonists (reviewed in Erickson and Graves, 2001). Genetic variation was also found in mice for drug response and cyclic AMP (Markovac and Erickson, 1983). This variation paralleled genetic variation in membrane methyl transferase and 2 genetic components were identified in recombinant inbred lines (Markovac and Erickson, 1985a). Although not cosegregating among RI lines, a robust effect of the *H-2* locus was found using the more powerful tool of congenic lines (see Chapter 7; Markovac and Erickson, 1985a,b). Another example in mice was the knockout of the 5-HT1B serotonin receptor which resulted in the KO males showing increased aggressiveness to foreign males (Sadou et al., 1994).

In conclusion, although pharmacogenetics started with, and was motivated by studies of human variation related to adverse drug events and, sometimes, apparent lack of efficacy, most of the variants had potential mouse models. The small size of mice made them less favorable for studies of drug kinetics, necessitating multiple samples over relatively short periods of time, and rats were frequently preferred. As the genetics of *Rattus norvegicus* developed towards the end of the Century, and transgenesis and knockouts were obtained in this species, it was frequently the favored model for such detailed drug turnover studies.

Note

1. My extensive involvement in studies of mouse N-acetyltransferase started with my collaboration with the "guru" of N-acetyltransferases, Wendell Weber, at the University of Michigan to create these congenics.

References

Ang, H.L., Deltour, L., Hayamizu, T.F., Zgombic-Knight, M., Duester, G., 1996. Retinoic acid synthesis in mouse embryos during gastrulation and craniofacial development linked to class IV alcohol dehydrogenase gene expression. J. Biol. Chem. 271, 9234–9256.

Arias, I.M., Gartner, L.M., Cohen, M., Ben-Ezzer, J., Levi, A.J., 1969. Chronic nonhemolytic unconjugated hyperbilirubinemia with glucuronyl transferase deficiency: clinical, biochemical, pharmacologic and genetic evidence for heterogeneity. Am. J. Med. 47, 395–409.

Atlas, S.A., Zweier, J.L., Nebert, D.W., 1980. Genetic differences in phenytoin pharmacokinetics in vivo clearance and in vitro metabolism among inbred strains of mice. Dev Pharmacol Ther 1, 281–304.

Balak, K.J., Keith, R.H., Felder, M.R., 1982. Genetic and developmental regulation of mouse liver alcohol dehydrogenase. J. Biol. Chem. 257, 15000–15007.

Belknap, J.K., MacInnes, J.W., McClearn, G.E, 1972. Ethanol sleep times and hepatic alcohol and aldehyde dehydrogenase activities in mice. Physiol. Behav 9, 453–457.

Betters, J.L., Yu, L., 2010. NPC1L1 and cholesterol transport. FEBS Lett. 584, 2740–2747.

Blakeslee, A.F., 1918. Unlike reaction of different individuals to fragrance in verbena flowers. Science 48, 298–299.

Bonner, J.J., Slavkin, H.C., 1975. Cleft palate susceptibility linked to histocompatibility-2 (H-*2*) in the mouse. Immunogenetics 2, 213–218.

Borst, P., Elferink, R., 2002. Mammalian ABC transporters in health and disease. Annu. Rev. Biochem. 71, 537–592.

Buckley, D.B., Klaassen, C.D., 2007. Tissue- and gender-specific mRNA expression of UDP-glucuronosyltransferases (UGTs) in mice. Drug Metab. Dispos. 35, 121–127.

Butley, M.S., Erickson, R.P., Pratt, W.B., 1978. Hepatic glucocorticoid receptors and the H-2 locus. Nature 275, 136–138.

Calla, B., Noble, K., Johnson, R.M., Walden, K.K.O., Schuler, M.A., Robertson, H.M., Berenbaum, M.R., 2017. Cytochrome P450 diversification and hostplant utilization patterns in specialist and generalist moths: Birth, death and adaptation. Mol. Ecol. 26. doi:10.1111/mec.14348 doi.org/.

Cao, W., Strnatka, D., McQueen, C.A., Hunter, R.J., Erickson, R.P., 2010. N-acetyltransferase 2 activity and folate levels. Life Sci 86, 103–106.

Cao, W., Chau, B., Hunter, R., Strnatka, D., McQueen, C.A., Erickson, R.P., 2005. Only low levels of exogenous N-acetyltransferase can be achieved in transgenic mice. Pharmocogenomics J. 5, 255–261.

Chan, K., Kan, Y.W., 1999. Nrf2 is essential for protection against acute pulmonary injury in mice. Proc. Natl. Acad. Sci. U.S.A. 96, 12731–12736.

Chauhan, D.P., Miller, M.S., Owens, I.S., Anderson, L.M., 1991. Gene Expression, Ontogeny and Transplacental Induction of Hepatic UDP-Glucuronosyl Transferase Activity in Mice. Develop. Pharm. Ther. 16, 139–149.

Chhoudhary, D., Jansson, J., Schenkman, J.B., Sarfarazi, M., Stoilov, I., 2003. Comparative expression profiling of 40 mouse cytochrome P450 genes in embryonic and adult tissues. Arch. Biochem. Biophys. 414, 91–100.

Davis, W.M., King, W.T., 1967. Pharmacogenetic factor in the convulsive responses of mice to flurothyl. Experientia 23, 214–215.

Erickson, R.P., 1998. From "magic bullet" to "specially engineered shotgun loads": the new genetics and the need for individualized pharmacotherapy. BioEssays 20, 683–685.

Erickson, R.P., Bhattacharyya, A., Hunter, R.J., Heidenreich, R.A., Cherrington, N.J., 2005. liver disease with altered bile acid transport in Niemann-Pick C mice on a high fat, 1% cholesterol diet. Amer. J. Physiol.: Liver GI 289, G300–G307.

Erickson, R.P., Butley, M.S., Sing, C.F., 1979. H-2 and non-H-2 determined strain variation in palatal shelf and tongue adenosine 3':5' cyclic monophosphate. J Immunogenet 6, 253–262.

Erickson, R.P., Graves, P.E., 2001. Genetic variation in β-adrenergic receptors and their relationship to susceptibility for asthma and therapeutic response. Drug Metab. Dispos. 29, 1–5.

Erickson, R.P., Karolyi, I.J., Diehl, S., 2005. Correlation of susceptibility to 6-aminonicontinamide and hydrocortisone-induced cleft palate. Life Sci. 76, 2071–2078.

Erickson, R.P., McQueen, C.A., Chau, B., Gokhale, V., Uchiyam, M., Toyoda, A., Ejima, F., Maho, N., Sakaki, Y., Gondo, Y., 2008a. An N-ethyl-N-nitrosourea-induced Mutation in N-acetyltransferase 1 in Mice. Biochem. Biophys. Res. Commun. 370, 285–288.

Erickson, R.P., Cao, W., Acuna, D.K., Strnatka, D.W., Hunter, R.J., Chau, B.T., Wakefield, L.V., Sim, E., McQueen, C.A., 2008b. Confirmation of the Role of N-acetyltransferase 2 in teratogen-induced cleft palate using transgenics and knockouts. Mol. Reprod. Dev. 75, 1071–1076.

Eriksson, K., Pikkarainen, P.H., 1968. Differences between the sexes in voluntary alcohol consumption and liver ADH-activity in inbred strains of mice. Metab 17, 1037–1042.

Estrada-Rogers, I., Levy, G.N., Webber, W.W., 1998. Substrate selectivity of mouse N-actyltransferases 1,2, and 3 expressed in COS-1 cells. Drug Metabol. Dispos. 26, 502–505.

Evans, D.A.P., Manley, K.A., Mc Cusick, V.A., 1960. Genetic control of isoniazid metabolism in man. Br. Med. J. 2, 485–490.

Fraser, F.C., Fainstat, T.D., 1951. The production of congenital defects in the offspring of pregnant mice treated with cortisone. Pediatrics 8, 527–533.

Fulcher, K.D., Welch, J.E., Klapper, D.G., O'Brien, D.A., Eddy, E.M., 1995. Identification of a unique μ-class gluta-thione S-transferase in mouse spermatogenic cells. Mol. Reprod. Dev. 42, 415–424.

Garrod, A., 1896. Notes on the occurrence of large quantities of haematoporphyrin in the urine of patients taking sulphonal. J. Path. Bact. 3, 434–448.

Ghatak, A., Faheem, O., Thompson, P.D., 2010. The genetics of statin-induced myopathy. Atherosclerosis 210, 337–343.

Ghersi-Egea, J.F., Walther, B., Decolin, D., Minn, A., Siest, G., 1987. The activity of 1-naphthol-UDP-glucuronosyl-transferase in the brain. Neuropharmacology 26, 367–372.

Giera, S., Braeuning, A., Kohle, C., Bursch, W., Metzger, U., Buchman, A., Schwarz, 2010. Wnt/β-Catenin Signaling Activates and Determines Hepatic Zonal Expression of Glutathione S-Transferases in Mouse Liver. Toxicol. Sci., 115.

Goldman, A.S., Shapiro, B.H., Katsumata, M., 1983. Human foetal palatal corticoid receptors and teratogens for cleft palate. Nature 272, 464–466.

Gilman, A., 1987. G proteins: transducers of receptor-generated signals. Ann. Rev. Biochem. 56, 615–649.

Godfrey, P., Rahal, J.O., Beamer, W.G., Copeland, N.G., Jenkins, N.A., Mayo, K.E., 1993. GHRH receptor of *little* mice contains a missense mutation in the extracellular domain that disrupts receptor function. Nature Genet 4, 227–232.

Harper, K.J., Erickson, R.P., 1983. Ionic effects on strain differences in hepatic cytosolic glucocorticoid receptor levels in mice. Teratology 27, 43–49.

Hayes, J.D., Chanas, S.A., Henderson, C.J., McMahon, M., Sun, C., Moffat, G.J., Wolf, C.R., Yamamoto, M., 2000. The Nrf2 transcription factor contributes both to the basal expression of glutathione S-transferases in mouse liver and to their induction by the chemopreventive synthetic antioxidants, butylated hydroxyanisole and ethoxyquin. Biochem. Soc. Trans. 28, 33–41.

Hayes, K.R., Young, B.M., Pletcher, M.T., 2009. Expression Quantitative Trait Loci Mapping Identifies New Ge-netic Models of Glutathione *S*-Transferase Variation. Drug Metab. Disposl 37, 1269–1276.

Herr, T.M., Bielinski, S.J., Bottinger, E., Brautbar, A., Brilliant, M.25 more, 2015. Practical considerations in ge-nomic decision support: The eMERGE experience. J Pathol Inform 28 (6), 50. doi:10.4103/2153-3539.165999.

Heydel, J.M., Holsztynska, E.J., Legendre, A., Thiebaud, N., Artur, Y., Le Bon, A.M., 2010. UDP-glucuronosyl-transferases (UGTs) in neuro-olfactory tissues: expression, regulation, and function. Drug Metab. Rev. 42, 74–97.

Hosoda, K., Hammer, R.E., Richardson, J.A., Baynash, A.G., Cheung, J.D., Giaid, A., Yanagisawa, M., 1994. Tar-geted and natural (piebald-lethal) mutations of endothelin-B receptor gene produce megacolon associated with spotted coat color in mice. Cell 79, 1267–1276.

Hsieh, T.-Y., Shiu, T.-Y., Huang, S.-M., Lin, H.-H., Lee, T.-C., Chen, P.-J., Chu, H.-C., Chang, W.-K., Jeng, K.-S., Lai, M.M.C., Chao, Y.-C., 2007. Molecular pathogenesis of Gilbert's syndrome: decreased TATA-binding protein binding affinity of UGT1A1 gene promoter. Pharmacogenet. Genomics 17, 229–236.

Huff, S.D., 1962. A genetically controlled response to the drug chlorpromazine. Abstract. Genetics 47, 962.

Jiang, Z., Asplin, J.R., Evan, A.P., Rajendran, V.M., Velasquez, H., Nottoli, T.P., Binder, H.J., Aronson, P.W., 2006. Calcium oxalate urolithiasis in mice lacking anion transporter Slc26a6. Nat. Genet. 38, 474–478.

Juriloff, D.M., 1995. Genetic analysis of the construction of the AEJ.A congenic strain indicates that non-syndromic CL(P) in the mouse is caused by two loci with epistatic interaction. J. Craniofac. Genet. Dev. Biol. 15, 1–12.

Juriloff, D.M., Mah, D.G., 1995. The major locus for multifactorial nonsyndromic cleft lip maps to mouse chromo-some 11. Mamm. Genome. 6, 63–69.

Juriloff, D.M., Harris, M.J., Brown, C.J., 2001. Unravelling the complex genetics of cleft lip in the mouse model. Mamm. Genome 12, 426–435.

Juriloff, D.M., Harris, M.J., Dewell, S.L., 2004. A digenic cause of cleft lip in A-strain mice and definition of candi-date genes for the two loci. Birth Defects Res. A Clin. Mol. Teratol. 70, 509–518.

Juriloff, D.M., Harris, M.J., Dewell, S.L., Brown, C.J., Mager, D.L., Gagnier, L., Mah, D.G., 2005. Investigations of the genomic region that contains the clf1 mutation, a causal gene in multifactorial cleft lip and palate in mice. Birth Defects Res. A Clin. Mol. Teratol. 73, 103–113.

Juriloff, D.M., Harris, M.J., McMahon, A.P., Caroll, T.J., Lidral, A.C., 2006. Wnt9b is the mutated gene involved in multifactorial nonsyndromic cleft lip with or without cleft palate in A/WySn mice, as confirmed by a genetic complementation test. Birth Defects Res. A Clin. Mol. Teratol. 76, 574–579.

Kalow, W., Gunn, D.R., 1957. The relationship between dose of succinylcholine and duration of apnea in man. J. Pharmacol. Exp. Ther. 120, 203–214.

Kaminsky, L.S., Zhiang, Z.Y., 1997. Human P450 metabolism of warfarin. Pharmacol Ther 73, 67–74.

Karolyi, J., Erickson, R.P., Liu, S., Killewald, L., 1990. Major effects on teratogen-induced facial clefting in mice determed by a single genetic region. Genetics 126, 201–205.

Lamb, J.G., Straub, P., Tukey, R.H., 1994. Cloning and characterization of cDNAs encoding mouse Ugt1.6 and rabbit UGT1.6: differential induction by 2,3,7,8-tetrachlorodibenzo-p-dioxin. Biochemistry 33, 10513–10520.

Lan, Y., Ryan, R.C., Zhang, Z., Bullard, S.A., Bush, J.O., Maltby, K.M., Lidral, A.C., Jiang, B., 2006. Expression of Wnt9b and activation of canonical Wnt signaling during midfacial morphogenesis in mice. Dev. Dyn. 235, 1448–1454.

Leach, K.L., Erickson, R.P., Pratt, W.B., 1982. The endogenous heat-stable glucocorticoid receptor stabilizing factor and the H-2 locus. J. Steroid Biochem. 17, 121–123.

Levy, G.N., Martell, K.J., de Leon, J.H., Weber, W.W., 1990. Metabolic, molecular genetic and toxicological aspects of the acetylation polymorphism in inbred mice. Pharmacogenetics 2, 197–206.

Link, R.E., Stevens, M.S., Kulatunga, M., Scheinin, M., Barsh, G.S., Kobilka, B.K., 1995. Targeted inactivation of the gene encoding the alpha 2c-adrenoceptor homolog. Mol. Pharmacol. 48, 48–55.

Makary, M., Kim, H.L., Safe, S., Womack, J., Ivie, W., 1989. Induction of glutathione S-transferases in genetically inbred male mice by dietary ethoxyquin hydrochloride. Comp. Biochem. Physiol. C Comp. Pharmacol. Toxicol. 92, 171–174.

Mansier, P., Médigue, C., Charlotte, N., Vermeiren, C., Coraboeuf, E.10 more, 1996. Decreased heart rate variability in transgenic mice overexpressing atrial beta 1-adrenoceptors. Am. J. Physiol. Heart and Circ. 271, H1464–H1472.

Markovac, J., Erickson, R.P., 1983. Genetic variation in β-adrenergic receptors in mice: A magnesium effect determined by a single gene. Genet. Res 42, 159–168.

Markovac, J., Erickson, R.P., 1985a. The genetics of hormone-induced cyclic AMP production and phospholipid N-methylation in inbred strains of mice. Genet. Res. 45, 167–177.

Markovac, J., Erickson, R.P., 1985b. A component of genetic variation among mice in activity of transmembrane methyltransferase I determined by the H-2 region. Biochem. Pharmacol. 34, 3421–3425.

Mattano, S.S., Erickson, R.P., Nesbitt, M., Weber, W.W., 1988. Linkage of NAT and Es-1 in the mouse and development of strains congenic for N-acetyltransferase. J. Hered. 79, 430–433.

McLellan, L.I., Hayes, J.D., 1987. Sex-specific constitutive expression of the pre-neoplastic marker glutathione S-transferase, YfYf, in mouse liver. Biochem. J. 245, 399–406.

McQueen, C.A., Chau, B., Erickson, R.P., Tjalkens, R.B., Philbert, M.A., 2003. The effects of genetic variation in N-acetyltransferases on 4-aminobiphenyl genotoxicity in mouse liver. Chem. Biol. Interact. 146, 51–60.

Meehan, R.R., Speed, R.M., Gosden, J.R., Rout, D., Hutton, J.J., Taylor, B.A., Hilkens, J., Kroezen, V., Hilgers, J., Adesnik, M., et al., 1988. Chromosomal organization of the cytochrome P450-2C gene family in the mouse: a locus associated with constitutive aryl hydrocarbon hydroxylase. Proc. Nat'l Acad. Sci.,U.S.A. 85, 2662–2666.

Meier, H., Allen, R.C., Hoag, W.G., 1962. Spontaneous hemorrhagic diathesis in inbred mice due to single or multiple "prothrombin-complex" deficiencies. Blood 19, 501–514.

Motulsky, A.G., 1957. Drug reactions, enzymes and biochemical genetics. J. Amer. Med. Assoc. 165, 835–837.

Nebert, D.W., 1999. Pharmacogenetics and pharmacogenomics: why is this relevant to the clinical geneticist? Clin. Genet. 56, 247–258.

Nerurkar, P.V., Schut, H.A.J., Anderson, L.M., Riggs, C.W., Snyderwine, E.G., Thorgeirsson, S.S., Weber, W.W., Rice, J.M., Levy, G.N., 1995. DNA adducts of 2-amino-3-methylimidazo[4,5f]quinoline (IQ) in colon, bladder, and kidney of congenic mice differing in AH responsiveness and N-acetyltransferase. Canc. Res. 55, 3043–3049.

Newton, J.M., Orlow, S.J., Barsh, G.S., 1996. Isolation and characterization of a mouse homolog of the X-linked ocular albinism (OA1) gene. Genomics 37, 219–225.

Nicol, C.J., Harrison, M.L., Laposa, R.R., Gimelshtein, I.I., Wells, P.G., 1995. A teratologic suppressor role for p53 in benzo[a]pyrene-treated transgenic p53-deficient mice. Nature Genet 10, 18–187.

Ono, S., Stenius, C., Christian, L., Harris, C., Ivey, C., 1970. More about the testosterone induction of kidney alcohol dehydrogenase activity in the mouse. Biochem. Genet. 4, 565–577.

Otterness, D.M., Weinshilboum, R.M., 1987. Mouse thiopurine methyltransferase pharmacogenetics: monogenic inheritance. J. Pharmacol. Exp. Ther. 240, 817–824.

Otterness, D.M., Weinshilboum, R.M., 1987. Mouse thiopurine methyltransferase pharmacogenetics: biochemical studies and recombinant inbred strains. J. Pharmacol. Exp. Ther. 243, 180–186.

Owen, I.S., 1977. Genetic regulation of UDP-glucuronosyltransferase induction by polycyclic aromatic compounds in mice. Co-segregation with aryl hydrocarbon (benzo(alpha)pyrene) hydroxylase induction. J. Biol. Chem. 252, 2827–2833.

Pairitz, G., Erickson, R.P., Schultz, J., Sing, C.J., 1985. Failure to detect association of isolated cleft palate with HLA antigens. J. Immunogenet. 12, 259–262.

Petzold, G.C., Hagiwara, A., Murthy, V.N., 2009. Serotonergic modulation of odor input to the mammalian olfactory bulb. Nat. Neurosci. 12, 784–791.

Poland, A., Glover, E., Taylor, B.A., 1987. The Murine Ah Locus: A New Allele and Mapping to Chromosome 12. Mol. Pharmacol. 32, 471–478.

Rasmussen-Torvik, L.J., Stallings, S.C., Gordon, A.S., Almoguera, B., Basford, M.A., and 48 more, 2014. Design and anticipated outcomes of the eMERGE-PGx project: a multicenter pilot for preemptive pharmacogenomics in electronic health record systems. Clin Pharmacol Ther 96, 482–489.

Rawles, L.A., Acuna, D., Erickson, R.P., 2007. The role of multiple drug resistance proteins in fetal and/or placental protection against teratogen-induced orofacial clefting. Mol. Reprod. Dev. 74, 1483–1489.

Robinson, J.R., Nebert, D.W., 1974. Genetic expression of aryl hydrocarbon hydroxylase induction, Presence or absence of association with zoxazolamine, diphenylhydantoin, and hexobarbital metabolism. Mol. Pharmacol. 10, 484–493.

Rushmore, T.H., Morton, M.R., Pickett, C.B., 1991. The antioxidant responsive element. Activation by oxidative stress and identification of the DNA consensus sequence required for functional activity. J. Biol. Chem. 266, 11632–11639.

Sadou, F., Amara, D.A., Dierich, A., LeMeur, M., Ramboz, S., Segu, L., Buhot, M.-C., Hen, R., 1994. Enhanced aggressive behavior in mice lacking 5-HT1B receptor. Science 265, 1875–1878.

Schwindinger, W.S., Giger, K.E., Betz, K.S., Stauffer, A.M., Sunderlin, E.M., Sim-Selley, L.J., Selley, D.E., Bronson, S.K., Robishaw, J.D., 2004. Mice with deficiency of G protein gamma3 are lean and have seizures. Mol. Cell Biol. 24, 7758–7768.

Shaw, G.M., Lammer, E.J., Wasserman, C.R., O'Malley, C.D., Tolarova, M.M., 1995. Risks of orofacial clefts in children born to women using multivitamins containing folic acid periconceptionally. Lancet 346, 393–396.

Sim, E., Pinter, K., Mushtaq, A., Upton, A., Sandy, J., Bhakta, S., Noble, M., 2003. Arylamine N-acetyltransferases: a pharmacogenomic approach to drug metabolism and endogenous function. Biochem. Soc. Trans. 31, 615–619.

Slater, V., Ahmed, M., Benedetto, C., Cheeseman, K., Packer, J.E., Dianzani, M.U., 1980. Electron transfer reactions in endoplasmic reticulum: Free radical production, lipid peroxidation, covalent binding and cell division. Int. J. Quantum Chem. S7, 347–356.

Slijepcevic, D., Kaufman, C., Wichers, C.G.K., Gilglioni, E.H., Lempp, F.A., Duijst, S., de Waart, D.R., Oude Elferink, R.P.J., Mier, W., Stieger, B., Beuers, U., Urban, S., van de Graaf, S.F.J., 2015. Impaired uptake of conjugated bile acids and hepatitis b virus pres1-binding in na(+)-taurocholate cotransporting polypeptide knockout mice. Hepatology 62, 207–219.

Snyder, L.H., 1932. Studies in Human Inheritance. IX. The inheritance of taste deficiency in man. Ohio J. Sci. 32, 436–440.

Thomas, P.E., Hutton, J.J., 1973. Genetics of aryl hydrocarbon hydroxylase induction in mice: additive inheritance in crosses between C3H-HeJ and DBA-2J. Biochem. Genet. 8, 249–257.

Thorgeirsson, S.S., Nebert, D.W., 1977. The Ah locus and the metabolism of chemical carcinogens and other foreign compounds. Adv. Cancer Res. 25, 149–193.

Uchihashi, S., Nishikawa, M., Sakaki, T., Ikushiro, S.-I., 2012. The critical role of amino acid residue at position 117 of mouse UDP-glucuronosyltransfererase 1a6a and 1a6b in resveratrol glucuronidation. J. Biochem. 152, 331–340.

Ullrich, V., Kremers, P., 1977. Multiple forms of cytochrome P450 in the microsomal monooxigenase system. Arch Toxicol 39, 41–50.

Vessell, E.S., 1967. Induction of drug-metabolizing enzymes in liver microsomes of mice and rats by softwood bedding. Science 157, 1057–1058.

Vessell, E.S., 1968. Factors altering the response of mice to hexobarbital. Pharm 1, 81–97.

Ward, A., Summers, M.J., Sim, E., 1995. Purification of recombinant human N-acetyltransferase type 1 (NAT1) expressed in E. coli and characterization of its potential role in folate metabolism. Biochem. Pharmacol. 49, 1759–1767.

Weinshilboum, R.M., Wang, L., 2017. Pharmacogenomics: Precision Medicine and Drug Response. Mayo Clin Proc 92, 1711–1722.

Wettschureck, N., Moers, A., Offermanns, S., 2004. Mouse models to study G-protein-mediated signaling. Pharmacol. Ther. 101, 75–89.

Whitelaw, E., Martin, D.I., 2001. Retrotransposons as epigenetic mediators of phenotypic variation in mammals. Nat. Genet. 27, 361–365.

Wolff, G.L., Gaylor, D.W., Frith, C.H., Suber, R.L., 1983. Controlled genetic variation in a subchronic toxicity assay: susceptibility to induction of bladder hyperplasia in mice by 2-acetylaminofluorene. J. Toxicol. Environ. Health 12, 255–265.

Wolff, G.L., Rodell, R.L., Moore, S.R., Cooney, A.C., 1998. Maternal epigenetics and methyl supplements affect agouti gene expression in Avy/a mice. FASEB J 12, 949–957.

Xu, Y.-Q., Zhang, D., Jin, T., Cai, D.-J., Wu, Q., Lu, Y., Liu, J., Klaassen, C.D., 2012. Diurnal variation of hepatic antioxidant gene expression in mice. PLOS One 7, e44237.

Book Bibliography

Anfinsen, C.B., 1959. The Molecular Basis of Evolution. John Wiley & Sons, Inc., New York.

Aragon, L., 1964. A History of the USSR: from Lenin to Khrushchev. Translated by Patrick O'Brien. David McKay Company, Inc., New York.

Baltzer, F., 1967. Theodor Boveri: Life and Work of a Great Biologist, 1862-1915. Translated from the German by Dorothea Rudnick. Universitiy of California Press, Berkeley and Los Angeles.

Bateson, W., 1894. Materials for the Study of Variation, treated with special regard to discontinuity in the origin of species. Cambridge University Press, Cambridge, UK.

Bateson, W., 1908. The Methods and Scope of Genetics. Cambridge University Press, Cambridge, UK.

Bateson, W., Mendel, G., 1909. Mendel's Principles of Heredity. A Defense with a translation of Mendel's Original Papers on Hybridization. Cambridge University Press, Cambridge, London.

Beatty, R.A., Glueckschon-Waelsch, S, 1972. The Genetics of the Spermatozoon, Proceedings of an International Symposium held at the University of Edinburgh, Scotland on August 16-20, 1971. Edinburgh, New York. University of Edinburgh.

Bennett, J.H. (Ed.), 1983. Natural Selection, Heredity, and Eugenics: Including selected correspondence of R.A. Fisher with Leonard Darwin and others. Clarendon Press, Oxford.

Benton, J., 1990. Naomi Mitchison: A Century of Experiment in Life and Letters. Pandora, London.

Boyd, B., 1990. Vladimir Nabokov: The Russian Years. Princeton University Press, Princeton, New Jersey.

Brent, L., 1996. A History of Transplantation Immunology. Academic Press, San Diego, London, Boston, New York, Sydney, Toronto, Tokyo.

Brewer, G.J., 1970. An Introduction to Isozyme Techniques, with a contribution by Charles F. Sing. Acadmic Press, New York and London.

Bull, James J, 1983. Evolution of Sex Determining Mechanisms. The Benjamin/Cummings publishing Company, London, Amsterdam, Don Mills, Ontario, Sydney, Tokyo.

Burt, A., Trivers, R., 2006. Genes in Conflict: The Biology of Selfish Genetic Elements. Harvard University Press, Cambridge, Mass.

Calder, J., 1997. The Nine Lives of Naomi Mitchison. Virago Press, London.

Carroll, S.B., 2013. Brave Genius: A Scientist, A Philosopher, and Their Daring Adventures From the French Resistance to the Nobel Prize. Crown Publishing, New York.

Chomard-Lexa, A., 2004. Lucien Cuenot: L'intution naturaliste. L'Hermattan, Paris.

Clayton, J., 2014. NRC National Institute for Medical Research: A Century of Science for Health. MRC press, London (ISBN-978-0-9572650-4-6).

Crosby, A.W., 1986. Ecological Imperialism: The Biological Expansion of Europe. Cambridge University Press, Cambridge, UK, pp. 900–1900.

Cuenot, L., 1925. *L'Adaptation*. Encyclopedie Scientifique. Librairie Octave Doin, Gaston Doin, Editeur, Paris.

Darwin, F., 1896. The Life and Letters of Charles Darwin, 2 vol.s. D. Appleton and Company, New York.

De Waal, E., 2010. The Hare with Amber Eyes: A Hidden Inheritance. Picador, Farrar, Straus and Giroux, New York.

Dobzhansky, T., 1937. Genetics and the Origin of Species. Columbia University Press, New York: Morningside Heights.

Dubos, R.J., 1976. Professor, the Institute, and DNA. The Rockefeller University Press, New York.

Duboule, D., 1994. Guidebook to the Homeobox Genes. Sambook and Tooze Publication at Oxford University Press, Oxford.

Dunn, L.C., 1965. A short history of genetics: The development of some main lines of thought :1864-1939. McGraw-Hill, New York, NY.

Edidin, M., Johnson, M.H., 1977. Immunobiology of Gametes. Clinical and Experimental Immunoreproduction. Cambridge University Press, Cambridge.

Ephrussi, B., 1972. Hybridization of Somatic Cells. Princeton University Press, Princeton, New Jersey.

Epstein, C.J., 1986. The Consequence of Chromosome Imbalance: Principles, Mechanisms, and Models. Cambridge University Press, Cambridge, London, New York, New Rochelle, Melbourne, Sydney.

Erickson, R.P., Anthony, Wynshaw-Boris (Eds.), 2016. Epstein's Inborn Errors of Development: The molecular basis of clinical disorders of morphogenesis, 3rd ed. Oxford University Press, NY, Oxford.

Festing, M.F.W, 1979. Inbred Strains in Biomedical Research. Oxford University Press, New York.

Fisher, R.A., 1930. The Genetical Theory of Natural Selection. Oxford University Press, London, New York.

Fleck, L., 1979. Genesis and Development of a Scientific Fact. University of Chicago Press, Chicago Translated from the German edition of 1935 by F. Bradley and T.J. Trenn, edited by T.J. Trenn and R.K. Merton.

Forrest, D.W., 1974. Francis Galton: The Life and Work of a Victorian Genius. Taplinger Publishing Co., Inc., New York.

Fortey, R., 2001. Trilobite: Eyewitness to Evolution. Random House, Inc., New York.

Foster, H.L., Small, J.D., Fox, J.G., 1981. The Mouse in Biomedical Research. Vol. I, History, Genetics, and Wild Mice. Academic Press, Inc., New York.

Franklin, A., Edwards, A.W.F., Fairbanks, D., Hartl, D.L., 2008. Ending the Mendel-Fisher Controversy. University of Pittsburgh Press, Pittsburgh, Pennsylvania.

Garrod, A.E., 1923. Inborn Errors of Metabolism, 2nd Edition. Henry Frowde and Hodder & Stoughton, London.

Hazen, G.W., 1926. The Japanese Waltzing Mouse: Its origin, heredity and relation to the genetic characters of other varieties of mice. Part III. Publication No. 337 of the Carnegie Institution of Washington, pp. 83–138.

Gilbert, S.F., 1991. A Conceptual History of Modern Embryology, Vol. 7 of Developmental Biology: A Comprehensive synthesis. Plenum Press, New York.

Goldschmidt, R.B., 1960. In and Out of the Ivory Tower—The Autobiography of Richard B. Goldschmidt. University of Washington Press, Seattle.

Goodfellow, P., 1984. Genetic Analysis of the Cell Surface. Receptors and Recognition Series B16. Chapman and Hall, London.

Green, E.L., 1966. Biology of the laboratory Mouse, 2nd. Ed. McGraw-Hill Book Company, New York.

Green, M.C., 1981. editor for the International Committee on Standardized Genetic Nomenclature for Mice. Genetic Variants and Strains of the Laboratory Mouse. Gustav Fischer Verlag, Stuttgart.

Gruneberg, H., 1943. The genetics of the mouse. Cambridge University Press, Cambridge, UK.

Gruneberg, H., 1952. The genetics of the mouse, 2nd edition. Martinus Nijhoof, The Hague.

Gruneberg, H., 1963. The pathology of development. Wiley, New York.

Guenet, J.L., Benavides, F., Panthier, J.-J., 2015. Montagutelli. Genetics of the Mouse. Springer, Heidelberg.

Gwatkin, Ralph B.L., 1993. Genes in Mammalian Reproduction. Wiley-Liss, New York.

Hadorn, E., 1955. English translation, 1961, Developmental Genetics and Lethal Factors. Methuen & Co. Ltd. John Wiley & Sons, Inc., LondonNew York.

Haldane, J.B.S., 1930. Enzymes. Langman, Green and Company 1965. MIT Press, LondonBoston, paper back ed.

Haldane, J.B.S., 1954. The Biocchemistry of Genetics. George Allen and Unwin, London.

Haldane, J.S., 1985. On being the right size and other essays. Oxford University Press, Oxford Ed. By J. M. Smith.

Haldane, J.S., 1919. The New Physiology and Other Addresses. Charles Griffin and Company, London.

Haldane, J.S., 1929. The Sciences and Philosophy: Gifford lectures. University of Glasgow, Doubleday, Doran and Company, Garden City, N Y, pp. 1927–1928.

Harris, H., 1959. Human Biochemical Genetics. Cambridge University Press, Cambridge, New York, London.

Harwood, J., 1993. Styles of Scientific Thought: The German Genetics Community 1900-1933. The University of Chicago Press, Chicago, Ill.

Henig, R.M., 2000. The Monk in the Garden: The Lost and Found Genius of Gregor Mendel, the Father of Genetics. Houghton Mifflin Co., Boston, New York.

Hodder, Ian, 2006. The Leopard's Tale: Revealing the mysteries of Catalhoyuk. Thames and Hudson, London.

Hornblum, A.M., Newman, J.L., Dober, G.J, 2013. Against their will: The secret history of medical experimentation on children in cold war America. Putnam MacMillann, New York, London.

Huxley, J., 1942. Evolution, The Modern Synthesis. Harper & Brothers, New York and London.

Hunter, R.F.H, 1995. Sex Determination, Differentiation and Intersexuality in Placental Mammals. Cambridge University Press, Cambridge.

Jacob, F., 1973. The logic of life. Pantheon Books, New York.

Jacob, F., 1982. The possible and actual. University of Washington Press, Seattle.

Jensen, Marius B., 2000. The Making of Modern Japan. The Belnap Press of the Harvard University Press, Cambridge, MA.

Judson, H.F., 1979. The eighth day of creation. Simon and Schuster, New York.

Kassow, S.D., 1989. Students, Professors, and the State in Tsarist Russia. University of California Press, Berkeley, CA.

Keeler, C.E., 1931. The Laboratory Mouse: Its Origin, Heredity, and Culture. Harvard University Press, Cambridge, MA.

King, C., 2019. Gods of the Upper Air: How a circle of renegade anthropologists reinvented race, sex, and gender in the twentieth century. Doubleday, New York.

Klein, J., 1975. Biology of the Mouse Histocompatibility-2 Complex. Springer-Verlag, New York, Heidelberg, Berlin.

Klein, J., 1986. Natural History of the Major Histocompatibility Complex. John Wiley & Sons, New York.

Koestler, A., 1971. The Case of the Midwife Toad. Random House, New York.

Kohler, R.E., 1994. Lords of the Fly: Drosophila Genetics and the Experimental life. Univ. of Chicago Press, Chicago.

Kuhn, T.S., 1970. The Structure of Scientific Revolutions, 2nd. Ed. University of Chicago Press, Chicago.

Lauber, P., 1971. Of Man and Mouse: How House Mice Became Laboratory Mice. The Viking Press, New York [book for youth].

Lear, J., 1978. Recombinant DNA, The Untold Story. Crown Publishers, Inc., New York.

Leroi, A.M., 2014. The Lagoon: How Aristotle Invented Science. Viking Press, New York, New York.

Lewontin, R.C., 1974. The Genetic Basis of Evolutionary Change 1974. Columbia University Press.

Ludmerer, K.M., 1972. Genetics and American Society: A Historical Appraisal. John Hopkins University Press, Baltimore, MD.

Manco, Jean, 2015. Ancestral Journeys: The Peopling of Europe from the First Venturers to the Vikings, 2nd Ed. Thames and Hudson,Ltd., London.

Mayr, E., 1942. Systematics and the Origin of Species (from the viewpoint of a zoologist). Columbia University Press, New York, NY.

Mayr, E., 1982. The Growth of Biological Thought: Diversity, Evolution, and Inheritance. The Belnap Press of Harvard University Press, Cambridge, MA.

McLaren, A., 1976. Mammalian Chimaeras. Cambridge university Press, Cambridge, London, New York, Melbourne.

Medvedev, Z., 1969. The Rise and Fall of T. D. Lysenko. Columbia University Press, New York.

Macholan, M., Baird Stuard, J.E., Munclinger, P., Pialek, J., 2012. Evolution of the House Mouse. Cambridge University Press, London, New York.

Mitchison, N., 1979. You May Well Ask: A Memoir 1920—1940. Victor Gallancz Ltd., London.

Mittwoch, U., 1973. Genetis of Sex Determination. Academic Press, New York and London.

Morgan, T.H., Sturtevant, A.H., Muller, H.J., Bridges, C.B., 1915. The Mechanism of Mendelian Heredity. Henry Holt and Co., New York.

Moriwaki, K., Shiroishi, T., Yonekawa, H., 1994. Genetics in Wild Mice: Its Application to Biomedical Research. Japan Scientific Societies Press, Tokyo.

Morris III, H.C., 1978. Origins of Inbred Mice. Academic Press, New York.

Moscona, A.A., Monroy, A., 1970. Current Topics in Molecular Biology 5. Academic press, New York.

Mukherjee, S., 2011. The Emperor of all Maladies, A Biography of Cancer. Harper Colllins, London.

Needham, J., 1934. A History of Embryology. Cambridge University Press, London.

Needham, J., 1942. Biochemistry and Morphogenesis. Cambridge University Press, Cambridge, U K.

Neel, J.V., Schull, W.J., 1991. The Children of Atomic Bomb Survivors: A Genetic Study. National Academy Press, Washington, D.C.

Neel, James V., 1994. Physician to the Gene Pool: Genetic Lessons and Other Stories. John Wiley and Sons, New York.

Ohmo, S., 1967. Sex chromosomes and Sex-linked Genes. Springer-Verlage, Berlin Heidelber, new York.

Overbye, D., 2000. Einstein in Love, A Scientific Romance. Viking Penguin. New York, N.Y.

Peterson, E.L., 2010. Finding Mind, Form, Organism, and Person in a Reductionist Age: The challenge of Gregory Bateson and C.H. Waddington to Biological and Anthropological Orthodoxy, 1924-1980. University of Notre Dame, ProQuest Dissertations Publishing, 3441562.

Pinsky, L., Erickson, R.P., Schimke, R.N., 1999. Genetic Disorders of Human Sexual Development. Oxford University Press, New York, Oxford.

Portugal, F.H., Cohen Jack, S., 1977. A Century of DNA. MIT, Press, Cambridge, Mass.

Provine, W.B., 1986. Sewall Wright and Evolutionary Biology. The University of Chicago Press, Chicago.

Rader, K., 2004. Making Mice: Standardizing Animals for American Biomedical Research, 1900-1955. Princeton University Press, Princeton, NJ.

Rainger, R., Benson, K.R., Maienscheinn, J., 1988. The American Development of Biology. University of Pennsylvania Press, Philadelphia.

Reed, K.C., Graves, J.A.M., 1993. Sex Chromosomes and Sex-determining Genes. Harwood Academic Publishers, Camberwell, Australia.

Ruse, M., 1996. Monad to Man: The concept of progress in Evolutionary Biology. Harvard university Press, Cambridge, Mass., London, England.

Sapp, J., 1987. Beyond the Gene: Cytoplasmic Inheritance and the Struggle of Authority in the Field of Heredity. Oxford University Press, New York.

Sayre, A., 1975. Rosalind Franklin & DNA. W. W. Norton & Company, Inc., New York.

Schull, W.J., 1990. Song Among the Ruins. Harvard University Press, Cambridge, Mass.

Searle, A.G., 1968. Comparitive Genetics of Coat Colour in Mammals. Logos press, Academic Press, New York.

Shine, I., Wrobel, S., 1976. Thomas Hunt Morgan: Pioneer of Genetics. The University Press of Kentucky, Lexington, KY.

Silver, Lee, 1995. Mouse Genetics: Concepts and Applications. Oxford University Press, New York, Oxford.

Silvers, W.K., 1979. The Coat Colors of Mice: A Model for Mammalian Gene Action and Interaction. Springer-Verlag, New York.

Snell, G.D., Jean, D., Nathenson, S., 1976. Histocompatibility. Academic Press, New York, San Francisco, London.

Snyder, L.J, 2015. Eye of the Beholder: Johannes Vermeer, Antoni van Leeuwenhoek, and the Reinvention of Seeing. W. W. Norton and Company, New York, London.

Stubbe, H., 1972. History of Genetics: From Prehistoric Times to the Rediscovery of Mendel's Laws, translated by T. R. W. Waters form the 2nd Ed. (1965). The MIT Press, Cambridge, MA.

Sturtevant, A.H., 1965. A History of Genetics. Harper and Row, New York.

Subtelny, S., Abbott, U.K. (Eds.), 1981. Levels of Genetic Control in Development. Alan R. Liss, Inc., New York.

Tierney, P., 2000. Darkness in El Dorado: How Scientists and Journalists Devastated the Amazon. W W Norton & Co., New York.

Waddington, C.H., 1956. Principles of embryology. George Allen & Unwin, Ltd., London.

Watson, J.D., 1968. The Double Helix. Atheneum, New York.

Weber, W.W., 1997. Pharmacogenetics. Oxford University Press, New York, Oxford.

Weismann, A., 1893. The Germ-Plasm: A Theory of Heredity. Wlater Scottt, Ltd., London Translated by W N Parker and H Ronnfeldt.

Welsome, E., 1999. The Plutonium files: America's secret medical experiments in the cold war. Dial Press, New York.

Wilkins, A.S., 1986. Genetic Analysis of Development. John Wiley & Sons, New York, Chichester, Brisbane, Toronto, Singapore.

Wilson, E.B., 1898. The Cell in Development and Inheritance. The MacMillan Company, London.

Zernicka-Goetz, M., Highfield, R., 2020. The Dance of Life: The New Science of How a Single Cell Becomes a Human Being. Basic Books, New York.

Index

Page numbers followed by "*f*" and "*t*" indicate, figures and tables respectively.